CHARLES PROTEUS STEINMETZ

Lectures on Electrical Engineering

EDITED BY PHILIP L. ALGER
CONSULTING PROFESSOR OF ELECTRICAL ENGINEERING
RENSSELAER POLYTECHNIC INSTITUTE

IN THREE VOLUMES

II

Electric Waves and Impulses

DOVER PUBLICATIONS, INC.
NEW YORK

Published in Canada by General Publishing Company, Ltd., 30 Lesmill Road, Don Mills, Toronto, Ontario.
Published in the United Kingdom by Constable and Company, Ltd., 10 Orange Street, London WC 2.

This Dover edition, first published in 1971, is a reprinting, in three volumes, of selected portions of the Steinmetz Electrical Engineering Library, originally published between 1915 and 1920 in nine volumes by the McGraw-Hill Book Company, Inc. The correspondence between the present edition and the original set is indicated on page vi of the General Preface.

International Standard Book Number: 0-486-62515-X
Library of Congress Catalog Card Number: 70-137004

Manufactured in the United States of America
Dover Publications, Inc.
180 Varick Street
New York, N. Y. 10014

GENERAL PREFACE

Steinmetz is well described by his adopted middle name, Proteus, for the range of his interests and accomplishments was vast. To his friends he was known simply as "the Doctor," but by others he was referred to variously as the wizard of Schenectady, forger of thunderbolts, and latter-day Vulcan. Harvard University called him "the foremost expert in applied electricity of this country," and he was a prolific writer on both technical and general subjects.

But it is as a teacher that he is best known, and with good reason. Steinmetz had a remarkable ability to explain things. He liked to talk with children, telling them why the stars twinkle, what makes the wind blow, and what makes flowers bloom. He was able to explain engineering problems in the same way, clearly and simply; he would resolve them into distinct, readily understood elements, and then solve them step by step, using approximations freely in order to emphasize method rather than detailed calculation.

In Steinmetz's day alternating-current phenomena were considered very mysterious, and there were very few textbooks available to explain them. Thus, a whole generation of engineers eagerly read the nine volumes of his Union College lectures and monographs published by McGraw-Hill, some of which went into five editions. Steinmetz realized that many engineers did not have the time or inclination for a rigorous mathematical presentation, and therefore endeavored in this series "to give a descriptive exposition of the physical nature and meaning, the origin and effects, of these phenomena, with the use of very little and only the simplest form of mathematics."

In view of the remarkable clarity of Steinmetz's explanations, and the breadth of his treatment of fundamental concepts, his writings of fifty years ago still offer one of the best ways of getting a basic electrical engineering education. They are also of historical interest: Steinmetz's use of the complex number $a + jb$ for alternating-current circuit calculations was just one of many unique contributions to electrical engineering recorded in these books.

This Dover edition of the lectures, in three volumes, includes all the material still of use to present-day students of engineering, with the exception of Volume 2 of the McGraw-Hill set, which is already available in both hardcover and paperback in a new edition revised by the present writer (*Mathematics for Science and Engineering*, by Philip L. Alger, McGraw-Hill, Second Edition, 1969). The relationship between the new Dover edition and the 9-volume McGraw-Hill edition of the lectures is:

Dover Edition	McGraw-Hill Edition
Volume I: *Elements of Electrical Engineering*	Volume 3, pp. 1–120; Volume 4, pp. 1–206; Volume 5, pp. 1–228
Volume II: *Electric Waves and Impulses*	Volume 4, pp. 341–465; Volume 7, complete; Volume 8, pp. 3–219; Volume 4, pp. 466–473 (Appendix)
Volume III: *Transient Electric Phenomena*	Volume 8, pp. 223–683

Omitted from the Dover edition are Volume 1, which consists of general lectures too diffuse to be of present-day interest, Volumes 6 and 9, and parts of the other volumes which deal with rotating machines, with more complex theories, or with apparatus now obsolete.

These lectures present concepts usually studied by electrical engineers in the first three years of college. They should be readily understood by any intelligent technician or engineering apprentice, so that by his own efforts he can gain a working knowledge of electrical engineering.

Some necessary changes have been made in the original text: a number of misprints have been silently emended, and cross references have been altered so as to be useful to the reader of these three new volumes. Where footnotes have been added they are differentiated from Steinmetz's own by the use of brackets.

Steinmetz occasionally refers the reader to his book *Engineering Mathematics* for a more detailed mathematical explanation than that given in the text. As mentioned above, this book is now available in a revised edition.

The reader should take note of the few changes in terminology that have occurred since Steinmetz's day:

1. Condensers are now called *capacitors*, condensance is

called *capacitance*, and condensive reactance is known as *capacitive reactance*.

2. A vector which indicates the phase position of a sinusoidally varying quantity is now known as a *phasor*; Steinmetz refers to these simply as vectors.

3. Wave screens are now called *wave filters*.

Philip L. Alger

Schenectady, New York
March, 1969

The books listed below will be useful as supplements to this work.

Alger, Philip L.: *Mathematics for Science and Engineering* (Based on *Engineering Mathematics* by C. P. Steinmetz), Second Edition; McGraw-Hill, 1969.

———: *The Nature of Induction Machines*; Gordon and Breach Science Publishers, 1965.

Bewley, L. V.: Alternating Current Machinery; The Macmillan Co., 1949.

Clement, P. R. and W. C. Johnson: *Electrical Engineering Science*; McGraw-Hill, 1960.

Concordia, C.: *Synchronous Machines*; John Wiley, 1951.

Dow, W. G.: *Fundamentals of Engineering Electronics*; John Wiley, 1952.

Fano, Chu, and Adler: *Electromagnetic Fields, Energy and Forces*; John Wiley, 1960.

Fitzgerald, A. E. and C. Kingsley, Jr.: *Electric Machinery*, Second Edition; McGraw-Hill, 1961.

Guillemin, E. A.: *Introductory Circuit Theory*; John Wiley, 1953.

Ham, J. M. and G. R. Slemon: *Scientific Basis of Electrical Engineering*; John Wiley, 1961.

Ku, Y. H.: *Electric Energy Conversion*; Ronald Press, 1959.

Lewis, W. W. and C. F. Goodheart: *Basic Electric Circuit Theory*; Ronald Press, 1958.

Majmudar, Harit: *Electromechanical Energy Converters*; Allyn and Bacon, 1965.

Messerle, H. K.: *Dynamic Circuit Theory*; Pergamon Press, 1965.

Scott, W. T.: *The Physics of Electricity and Magnetism*; John Wiley, 1959.

Slemon, G. R.: *Magnetoelectric Devices*; John Wiley, 1966.

Veinott, C. G.: *Fractional Horsepower Electric Motors*, Second Edition; McGraw-Hill, 1948.

Zimmerman, H. J. and S. J. Mason: *Electric Circuit Theory*; John Wiley, 1959.

The following books provide more information about the life and thought of C. P. Steinmetz.

Alger, Philip L. and E. Caldecott: *Steinmetz—The Philosopher*; Mohawk Development Service, Inc., P.O. Box 7, Bellevue Station, Schenectady, New York 12306.

Miller, Floyd: *The Electrical Genius of Liberty Hall*; McGraw-Hill, 1962.

Miller, J. A.: *Modern Jupiter*; The American Society of Mechanical Engineers, 1958.

PREFACE TO VOLUME II

The first part of this volume considers the shapes and distortions of voltage and current waves, and then discusses polyphase systems at some length, including the effects of unbalance, and methods of metering.

The second part deals with transient phenomena, analyzed on the basis of energy storage and decay. Transients in both d-c and a-c circuits are studied, and several chapters are devoted to traveling waves and oscillations on transmission lines. In the final chapter formulas are derived for calculating inductance and capacitance of parallel conductors.

The third part is devoted to a detailed mathematical analysis of many kinds of transients, including the buildup of fields in d-c machines, the charging of capacitors, and the short-circuit currents of alternators. Many examples are worked out numerically, with illustrative oscillograms and charts.

The algebra of complex quantities necessary for a-c circuit calculations is explained in an appendix.

Philip L. Alger

Schenectady, New York
March, 1969

CONTENTS FOR VOLUME II

PART I
ALTERNATING CURRENT PHENOMENA

General Waves

CHAPTER I. DISTORTION OF WAVE-SHAPE, AND ITS CAUSES

CHAPTER II. EFFECTS OF HIGHER HARMONICS

PART II
ELECTRIC DISCHARGES, WAVES AND IMPULSES

PART III
TRANSIENTS IN TIME

Chapter I. The Constants of the Electric Circuit

Chapter II. Introduction

PART I

ALTERNATING-CURRENT PHENOMENA

GENERAL WAVES

CHAPTER I

DISTORTION OF WAVE-SHAPE AND ITS CAUSES

1. In discussing alternating currents and alternating e.m.fs., the usual procedure is to consider them as sine waves or as replaced by their equivalent sine waves.

While this is sufficiently exact in most cases, under certain circumstances the deviation of the wave from sine shape becomes of importance, and with certain distortions it may not be possible to replace the distorted wave by an equivalent sine wave, since the angle of phase displacement of the equivalent sine wave becomes indefinite. Thus it becomes desirable to investigate the distortion of the wave, its causes and its effects.

Since, as stated before, any alternating wave can be represented by a series of sine functions of odd orders, the investigation of distortion of wave-shape resolves itself in the investigation of the higher harmonics of the alternating wave.

In general we have to distinguish between higher harmonics of e.m.f. and higher harmonics of current. Both depend upon each other in so far as with a sine wave of impressed e.m.f. a distorting effect will cause distortion of the current wave, while with a sine wave of current passing through the circuit, a distorting effect will cause higher harmonics of e.m.f.

2. In a conductor revolving with uniform velocity through a uniform and constant magnetic field, a sine wave of e.m.f. is generated. In a circuit with constant resistance and constant reactance, this sine wave of e.m.f. produces a sine wave of current. Thus distortion of the wave-shape or higher harmonics may be due to lack of uniformity of the velocity of the revolving conductor; lack of uniformity or pulsation of the magnetic field; pulsation of the resistance or pulsation of the reactance.

The first two cases, lack of uniformity of the rotation or of the magnetic field, cause higher harmonics of e.m.f. at open circuit. The last, pulsation of resistance and reactance, causes higher harmonics only when there is current in the circuit, that is, under load.

Lack of uniformity of the rotation is hardly ever of practical interest as a cause of distortion, since in alternators, due to mechanical momentum, the speed is always very nearly uniform during the period. A periodic pulsation of speed may occur in low speed singlephase machines.

Thus as causes of higher harmonics remain:

1st. Lack of uniformity and pulsation of the magnetic field, causing a distortion of the generated e.m.f. at open circuit as well as under load.

2d. Pulsation of the reactance, causing higher harmonics under load.

3d. Pulsation of the resistance, causing higher harmonics under load also.

Taking up the different causes of higher harmonics, we have:

Lack of Uniformity and Pulsation of the Magnetic Field.

3. Since most of the alternating-current generators contain definite and sharply defined field-poles covering in different types different proportions of the pitch, in general the magnetic flux interlinked with the armature coil will not vary as a sine wave, of the form

$$\Phi \cos \beta,$$

but as a complex harmonic function, depending on the shape and the pitch of the field-poles and the arrangement of the armature conductors. In this case the magnetic flux issuing from the field-pole of the alternator can be represented by the general equation,

$$\begin{aligned}\Phi = A_0 + A_1 \cos \beta + A_2 \cos 2\beta + A_3 \cos 3\beta + \dots \\ + B_1 \sin \beta + B_2 \sin 2\beta + B_3 \sin 3\beta + \dots\end{aligned}$$

If the reluctance of the armature is uniform in all directions, so that the distribution of the magnetic flux at the field-pole face does not change by the rotation of the armature, the rate of cutting magnetic flux by an armature conductor is Φ, and the e.m.f. generated in the conductor thus equal thereto in wave-shape. As a rule A_0, A_2, A_4 . . . B_2, B_4 equal zero; that is, successive field-poles are equal in strength and distribution of magnetism, but of opposite polarity. In some types of machines, however, especially inductor alternators, this is not the case.

The e.m.f. generated in a full-pitch armature turn—that is, armature conductor and return conductor distant from former by the pitch of the armature pole (corresponding to the distance from field-pole center to pole center)—is

$$\begin{aligned}\delta e &= \Phi_0 - \Phi_{180} \\ &= 2\,\{A_1 \cos\beta + A_3 \cos 3\,\beta + A_5 \cos 5\,\beta + \dots \\ &\quad + B_1 \sin\beta + B_3 \sin 3\,\beta + B_5 \sin 5\,\beta + \dots\}\end{aligned}$$

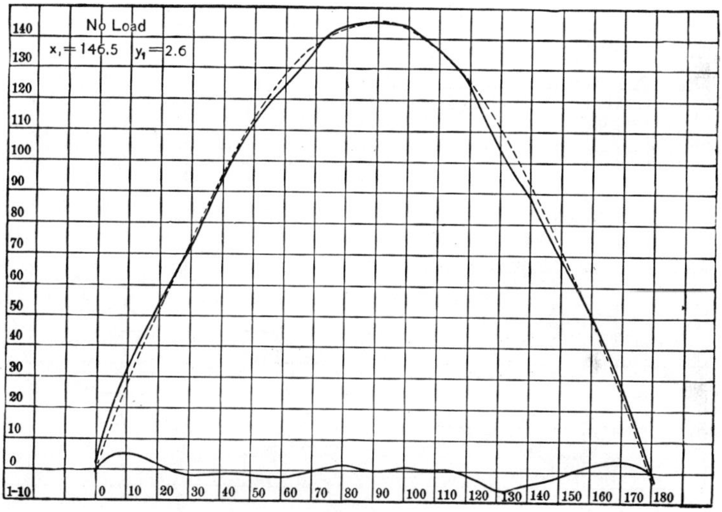

FIG. 1.

Even with an unsymmetrical distribution of the magnetic flux in the air-gap, the e.m.f. wave generated in a full-pitch armature coil is symmetrical, the positive and negative half-waves equal, and correspond to the mean flux distribution of adjacent poles. With fractional pitch-windings—that is, windings whose turns cover less than the armature pole-pitch—the generated e.m.f. can be unsymmetrical with unsymmetrical magnetic field, but as a rule is symmetrical also. In unitooth alternators the total generated e.m.f. has the same shape as that generated in a single turn.

With the conductors more or less distributed over the surface of the armature, the total generated e.m.f. is the resultant of several e.m.fs. of different phases, and is thus more uniformly varying; that is, more sinusoidal, approaching sine shape to within 3 per cent. or less, as for instance the curves Fig. 1 and Fig. 2 show, which represent the no-load and full-load wave of e.m.f. of a three-phase multitooth alternator. The principal term of these harmonics is the third harmonic, which consequently appears more or less in all alternator waves. As a rule these harmonics can be considered together with the harmonics due to the varying reluctance of the magnetic circuit.

In iron-clad alternators with few slots and teeth per pole, the passage of slots across the field-poles causes a pulsation of the

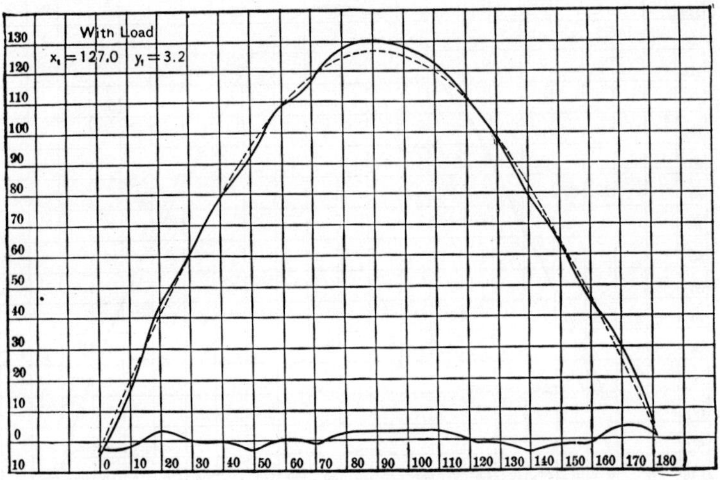

FIG. 2.

magnetic reluctance, or its reciprocal, the magnetic reactance of the circuit. In consequence thereof the magnetism per field-pole, or at least that part of the magnetism passing through the armature, will pulsate with a frequency $2\,\gamma$, if γ = number of slots per pole.

Thus, in a machine with one slot per pole the instantaneous magnetic flux interlinked with the armature conductors can be expressed by the equation

$$\phi = \Phi \cos \beta \left\{1 + \epsilon \cos \left[2\,\beta - \theta\right]\right\}$$

where Φ = average magnetic flux,

ϵ = amplitude of pulsation,

and θ = phase of pulsation.

In a machine with γ slots per pole, the instantaneous flux interlinked with the armature conductors will be

$$\phi = \Phi \cos \beta \left\{1 + \epsilon \cos [2 \gamma\beta - \theta]\right\},$$

if the assumption is made that the pulsation of the magnetic flux follows a simple sine law, as first approximation.

In general the instantaneous magnetic flux interlinked with the armature conductors will be

$$\phi = \Phi \cos \beta \left\{1 + \epsilon_1 \cos (2 \beta - \theta_1) + \epsilon_2 \cos (4 \beta - \theta_2) + \ . \ . \ . \right\},$$

where the term ϵ_γ is predominating, if γ = number of armature slots per pole. This general equation includes also the effect of lack of uniformity of the magnetic flux.

In case of a pulsation of the magnetic flux with the frequency, 2γ, due to an existence of γ slots per pole in the armature, the instantaneous value of magnetism interlinked with the armature coil is

$$\phi = \Phi \cos \beta \left\{1 + \epsilon \cos [2 \gamma\beta - \theta]\right\}.$$

Hence the e.m.f. generated thereby,

$$\begin{aligned} e &= - n \frac{d\phi}{dt} \\ &= - \sqrt{2}\pi f \Phi \frac{d}{d\beta} \left\{\cos \beta \left(1 + \epsilon \cos [2 \gamma\beta - \theta]\right)\right\}. \end{aligned}$$

And, expanded,

$$e = \sqrt{2}\, \pi f n \Phi \left\{ \sin \beta + \epsilon \frac{2\gamma - 1}{2} \sin [(2\gamma - 1)\beta - \theta] + \epsilon \frac{2\gamma + 1}{2} \sin [(2\gamma + 1)\beta - \theta] \right\}.$$

Hence, the pulsation of the magnetic flux with the frequency, 2γ, as due to the existence of γ slots per pole, introduces two harmonics, of the orders $(2\gamma - 1)$ and $(2\gamma + 1)$.

4. If $\gamma = 1$ it is

$$e = \sqrt{2}\, \pi f n \Phi \left\{ \sin \beta + \frac{\epsilon}{2} \sin (\beta - \theta) + \frac{3\epsilon}{2} \sin (3\beta - \theta) \right\};$$

that is, in a unitooth single-phaser a pronounced triple harmonic may be expected, but no pronounced higher harmonics.

Fig. 3 shows the wave of e.m.f. of the main coil of a monocyclic alternator at no load, represented by,

$$e = E \{ \sin \beta - 0.242 \sin (3 \beta - 6.3) - 0.046 \sin (5 \beta - 2.6) + 0.068 \sin (7 \beta - 3.3) - 0.027 \sin (9 \beta - 10.0) - 0.018 \sin (11 \beta - 6.6) + 0.029 \sin (13 \beta - 8.2) \};$$

hence giving a pronounced triple harmonic only, as expected.

If $\gamma = 2$, it is,

$$e = \sqrt{2} \pi f n \Phi \left\{ \sin \beta + \frac{3 \epsilon}{2} \sin (3 \beta - \theta) + \frac{5 \epsilon}{2} \sin (5 \beta - \theta) \right\},$$

the no-load wave of a unitooth quarter-phase machine, having pronounced triple and quintuple harmonics.

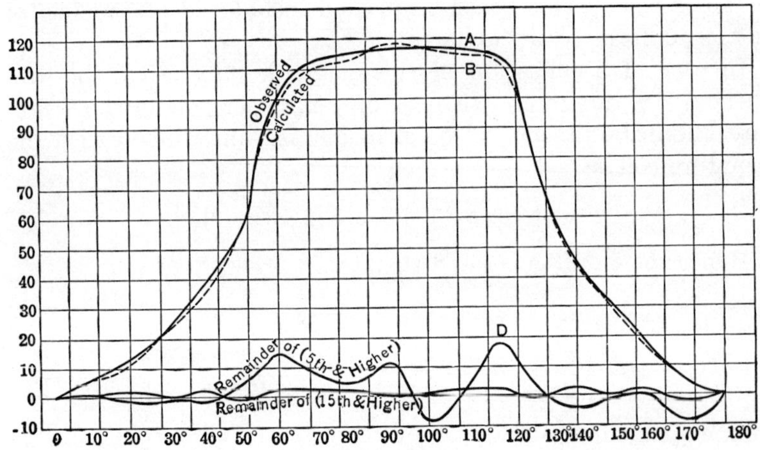

FIG. 3.—No-load of e.m.f. of unitooth monocyclic alternator.

If $\gamma = 3$, it is,

$$e = \sqrt{2} \pi f n \Phi \left\{ \sin \beta + \frac{5 \epsilon}{2} \sin (5 \beta - \theta) + \frac{7 \epsilon}{2} \sin (7 \beta - \theta) \right\}.$$

That is, in a unitooth three-phaser, a pronounced quintuple and septuple harmonic may be expected, but no pronounced triple harmonic.

Fig. 4 shows the wave of e.m.f. of a unitooth three-phaser at no-load, represented by

$$e = E \{ \sin \beta - 0.12 \sin (3 \beta - 2.3) - 0.23 \sin (5 \beta - 1.5) + 0.134 \sin (7 \beta - 6.2) - 0.002 \sin (9 \beta + 27.7) - 0.046 \sin (11 \beta - 5.5) + 0.031 \sin (13 \beta - 61.5) \}.$$

Thus giving a pronounced quintuple and septuple and a

lesser triple harmonic, probably due to the deviation of the field from uniformity, as explained above, and deviation of the pulsation of reluctance from sine-shape. In some especially favorable cases, harmonics as high as the 35th and 37th have been observed, caused by pulsation of the reluctance, and even still higher harmonics.

In general, if the pulsation of the magnetic reactance is denoted by the general expression

$$1 + \sum_{1}^{\infty} {}_{\gamma}\epsilon_{\gamma}r \cos (2\,\gamma\beta - \theta_{\gamma}),$$

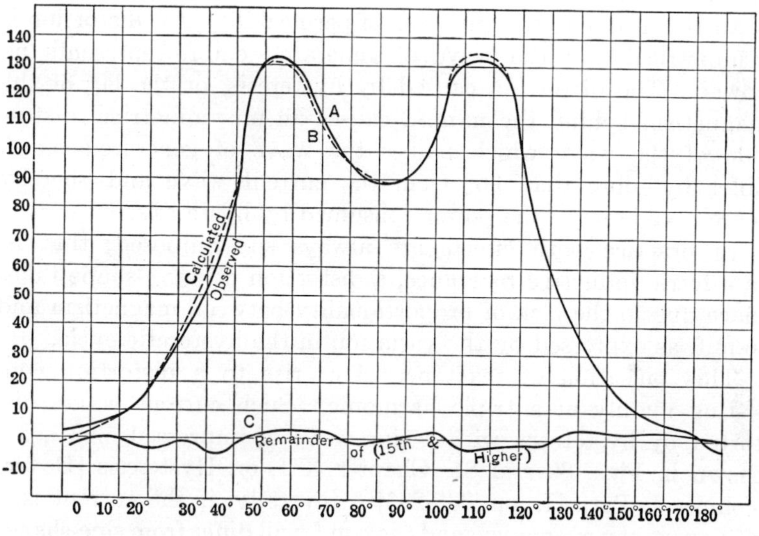

FIG. 4. —No-load wave of e.m.f. of unitooth three-phase alternator.

the instantaneous magnetic flux is

$$\phi = \Phi \cos\beta \left\{ 1 + \sum_{1}^{\infty}\gamma\epsilon_{\gamma} \cos (2\,\gamma\beta - \theta_{\gamma}) \right\}$$

$$= \Phi \left\{ \cos\beta + \frac{\epsilon_1}{2} \cos (\beta - \theta_1) + \sum_{1}^{\infty}\gamma \left[\frac{\epsilon_{\gamma}}{2} \cos [(2\,\gamma + 1)\beta - \theta_{\gamma}] + \frac{\epsilon_{\gamma+1}}{2} \cos [(2\,\gamma + 1)\beta - \theta_{\gamma+1}] \right] \right\};$$

hence, the e.m.f.,

$$e = \sqrt{2}\,\pi f n \Phi \left\{ \sin\beta + \frac{\epsilon_1}{2} \sin (\beta - \theta_1) + \sum_{1}^{\infty}\gamma \frac{2\,\gamma + 1}{2} [\epsilon_{\gamma} \sin [(2\,\gamma + 1)\beta - \theta_{\gamma}] + \epsilon_{\gamma+1} \sin [(2\,\gamma + 1)\beta - \theta_{\gamma+1}]] \right\}.$$

With the general adoption of distributed fractional pitch armature windings, such pronounced wave shape distortions as shown by the unitooth alternators shown as illustrations, have become infrequent.

Pulsation of Reactance

5. The main causes of a pulsation of reactance are magnetic saturation and hysteresis, and synchronous motion. Since in an iron-clad magnetic circuit the magnetism is not proportional to the m.m.f., the wave of magnetism and thus the wave of e.m.f. will differ from the wave of current. As far as this distortion is due to the variation of permeability, the distortion is symmetrical and the wave of generated e.m.f. represents no power. The distortion caused by hysteresis, or the lag of the magnetism behind the m.m.f., causes an unsymmetrical distortion of the wave which makes the wave of generated e.m.f. differ by more than 90° from the current wave and thereby represents power—the power consumed by hysteresis.

In practice both effects are always superimposed; that is, in a ferric inductive reactance, a distortion of wave-shape takes place due to the lack of proportionality between magnetism and m.m.f. as expressed by the variation in the hysteretic cycle.

This pulsation of reactance gives rise to a distortion consisting mainly of a triple harmonic. Such current waves distorted by hysteresis, with a sine wave of impressed e.m.f., are shown in Figs. 80 and 81, Chapter XII, on Hysteresis [Dover, Volume I, Part II, pp. 235-236]. Inversely, if the current is a sine wave, the magnetism and the e.m.f. will differ from sine-shape.

For further discussion of this distortion of wave-shape by hysteresis, Chapter XII may be consulted.

6. Distortion of wave-shape takes place also by the pulsation of reactance due to synchronous rotation, as discussed in the chapter on "Reaction Machines," in *Theory and Calculation of Electrical Apparatus* [McGraw-Hill edition, volume 6].

With a sine wave of e.m.f., distorted current waves result.

Inversely, if a sine wave of current,

$$i = I \cos \beta,$$

exists through a circuit of synchronously varying reactance, as for instance, the armature of a unitooth alternator or synchronous motor—or, more general, an alternator whose armature reluctance is different in different positions with regard to the field-poles—and the reactance is expressed by

$$X = x\,\{1 + \epsilon \cos(2\beta - \theta)\};$$

or, more general,

$$X = x\left\{1 + \sum_1^\infty {}_\gamma \epsilon_\gamma \cos(2\gamma\beta - \theta_\gamma)\right\};$$

the wave of magnetism is

$$\phi = \frac{X}{2\pi f n}\cos\beta = \frac{x}{2\pi f n}\left\{\cos\beta + \sum_1^\infty {}_\gamma \epsilon_\gamma \cos\beta\cos(2\gamma\beta - \theta_\gamma)\right\}$$

$$= \frac{x}{2\pi f n}\left\{\cos\beta + \frac{\epsilon_1}{2}\cos(\beta - \theta_1) + \sum_1^\infty {}_\gamma\left[\frac{\epsilon_\gamma}{2}\cos[(2\gamma + 1)\beta - \theta_\gamma] + \frac{\epsilon_{2\gamma+1}}{2}\cos[(2\gamma+1)\beta - \theta_{\gamma+1}]\right]\right\};$$

hence the wave of generated e.m.f.,

$$e = -n\frac{d\phi}{dt} = -2\pi f n\frac{d\phi}{d\beta}$$

$$= x\left\{\sin\beta + \frac{\epsilon_1}{2}\sin(\beta - \theta_1) + \sum_1^\infty {}_\gamma \frac{2\gamma + 1}{2}\left[\epsilon_\gamma \sin[(2\gamma + 1)\beta - \theta_\gamma] + \epsilon_{2\gamma+1}\sin[(2\gamma + 1)\beta - \theta_{\gamma+1}]\right]\right\};$$

that is, the pulsation of reactance of frequency, 2γ, introduces two higher harmonics of the order $(2\gamma - 1)$ and $(2\gamma + 1)$.

If

$$X = x\{1 + \epsilon\cos(2\beta - \theta)\},$$

it is

$$\phi = \frac{x}{2\pi f n}\left\{\cos\beta + \frac{\epsilon}{2}\cos(\beta - \theta) + \frac{\epsilon}{2}\cos(3\beta - \theta)\right\};$$

$$e = x\left\{\sin\beta + \frac{\epsilon}{2}\sin(\beta - \theta) + \frac{3\epsilon}{2}\sin(3\beta - \theta)\right\}.$$

Since the pulsation of reactance due to magnetic saturation and hysteresis is essentially of the frequency, $2f$—that is, describes a complete cycle for each half-wave of current—this shows why the distortion of wave-shape by hysteresis consists essentially of a triple harmonic.

The phase displacement between e and i, and thus the power consumed or produced in the electric circuit, depends upon the angle, θ, as discussed before.

7. In case of a distortion of the wave-shape by reactance, the distorted waves can be replaced by their equivalent sine waves, and the investigation with sufficient exactness for most cases be carried out under the assumption of sine waves, as done in the preceding chapters.

Similar phenomena take place in circuits containing polarization cells, leaky condensers, or other apparatus representing a synchronously varying negative reactance. Possibly dielectric hysteresis in condensers causes a distortion similar to that due to magnetic hysteresis.

Inversely, at very high voltages, where corona appears on the conductors, with a sine wave of impressed voltage, a distortion of the capacity current wave occurs, which is largely effective, but partly reactive due to the increase of capacity under corona.

Pulsation of Resistance

8. To a certain extent the investigation of the effect of synchronous pulsation of the resistance coincides with that of reactance; since a pulsation of reactance, when unsymmetrical with regard to the current wave, introduces a power component which can be represented by an "effective resistance."

Inversely, an unsymmetrical pulsation of the ohmic resistance introduces a wattless component, to be denoted by "effective reactance."

A typical case of a synchronously pulsating resistance is represented in the alternating arc.

The apparent resistance of an arc depends upon the current through the arc; that is, the apparent resistance of the arc = $\frac{\text{potential difference between electrodes}}{\text{current}}$, is high for small currents, low for large currents. Thus in an alternating arc the apparent resistance will vary during every half-wave of current between a maximum value at zero current and a minimum value at maximum current, thereby describing a complete cycle per half-wave of current.

Let the effective value of current through the arc be represented by I.

Then the instantaneous value of current, assuming the current wave as sine wave, is represented by

$$i = I \sqrt{2} \sin \beta;$$

and the apparent resistance of the arc, in first approximation, by

$$R = r(1 + \epsilon \cos 2\,\beta);$$

thus the potential difference at the arc is

$$e = iR = I\sqrt{2}\,r \sin \beta\,(1 + \epsilon \cos 2\,\beta)$$
$$= rI\sqrt{2}\left\{\left(1 - \frac{\epsilon}{2}\right) \sin \beta + \frac{\epsilon}{2} \sin 3\,\beta\right\}.$$

Hence the effective value of potential difference,

$$E = rI\sqrt{\left(1 - \frac{\epsilon}{2}\right)^2 + \frac{\epsilon^2}{4}}$$
$$= rI\sqrt{1 - \epsilon + \frac{\epsilon^2}{2}},$$

and the apparent resistance of the arc,

$$r_0 = \frac{E}{I} = r\sqrt{1 - \epsilon + \frac{\epsilon^2}{2}}.$$

The instantaneous power consumed in the arc is

$$ie = 2\,rI^2\left\{\left(1 - \frac{\epsilon}{2}\right) \sin^2 \beta + \frac{\epsilon}{2} \sin \beta \sin 3\,\beta\right\}.$$

Hence the effective power,

$$P = rI^2\left(1 - \frac{\epsilon}{2}\right).$$

The apparent power, or volt-amperes consumed by the arc,

$$IE = rI^2\sqrt{1 - \epsilon + \frac{\epsilon^2}{2}}.$$

Thus the power-factor of the arc,

$$p = \frac{P}{IE} = \frac{1 - \frac{\epsilon}{2}}{\sqrt{1 - \epsilon + \frac{\epsilon}{2}}};$$

that is, less than unity.

9. We find here a case of a circuit in which the power-factor —that is, the ratio of watts to volt-amperes—differs from unity without any displacement of phase; that is, while current and e.m.f. are in phase with each other, but are distorted, the alternating wave cannot be replaced by an equivalent sine wave, since the assumption of equivalent sine wave would introduce a phase displacement,

$$\cos \theta = p$$

of an angle, θ, whose sign is indefinite.

As an example are shown, in Fig. 5, for the constants, $I = 12$, $r = 3$, $\epsilon = 0.9$, the resistance,

$$R = 3\ (1 + 0.9 \cos 2\ \beta);$$

the current,

$$i = 17 \sin \beta;$$

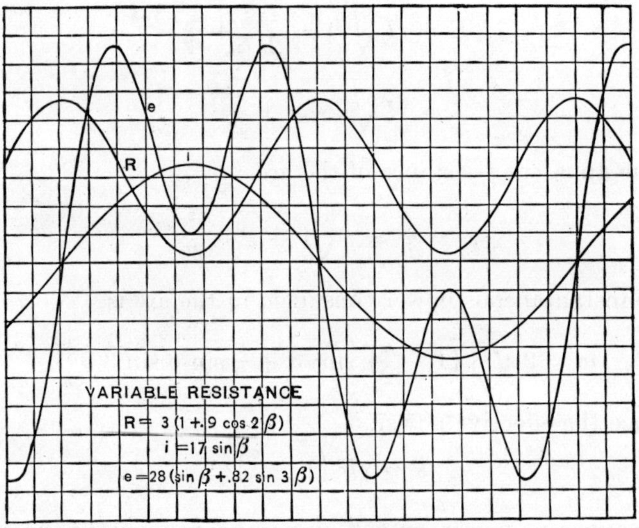

FIG. 5. —Periodically varying resistance.

the potential difference,

$$e = 28\ (\sin \beta + 0.82 \sin 3\ \beta).$$

In this case the effective e.m.f. is

$$E = 25.5;$$

the apparent resistance,

$$r_0 = 2.13;$$

the power,

$$P = 244;$$

the apparent power,

$$EI = 307;$$

the power-factor,

$$p = 0.796.$$

As seen, with a sine wave of current the e.m.f. wave in an alternating arc will become double-peaked, and rise very abruptly near the zero values of current. Inversely, with a sine wave of e.m.f. the current wave in an alternating arc will become peaked, and very flat near the zero values of e.m.f.

10. In reality the distortion is of more complex nature, since the pulsation of resistance in the arc does not follow a simple sine law of double frequency, but varies much more abruptly near the zero value of current, making thereby the variation of e.m.f. near the zero value of current much more abruptly, or, inversely, the variation of current more flat.

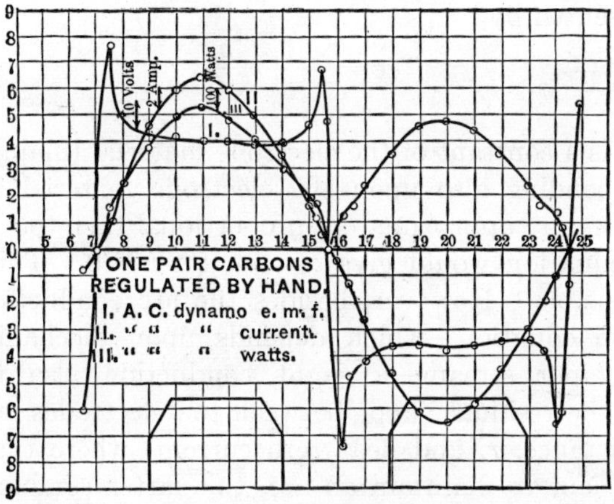

FIG. 6. —Electric arc.

A typical wave of potential difference, with an approximate sine wave of current through the arc, is given in Fig. 6 .[1]

11. The value of ϵ, the amplitude of the resistance pulsation, largely depends upon the nature of the electrodes and the steadiness of the arc, and with soft carbons and a steady arc is small, and the power-factor, p, of the arc near unity. With hard carbons and an unsteady arc, ϵ rises greatly, higher harmonics appear in the pulsation of resistance, and the power-factor, p, falls, being in extreme cases even as low as 0.6. Especially is this the case with metal arcs.

This double-peaked appearance of the voltage wave, as shown by Figs. 5 and 6 , is characteristic of the arc to such an extent

[1] From American Institute of Electrical Engineers, Transactions, 1890, p. 376. Tobey and Walbridge, on the Stanley Alternate Arc Dynamo.

that when in the investigation of an electric circuit by oscillograph such a wave-shape is found, the existence of an arc or arcing ground somewhere in the circuit may usually be suspected. This is of importance as in high-voltage systems arcs are liable to cause dangerous voltages.

The pulsation of the resistance in an arc, as shown in Fig. 6 for hard carbons, is usually very far from sinusoidal, as assumed in Fig. 5. It is due to the feature of the arc that the voltage consumed in the arc flame decreases with increase of current—approximately inversely proportional to the square root of the current—and so is lowest at maximum current.

Approximately, the volt-ampere characteristic of the arc can be represented by,

$$e = e_0 + \frac{c}{\sqrt{i}}, \qquad (1)$$

where e_0 is a constant of the electrode material (mainly), c a constant depending also upon the electrode material and on the arc length, and approximately proportional thereto.

This equation would give $e = \infty$, for $i = 0$. This obviously is not feasible. However, besides the arc conduction as given by above equation—which depends upon mechanical motion of the vapor stream—a slight conduction also takes place through the residual vapor between the electrodes, as a path of high resistance, r, and near zero current, where the voltage is not sufficient to maintain an arc, this latter conduction carries the current.

The characteristic of the alternating-current arc therefore consists of the combination of two curves: the arc characteristic, (1), and the resistance characteristic,

$$e = ri. \qquad (2)$$

The phenomenon then follows that curve which gives the lowest voltage; that is, for high values of current, is represented by equation (1), for low values of current, by equation (2).

12. As an example are shown in Fig. 7 the calculated curves of an alternating arc between hard carbons (or carbides), for the constants,

$$e_0 = 30 \text{ volts},$$
$$c = 40,$$
$$r = 70 \text{ ohms}.$$

The curve I represents the arc conduction, following equation (1),

$$e = 30 + \frac{40}{\sqrt{i}},$$

and the curve II represents the conduction through the (stationary) residual vapor, by equation (2), near the zero points, A and D, of the current,

$$e = 70\, i.$$

As seen, from A to B the voltage varies approximately proportionally with the current. At B the arc starts, and the vol-

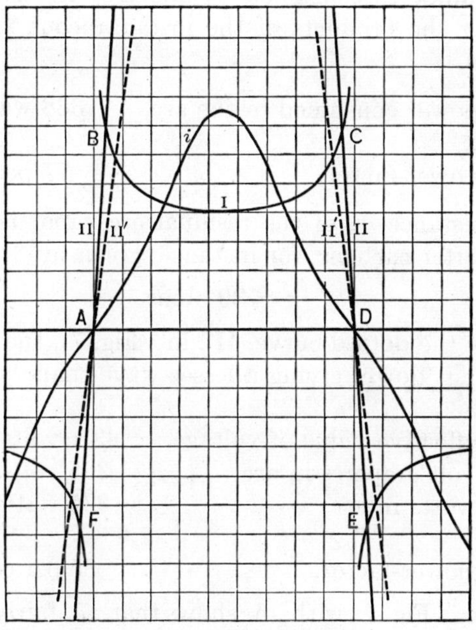

FIG. 7.

tage drops with the further increase of current, and then rises again with the decreasing current, until at C, the intersection point between curves I and II, the arc extinguishes and the voltage follows curve II, until at E the arc starts again. The two sharp peaks of the curve thus represent respectively the starting and the extinction of the arc.

Since the high values of voltage near zero current lower and the low values of voltage near maximum current raise the value of the current, the current wave does not remain a sine wave, if the arc voltage is an appreciable part of the total voltage, but the current wave becomes peaked, with flat zero, as expressed approximately by a third harmonic in phase with the fundamental. The current wave in Fig. 7 so has been assumed as

$$i = 13 \cos \phi + 2 \cos 3 \phi.$$

From Fig. 7 follows:

effective value of current, 9.30 amp.,
effective value of voltage, 47.2 volts;

hence, volt-amperes consumed by the arc, 439 volt-amp.; and, by averaging the products of the instantaneous values of volts and amperes,

power consumed in the arc, 388 watts;

hence,

power-factor, 77 per cent.

If the resistance, r, of the residual arc-vapor is lower, as by the use of softer carbons, for instance, given by

$$r = 30 \text{ ohms},$$

as shown by the dotted curve, II′, in Fig. 7 , the voltage peaks are greatly cut down, giving a lesser wave-shape distortion, and so,

effective value of voltage, 43.1 volts,
volt-amperes in arc, 395 volt-amp.,
watts in arc, 335 watts,

hence,

power-factor, 85 per cent.

Comparing Fig. 7 with 6 shows that 7 fairly well approximates 6 , except that in Fig. 6 the second peak is lower than the first. This is due to the lower resistance, r, of the residual vapor immediately after the passage of the arc than before the starting of the arc. Fig. 6 also shows a decrease of resistance, r, immediately before starting, or after extinction of the arc, which may be represented by some expression like

$$r = r_0 i^{-b},$$

where

$$b < 1,$$

but which has not been considered in Fig. 7.

The softer the carbons, the more is the latter effect appreciable and the peaks rounded off, thus causing the curve to approach the appearance of Fig. 176, while with metal arcs, where r is very high, the peaks, especially the first, become very sharp and high, frequently reaching values of several thousand volts.

For further discussion on the effect of the arc, see "Theory of Electric Circuits" Dover edition, Volume I, Part III.

13. One of the most important sources of wave-shape distortion is the presence of iron in a magnetic circuit. The magnetic induction in iron, and therewith the magnetic flux, is not proportional to the magnetizing force or the exciting current, but the magnetic induction and the magnetizing force are related to each other by the hysteresis cycle of the iron, as discussed in Chapter XII, [Dover, Volume I, Part II]. In an iron-clad magnetic circuit, the magnetic flux and the current, therefore, cannot both be sine waves; if the magnetic flux and therefore the generated e.m.f. are sine waves, the current differs from sine wave-shape, while if a sine wave of current is sent through the circuit, the magnetic flux and the generated e.m.f. cannot be sine waves.

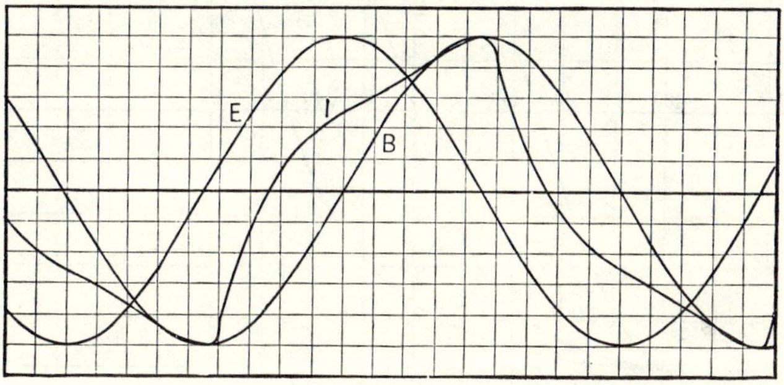

FIG. 8.

A. Sine Wave of Voltage

Let a sine wave of e.m.f. be impressed upon an iron-clad reactance coil, or a primary coil of a transformer with open secondary circuit. Neglecting the ohmic resistance of the circuit, that is, assuming the generated e.m.f. as equal or practically equal to the impressed e.m.f., the voltage consumed by the generated e.m.f. and therewith the magnetic flux are sine waves, as represented by E and B in Fig. 8. The current which produces this magnetic flux, B, and so the voltage, E, then is derived point by point from B, by the hysteresis cycle of the iron. With the hysteresis cycle given in Fig. 9, the current then has the wave-shape given as I in Fig. 8, that is, greatly differs from a sine wave. This distortion of the current wave is mainly due to the bend of the magnetic characteristic, that is, the magnetic saturation, and not to the energy loss or the area of the curve. This is seen by resolving the current wave, I, into two components: an energy component, i', in phase with

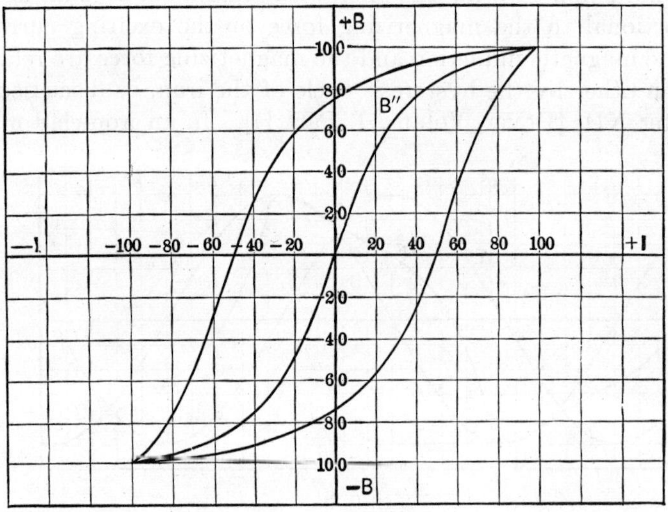

FIG. 9.

the e.m.f., $e = E \sin \phi$, and a wattless component, i'', in quadrature with E, and in phase with B. These components are calculated as

$$i' = \tfrac{1}{2} \{i_\phi + i_{\pi-\phi}\},$$

and

$$i'' = \tfrac{1}{2} \{i_\phi - i_{\pi-\phi}\},$$

where i_ϕ and $i_{\pi-\phi}$ are the instantaneous values of the current, I, at the angles ϕ and $\pi - \phi$, respectively.

These components, the *hysteresis power current*, i', and the reactive *magnetizing current*, i'', are plotted in Fig. 10 and show that i' is nearly a sine wave, while i'' is greatly distorted and peaked.

The total current, I, derived by the hysteresis cycle, Fig. 9, from the magnetic flux,

$$B = B_0 \cos \phi,$$

can be resolved into an infinite series of harmonic waves, that is, a trigonometric or Fourier series of the form:

$$\begin{aligned} i = a_1 \cos \phi + a_3 \cos 3\phi + a_5 \cos 5\phi + \dots + a_n \cos n\phi + \dots \\ + b_1 \sin \phi + b_3 \sin 3\phi + b_5 \sin 5\phi + \dots + b_n \sin n\phi + \dots \end{aligned}$$

or of the form:

$$\begin{aligned} i = c_1 \cos (\phi - \theta_1) + c_3 \cos (3\phi - \theta_3) + c_5 \cos (5\phi - \theta_5) \\ + \dots + c_n \cos (n\phi - \theta_n) + \dots \end{aligned}$$

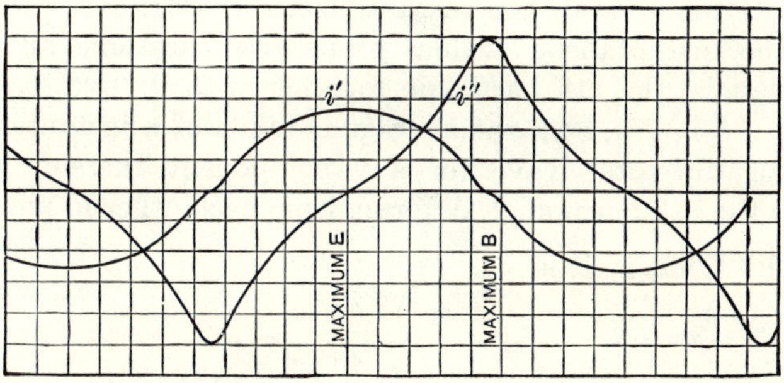

FIG. 10.

where

$$c_n = \sqrt{a_n^2 + b_n^2},$$

$$\tan \theta_n = \frac{b_n}{a_n}.$$

The coefficients a_n and b_n are determined by the definite integrals:[1]

$$a_n = \frac{2}{\pi}\int_0^\pi i \cos n\phi d\phi = 2 \times avg\,(i \cos n\phi)_0^\pi,$$

$$b_n = \frac{2}{\pi}\int_0^\pi i \sin n\phi d\phi = 2 \times avg\,(i \sin n\phi)_0^\pi;$$

that is, by multiplying the instantaneous values of i, as given numerically, by $\cos n\phi$ and $\sin n\phi$, respectively, and then averaging.

[1] See "Engineering Mathematics."

Just as in most investigations dealing with alternating currents, not the fundamental sine wave, but the fundamental sine wave together with all its higher harmonics, that is, the total wave, is of importance; so also when dealing with the higher harmonics, frequently not the individual higher harmonic sine wave is of importance, but the higher harmonic together with all of its higher harmonics. For instance, when dealing with the disturbances caused by the third harmonic in a three-phase system, the third harmonic together with all its higher harmonics or overtones, as the ninth, fifteenth, twenty-first, etc., comes in consideration, that is, all the components which repeat after one-third cycle. The higher harmonic then appears as a distorted wave, including its higher harmonics.

To determine, from the instantaneous values of a distorted wave, the instantaneous values of its nth harmonic distorted wave, that is, the nth harmonic together with its overtones, of order 3 n, 5 n, 7 n, etc., the average is taken of n instantaneous values of the total wave (or any component thereof, which includes the nth harmonic). differing from each other in phase by $\frac{1}{n}$ period. That is, it is

$$i_n = \sum_{o}^{n-1} {}^{\kappa}\, i_{\phi + \frac{2\kappa\pi}{n}}.$$

This method is based on the relations:

$$\sum_{o}^{n-1} {}^{\kappa} \cos\left(m\phi + \frac{2\,\kappa\pi}{n}\right) = n \cos m\phi,$$

$$\sum_{o}^{n-1} {}^{\kappa} \sin\left(m\phi + \frac{2\,\kappa\pi}{n}\right) = n \sin m\phi,$$

if $m = n$ or if m is a multiple of n; otherwise these sums $= 0$, where m and n are integer numbers.

14. In this manner the wave of exciting current, I, of Fig. 8 is resolved, in Fig. 11 , into the fundamental sine wave, i_1, and the higher harmonics, i_3, i_5, i_7, which are general waves, that is, include their higher harmonics.

Analytically, it can be represented by

$$\begin{aligned} i = {} & 8.857 \cos(\phi + 37.6^\circ) + 1.898 \cos 3\,(\phi + 4.1^\circ) \\ & + 0.585 \cos 5\,(\phi - 1.7^\circ) + 0.319 \cos 7\,(\phi - 3.2^\circ) \\ & + 0.158 \cos 9\,(\phi - 2.5^\circ) + \ldots \end{aligned}$$

where $B = 10{,}000 \cos \phi$ is the wave of magnetic induction.

The equivalent sine wave of above current wave is

$$i_0 = 9.104 \cos (\phi - 36.3°).$$

In this case of the distortion of a current wave by an iron-clad reactance coil or transformer, with a sine wave of impressed e.m.f., it is, from the above equation of the current wave,

Effective value of the total current	6.423
Effective value of its fundamental sine wave . . .	6.27
Effective value of the sum of all its higher harmonics	1.43.

That is, the effective value of all the harmonics is 22.3 per cent. of the effective value of the total current.

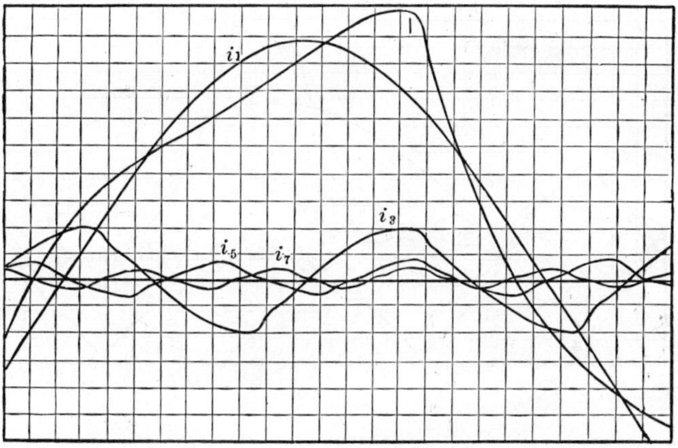

FIG. 11.

B. Sine Wave of Current

15. If a sine wave of current exists through an iron-clad magnetic circuit, as, for instance, an iron-clad reactance coil or transformer connected in series to a circuit traversed by a sine wave, the potential difference at the terminals of the reactance cannot be a sine wave, but contains higher harmonics.

From the sine wave of current

$$i = I \cos \phi,$$

follows by the hysteresis cycle, Fig. 9, the wave of magnetism. This is not a sine wave, but hollowed out on the rising, humped on the decreasing side, that is, has a distortion about opposite from that of the current wave in Fig. 8 ; the wave of magnetism has the maximum at the same angle, ϕ, as the current, but passes the zero much later than the current.

From the wave of magnetism follows the wave of generated e.m.f., and so (approximately, that is, neglecting resistance) of terminal voltage, e, at the reactance, since e is proportional to $\frac{dB}{d\phi}$. It is plotted as E in Fig. 12 , and resolved into its harmonics in the same manner as the current wave in A.

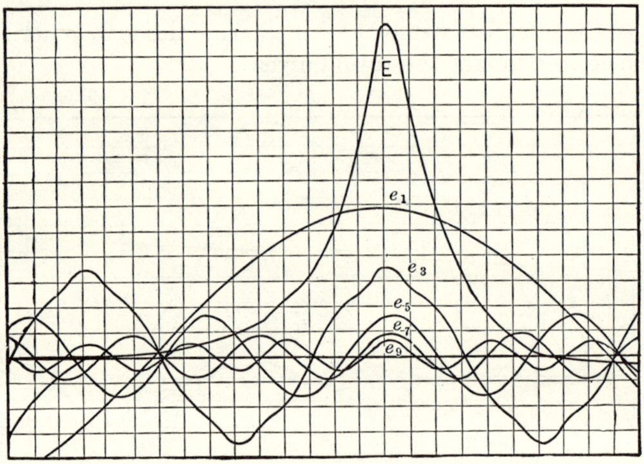

FIG. 12.

As seen, with a sine wave of current traversing an iron-clad reactance, the e.m.f. wave is very greatly distorted, and the maximum value of the distorted e.m.f. wave is more than twice the maximum of its fundamental sine wave.

Denoting the current wave by,

$$i = 10 \sin (\phi + 30°),$$

the e.m.f. wave in Fig. 183 is represented by

$$\begin{aligned} e = {} & 11.67 \cos (\phi + 2.5°) + 6.64 \cos 3 (\phi - 1.13°) \\ & + 3.24 \cos 5 (\phi - 2.4°) + 1.8 \cos 7 (\phi - 1.53°) + \\ & 1.16 \cos 9 (\phi - 0.5°) + 0.80 \cos 11 (\phi - 2°) \\ & + 0.53 \cos 13 (\phi - 2°) + 0.19 \cos 15 (\phi - 1°) + \dots \end{aligned}$$

that is, all the harmonics are nearly in phase with each other, so accounting for the very steep peak. It is

Effective value of total wave	9.91
Effective value of its fundamental sine wave . . .	8.25
Effective value of the sum of all its higher harmonics	5.48

that is, the effective value of all the higher harmonics is 55.3 per cent. of the effective value of the total wave.

The impedance of this iron-clad reactance, with a sine wave current of 7.07 effective, so is

$$z = \frac{9.91}{7.07} = 1.40,$$

while the same reactance, with a sine wave e.m.f. of 7.07 effective, in A, gives the impedance,

$$z = \frac{7.07}{6.42} = 1.10.$$

The conclusion is that an iron-clad magnetic circuit is not suitable for a reactor, since even below saturation (as above assumed) it produces very great wave-shape distortion.

As discussed before, the insertion of even a small air-gap into the magnetic circuit makes the current wave nearly coincide in phase and in shape with the wave of magnetism.

C. Three-phase Circuits

16. The wave-shape distortion in an iron-clad magnetic circuit has an important bearing on transformer connections in three-phase circuits.

The e.m.fs. and the currents in a three-phase system are displaced from each other in phase by one-third of a period or 120°. Their third harmonics, therefore, differ by 3 × 120°, or a complete period, that is, are in phase with each other. That is, whatever third harmonics of e.m.f. and of current may exist in a three-phase system must be in phase with each other in all three phases, or, in other words, for the third harmonics the three-phase system is single-phase.

The sum of the three e.m.fs. between the lines of a three-phase system (Δ voltages) is zero. Since their third harmonic would be in phase with each other, and so add up, it follows:

The voltages between the lines of a three-phase system, or Δ voltages, cannot contain any third harmonic or its overtones (ninth, fifteenth, twenty-first, etc., harmonics).

Since in a three-wire, three-phase system the sum of the three currents in the line is zero, but their third harmonics would be in phase with each other, and their sum, therefore, not zero, it follows:

The currents in the lines of a three-wire, three-phase system, or Y currents, cannot contain any third harmonic.

Third harmonics, however, can exist in the Y voltage or voltage between line and neutral of the system, and since the third harmonics are in phase with each other, in this case, a potential difference of triple frequency exists between the neutral of the system and all three phases as the other terminal, that is, the whole system pulsates against the neutral at triple frequency.

Third harmonics can also exist in the currents between the lines, or Δ currents. Since the two currents from one line to the other two lines are displaced 60° from each other, their third harmonics are in opposition and, therefore, neutralize. That is, the third harmonics in the Δ currents of a three-phase system do not exist in the Y currents in the lines, but exist only in a local closed circuit.

Third harmonics can exist in the line currents in a four-wire, three-phase system, as a system with grounded neutral. In this case the third harmonics of currents in the lines return jointly over the fourth or neutral wire, and even with balanced load on the three phases, the neutral wire carries a current which is of triple frequency.

17. With a sine wave of impressed e.m.f. the current in an iron-clad circuit, as the exciting current of a transformer, must contain a strong third harmonic, otherwise the e.m.f. cannot be a sine wave. Since in the lines of a three-phase system the third harmonics of current cannot exist, interesting wave-shape distortions thus result in transformers, when connected to a three-phase system in such a manner that the third harmonic of the exciting current would have to enter the line as Y current, and so is suppressed.

For instance, connecting three iron-clad reactors, as the primary coils of three transformers—with their secondaries open-circuited—in star or Y connection into a three-phase system, with a sine wave of e.m.f., e, impressed upon the lines. Normally, the voltage of each transformer should be a sine wave also, and equal $\frac{e}{\sqrt{3}}$. This, however, would require that the current taken by the transformer as exciting current contains a third harmonic. As such a third harmonic cannot exist in a three-phase circuit, the wave of magnetism cannot be a sine wave, but must contain a third harmonic, about opposite to that which was suppressed in the exciting current. The e.m.f. generated by this magnetism, and therewith the potential difference at the transformer or Y voltage, therefore, must also contain a third harmonic, and its overtones, three times as great as that of the magnetism, due to the triple frequency.

With three transformers connected in Y into a three-phase system with open secondary circuit, we have, then, with a sine wave of e.m.f. impressed between the three-phase lines, the conditions:

The voltage at the transformers, or Y voltage, cannot be a sine wave, but must contain a third harmonic and its overtones, but can contain no other harmonics, since the other harmonics, as the fifth, seventh, etc., would not eliminate by combining two Y voltages to the Δ voltage or line voltage, and the latter was assumed as sine wave.

The exciting current in the transformers cannot contain any third harmonic or its overtones, but can contain all other harmonics.

The magnetic flux is not a sine wave, but contains a third harmonic and its overtones, corresponding to those of the Y voltage, but contains no other harmonics, and is related to the exciting current by the hysteresis cycle.

Herefrom then the wave-shapes of currents, magnetism and voltage can be constructed. Obviously, since the relation between current and magnetism is merely empirical, given by the hysteresis cycle, this cannot be done analytically, but only by the calculation or construction of the instantaneous values of the curves.

18. For the hysteresis cycle in Fig. 9 , and for a system of transformers connected in Y, with open secondary circuit, into a three-phase system with a sine wave of e.m.f. between the lines, the curves of exciting current, magnetic flux and voltage per transformer, or between lines and neutral, are constructed in Fig. 13 .

i is the exciting current of the transformer, and contains all the harmonics, except the third and its multiples. It is given by the equation:

$$i = 8.28 \sin (\phi + 30.8°) - 0.71 \sin (5\,\phi - 17.2°) + \ldots$$

B is the magnetic flux density in the transformer. It contains only the third harmonic and its multiples, but no other harmonics, and is given by the equation:

$$B = 10.0 \sin \phi + 1.38 \sin (3\phi - 9.2°) + 0.045 \sin 9\phi + \ldots$$

e is the potential difference of the transformer terminals, or voltage between the three-phase lines and the transformer neutral. It contains the third harmonic and its multiples, but no other harmonics, and is given by the equation:

$$e = 10.0 \cos \phi + 4.14 \cos (3\phi - 9.2°) + 0.405 \cos 9\phi + \ldots$$

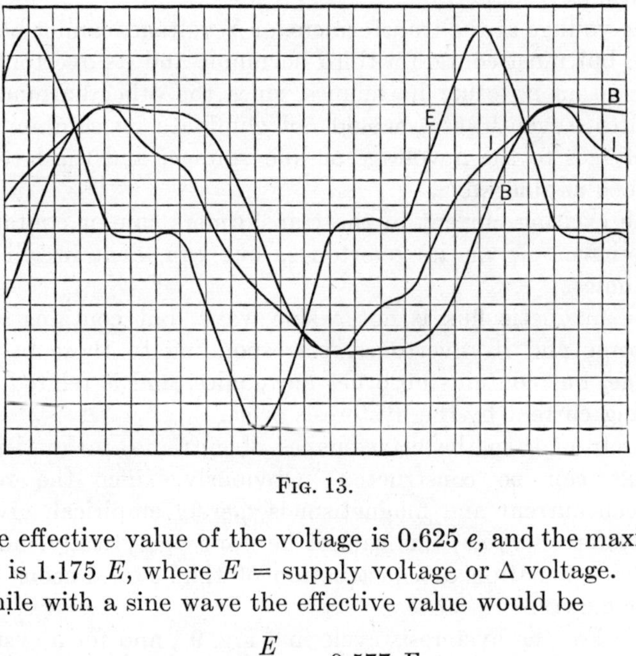

Fig. 13.

The effective value of the voltage is $0.625\, e$, and the maximum value is $1.175\, E$, where E = supply voltage or Δ voltage.

While with a sine wave the effective value would be

$$\frac{E}{\sqrt{3}} = 0.577\, E,$$

and the maximum value

$$\frac{E\sqrt{2}}{\sqrt{3}} = 0.815\, E,$$

that is, by the suppression of the third harmonic of exciting current in the three-phase system, the effective value of the voltage per transformer, or voltage between three-phase lines and neutral (or ground, if the neutral is grounded) has been increased by

8.5 per cent., the maximum value by 44.6 per cent., and the voltage wave has become very peaked, by a pronounced third harmonic of an effective value of 0.24 E—that is, 38.5 per cent. of the effective value of the total wave.

The very high peak of e.m.f. produced by this wave-shape distortion is liable to be dangerous in high-potential, three-phase systems by increasing the strain on the insulation between lines and ground, and leading to resonance phenomena with the third harmonic.

The maximum value of the distorted wave of magnetism is 8.89, while with a sine wave it would be 10.0, that is, the maximum of the wave of magnetism has been reduced by 11.1 per cent., and the core loss of the transformer so by about 17 per cent.

19. Assuming now that in such transformers, connected with their primaries in Y into a three-phase circuit, the secondaries are connected in Δ. The third harmonics of e.m.f., generated in the three transformer secondaries, then are in series in short-circuit, thus produce a local current in the secondary transformer triangle. This current is of triple frequency, and hence supplies the third harmonic of exciting current, which was suppressed in the primary, and thereby eliminates the third harmonic of magnetism and of e.m.f., which results from the suppression of the third harmonic of exciting current, and so limits itself. That is, connecting the transformer secondaries in Δ, the wave-shape distortion disappears, and voltage and magnetism are again sine waves, and the exciting current is that corresponding to a sine wave of magnetism, except that it is divided between primary and secondary; the third harmonic of the exciting current does not exist in the primary, but is produced by induction in the secondary circuit. Obviously, in this case the magnetic flux and the voltage are not perfect sine waves, but contain a slight third harmonic, which produces in the secondary the triple-frequency exciting current.

If the primary neutral of the transformers is connected to a fourth wire, in a four-wire, three-phase system or three-phase system with grounded neutral, and this fourth wire leads back to the generator neutral, or a neutral of a transformer in which the triple-frequency current can exist, that is, in which the secondary is connected in Δ, the wave-shape distortion also disappears.

It follows herefrom that in the three-phase system attention must be paid to provide a path for the third harmonic of the transformer exciting current, either directly or inductively, otherwise a serious distortion of the e.m.f. wave of the transformers occurs.

CHAPTER II

EFFECTS OF HIGHER HARMONICS

20. To elucidate the variation in the shape of alternating waves caused by various harmonics, in Figs. 14 and 15 are shown the wave-forms produced by the superposition of the

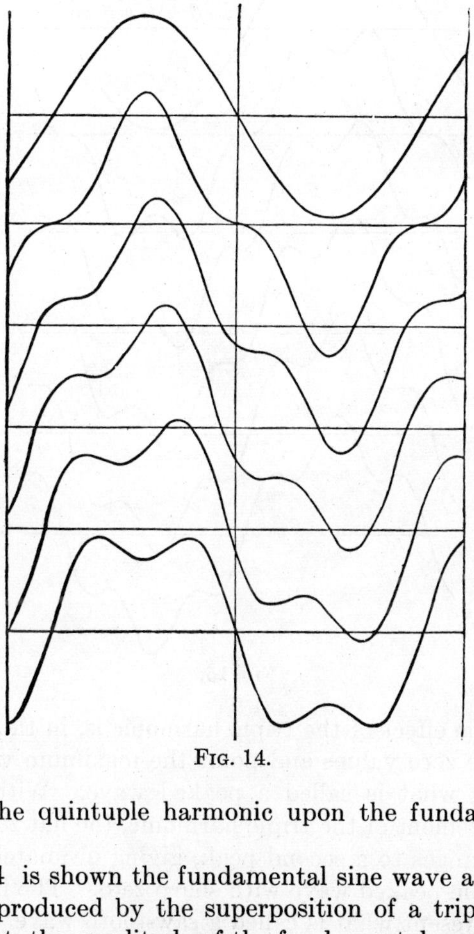

Fig. 14.

triple and the quintuple harmonic upon the fundamental sine wave.

In Fig. 14 is shown the fundamental sine wave and the complex waves produced by the superposition of a triple harmonic of 30 per cent. the amplitude of the fundamental, under the relative phase displacments of 0°, 45°, 90°, 135°, and 180°, represented by the equations:

$$\sin\beta$$
$$\sin\beta - 0.3\sin 3\beta$$
$$\sin\beta - 0.3\sin(3\beta - 45°)$$
$$\sin\beta - 0.3\sin(3\beta - 90°)$$
$$\sin\beta - 0.3\sin(3\beta - 135°)$$
$$\sin\beta - 0.3\sin(3\beta - 180°).$$

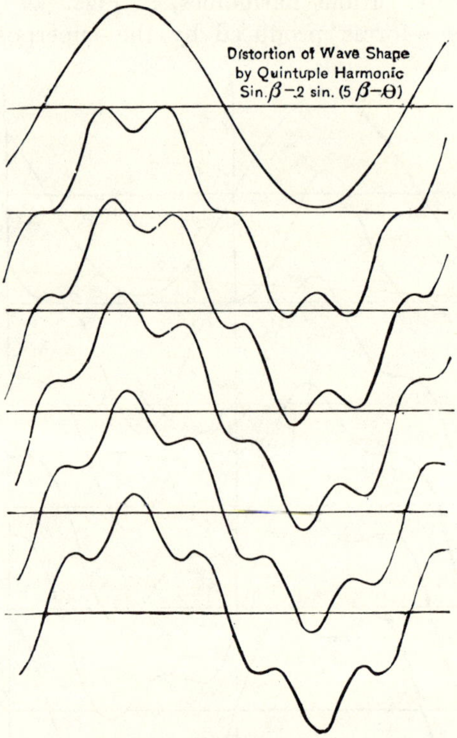

FIG. 15.

As seen, the effect of the triple harmonic is, in the first figure, to flatten the zero values and point the maximum values of the wave, giving what is called a peaked wave. With increasing phase displacement of the triple harmonic, the flat zero rises and gradually changes to a second peak, giving ultimately a flat-top or even double-peaked wave with sharp zero. The intermediate positions represent what is called a saw-tooth wave.

In Fig. 15 are shown the fundamental sine wave and the

complex waves produced by superposition of a quintuple harmonic of 20 per cent. the amplitude of the fundamental, under the relative phase displacement of 0°, 45°, 90°, 135°, 180°, represented by the equations:

$$\sin \beta$$
$$\sin \beta - 0.2 \sin 5 \beta$$
$$\sin \beta - 0.2 \sin (5 \beta - 45°)$$
$$\sin \beta - 0.2 \sin (5 \beta - 90°)$$
$$\sin \beta - 0.2 \sin (5 \beta - 135°)$$
$$\sin \beta - 0.2 \sin (5 \beta - 180°).$$

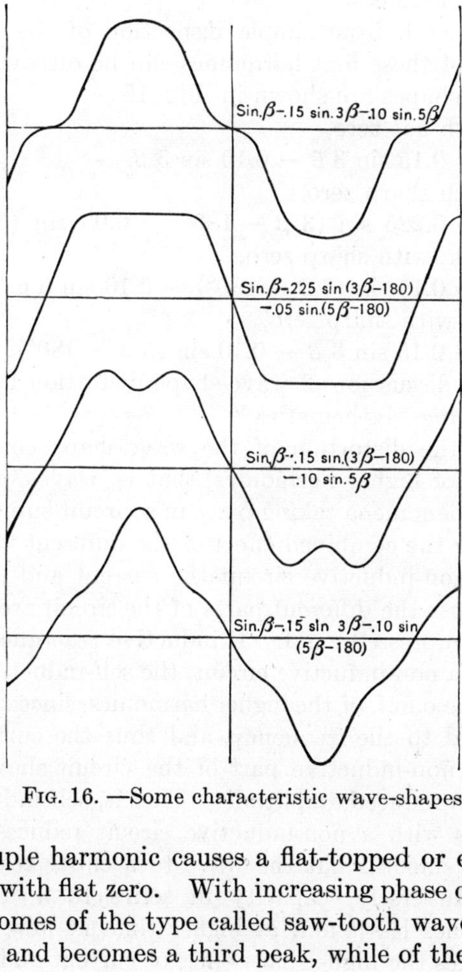

FIG. 16. —Some characteristic wave-shapes.

The quintuple harmonic causes a flat-topped or even double-peaked wave with flat zero. With increasing phase displacement the wave becomes of the type called saw-tooth wave also. The flat zero rises and becomes a third peak, while of the two former peaks, one rises, the other decreases, and the wave gradually changes to a triple-peaked wave with one main peak, and a sharp zero.

As seen, with the triple harmonic, flat top or double peak coincides with sharp zero, while the quintuple harmonic flat top or double peak coincides with flat zero.

Sharp peak coincides with flat zero in the triple, with sharp zero in the quintuple harmonic. With the triple harmonic, the saw-tooth shape appearing in case of a phase difference between fundamental and harmonic is single, while with the quintuple harmonic it is double.

Thus in general, from simple inspection of the wave-shape, the existence of these first harmonics can be discovered. Some characteristic shapes are shown in Fig. 16.

Flat top with flat zero,

$$\sin \beta - 0.15 \sin 3\beta - 0.10 \sin 5\beta.$$

Flat top with sharp zero,

$$\sin \beta - 0.225 \sin (3\beta - 180^\circ) - 0.05 \sin (5\beta - 180^\circ).$$

Double peak, with sharp zero,

$$\sin \beta - 0.15 \sin (3\beta - 180^\circ) - 0.10 \sin 5\beta.$$

Sharp peak with sharp zero,

$$\sin \beta - 0.15 \sin 3\beta - 0.10 \sin (5\beta - 180^\circ).$$

For further discussion of wave-shape distortion by harmonics see "Engineering Mathematics."

21. Since the distortion of the wave-shape consists in the superposition of higher harmonics, that is, waves of higher frequency, the phenomena taking place in a circuit supplied by such a wave will be the combined effect of the different waves.

Thus in a non-inductive circuit the current and the potential difference across the different parts of the circuit are of the same shape as the impressed e.m.f. If inductive reactance is inserted in series with a non-inductive circuit, the self-inductive reactance consumes more e.m.f. of the higher harmonics, since the reactance is proportional to the frequency, and thus the current and the e.m.f. in the non-inductive part of the circuit show the higher harmonics in a reduced amplitude. That is, self-inductive reactance in series with a non-inductive circuit reduces the higher harmonics or smooths out the wave to a closer resemblance to sine-shape. Inversely, capacity in series to a non-inductive circuit consumes less e.m.f. at higher than at lower frequency, and thus makes the higher harmonics of current and of potential difference in the non-inductive part of the circuit more pronounced—intensifies the harmonics.

Self-induction and capacity in series may cause an increase of voltage due to complete or partial resonance with higher harmonics, and a discrepancy between volt-amperes and watts, without corresponding phase displacement, as will be shown hereafter.

22. In long-distance transmission over lines of noticeable inductive and condensive reactance, rise of voltage due to resonance may occur with higher harmonics, as waves of higher frequency, while the fundamental wave is usually of too low a frequency to cause resonance.

An approximate estimate of the possible rise by resonance with various harmonics can be obtained by the investigation of a numerical example. Let in a long-distance line, fed by step-up transformers at 60 cycles,

The resistance drop in the transformers at full-load = 1 per cent.
The reactance voltage in the transformers at full-load = 5 per cent. with the fundamental wave.
The resistance drop in the line at full-load = 10 per cent.
The reactance voltage in the line at full-load = 20 per cent. with the fundamental wave.
The capacity or charging current of the line = 20 per cent. of the full-load current, I, at the frequency of the fundamental.

The line capacity may approximately be represented by a condenser shunted across the middle of the line. The e.m.f. at the generator terminals, E, is assumed as maintained constant.

The e.m.f. consumed by the resistance of the circuit from generator terminals to condenser is

$$Ir = 0.06\,E,$$

or,

$$r = 0.06\,\frac{E}{I}.$$

The reactance e.m.f. between generator terminals and condenser is, for the fundamental frequency,

$$Ix = 0.15\,E,$$

or,

$$x = 0.15\,\frac{E}{I};$$

thus the reactance corresponding to the frequency $(2k-1)f$ of the higher harmonic is

$$x(2k-1) = 0.15(2k-1)\frac{E}{I}.$$

The capacity current at fundamental frequency is,

$$i = 0.2\,I;$$

hence, at the frequency $(2k-1)f$,

$$i = 0.2(2k-1)e'\frac{I}{E},$$

if

e' = e.m.f. of the $(2k-1)$th harmonic at the condenser,
e = e.m.f. of the $(2k-1)$th harmonic at the generator terminals.

The e.m.f. at the condenser is

$$e' = \sqrt{e^2 - i^2r^2} + ix(2k-1);$$

hence, substituted,

$$a = \frac{e'}{e} = \frac{1}{\sqrt{1 - 0.059856(2k-1)^2 + 0.0009(2k-1)^4}},$$

the rise of voltage by inductive and condensive reactance.

Substituting,

	k =	1	2	3	4	5	6
or,	$2k-1$ =	1	3	5	7	9	11
and	a =	1.03	1.36	3.76	2.18	0.70	0.38

That is, the fundamental will be increased at open circuit by 3 per cent., the triple harmonic by 36 per cent., the quintuple harmonic by 276 per cent., the septuple harmonic by 118 per cent., while the still higher harmonics are reduced.

The maximum possible rise will take place for

$$\frac{da}{d(2k-1)} = 0, \text{ or, } 2k-1 = 5.77;$$

that is, at a frequency $f = 346$, and $a = 14.4$.

That is, complete resonance will appear at a frequency between quintuple and septuple harmonic, and would raise the voltage at this particular frequency 14.4-fold.

If the voltage shall not exceed the impressed voltage by more than 100 per cent., even at coincidence of the maximum of the harmonic with the maximum of the fundamental,

the triple harmonic must be less than 70 per cent. of the fundamental,
the quintuple harmonic must be less than 26.5 per cent. of the fundamental,
the septuple harmonic must be less than 46 per cent. of the fundamental.

The voltage will not exceed twice the normal, even at a frequency of complete resonance with the higher harmonic, if none of the higher harmonics amounts to more than 7 per cent. of the fundamental. Herefrom it follows that the danger of resonance in high-potential lines is frequently overestimated, since the conditions assumed in this example are rather more severe than found in lines of moderate length, the capacity current of such line very seldom reaching 20 per cent. of the main current.

23. The power developed by a complex harmonic wave in a non-inductive circuit is the sum of the powers of the individual harmonics. Thus if upon a sine wave of alternating e.m.f. higher harmonic waves are superposed, the effective e.m.f. and the power produced by this wave in a given circuit or with a given effective current are increased. In consequence hereof alternators and synchronous motors of iron-clad unitooth construction —that is, machines giving waves with pronounced higher harmonics—may give with the same number of turns on the armature, and the same magnetic flux per field-pole at the same frequency, a higher output than machines built to produce sine waves.

24. This explains an apparent paradox:

If in the three-phase star-connected generator with the magnetic field constructed as shown diagrammatically in Fig. 17 the magnetic flux per pole $= \Phi$, the number of turns in series per circuit $= n$, the frequency $= f$, the e.m.f. between any two collector rings is

$$E = \sqrt{2}\, \pi f\, 2\, n\Phi\, 10^{-8},$$

since $2\,n$ armature turns simultaneously interlink with the magnetic flux, Φ.

The e.m.f. per armature circuit is

$$e = \sqrt{2}\, \pi f n\Phi\, 10^{-8};$$

hence the e.m.f. between collector rings, as resultant of two e.m.fs., e, displaced by 60° from each other, is

$$E = e\sqrt{3} = \sqrt{2}\, \pi f \sqrt{3}\, n\Phi\, 10^{-8},$$

while the same e.m.f. was found from the number of turns, the magnetic flux, and the frequency by direct calculation to be equal to $2e$; that is, the two values found for the same e.m.f. have the proportion $\sqrt{3}:2 = 1:1.154$.

This discrepancy is due to the existence of more pronounced higher harmonics in the wave e than in the wave $E = e \times \sqrt{3}$, which have been neglected in the formula

$$e = \sqrt{2}\,\pi f n \Phi 10^{-8}.$$

Hence it follows that, while the e.m.f. between two collector rings in the machine shown diagrammatically in Fig. 17 is only

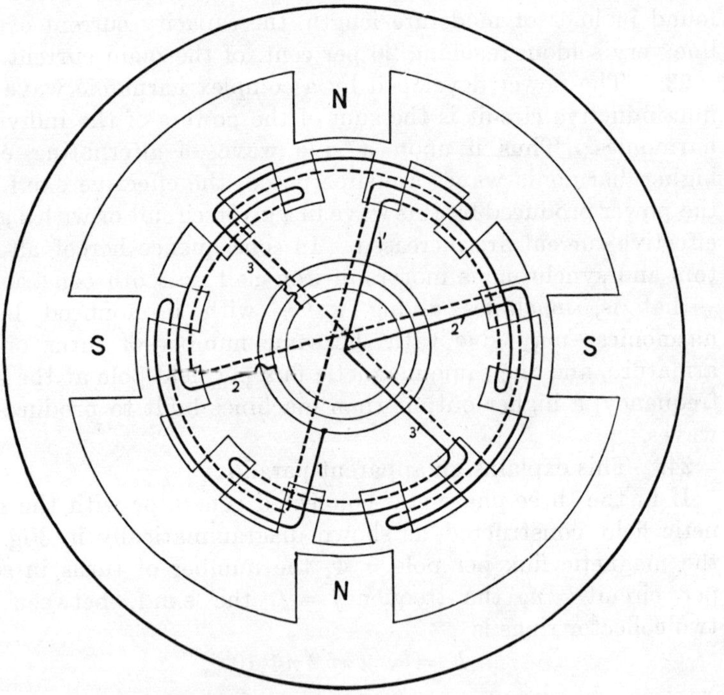

FIG. 17. —Three-phase star-connected alternator.

$e \times \sqrt{3}$, by massing the same number of turns in one slot instead of in two slots, we get the e.m.f. $2e$, or 15.4 per cent. higher e.m.f., that is, larger output.

It follows herefrom that the distorted e.m.f. wave of a unitooth alternator is produced by lesser magnetic flux per pole—that is, in general, at a lesser hysteretic loss in the armature or at higher efficiency—than the same effective e.m.f. would be produced with the same number of armature turns if the magnetic disposition were such as to produce a sine wave.

25. Inversely, if such a distorted wave of e.m.f. is impressed upon a magnetic circuit, as, for instance, a transformer, the wave of magnetism in the primary will repeat in shape the wave of magnetism interlinked with the armature coils of the alternator, and consequently with a lesser maximum magnetic flux the same effective counter e.m.f. will be produced, that is, the same power converted in the transformer. Since the hysteretic loss in the transformer depends upon the maximum value of magnetism, it follows that the hysteretic loss in a transformer is less with a distorted wave of a unitooth alternator than with a sine wave.

26. From another side the same problem can be approached: If upon a transformer a sine wave of e.m.f. is impressed, the wave of magnetism will be a sine wave also. If now upon the sine wave of e.m.f. higher harmonics, as sine waves of triple, quintuple, etc., frequency are superposed in such a way that the corresponding higher harmonic sine waves of magnetism do not increase the maximum value of magnetism, or even lower it by a coincidence of their negative maxima with the positive maximum of the fundamental, in this case all the power represented by these higher harmonics of e.m.f. will be transformed without an increase of the hysteretic loss, or even with a decreased hysteretic loss.

Obviously, if the maximum of the higher harmonic wave of magnetism coincides with the maximum of the fundamental, and thereby makes the wave of magnetism more pointed, the hysteretic loss will be increased more than in proportion to the increased power transformed, *i.e.*, the efficiency of the transformer will be lowered.

That is, some distorted waves of e.m.f. are transformed at a lesser, some at a larger, hysteretic loss than the sine wave, if the same effective e.m.f. is impressed upon the transformer.

The unitooth alternator wave and the first wave in Fig. 14 belong to the former class; the waves derived from continuous-current machines, tapped at two equidistant points of the armature, frequently, to the latter class.

27. Regarding the loss of energy by Foucault or eddy currents, this loss is not affected by distortion of wave-shape, since the e.m.f. of eddy currents, like the generated e.m.f., is proportional to the secondary e.m.f.; and thus at constant impressed primary e.m.f. the power consumed by eddy currents bears a constant relation to the output of the secondary circuit, as obvious, since the division of power between the two secondary circuits—the eddy-current circuit and the useful or consumer circuit—is unaffected by wave-shape or intensity of magnetism.

In high-potential lines, distorted waves whose maxima are very high above the effective values, as peaked waves, are objectionable by increasing the strain on the insulation. The striking-distance of an alternating voltage depends upon the maximum value, except at extremely high frequencies, such as oscillating discharges. In the latter, the very short duration of the voltage peak may reduce the disruptive strength, as dielectric disruption requires energy, that is, not only voltage, but time also.

CHAPTER III

SYMBOLIC REPRESENTATION OF GENERAL ALTERNATING WAVES

28. The vector representation,

$$A = a^1 + ja^{11} = a (\cos \theta + j \sin \theta)$$

of the alternating wave,

$$A = a_0 \cos (\phi - \theta)$$

applies to the sine wave only.

The general alternating wave, however, contains an infinite series of terms, of odd frequencies,

$$A = A_1 \cos (\phi - \theta_1) + A_3 \cos (3 \phi - \theta_3) + A_5 \cos (5 \phi - \theta_5) +$$

thus cannot be directly represented by one complex vector quantity.

The replacement of the general wave by its equivalent sine wave, as before discussed, that is, a sine wave of equal effective intensity and equal power, while sufficiently accurate in many cases, completely fails in other cases, especially in circuits containing capacity, or in circuits containing periodically (and in synchronism with the wave) varying resistance or reactance (as alternating arcs, reaction machines, synchronous induction motors, oversaturated magnetic circuits, etc.).

Since, however, the individual harmonics of the general alternating wave are independent of each other, that is, all products of different harmonics vanish, each term can be represented by a complex symbol, and the equations of the general wave then are the resultants of those of the individual harmonics.

This can be represented symbolically by combining in one formula symbolic representations of different frequencies, thus,

$$\dot{A} = \sum_{1}^{\infty}{}_{2n-1} (a_n{}^1 + j_n a_n{}^{11}),$$ [1]

[1] The index $2n - 1$ in the Σ sign denotes that only the odd values of n are considered. If the wave contained even harmonics, the even values of n would also be considered, and the index in the Σ sign would be n.

where

$$j_n = \sqrt{-1},$$

and the index of the j_n merely denotes that the j's of different indices, n, while algebraically identical, physically represent different frequencies, and thus cannot be combined.

The general wave of e.m.f. is thus represented by

$$\dot{E} = \sum_1^\infty {}_{2n-1} (e_n^1 + j_n e_n^{11}),$$

the general wave of current by

$$\dot{I} = \sum_1^\infty {}_{2n-1} (i_n^1 + j_n i_n^{11}).$$

If

$$Z_1 = r + j\,(x_m + x_0 + x_c)$$

is the impedance of the fundamental harmonic, where

x_m is that part of the reactance which is proportional to the frequency (inductance, etc.).

x_0 is that part of the reactance which is independent of the frequency (mutual inductance, synchronous motion, etc.).

x_c is that part of the reactance which is inversely proportional to the frequency (capacity, etc.).

The impedance for the nth harmonic is

$$Z = r + j_n\left(n x_m + x_0 + \frac{x_c}{n}\right).$$

This term can be considered as the general symbolic expression of the impedance of a circuit of general wave-shape.

Ohm's law, in symbolic expression, assumes for the general alternating wave the form

$$\dot{I} = \frac{\dot{E}}{Z} \text{ or, } \sum_1^\infty {}_{2n-1} (i_n^1 + j_n i_n^{11}) = \sum_1^\infty {}_{2n-1} \frac{e_n^1 + j_n e_n^{11}}{r + j_n\left(n x_m + x_0 + \frac{x_c}{n}\right)};$$

$$\dot{E} = \dot{I}Z \text{ or, } \sum_1^\infty {}_{2n-1} (e_n^1 + j_n e_n^{11}) = \sum_1^\infty {}_{2n-1} \left[r + j_n\left(n x_m + x_0 + \frac{x_c}{n}\right)\right] (i_n^1 + j_n i_n^{11});$$

$$Z = \frac{\dot{E}}{\dot{I}} \text{ or } Z_n = r + j_n\left(n x_m + x_0 + \frac{x_c}{n}\right) = \frac{e_n + j_n e_n^{11}}{i_n^1 + j_n i_n^{11}}.$$

The symbols of multiplication and division of the terms, $\dot{E}$, $\dot{I}$, Z, thus represent, not algebraic operation, but multiplication and division of corresponding terms of $\underset{\cdot}{E}$, $\underset{\cdot}{I}$, $\underset{\cdot}{Z}$, that is, terms of the same index, n, or, in algebraic multiplication and division of the series, $\underset{\cdot}{E}$, $\underset{\cdot}{I}$, all compound terms, that is, terms containing two different n's, vanish.

29. The effective value of the general wave,
$a = A_1 \cos(\phi - \theta_1) + A_3 \cos(3\phi - \theta_3) + A_5 \cos(5\phi - \theta_2) + ..$
is the square root of the sum of mean squares of individual harmonics,

$$A = \sqrt{\tfrac{1}{2}\{A_1^2 + A_3^2 + A_5^2 + \dots\}}.$$

Since, as discussed above, the compound terms of two different indices, n, vanish, the absolute value of the general alternating wave,

$$\underset{\cdot}{A} = \sum_1^\infty {}_{2n-1} \frac{a_n^1 + j_n a_n^{11}}{b_n^1 + j_n b_n^{11}},$$

is thus,

$$A = \sqrt{\sum_1^\infty {}_{2n-1} \frac{{a_n^1}^2 + {a_n^{11}}^2}{{b_n^1}^2 + {b_n^{11}}^2}},$$

which offers an easy means of reduction from symbolic to absolute values.

Thus, the absolute value of the e.m.f.,

$$\underset{\cdot}{E} = \sum_1^\infty {}_{2n-1} (e_n^1 + j_n e_n^{11}),$$

is

$$E = \sqrt{\sum_1^\infty {}_{2n-1} ({e_n^1}^2 + {e_n^{11}}^2)},$$

the absolute value of the current,

$$\underset{\cdot}{I} = \sum_1^\infty {}_{2n-1} (i_n^1 + j_n i_n^{11}),$$

is

$$I = \sqrt{\sum_1^\infty {}_{2n-1} ({i_n^1}^2 + {i_n^{11}}^2)}.$$

30. The double frequency power (torque, etc.) equation of the general alternating wave has the same symbolic expression as with the sine wave,

$$
\begin{aligned}
P &= [\dot{E}\dot{I}] \\
&= P^1 + jP^j \\
&= [\dot{E}\dot{I}]^1 + j[EI]j \\
&= \sum_1^\infty {}_{2n-1} (e_n{}^1 i_n{}^1 + e_n{}^{11} i_n{}^{11}) + \sum_1^\infty {}_{2n-1}\, j_n\, (e_n{}^{11} i_n{}^1 - e_n{}^1 i_n{}^{11}),
\end{aligned}
$$

where

$$P^1 = [\dot{E}\dot{I}]^1 = \sum_1^\infty {}_{2n-1} (e_n{}^1 i_n{}^1 + e_n{}^{11} i_n{}^{11}),$$

$$P^j = [\dot{E}\dot{I}]^j = \sum_1^\infty {}_{2n-1} \frac{j_n}{j} (e_n{}^{11} i_n{}^1 - e_n{}^1 i_n{}^{11}).$$

The j_n enters under the summation sign of the reactive or "wattless power," P^j, so that the wattless powers of the different harmonics cannot be algebraically added.

Thus,

The total "true power" of a general alternating-current circuit is the algebraic sum of the powers of the individual harmonics.

The total "reactive power" of a general alternating-current circuit is not the algebraic, but the absolute sum of the wattless powers of the individual harmonics.

Thus, regarding the reactive power as a whole, in the general alternating circuit no distinction can be made between lead and lag, since some harmonics may be leading, others lagging.

The apparent power, or total volt-amperes, of the circuit is

$$P_a = EI = \sqrt{\sum_1^\infty {}_{2n-1} (e_n{}^{1^2} + e_n{}^{11^2}) \sum_1^\infty {}_{2n-1} (i_n{}^{1^2} + i_n{}^{11^2})}.$$

The power-factor of the circuit is,

$$p = \frac{P^1}{P_a} = \frac{\sum_1^\infty {}_{2n-1} (e_n{}^1 i_n{}^1 + e_n{}^{11} i_n{}^{11})}{\sqrt{\sum_1^\infty {}_{2n-1} (e_n{}^{1^2} + e_n{}^{11^2}) \sum_1^\infty {}_{2n-1} (i_n{}^{1^2} + i_n{}^{11^2})}}.$$

The term "inductance factor," however, has no meaning any more, since the reactive powers of the different harmonics are not directly comparable.

The quantity

$$q_0 = \sqrt{1 - p^2}$$

has no physical significance, and is not $\dfrac{\text{reactive power}}{\text{total apparent power}}$

The term

$$\frac{P^i}{EI} = \sum_1^{\infty} {}_{2n-1} \frac{j_n}{j} \frac{e_n^{11} i_n^1 - e_n^{11} i_n^1}{EI}$$

$$= \sum_1^{\infty} {}_{2n-1} \frac{j_n}{j} q_n,$$

where

$$q_n = \frac{e_n^{11} i_n^1 - e_n^1 i_n^{11}}{EI},$$

consists of a series of inductance factors, q_n, of the individual harmonics.

As a rule, if

$$q^2 = \sum_1^{\infty} {}_{2n-1} q_n^2,$$

$$p^2 + q^2 < 1,$$

for the general alternating wave, that is, q differs from

$$q_0 = \sqrt{1 - p^2}.$$

The complex quantity,

$$V = \frac{\dot{P}}{P_a} = \frac{[\dot{E}\dot{I}]}{EI} = \frac{[\dot{E}\dot{I}]^1 + j[\dot{E}\dot{I}]^j}{EI}$$

$$= \frac{\sum_1^{\infty} {}_{2n-1} (e_n^1 i_n^1 + e_n^{11} i_n^{11}) + \sum_1^{\infty} {}_{2n-1} j_n (e_n^{11} i_n^1 - e_n^1 i_n^{11})}{\sqrt{\sum_1^{\infty} {}_{2n-1} ({e_n^1}^2 + {e_n^{11}}^2) \sum_1^{\infty} {}_{2n-1} ({i_n^1}^2 + {i_n^{11}}^2)}}$$

$$= p + \sum_1^{\infty} {}_{2n-1} j_n q_n,$$

takes in the circuit of the general alternating wave the same position as power-factor and inductance factor with the sine wave.

$\underset{\cdot}{V} = \frac{\dot{P}}{P_a}$ may be called the "*circuit-factor.*"

It consists of a real term, p, the power-factor, and a series of imaginary terms, $j_n q_n$, the inductance factors of the individual harmonics.

The absolute value of the circuit-factor,

$$v = \sqrt{p^2 + \sum_1^{\infty} {}_{2n-1} q_n^2},$$

as a rule, is < 1.

31. Some applications of this symbolism will explain its mechanism and its usefulness more fully.

First Example.—Let the e.m.f.,

$$\dot{E} = \sum_{1}^{5} {}_{2n-1} (e_n{}^1 + j_n e^{11}),$$

be impressed upon a circuit of the impedance,

$$Z = r + j_n \left(n x_m - \frac{x_c}{n} \right) = 10 + j_n \left(10n - \frac{30}{n} \right)$$

that is, containing resistance, r, inductive reactance x_m and condensive reactance x_c in series.

Let

$$e_1{}^1 = 720 \qquad e_1{}^{11} = -540$$
$$e_3{}^1 = 283 \qquad e_3{}^{11} = 283$$
$$e_5{}^1 = -104 \qquad e_5{}^{11} = -138$$

or,

$$e_1 = 900 \qquad \tan \theta_1 = 0.75$$
$$e_3 = 400 \qquad \tan \theta_3 = -1.0$$
$$e_5 = 173 \qquad \tan \theta_5 = -1.33$$

It is thus in symbolic expression,

$$Z_1 = 10 - 80 j_1 \qquad z_1 = 80.6$$
$$Z_3 = 10 \qquad z_3 = 10.0$$
$$Z_5 = 10 + 32 j_5 \qquad z_5 = 33.5,$$

and e.m.f.,

$$\dot{E} = (720 - 540 j_1) + (283 + 283 j_3) + (-104 - 138 j_5),$$

or, absolute,

$$E = 1000,$$

and current,

$$\dot{I} = \frac{\dot{E}}{Z} = \frac{720 - 540 j_1}{10 - 80 j_1} + \frac{283 + 283 j_3}{10} + \frac{-104 - 138 j_5}{10 + 32 j_5}$$
$$= (7.76 + 8.04 j_1) + (28.3 + 28.3 j_3) + (-4.86 + 1.73 j_5)$$

or, absolute,

$$I = 41.85,$$

of which is of fundamental frequency,

$$I_1 = 11.15$$

of triple frequency,

$$I_3 = 40$$

of quintuple frequency,

$$I_5 = 5.17.$$

The total apparent power of the circuit is

$$P_a = EI = 41{,}850.$$

The true power of the circuit is,

$$P^1 = [EI]^1 = 1240 + 16{,}000 + 270,$$
$$= 17{,}510,$$

the reactive power,

$$jP^j = j[EI]^j = -10{,}000\, j_1 + 850\, j_5;$$

thus, the total power,

$$\dot{P} = 17{,}510 - 10{,}000\, j_1 + 850\, j_5.$$

That is, the reactive power of the first harmonic is leading, that of the third harmonic zero, and that of the fifth harmonic lagging.

$$17{,}510 = I^2r, \text{ as obvious.}$$

The circuit-factor is,

$$\dot{V} = \frac{\dot{P}}{Pa} = \frac{[\dot{E}\dot{I}]}{EI}$$
$$= 0.418 - 0.239\, j_1 + 0.0203\, j_5,$$

or, absolute,

$$v = \sqrt{0.418^2 + 0.239^2 + 0.0203^2}.$$
$$= 0.482.$$

The power-factor is

$$p = 0.418.$$

The inductance factor of the first harmonic is $q_1 = -0.239$, that of the third harmonic $q_3 = 0$, and of the fifth harmonic $q_5 = +0.0203$.

Considering the waves as replaced by their equivalent sine waves, from the sine wave formula,

$$p^2 + q_0^2 = 1,$$

the inductance factor would be,

$$q_0 = 0.914,$$

and the phase angle,

$$\tan\theta = \frac{q_0}{p} = \frac{0.914}{0.418} = 2.8, \qquad \theta = 65.4°,$$

giving apparently a very great phase displacement, while in reality, of the 41.85 amp. total current, 40 amp. (the current of the third harmonic) are in phase with their e.m.f.

We thus have here a case of a circuit with complex harmonic waves which cannot be represented by their equivalent sine waves. The relative magnitudes of the different harmonics in the wave of current and of e.m.f. differ essentially, and the circuit has simultaneously a very low power-factor and a very low inductance factor; that is, a low power-factor exists without corresponding phase displacement, the circuit-factor being less than one-half.

Such circuits, for instance, are those including alternating arcs, reaction machines, synchronous induction motors, reactances with over-saturated magnetic circuit, high potential lines in which the maximum difference of potential exceeds the corona voltage, polarization cells and in general electrolytic conductors above the dissociation voltage of the electrolyte, etc. Such circuits cannot correctly, and in many cases not even approximately, be treated by the theory of the equivalent sine waves, but require the symbolism of the complex harmonic wave.

32. *Second Example.*—A condenser of capacity, $C_0 = 20$ mf. is connected into the circuit of a 60-cycle alternator giving a wave of the form,

$$e = E(\cos\phi - 0.10\cos 3\,\phi - 0.08\cos 5\,\phi + 0.06\cos 7\,\phi),$$

or, in symbolic expression,

$$\dot{E} = e(1_1 - 0.10_3 - 0.08_5 + 0.06_7).$$

The synchronous impedance of the alternator is

$$Z_0 = r_0 + j_n n x_0 = 0.3 + 5\,nj_n.$$

What is the apparent capacity, C, of the condenser (as calculated from its terminal volts and amperes) when connected directly with the alternator terminals, and when connected thereto through various amounts of resistance and inductive reactance?

The condensive reactance of the condenser is

$$x_c = \frac{10^6}{2\,\pi f C_0} = 132 \text{ ohms},$$

or, in symbolic expression,

$$-\,j_n\frac{x_c}{n} = -\,\frac{132}{n}\,j_n.$$

Let $Z_1 = r + j_n nx$ = impedance inserted in series with the condenser.

The total impedance of the circuit is then

$$Z = Z_0 + Z_1 - j_n \frac{x_c}{n} = (0.3 + r) + j_n \left([5 + x]\, n - \frac{132}{n}\right).$$

The current in the circuit is

$$\dot{I} = \frac{\dot{E}}{Z} = e\left[\frac{1}{(0.3 + r) + j\,(x - 127)} - \frac{0.1}{(0.3 + r) + j_3\,(3\,x - 29)} - \frac{0.08}{(0.3 + r) + j_5\,(5\,x - 1.4)} + \frac{0.06}{(0.3 + r) - j_7\,(7\,x + 16.1)}\right],$$

and the e.m.f. at the condenser terminals,

$$\dot{E}_1 = -j_n \frac{x_c \dot{I}}{n} = -e\left[\frac{132\,j_1}{(0.3 + r) + j_1\,(x - 127)} - \frac{4.4\,j_3}{(0.3+r)+j_3(3\,x-29)} - \frac{2.11\,j_5}{(0.3 + r) + j_5\,(5\,x - 1.4)} + \frac{1.13\,j_7}{(0.3 + r) + j_7\,(7\,x + 16.1)}\right];$$

thus the apparent condensive reactance of the condenser is

$$x_1 = \frac{E_1}{I},$$

and the apparent capacity,

$$C = \frac{10^6}{2\,\pi f x_1}.$$

(*a*) $x = 0$: Resistance, r, in series with the condenser. Reduced to absolute values it is

$$\frac{1}{x_1^2} = \frac{\dfrac{1}{(0.3+r)+16129} + \dfrac{0.01}{(0.3+r)^2+841} + \dfrac{0.0064}{(0.3+r)^2+1.96} + \dfrac{0.0036}{(0.3+r)^2+259}}{\dfrac{16129}{(0.3+r)^2+16129} + \dfrac{19.4}{(0.3+r)^2+841} + \dfrac{4.45}{(0.3+r)^2+1.96} + \dfrac{1.28}{(0.3+r)^2+259}}.$$

(*b*) $r = 0$: Inductive reactance, x, in series with the condenser. Reduced to absolute values it is

$$\frac{1}{x_1^2} = \frac{\dfrac{1}{0.09+(x-127)^2} + \dfrac{0\,.01}{0.09+(3\,x-29)^2} + \dfrac{0.0064}{0.09+(5\,x-1.4)^2} + \dfrac{0.0036}{0.09+(7\,x+16.1)^2}}{\dfrac{16129}{0.09+(x-127)^2} + \dfrac{19.4}{0.09+(3\,x-29)^2} + \dfrac{4.45}{0.09+(5\,x-1.4)^2} + \dfrac{1.28}{0.09+(7\,x+16.1)^2}}.$$

From $\frac{1}{{x_1}^2}$ are derived the values of apparent capacity,

$$C = \frac{10^6}{2\pi f x_1},$$

and plotted in Fig. 18 for values of r and x respectively varying from 0 to 22 ohms.

As seen, with neither additional resistance nor reactance in series to the condenser, the apparent capacity with this generator wave is 84 mf., or 4.2 times the true capacity, and gradually decreases with increasing series resistance, to C = 27 mf. = 1.35 times the true capacity at r = 13.2 ohms, or one-tenth the true capacity reactance. With r = 132 ohms, or with an additional resistance equal to the condensive reactance, C = 20.2 mf. or only

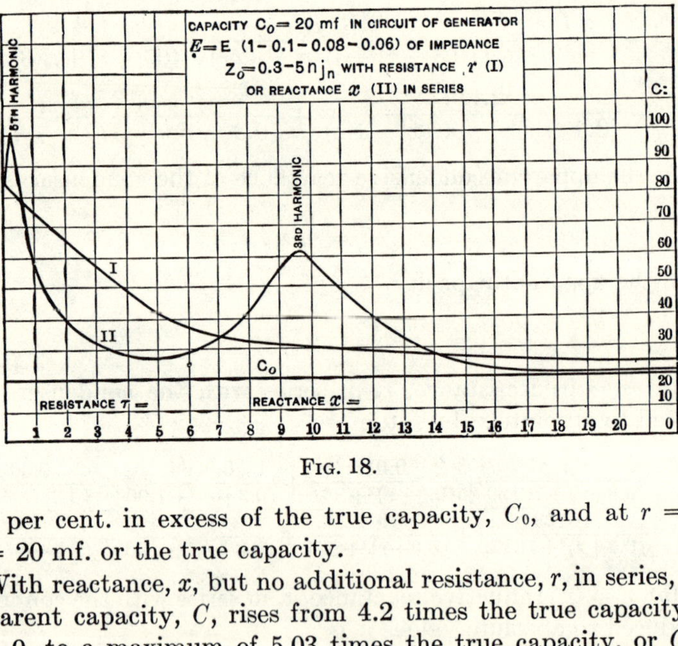

Fig. 18.

one per cent. in excess of the true capacity, C_0, and at $r = \infty$, C = 20 mf. or the true capacity.

With reactance, x, but no additional resistance, r, in series, the apparent capacity, C, rises from 4.2 times the true capacity at x = 0, to a maximum of 5.03 times the true capacity, or C = 100.6 mf. at x = 0.28, the condition of resonance of the fifth harmonic, then decreases to a minimum of 27 mf., or 35 per cent. in excess of the true capacity, rises again to 60.2 mf., or 3.01 times the true capacity at x = 9.67, the condition of resonance with the third harmonic, and finally decreases, reaching 20 mf., or the true capacity at x = 132, or an inductive reactance equal to the condensive reactance.

It thus follows that the true capacity of a condenser cannot even approximately be determined by measuring volts and amperes if there are any higher harmonics present in the generator wave, except by inserting a very large resistance or reactance in series to the condenser.

33. *Third Example.*—An alternating-current generator of the wave,

$$E_0 = 2000\,[1_1 + 0.12_3 - 0.23_5 - 0.13_7],$$

and of synchronous impedance,

$$Z_0 = 0.3 + 5\,nj_n,$$

feeds over a line of impedance,

$$Z_1 = 2 + 4\,nj_n,$$

a synchronous motor of the wave,

$$\dot{E}_1 = 2250\,[(\cos\theta - j_1 \sin\theta) + 0.24\,(\cos 3\,\theta - j_3 \sin 3\,\theta)],$$

and of synchronous impedance,

$$Z_2 = 0.3 + 6\,nj_n.$$

The total impedance of the system is then,

$$Z = Z_0 + Z_1 + Z_2 = 2.6 + 15\,nj_n,$$

thus the current,

$$\dot{I} = \frac{\dot{E}_0 - \dot{E}_1}{Z}$$

$$= \frac{2000 - 2250\cos\theta + 2250\,j_1\sin\theta}{2.6 + 15\,j_1} + \frac{240 - 540\cos 3\,\theta + 540\,j_3 \sin 3\,\theta}{2.6 - 45\,j_3}$$

$$- \frac{460}{2.6 + 75\,j_5} - \frac{260}{2.6 + 105\,j_7}$$

$$= (a_1{}^1 - j_1 a_1{}^{11}) + (a_3{}^1 - j_3 a_3{}^{11}) + (a_5{}^1 - j_5 a_5{}^{11}) + (a_7{}^1 - j_7 a_7{}^{11});$$

where

$$\begin{aligned}
a_1{}^1 &= 22.5 - 25.2\cos\theta + 146\sin\theta,\\
a_3{}^1 &= 0.306 - 0.69\cos 3\,\theta + 11.9\sin 3\,\theta,\\
a_5{}^1 &= 0.213,\\
a_7{}^1 &= -\,0.061,\\
a_1{}^{11} &= 130 - 146\cos\theta - 25.2\sin\theta,\\
a_3{}^{11} &= 5.3 - 11.9\cos 3\,\theta - 0.69\sin 3\,\theta,\\
a_5{}^{11} &= -\,6.12,\\
a_7{}^{11} &= -\,2.48,
\end{aligned}$$

or, absolute,

first harmonic,

$$a_1 = \sqrt{a_1^{12} + a_1^{112}},$$

third harmonic,

$$a_3 = \sqrt{a_3^{12} + a_3} \ ,$$

fifth harmonic,

$$a_5 = 6.12,$$

seventh harmonic,

$$a_7 = 2.48,$$

$$I = \sqrt{a_1^2 + a_3^2 + a_5^2 + a_7^2};$$

while the total current of higher harmonics is

$$I_0 = \sqrt{a_3^2 + a_6^2 + a_7^2}.$$

The true input of the synchronous motor is

$$P^1 = [E_1 I]^1$$

$$= (2250\, a_1^1 \cos\theta + 2250\, a_1^{11} \sin\theta) + (540\, a_1^1 \cos 3\,\theta + 540\, a_3^{11} \sin 3\,\theta$$

$$= P_1^1 + P_3^1$$

$$P_1^1 = 2250\,(a_1^1 \cos\theta + a_1^{11} \sin\theta),$$

is the power of the fundamental wave,

$$P_3^1 = 540\,(a_3^1 \cos 3\,\theta + a_3^{11} \sin 3\,\theta),$$

the power of the third harmonic.

The fifth and seventh harmonics do not give any power, since they are not contained in the synchronous motor wave. Substituting now different numerical values for θ, the phase angle between generator e.m.f. and synchronous motor counter e.m.f., corresponding values of the currents, I, I_0, and the powers, P^1, P_1^1, P_3^1, are derived. These are plotted in Fig. 19 with the total current, I, as abscissas. To each value of the total current, I, correspond two values of the total power, P^1, a positive value plotted as Curve I—synchronous motor—and a negative value plotted as Curve II—alternating-current generator—. Curve III gives the total current of higher frequency, I_0, Curve IV the difference between the total current and the current of fundamental frequency, $I—I_1$, in percentage of the total current, I, and V the power of the third harmonic, P_3^1, in percentage of the total power, P^1.

Curves III, IV, and V correspond to the positive or synchronous motor part of the power curve, P^1. As seen, the increase of current due to the higher harmonics is small, and entirely disappears at about 180 amp. The power of the third harmonic is positive, that is, adds to the power of the synchronous motor

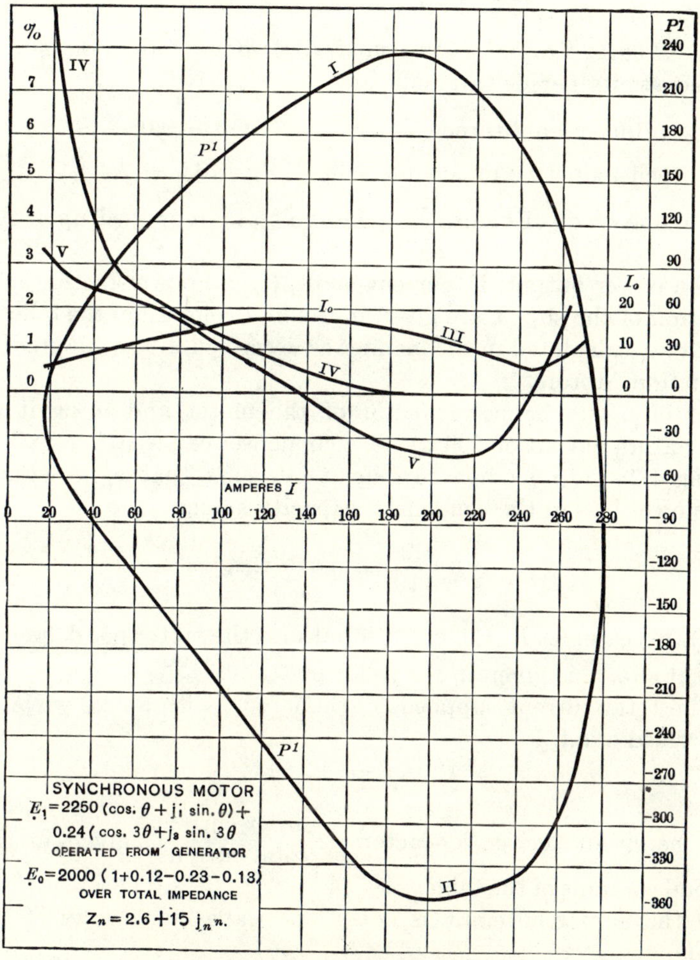

FIG. 19. —Synchronous motor.

up to about 140 amp. or near the maximum output of the motor, and then becomes negative.

It follows herefrom that higher harmonics in the e.m.f. waves of generators and synchronous motors do not represent a mere waste of current, but may contribute more or less to the output of the motor. Thus at 75 amp. total current, the percentage of increase of power due to the higher harmonic is equal to the increase of current, or in other words the higher harmonics of current do work with the same efficiency as the fundamental wave.

34. *Fourth Example.*—In a small three-phase induction motor, the constants per delta circuit are

primary admittance $Y = 0.002 - 0.03\,j,$

self-inductive impedance $Z_0 = Z_1 = 0.6 + 2.4\,j,$:

and a sine wave of e.m.f., $e_0 = 110$ volts, is impressed upon the motor.

The power output, P, current input, I_s, and power-factor, p, as function of the slip, s, are given in the first columns of the following table, calculated in the manner as described in the chapter on Induction Motors.

To improve the power-factor of the motor and bring it to unity at an output of 500 watts, a condenser capacity is required giving 4.28 amp. leading current at 110 volts, that is, neglecting the power loss in the condenser, capacity susceptance

$$\frac{4.28}{110} = 0.039.$$

In this case, let I_s = current input into the motor per delta circuit at slip s, as given in the following table.

The total current supplied by the circuit with a sine wave of impressed e.m.f. is

$$\dot{I}' = \dot{I}_s + 4.28\,j,$$

and herefrom the power-factor $= \dfrac{\text{power current}}{\text{total current}}$, given in the second columns of the table.

If the impressed e.m.f. is not a sine wave but a wave of the shape,

$$\dot{E}_0 = e_0(1_1 + 0.12_3 - 0.23_5 - 0.134_7),$$

to give the same output, the fundamental wave must be the same: $e_0 = 110$ volts, when assuming the higher harmonics in the motor as wattless, that is,

$$\dot{E}_0 = 110_1 + 13.2_3 - 25.3_5 - 14.7_7 = e_0 + \dot{E}_0{}^1,$$

where $\dot{E}_0^1 = 13.2_3 - 25.3_5 - 14.7_7$

$= $ component of impressed e.m.f. of higher frequency.

The effective value is

$$E_0 = 114.5 \text{ volts.}$$

The condenser admittance for the general alternating wave is

$$Y_c = 0.039\, nj_n.$$

Since the frequency of rotation of the motor is small compared with the frequency of the higher harmonics, as total impedance of the motor for these higher harmonics can be assumed the stationary impedance, and by neglecting the resistance we have

$$Z^1 = nj_n(x_0 + x_1) = 4.8\, nj_n$$

The exciting admittance of the motor, for these higher harmonics, is, by neglecting the conductance,

$$Y^1 = -\frac{bj_n}{n} = -\frac{0.03\, j_n}{n},$$

and the higher harmonics of counter e.m.f.,

$$\dot{E}^1 = \frac{E_0^1}{2}.$$

Thus we have,

current input in the condenser,

$$\dot{I}_c = \dot{E}_0 Y_c = +\, 4.28\, j_1 + 1.54\, j_3 - 4.93\, j_5 - 4.02\, j_7;$$

high-frequency component of motor-impedance current,

$$\frac{\dot{E}_0^1}{\dot{Z}^1} = -\, 0.92\, j_3 + 1.06\, j_5 + 0.44\, j_7;$$

high-frequency component of motor-exciting current,

$$\dot{E}^1 Y^1 = \frac{\dot{E}_0^1 Y^1}{2} = -\, 0.07\, j_3 + 0.08\, j_5 + 0.03\, j_7:$$

thus, total high-frequency component of motor current,

$$\dot{I}_0^1 = \frac{\dot{E}_0^1}{Z^1} + \dot{E}^1 Y^1 = -\, 0.99\, j_3 + 1.14\, j_5 + 0.47\, j_7,$$

and total current, without condenser,

$$\underset{\cdot}{I_0} = \underset{\cdot}{I_s} + \underset{\cdot}{I_0^1} = \underset{\cdot}{I_s} - 0.99\, j_3 + 1.14\, j_5 + 0.47\, j_7,$$

with condenser,

$$\underset{\cdot}{I} = \underset{\cdot}{I_s} + \underset{\cdot}{I_0^1} - \underset{\cdot}{I_c} = \underset{\cdot}{I_s} + 4.28\, j_1 + 0.55\, j_3 - 3.79\, j_5 + 3.55\, j_7;$$

and herefrom the power-factor.

In the following table and in Fig. 20 are given the values of current and power-factor:

I. With sine wave of e.m.f., of 110 volts, and no condenser.
II. With sine wave of e.m.f., of 110 volts, and with condenser.
III. With distorted wave of e.m.f., of 114.5 volts, and no condenser.
IV. With distorted wave of e.m.f., of 114.5 volts, and with condenser.

TABLE

		I.			II.		III.		IV.	
s	P	I_s	I_s	p	I'	p	I_0	p	I	p
0.0	0	0.24+ 3.10 j	3.1	7.8	1.2	20.0	3.5	6.6	5.2	4.4
0.01	160	1.73+ 3.16 j	3.6	48.0	2.1	84.0	3.9	43.0	5.5	31.0
0.02	320	3.32+ 3.47 j	4.8	69.0	3.4	97.2	5.1	64.0	6.1	54.0
0.035	500	5.16+ 4.28 j	6.7	77.0	5.2	100.0	6.9	72.5	7.2	68.0
0.05	660	6.95+ 5.4 j	8.8	79.0	7.0	98.7	8.9	76.0	8.6	77.0
0.07	810	8.77+ 7.3 j	11.4	77.0	9.3	94.5	11.5	73.5	10.6	80.0
0.10	885	10.1 + 9.85 j	14.1	71.5	11.5	87.0	14.2	68.0	12.6	77.0
0.13	900	10.45+11.45 j	15.5	67.5	12.7	82.0	15.6	64.5	13.7	73.0
0.15	890	10.75+12.9 j	16.8	64.0	13.8	78.0	16.9	61.0	14.7	70.0

The curves II and IV with condenser are plotted in dotted lines in Fig. 20 As seen, even with such a distorted wave the current input and power-factor of the motor are not much changed if no condenser is used. When using a condenser in shunt to the motor, however, with such a wave of impressed e.m.f. the increase of the total current, due to higher-frequency currents in the condenser, is greater than the decrease, due to the compensation of lagging currents, and the power-factor is actually lowered by the condenser, over the total range of load up to overload, and especially at light load.

Where a compensator or transformer is used for feeding the condenser, due to the internal self-inductance of the compensator, the higher harmonics of current are still more accentuated, that is, the power-factor still more lowered.

In the preceding the energy loss in the condenser and compensator and that due to the higher harmonics of current in the motor has been neglected. The effect of this energy loss is a slight decrease of efficiency and corresponding increase of power-factor. The power produced by the higher harmonics has also been neglected; it may be positive or negative, according to the index

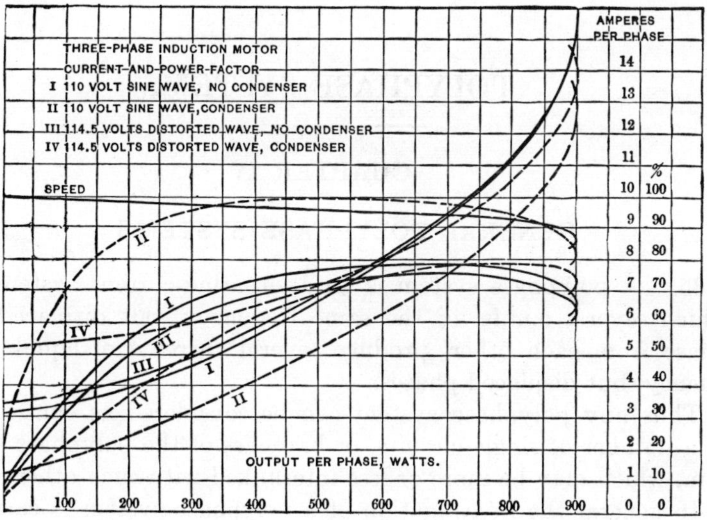

Fig. 20.

of the harmonic, and the winding of the motor primary. Thus, for instance, the effect of the triple harmonic is negative in the quarter-phase motor, zero in the three-phase motor, etc.; altogether, however, the effect of these harmonics is usually small.

POLYPHASE SYSTEMS

CHAPTER IV

GENERAL POLYPHASE SYSTEMS

35. A polyphase system is an alternating-current system in which several e.m.fs. of the same frequency, but displaced in phase from each other, produce several currents of equal frequency, but displaced phases.

Thus any polyphase system can be considered as consisting of a number of single circuits, or branches of the polyphase system, which may be more or less interlinked with each other.

In general the investigation of a polyphase system is carried out by treating the single-phase branch circuits independently.

Thus all the discussions on generators, synchronous motors, induction motors, etc., in the preceding chapters, apply to single-phase systems as well as polyphase systems, in the latter case the total power being the sum of the powers of the individual or branch circuits.

If the polyphase system consists of n equal e.m.fs. displaced from each other by $\frac{1}{n}$ of a period, the system is called a *symmetrical system*, otherwise an *unsymmetrical system*.

Thus the three-phase system, consisting of three equal e.m.fs. displaced by one-third of a period, is a symmetrical system. The quarter-phase system, consisting of two equal e.m.fs. displaced by 90°, or one-quarter of a period, is an unsymmetrical system.

36. The power in a single-phase system is pulsating; that is, the watt curve of the circuit is a sine wave of double frequency, alternating between a maximum value and zero, or a negative maximum value. In a polyphase system the watt curves of the different branches of the system are pulsating also. Their sum, however, or the total power of the system, may be either constant or pulsating. In the first case, the system is called a *balanced system,* in the latter case an *unbalanced system.*

The three-phase system and the quarter-phase system, with equal load on the different branches, are balanced systems; with unequal distribution of load between the individual branches both systems become unbalanced systems.

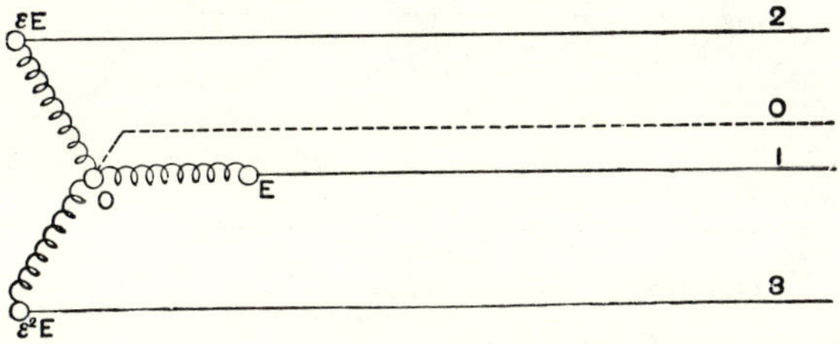

FIG. 21.

The different branches of a polyphase system may be either independent from each other, that is, without any electrical interconnection, or they may be interlinked with each other. In the first case the polyphase system is called an *independent system,* in the latter case an *interlinked system.*

The three-phase system with star-connected or ring-connected generator, as shown diagrammatically in Figs. 21 and 22, is an interlinked system.

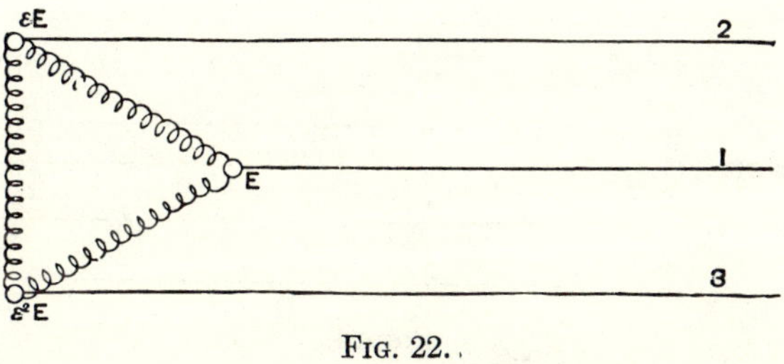

FIG. 22.

The four-phase system as derived by connecting four equidistant points of a continuous-current armature with four collector rings, as shown diagrammatically in Fig. 23, is an interlinked system also. The four-wire, quarter-phase system produced by a generator with two independent armature coils, or by two single-phase generators rigidly connected with each other in quadrature, is an independent system. As interlinked system, it is shown in Fig. 24, as star-connected, four-phase system.

37. Thus, polyphase systems can be subdivided into:

Symmetrical systems and unsymmetrical systems.

Balanced systems and unbalanced systems.

Interlinked systems and independent systems.

The only polyphase systems which have found practical application are:

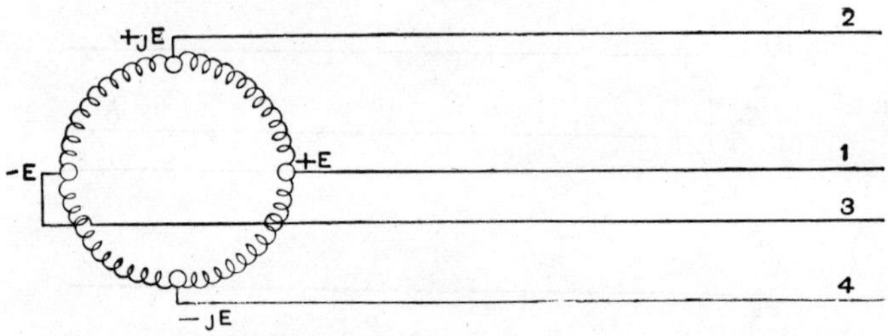

FIG. 23.

The three-phase system, consisting of three e.m.fs. displaced by one-third of a period, is used exclusively as interlinked system.

The quarter-phase system, consisting of two e.m.fs. in quadrature, and used with four wires, or with three wires, which may be either an interlinked system or an independent system.

The six-phase system, consisting of two three-phase systems in opposition to each other, and derived by transformation from

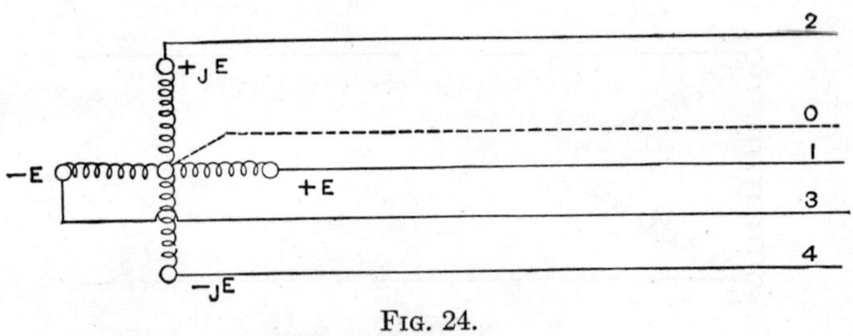

FIG. 24.

a three-phase system, in the alternating supply circuit of large synchronous converters.

The inverted three-phase system, consisting of two e.m.fs. displaced from each other by 60°, and derived from two phases of a three-phase system by transformation with two transformers, of which the secondary of one is reversed with regard to its primary (thus changing the phase difference from 120° to 180° − 120° = 60°) finds a limited application in low-tension distribution.

CHAPTER V

SYMMETRICAL POLYPHASE SYSTEMS

38. If all the e.m.fs. of a polyphase system are equal in intensity and differ from each other by the same angle of difference of phase, the system is called a symmetrical polyphase system.

Hence, a symmetrical n-phase system is a system of n e.m.fs. of equal intensity, differing from each other in phase by $\frac{1}{n}$ of a period:

$$\begin{aligned} e_1 &= E \sin \beta; \\ e_2 &= E \sin \left(\beta - \frac{2\pi}{n}\right); \\ e_3 &= E \sin \left(\beta - \frac{4\pi}{n}\right); \\ &\vdots \\ e_n &= E \sin \left(\beta - \frac{2(n-1)\pi}{n}\right). \end{aligned}$$

The next e.m.f. is, again,

$$e_1 = E \sin (\beta - 2\pi) = E \sin \beta.$$

In the vector diagram the n e.m.fs. of the symmetrical n-phase system are represented by n equal vectors, following each other under equal angles.

Since in symbolic writing rotation by $\frac{1}{n}$ of a period, or angle $\frac{2\pi}{n}$, is represented by multiplication with

$$\cos \frac{2\pi}{n} + j \sin \frac{2\pi}{n} = \epsilon,$$

the e.m.fs. of the symmetrical polyphase system are

$$\dot{E};$$

$$\dot{E} \left(\cos \frac{2\pi}{n} + j \sin \frac{2\pi}{n}\right) = \dot{E}\epsilon;$$

$$\dot{E}\left(\cos\frac{4\pi}{n}+j\sin\frac{4\pi}{n}\right)=\dot{E}\epsilon^2;$$

$$\dot{E}\left(\cos\frac{2(n-1)\pi}{n}+j\sin\frac{2(n-1)\pi}{n}\right)=\dot{E}\epsilon^{n-1}.$$

The next e.m.f. is again,

$$\dot{E}(\cos 2\pi+j\sin 2\pi)=\dot{E}\epsilon^n=\dot{E}.$$

Hence, it is

$$\epsilon=\cos\frac{2\pi}{n}+j\sin\frac{2\pi}{n}=\sqrt[n]{1}.$$

Or in other words:

In a symmetrical n-phase system any e.m.f. of the system is expressed by

$$\epsilon^i\dot{E};$$

where

$$\epsilon=\sqrt[n]{1}.$$

39. Substituting now for n different values, we get the different symmetrical polyphase systems, represented by

$$\epsilon^i\dot{E},$$

where

$$\epsilon=\sqrt[n]{1}=\cos\frac{2\pi}{n}+j\sin\frac{2\pi}{n}.$$

1. $n=1,\quad \epsilon=1,\quad \epsilon^i\dot{E}=\dot{E},$

the ordinary single-phase system.

2. $n=2,\quad \epsilon=-1,\quad \epsilon^i\dot{E}=\dot{E}$ and $-\dot{E}.$

Since $-\dot{E}$ is the return of $\dot{E}$, $n=2$ gives again the single-phase system.

3. $n=3,\quad \epsilon=\cos\frac{2\pi}{3}+j\sin\frac{2\pi}{3}=\frac{-1+j\sqrt{3}}{2}$

$$\epsilon^2=\frac{-1-j\sqrt{3}}{2}.$$

The three e.m.fs. of the three-phase system are

$$\epsilon^i\dot{E}=\dot{E},\quad\frac{-1+j\sqrt{3}}{2}\dot{E},\quad\frac{-1-j\sqrt{3}}{2}\dot{E}.$$

Consequently the three-phase system is the lowest symmetrical polyphase system.

4. $n = 4, \quad \epsilon = \cos \frac{2\pi}{4} + j \sin \frac{2\pi}{4} = j, \quad \epsilon^2 = -1, \quad \epsilon^3 = -j.$

The four e.m.fs. of the four-phase system are,

$$\epsilon^i \dot{E} = \dot{E}, \quad j\ddot{E}, \quad -\dot{E}, \quad -j\dot{E}.$$

They are in pairs opposite to each other,

$$\dot{E} \text{ and } -\dot{E}; \; j\dot{E} \text{ and } -j\dot{E}.$$

Hence can be produced by two coils in quadrature with each other, analogous as the two-phase system, or ordinary alternating current system, can be produced by one coil.

Thus the symmetrical quarter-phase system is a four-phase system.

Higher systems than the quarter-phase or four-phase system have not been very extensively used, and are thus of less practical interest. A symmetrical six-phase system, derived by transformation from a three-phase system, has found application in synchronous converters, as offering a higher output from these machines, and a symmetrical eight-phase system proposed for the same purpose.

40. . A characteristic feature of the symmetrical n-phase system is that under certain conditions it can produce a rotating m.m.f. of constant intensity.

If n equal magnetizing coils act upon a point under equal angular displacements in space, and are excited by the n e.m.fs. of a symmetrical n-phase system, a m.m.f. of constant intensity is produced at this point, whose direction revolves synchronously with uniform velocity.

Let

n' = number of turns of each magnetizing coil.
E = effective value of impressed e.m.f.
I = effective value of current.

Hence,

$F = n'I$ = effective m.m.f. of one of the magnetizing coils.

Then the instantaneous value of the m.m.f. of the coil acting in the direction, $\frac{2\pi i}{n}$, is

$$f_i = F\sqrt{2} \sin \left(\beta - \frac{2\pi i}{n}\right)$$

$$= n'I\sqrt{2} \sin \left(\beta - \frac{2\pi i}{n}\right).$$

The two rectangular space components of this m.m.f. are

$$f_i' = f_i \cos \frac{2\pi i}{n}$$
$$= n'I\sqrt{2} \cos \frac{2\pi i}{n} \sin \left(\beta - \frac{2\pi i}{n}\right).$$

and

$$f_i'' = f_i \sin \frac{2\pi i}{n}$$
$$= n'I\sqrt{2} \sin \frac{2\pi i}{n} \sin \left(\beta - \frac{2\pi i}{n}\right).$$

Hence the m.m.f. of this coil can be expressed by the symbolic formula

$$f_i = n'I\sqrt{2} \sin \left(\beta - \frac{2\pi i}{n}\right)\left(\cos \frac{2\pi i}{n} + j \sin \frac{2\pi i}{n}\right).$$

Thus the total or resultant m.m.f. of the n coils displaced under the n equal angles is

$$f = \sum_1^n{}^i f_i = n'I\sqrt{2} \sum_1^n{}^i \sin \left(\beta - \frac{2\pi i}{n}\right)\left(\cos \frac{2\pi i}{n} + j \sin \frac{2\pi i}{n}\right).$$

or, expanded,

$$f = n'I\sqrt{2}\left\{\sin\beta \sum_1^n{}^i \left(\cos^2 \frac{2\pi i}{n} + j \sin \frac{2\pi i}{n} \cos \frac{2\pi i}{n}\right)\right.$$
$$\left. - \cos\beta \sum_1^n{}^i \left(\sin \frac{2\pi i}{n} \cos \frac{2\pi i}{n} + j \sin^2 \frac{2\pi i}{n}\right)\right\}.$$

It is, however,

$$\cos^2 \frac{2\pi i}{n} + j \sin \frac{2\pi i}{n} \cos \frac{2\pi i}{n} = \tfrac{1}{2}\left(1 + \cos \frac{4\pi i}{n} + j \sin \frac{4\pi i}{n}\right)$$
$$= \tfrac{1}{2}(1 + \epsilon^{2i}),$$

$$\sin \frac{2\pi i}{n} \cos \frac{2\pi i}{n} + j \sin^2 \frac{2\pi i}{n} = \frac{j}{2}\left(1 - \cos \frac{4\pi i}{n} - j \sin \frac{4\pi i}{n}\right)$$
$$= \frac{j}{2}(1 - \epsilon^{2i});$$

and, since

$$\sum_1^n{}^i \epsilon^{2i} = 0, \qquad \sum_1^n{}^i \epsilon^{-2i} = 0,$$

it is,

$$f = \frac{nn'I\sqrt{2}}{2}(\sin\beta - j\cos\beta);$$

or,

$$f = \frac{nn'I}{\sqrt{2}}(\sin\beta - j\cos\beta)$$

$$= \frac{nF}{\sqrt{2}}(\sin\beta - j\cos\beta);$$

the symbolic expression of the m.m.f. produced by the n circuits of the symmetrical n-phase system, when exciting n equal magnetizing coils displaced in space under equal angles.

The absolute value of this m.m.f. is

$$F_0 = \frac{nn'I}{\sqrt{2}} = \frac{nF}{\sqrt{2}} = \frac{nF_{max}}{2}.$$

Hence constant and equal $\frac{n}{\sqrt{2}}$ times the effective m.m.f. of each coil or $\frac{n}{2}$ times the maximum m.m.f. of each coil.

The phase of the resultant m.m.f. at the time represented by the angle β is

$$\tan\theta = -\cot\beta;\ \text{hence}\ \theta = -\beta\frac{\pi}{2}.$$

That is, the m.m.f. produced by a symmetrical n-phase system revolves with constant intensity,

$$F_0 = \frac{nF}{\sqrt{2}},$$

and constant speed, in synchronism with the frequency of the system; and, if the reluctance of the magnetic circuit is constant, the magnetism revolves with constant intensity and constant speed also, at the point acted upon symmetrically by the n m.m.fs. of the n-phase system.

This is a characteristic feature of the symmetrical polyphase system.

41. In the three-phase system, $n = 3$, $F_0 = 1.5\ F_{max}$, where F_{max} is the maximum m.m.f. of each of the magnetizing coils.

In a symmetrical quarter-phase system, $n = 4$, $F_0 = 2\ F_{max}$, where F_{max} is the maximum m.m.f. of each of the four magnetizing coils, or, if only two coils are used, since the four-phase m.m.fs. are opposite in phase by two, $F_0 = F_{max}$, where F_{max} is the maximum m.m.f. of each of the two magnetizing coils of the quarter-phase system.

While the quarter-phase system, consisting of two e.m.fs. displaced by one-quarter of a period, is by its nature an unsymmetrical system, it shares a number of features—as, for instance, the ability of producing a constant-resultant m.m.f.—with the symmetrical system, and may be considered as one-half of a symmetrical four-phase system.

Such systems, consisting of one-half of a symmetrical system, are called *hemisymmetrical systems*.

CHAPTER VI

BALANCED AND UNBALANCED POLYPHASE SYSTEMS

42. If an alternating e.m.f.,

$$e = E\sqrt{2} \sin \beta,$$

produces a current,

$$i = I\sqrt{2} \sin (\beta - \theta),$$

where θ is the angle of lag, the power is

$$\begin{aligned} p = ei &= 2\, EI \sin \beta \sin (\beta - \theta) \\ &= EI\, (\cos \theta - \cos (2\, \beta - \theta)), \end{aligned}$$

and the average value of power,

$$P = EI \cos \theta.$$

Substituting this, the instantaneous value of power is found as

$$p = P\left(1 - \frac{\cos (2\, \beta - \theta)}{\cos \theta}\right).$$

Hence the power, or the flow of energy, in an ordinary single-phase, alternating-current circuit is fluctuating, and varies with twice the frequency of e.m.f. and current, unlike the power of a continuous-current circuit, which is constant,

$$p = ei.$$

If the angle of lag, $\theta = 0$, it is,

$$p = P(1 - \cos 2\, \beta);$$

hence the flow of energy varies between zero and $2\, P$, where P is the average flow of energy or the effective power of the circuit.

If the current lags or leads the e.m.f. by angle θ, the power varies between

$$P\left(1 - \frac{1}{\cos \theta}\right) \text{and } P\left(1 + \frac{1}{\cos \theta}\right),$$

that is, becomes negative for a certain part of each half-wave. That is, for a time during each half-wave, energy flows back into the generator, while during the other part of the half-wave the generator sends out energy, and the difference between both is the effective power of the circuit.

If $\theta = 90°$, it is

$$p = -EI \sin 2\beta;$$

that is, the effective power $P = 0$, and the energy flows to and fro between generator and receiving circuit.

Under any circumstances, however, the flow of energy in the single-phase system is fluctuating, at least between zero and a maximum value, frequently even reversing.

43. If in a polyphase system

$e_1, e_2, e_3, \ldots . = $ instantaneous values of e.m.f.;
$i_1, i_2, i_3, \ldots . = $ instantaneous values of current produced thereby,

the total power in the system is

$$p = e_1 i_1 + e_2 i_2 + e_3 i_3 + \ldots .$$

The average power is

$$P = E_1 I_1 \cos \theta_1 + E_2 I_2 \cos \theta_2 + \ldots .$$

The polyphase system is called a balanced system, if the flow of energy

$$p = e_1 i_1 + e_2 i_2 + e_3 i_3 + \ldots .$$

is constant, and it is called an unbalanced system if the flow of energy varies periodically, as in the single-phase system; and the ratio of the minimum value to the maximum value of power is called the *balance-factor of the system.*

Hence in a single-phase system on non-inductive circuit, that is, at no-phase displacement, the balance-factor is zero; and it is negative in a single-phase system with lagging or leading current, and becomes equal to -1 if the phase displacement is 90°—that is, the circuit is wattless.

44. Obviously, in a polyphase system the balance of the system is a function of the distribution of load between the different branch circuits.

A balanced system in particular is called a polyphase system, whose flow of energy is constant, if all the circuits are loaded equally with a load of the same character, that is, the same phase displacement.

45. All the symmetrical systems from the three-phase system upward are balanced systems. Many unsymmetrical systems are balanced systems also.

1. Three-phase system:

Let

$$e_1 = E\sqrt{2}\sin\beta, \qquad \text{and } i_1 = I\sqrt{2}\sin(\beta-\theta),$$
$$e_2 = E\sqrt{2}\sin(\beta-120), \qquad i_2 = I\sqrt{2}\sin(\beta-\theta-120),$$
$$e_3 = E\sqrt{2}\sin(\beta-240), \qquad i_3 = I\sqrt{2}\sin(\beta-\theta-240),$$

be the e.m.fs. of a three-phase system and the currents produced thereby.

Then the total power is

$$p = 2\,EI\{\sin\beta\sin(\beta-\theta) + \sin(\beta-120)\sin(\beta-\theta-120) + \sin(\beta-240)\sin(\beta-\theta-240)\}$$
$$= 3\,EI\cos\theta = P, \text{ or constant.}$$

Hence the symmetrical three-phase system is a balanced system.

2. Quarter-phase system:

Let

$$e_1 = E\sqrt{2}\sin\beta, \qquad i_1 = I\sqrt{2}\sin(\beta-\theta),$$
$$e_2 = E\sqrt{2}\cos\beta, \qquad i_2 = I\sqrt{2}\cos(\beta-\theta)$$

be the e.m.fs. of the quarter-phase system, and the currents produced thereby.

This is an unsymmetrical system, but the instantaneous value of power is

$$p = 2\,EI\,\{\sin\beta\sin(\beta-\theta) + \cos\beta\cos(\beta-\theta)\}$$
$$= 2\,EI\cos\theta = P, \text{ or constant.}$$

Hence the quarter-phase system is an unsymmetrical balanced system.

3. The symmetrical n-phase system, with equal load and equal phase-displacement in all n branches, is a balanced system. For, let

$$e_i = E\sqrt{2}\sin\left(\beta - \frac{2\,\pi i}{n}\right) = \text{e.m.f.};$$

$$i_i = I\sqrt{2}\sin\left(\beta - \theta - \frac{2\,\pi i}{n}\right) = \text{current};$$

the instantaneous value of power is

$$p = \sum_1^n {}^i e_i i_i$$

$$= 2\,EI \sum_1^n {}^i \sin\left(\beta - \frac{2\,\pi i}{n}\right) \sin\left(\beta - \theta - \frac{2\,\pi i}{n}\right)$$

$$= EI \left\{ \sum_1^n {}^i \cos\theta - \sum_1^n {}^i \cos\left(2\,\beta - \theta - \frac{4\,\pi i}{n}\right)\right\};$$

or $$p = nEI \cos\theta = P, \text{ or constant.}$$

46. An unbalanced polyphase system is the so-called inverted three-phase system, derived from two branches of a three-phase system by transformation by means of two transformers, whose secondaries are connected in opposite direction with respect to their primaries. Such a system takes an intermediate position between the Edison three-wire system and the three-phase system. It shares with the latter the polyphase feature, and with the Edison three-wire system the feature that the potential difference between the outside wires is higher than between middle wire and outside wire.

By such a pair of transformers the two primary e.m.fs. of 120° displacement of phase are transformed into two secondary e.m.fs., differing from each other by 60°. Thus in the secondary circuit the difference of potential between the outside wires is $\sqrt{3}$ times the difference of potential between middle wire and outside wire. At equal load on the two branches, the three currents are equal, and differ from each other by 120°, that is, have the same relative proportion as in a three-phase system. If the load on one branch is maintained constant, while the load of the other branch is reduced from equality with that in the first branch down to zero, the current in the middle wire first decreases, reaches a minimum value of $\frac{\sqrt{3}}{2} = 0.866$ of its original value, and then increases again, reaching at no-load the same value as at full-load.

The balance factor of the inverted three-phase system on non-inductive load is 0.333.

47. In Figs. 25 to 32 are shown the e.m.fs., as e and currents as i in full lines, and the power as p in dotted lines, for balance-factor, 0; balance-factor, − 0.333; balance-factor, + 1; balance-factor, + 1; balance-factor, + 1; balance-factor, + 1; balance-factor, + 0.333, and balance-factor, 0.

48. The flow of energy in an alternating-current system is a most important and characteristic feature of the system, and by its nature the systems may be classified into:

Monocyclic systems, or systems with a balance-factor zero or negative.

Polycyclic systems, with a positive balance-factor.

Balance-factor -1 corresponds to a wattless single-phase circuit, balance-factor zero to a non-inductive single-phase circuit, balance-factor $+1$ to a balanced polyphase system.

49. In polar coördinates the flow of energy of an alternating current system is represented by using the instantaneous value of power as radius vector, with the angle, β, corresponding to the time as amplitude, one complete period being represented by one revolution.

In this way the power of an alternating-current system is represented by a closed symmetrical curve, having the zero point as quadruple point. In the monocyclic systems the zero point is quadruple nodal point; in the polycyclic systems quadruple isolated point.

Thus these curves are sextics.

Since the flow of energy in any single-phase branch of the alternating-current system can be represented by a sine wave of double frequency,

$$p = P\left(1 + \frac{\sin (2\beta - \theta)}{\cos \theta}\right),$$

the total flow of energy of the system as derived by the addition of the powers of the branch circuits can be represented in the form

$$p = P(1 + \epsilon \sin (2\beta - \theta_0)).$$

This is a wave of double frequency also, with ϵ as amplitude of fluctuation of power.

This is the equation of the power characteristics of the system in polar coördinates.

50. To derive the equation in rectangular coördinates we introduce a substitution which revolves the system of coördinates by an angle, $\frac{\theta_0}{2}$, so as to make the symmetry axes of the power characteristic the coördinate axes.

$$p = \sqrt{x^2 + y^2},$$

$$\tan\left(\beta - \frac{\theta_0}{2}\right) = \frac{y}{x};$$

hence,

$$\sin(2\beta - \theta_0) = 2\sin\left(\beta - \frac{\theta_0}{2}\right)\cos\left(\beta - \frac{\theta_0}{2}\right) = \frac{2xy}{x^2 + y^2},$$

substituted,

$$\sqrt{x^2 + y^2} = P\left\{1 + \frac{2\epsilon xy}{x^2 + y^2}\right\},$$

or, expanded,

$$(x^2 + y^2)^3 - P^2(x^2 + y^2 + 2\epsilon xy)^2 = 0,$$

the sextic equation of the power characteristic.

Introducing

$$a = (1 + \epsilon)P = \text{maximum value of power},$$
$$b = (1 - \epsilon)P = \text{minimum value of power};$$

we have

$$P = \frac{a + b}{2},$$

$$\epsilon = \frac{a - b}{a + b};$$

hence, substituted, and expanded,

$$(x^2 + y^2)^3 - \frac{1}{4}\{a(x + y)^2 + b(x - y)^2\}^2 = 0,$$

the equation of the power characteristic, with the *main power axes*, a and b, and the balance-factor, $\frac{b}{a}$.

It is thus:

Single-phase. non-inductive circuit, $p = P(1 + \sin 2\theta)$, $b = 0$, $a = 2P$,

$$(x^2 + y^2)^3 - P^2(x + y)^4 = 0, \quad \frac{b}{a} = 0.$$

Single-phase circuit, 60° lag: $p = P(1 + 2\sin 2\theta)$, $b = -P$, $a = +3P$,

$$(x^2 + y^2)^3 - P^2(x^2 + y^2 + 4xy)^2 = 0, \quad \frac{b}{a} = -\frac{1}{3}.$$

Single-phase circuit, 90° lag: $p = EI\sin 2\theta$,

$$b = -EI, \quad a = +EI,$$

$$(x^2 + y^2)^3 - 4(EI)^2x^2y^2, \quad \frac{b}{a} = -1.$$

Three-phase non-inductive circuit, $p = P$, $b = 1$, $a = 1$,

$$x^2 + y^2 - P^2 = 0, \text{ circle. } \frac{b}{a} = +1.$$

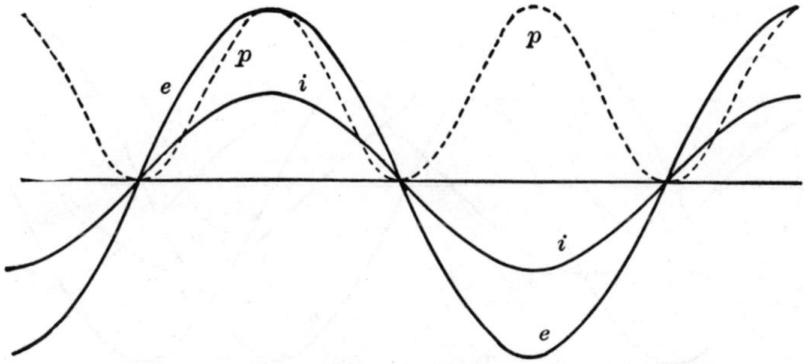

FIG. 25. —Single-phase, non-inductive circuit.

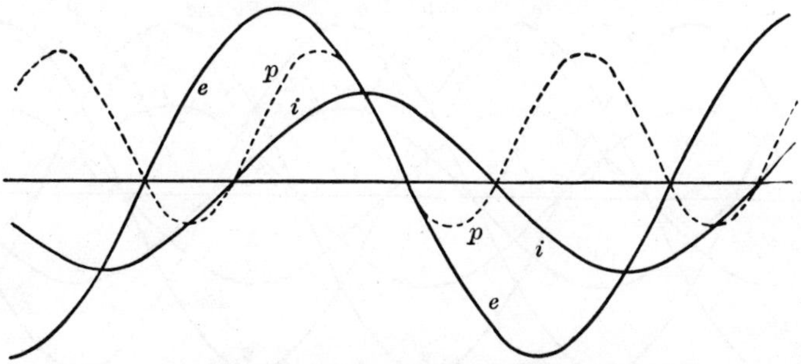

FIG. 26. —Single-phase, 60° lag.

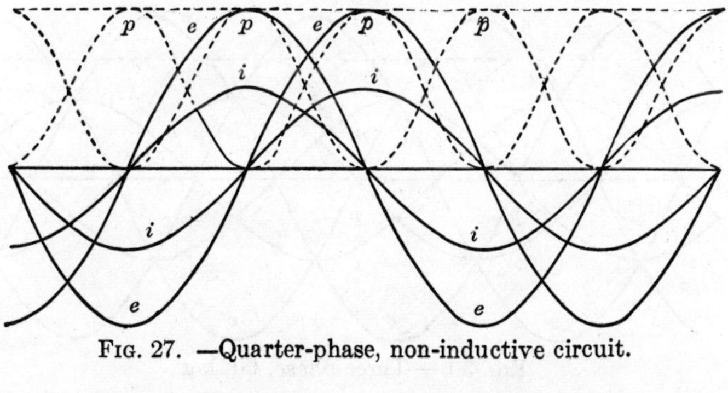

FIG. 27. —Quarter-phase, non-inductive circuit.

Three-phase circuit, 60° lag, $p = P$, $b = 1$, $a = 1$,

$$x^2 + y^2 - P^2 = 0, \text{ circle. } \frac{b}{a} = +1.$$

Quarter-phase non-inductive circuit, $p = P$, $b = 1$, $a = 1$,

$$x^2 + y^2 - P^2 = 0, \text{ circle. } \frac{b}{a} = +1.$$

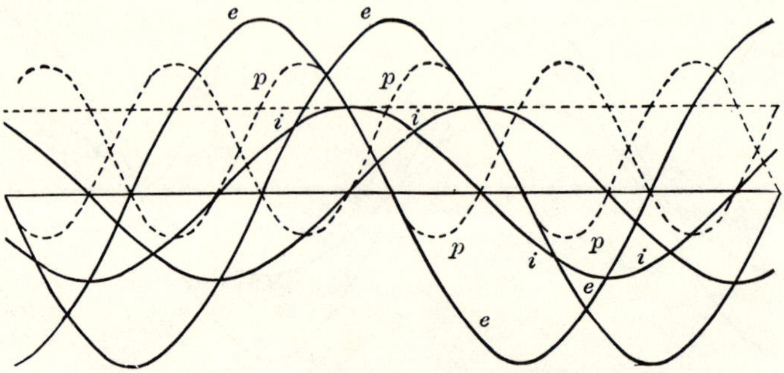

FIG. 28. —Quarter-phase, 60° lag.

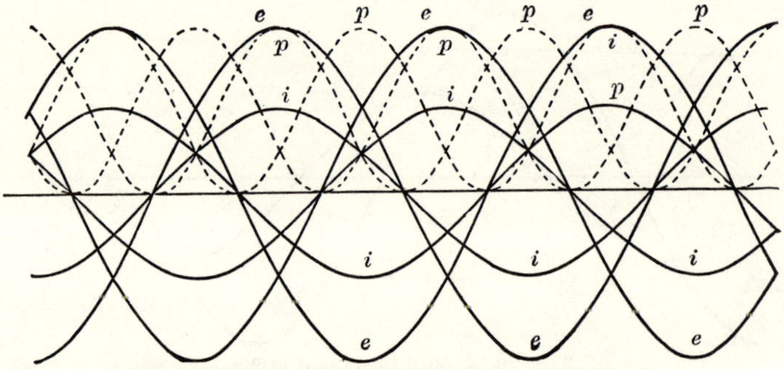

FIG. 29. —Three-phase, non-inductive circuit.

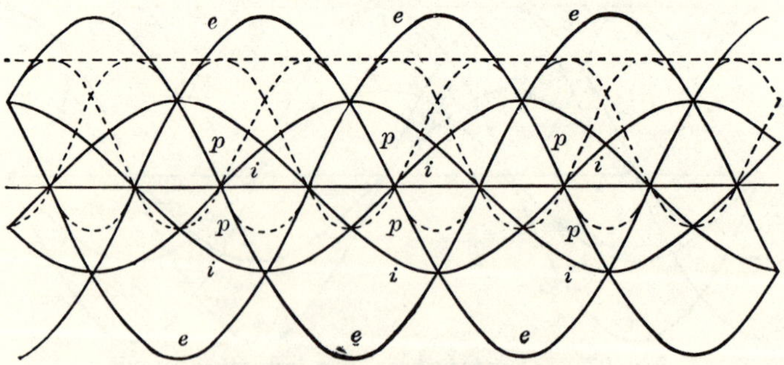

FIG. 30. —Three-phase, 60° lag.

Quarter-phase circuit, 60° lag, $p = P$, $b = 1$, $a = 1$,

$$x^2 + y^2 - P^2 = 0, \text{ circle. } \frac{b}{a} = +1.$$

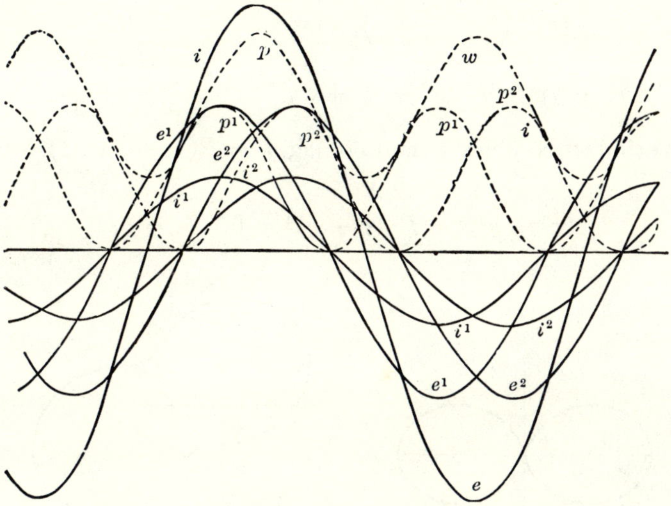

Fig. 31. —Inverted three-phase, non-inductive circuit.

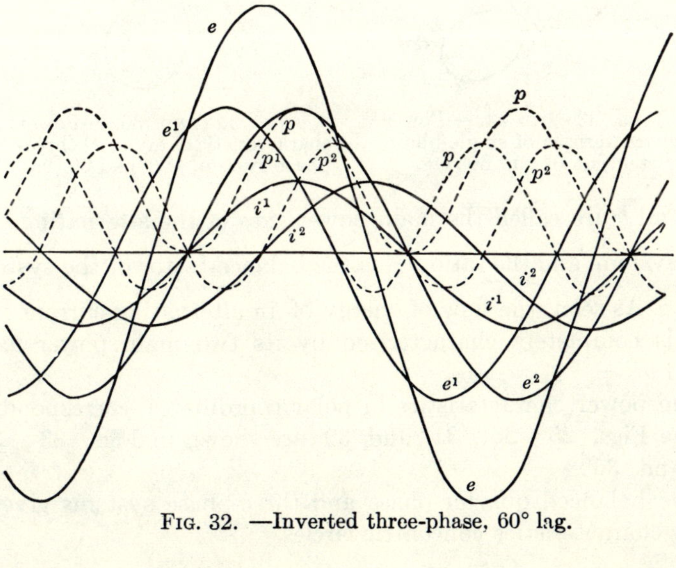

Fig. 32. —Inverted three-phase, 60° lag.

Inverted three-phase non-inductive circuit,

$$p = P\left(1 + \frac{\sin 2\,\theta}{2}\right),\ b = \frac{1}{2}P,\ a = \frac{3}{2}P,$$

$$(x^2 + y)^3 - P^2\,(x^2 + y^2 + xy)^2 = 0.\quad \frac{b}{a} = +\frac{1}{3}.$$

Inverted three-phase circuit 60° lag, $p = P(1 + \sin 2\,\theta)$, $b = 0$, $a = 2\,P$,

$$(x^2 + y^2)^3 - P^2\,(x + y)^4 = 0.\quad \frac{b}{a} = 0.$$

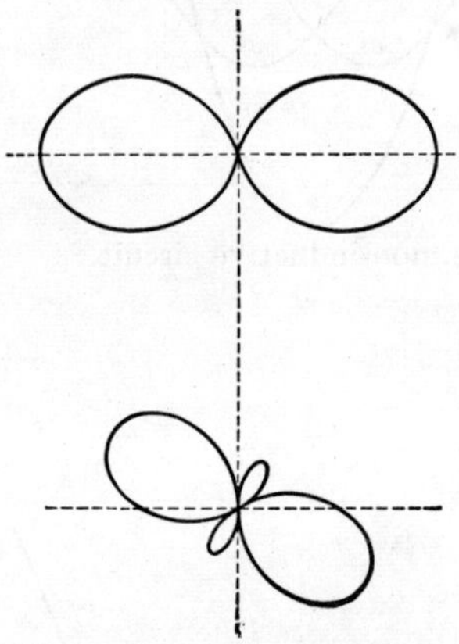

FIGS. 33 AND 34. —Power characteristic of single-phase system, at 0° and 60° lag.

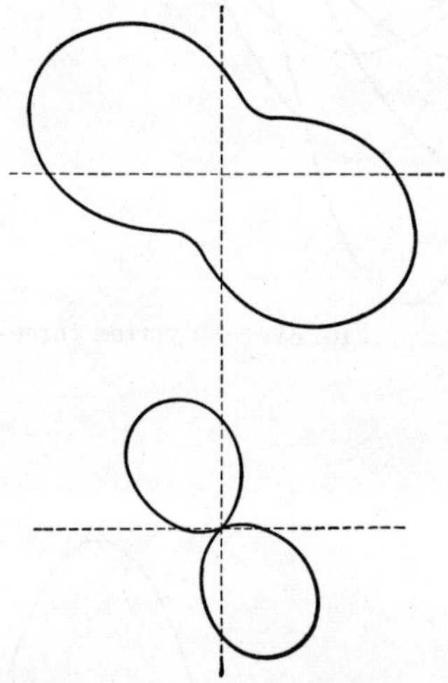

FIGS. 35 AND 36. —Power characteristic of inverted three-phase system, at 0° and 60° lag.

a and b are called the main power axes of the alternating-current system, and the ratio, $\frac{b}{a}$, is the balance-factor of the system.

51. As seen, the flow of energy of an alternating-current system is completely characterized by its two main power axes, a and b.

The power characteristics in polar coördinates, corresponding to the Figs. 25 , 26 , 31 and 32 are shown in Figs. 33 , 34, 35 and 36 .

The balanced quarter-phase and three-phase systems give as polar characteristics concentric circles.

CHAPTER VII

INTERLINKED POLYPHASE SYSTEMS

52. In a polyphase system the different circuits of displaced phases, which constitute the system, may either be entirely separate and without electrical connection with each other, or they may be connected with each other electrically, so that a part of the electrical conductors are in common to the different phases, and in this case the system is called an interlinked polyphase system.

Thus, for instance, the quarter-phase system will be called an independent system if the two e.m.fs. in quadrature with each other are produced by two entirely separate coils of the same, or different, but rigidly connected, armatures, and are connected to four wires which energize independent circuits in motors or other receiving devices. If the quarter-phase system is derived by connecting four equidistant points of a closed-circuit drum or ring-wound armature to the four collector rings, the system is an interlinked quarter-phase system.

Similarly in a three-phase system. Since each of the three currents which differ from each other by one-third of a period is equal to the resultant of the other two currents, it can be considered as the return circuit of the other two currents, and an interlinked three-phase system thus consists of three wires conveying currents differing by one-third of a period from each other, so that each of the three currents is a common return of the other two, and inversely.

53. In an interlinked polyphase system two ways exist of connecting apparatus into the system.

1. The *star connection*, represented diagrammatically in Fig. 37. In this connection the n circuits, excited by currents differing from each other by $\frac{1}{n}$ of a period, are connected with their one end together into a neutral point or common connection, which may either be grounded, or connected with other corresponding neutral points, or insulated.

In a three-phase system this connection is usually called a Y connection, from a similarity of its diagrammatical representation with the letter *Y*, as shown in Fig. 21.

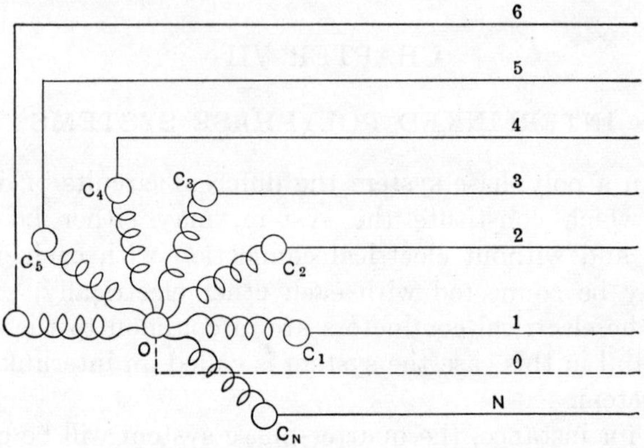

FIG. 37.

2. The *ring connection*, represented diagrammatically in Fig. 38, where the *n* circuits of the apparatus are connected with each other in closed circuit, and the corners or points of connection of adjacent circuits connected to the *n* lines of the polyphase

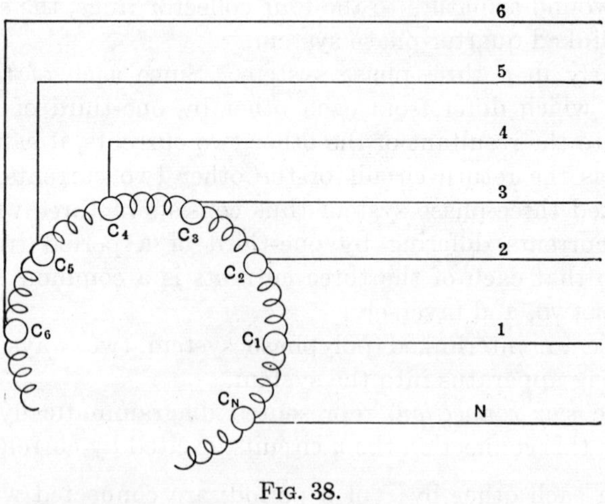

FIG. 38.

system. In a three-phase system this connection is called the delta (Δ) connection, from the similarity of its diagrammatic representation with the Greek letter delta, as shown in Fig. 22.

In consequence hereof we distinguish between star-connected and ring-connected generators, motors, etc., or in three-phase systems Y-connected and Δ-connected apparatus.

54. Obviously, the polyphase system as a whole does not differ, whether star connection or ring connection is used in the generators or other apparatus; and the transmission line of a symmetrical n-phase system always consists of n wires carrying currents of equal strength, when balanced, differing from each other in phase by $\frac{1}{n}$ of a period. Since the line wires radiate from the n terminals of the generator, the lines can be considered as being in star connection.

The circuits of all the apparatus, generators, motors, etc., can either be connected in star connection, that is, between one line and a neutral point, or in ring connection, that is, between two adjacent lines.

In general some of the apparatus will be arranged in star connection, some in ring connection, as the occasion may require.

55. In the same way as we speak of star connection and ring connection of the circuits of the apparatus, the terms star voltage and ring voltage, star current and ring current, etc., are used, whereby as star voltage or in a three-phase circuit Y voltage, the potential difference between one of the lines and the neutral point, that is, a point having the same difference of potential against all the lines, is understood; that is, the voltage as measured by a voltmeter connected into star or Y connection. By ring or delta voltage is understood the difference of potential between adjacent lines, as measured by a voltmeter connected between adjacent lines, in ring or delta connection.

In the same way the star or Y current is the current in a circuit from one line to a neutral point; the ring or delta current, the current in a circuit from one line to the next line.

The current in the transmission line is always the star or Y current, and the potential difference between the line wires, the ring or delta voltage.

Since the star voltage and the ring voltage differ from each other, apparatus requiring different voltages can be connected into the same polyphase mains, by using either star or ring connection.

56. If in a generator with star-connected circuits, the e.m.f. per circuit $= E$, and the common connection or neutral point is denoted by zero, the voltages of the n terminals are

$$\dot{E}, \epsilon \dot{E}, \epsilon^2 \dot{E} \ldots\ldots \epsilon^{n-1}\dot{E};$$

or in general, $\epsilon^i\dot{E}$,

at the ith terminal, where,

$$i = 0, 1, 2 \ldots\ldots n - 1, \quad \epsilon = \cos\frac{2\pi}{n} + j\sin\frac{2\pi}{n} = \sqrt[n]{1}.$$

Hence the e.m.f. in the circuit from the ith to the kth terminal is

$$\dot{E}_{ki} = \epsilon^k \dot{E} - \epsilon^i \dot{E} = (\epsilon^k - \epsilon^i)\dot{E}.$$

The e.m.f. between adjacent terminals i and $i + 1$ is

$$(\epsilon^{i+1} - \epsilon^i)\dot{E} = \epsilon^i(\epsilon - 1)\dot{E}.$$

In a generator with ring-connected circuits, the e.m.f. per circuit

$$\epsilon^i\dot{E},$$

is the ring e.m.f., and takes the place of

$$\epsilon^i(\epsilon - 1)\dot{E};$$

while the e.m.f. between terminal and neutral point, or the star e.m.f., is

$$\frac{\epsilon^i}{\epsilon - 1}\dot{E}.$$

Hence in a star-connected generator with the e.m.f. $\dot{E}$ per circuit, it is:

star e.m.f., $\epsilon^i \dot{E}$,
ring e.m.f., $\epsilon^i (\epsilon - 1)\dot{E}$,
e.m.f. between terminal i and terminal k, $(\epsilon^k - \epsilon^i)\dot{E}$.

In a ring-connected generator with the e.m.f., $\dot{E}$, per circuit, it is

star e.m.f., $\dfrac{\epsilon^i}{\epsilon - 1}\dot{E}$,
ring e.m.f., $\epsilon^i\dot{E}$,
e.m.f. between terminals i and k, $\dfrac{\epsilon^k - \epsilon^i}{\epsilon - 1}\dot{E}$.

In a star-connected apparatus, the e.m.f. and the current per circuit have to be the star e.m.f. and the star current. In a ring-connected apparatus the e.m.f. and current per circuit have to be the ring e.m.f. and ring current.

In the generator of a symmetrical polyphase system, if

$\epsilon^i E$ are the e.m.fs. between the n terminals and the neutral point, or star e.m.fs.

I_i = the currents issuing from terminals i over a line of the impedance, Z_i (including generator impedance in star connection), we have

voltage at end of line i,

$$\epsilon^i E - Z_i I_i,$$

and difference of potential between terminals k and i

$$(\epsilon^k - \epsilon^i)E - (Z_k I_k - Z_i I_i),$$

where I_i is the star current of the system, Z_i the star impedance.

The ring voltage at the end of the line between terminals i and k is E_{ik}, and

$$E_{ik} = - E_{ki}.$$

If now I_{ik} denotes the current from terminal i to terminal k, and Z_{ik} impedance of the circuit between terminal i and terminal k, where

$$I_{ik} = - I_{ki},$$

$$Z_{ik} = Z_{ki},$$

we have

$$E_{ik} = Z_{ik} I_{ik}.$$

If I_{io} denotes the current in the circuit from terminal i to a ground or neutral point, and Z_{io} is the impedance of this circuit between terminal i and neutral point, it is

$$E_{io} = \epsilon^i E - Z_i I_i = Z_{io} I_{io}.$$

57. We have thus, by Ohm's law and Kirchoff's law:

If $\epsilon_i E$ is the e.m.f. per circuit of the generator, between the terminal, i, and the neutral point of the generator, or the star e.m.f.

I_i = the current at the terminal, i, of the generator, or the star current.

Z_i = the impedance of the line connected to a terminal, i, of the generator, including generator impedance.

$\underset{\cdot}{E}_i$ = the e.m.f. at the end of line connected to a terminal, i, of the generator.

$\underset{\cdot}{E}_{ik}$ = the difference of potential between the ends of the lines, i and k.

$\underset{\cdot}{I}_{ik}$ = the current from line i to line k.

$\underset{\cdot}{Z}_{ik}$ = the impedance of the circuit between lines i and k.

$\underset{\cdot}{I}_{io}, \underset{\cdot}{I}_{ioo} \ldots\ldots$ = the current from line i to neutral points 0, 00, ...

$Z_{io}, Z_{ioo} \ldots\ldots$ = the impedance of the circuits between line i and neutral points 0, 00, ...

Then:

1. $\underset{\cdot}{E}_{ik} = -\underset{\cdot}{E}_{ki}, \; \underset{\cdot}{I}_{ik} = -\underset{\cdot}{I}_{ki} \; Z_{ik} = Z_{ki}, \; I_{io} = -\underset{\cdot}{I}_{oi},$
   $Z_{io} = Z_{oi},$ etc.
2. $\underset{\cdot}{E}_i = \epsilon^i \underset{\cdot}{E} - Z_i \underset{\cdot}{I}_i.$
3. $\underset{\cdot}{E}_i = Z_{io} \underset{\cdot}{I}_{io} = Z_{ioo} \underset{\cdot}{I}_{ioo} = \ldots\ldots$
4. $\underset{\cdot}{E}_{ik} = \underset{\cdot}{E}_k - \underset{\cdot}{E}_i = (\epsilon^k - \epsilon^i)\underset{\cdot}{E} - (Z_k \underset{\cdot}{I}_k - Z_i \underset{\cdot}{I}_i).$
5. $\underset{\cdot}{E}_{ik} = Z_{ik} \underset{\cdot}{I}_{ik}.$
6. $\underset{\cdot}{I}_i = \sum_0^n {}^k \underset{\cdot}{I}_{ik}.$

7. If the neutral point of the generator does not exist, as in ring connection, or is insulated from the other neutral points:

$$\sum_1^n {}^i \underset{\cdot}{I}_i = 0$$

$$\sum_1^n {}^i \underset{\cdot}{I}_{io} = 0;$$

$$\sum_1^n {}^i \underset{\cdot}{I}_{ioo} = 0, \text{ etc}$$

Where 0, 00, etc., are the different neutral points which are insulated from each other.

If the neutral point of the generator and all the other neutral points are grounded or connected with each other, we have,

$$\sum_1^n {}^i \underset{\cdot}{I}_i = \sum_1^n {}^i (\underset{\cdot}{I}_{io} + \underset{\cdot}{I}_{ioo} + \ldots\ldots)$$

$$= \sum_1^n {}^i \underset{\cdot}{I}_{io} + \sum_1^n {}^i \underset{\cdot}{I}_{ioo} + \ldots\ldots$$

If the neutral point of the generator or other neutral points are grounded, the system is called a grounded system. If the neutral points are not grounded, the system is an insulated polyphase system, and an insulated polyphase system with equalizing return, if all the neutral points are connected with each other.

8. The power of the polyphase system is

$$P = \sum_{1}^{n}{}^{i}\, \epsilon^{i} E I_{ii} \cos \theta_{i} \text{ at the generator,}$$

$$P = \sum_{0}^{n}{}^{i} \sum_{i}^{n}{}^{k}\, E_{ik} I_{ik} \cos \theta_{ik} \text{ in the receiving circuits.}$$

CHAPTER VIII

TRANSFORMATION OF POLYPHASE SYSTEMS

58. In transforming one polyphase system into another polyphase system, it is obvious that the primary system must have the same flow of energy as the secondary system, neglecting losses in transformation, and that consequently a balanced system will be transformed again into a balanced system, and an unbalanced system into an unbalanced system of the same balance-factor, since the transformer is not able to store energy, and thereby to change the nature of the flow of energy. The energy stored as magnetism amounts in a well-designed transformer only to a very small percentage of the total energy. This shows the futility of producing symmetrical balanced polyphase systems by transformation from the unbalanced single-phase system without additional apparatus able to store energy efficiently, as revolving machinery, etc.

Since any e.m.f. can be resolved into, or produced by, two components of given directions, the e.m.f. of any polyphase system can be resolved into components or produced from components of two given directions. This enables the transformation of any polyphase system into any other polyphase system of the same balance-factor by two transformers only.

59. Let E_1, E_2, E_3 . . . . be the e.m.fs. of the primary system which shall be transformed into

E'_1, E'_2, E'_3 . . . . the e.m.fs. of the secondary system.

Choosing two magnetic fluxes, $\overline{\Phi}$ and $\overline{\overline{\Phi}}$, of different phases, as magnetic circuits of the two transformers, which generate the e m.fs., $\overline{e}$ and $\overline{\overline{e}}$, per turn, by the law of parallelogram the e.m.fs., $E_1, E_2,$ . . . . can be resolved into two components, $\overline{E}_1$ and $\overline{\overline{E}}_1$, $\overline{E}_2$ and $\overline{\overline{E}}_2$, ... of the phases, $\overline{e}$ and $\overline{\overline{e}}$.

Then

$\overline{E}_1, \overline{E}_2,$ . . . . are the counter e.m.fs. which have to be generated in the primary circuits of the first transformer;

$\overline{\overline{E}}_1, \overline{\overline{E}}_2,$ . . . . the counter e.m.fs. which have to be generated in the primary circuits of the second transformer.

Hence

$\frac{\bar{E}_1}{\bar{e}}, \frac{\bar{E}_2}{\bar{e}}$ . . . . are the numbers of turns of the primary coils of the first transformer.

Analogously

$\frac{\bar{\bar{E}}_1}{\bar{\bar{e}}}, \frac{\bar{\bar{E}}_2}{\bar{\bar{e}}}$ . . . . are the numbers of turns of the primary coils in the second transformer.

In the same manner as the e.m.fs. of the primary system have been resolved into components in phase with $\bar{e}$ and $\bar{\bar{e}}$, the e.m.fs. of the secondary system, E'_1, E'_2, ... are produced from components, $\bar{E}'_1$, and $\bar{\bar{E}}'_1$, $\bar{E}'_2$, and $\bar{\bar{E}}'_2$ . . . . in phase with e and e, and give as numbers of secondary turns—

$\frac{\overline{E'_1}}{\bar{e}}, \frac{\overline{E'_2}}{\bar{e}}$ . . . . in the first transformer;

$\frac{\overline{\overline{E'_1}}}{\bar{\bar{e}}}, \frac{\overline{\overline{E'_2}}}{\bar{\bar{e}}}$ . . . . in the second transformer.

That means each of the two transformers, m and $\bar{\bar{m}}$, contains in general primary turns of each of the primary phases, and secondary turns of each of the secondary phases. Loading now the secondary polyphase system in any desired manner, corresponding to the secondary currents, primary currents will exist in such a manner that the total flow of energy in the primary polyphase system is the same as the total flow of energy in the secondary system, plus the loss of power in the transformers.

60. As an instance may be considered the transformation of the symmetrical balanced three-phase system,

$$E \sin \beta, \quad E \sin (\beta - 120), \quad E \sin (\beta - 240),$$

into an unsymmetrical balanced quarter-phase system,

$$E' \sin \beta, \quad E' \sin (\beta - 90).$$

Let the magnetic flux of the two transformers be chosen in quadrature

$$\Phi \cos \beta \text{ and } \Phi \cos (\beta - 90).$$

Then the e.m.fs. generated per turn in the transformers are

$$e \sin \beta \text{ and } e \sin (\beta - 90);$$

hence, in the primary circuit the first phase, $E \sin \beta$, will give, in the first transformer, $\frac{E}{e}$ primary turns; in the second transformer, 0 primary turns.

The second phase, $E \sin (\beta - 120)$, will give, in the first transformer, $\frac{-E}{2e}$ primary turns; in the second transformer, $\frac{E \times \sqrt{3}}{2e}$ primary turns.

The third phase, $E \sin (\beta - 240)$, will give, in the first transformer, $\frac{-E}{2e}$ primary turns; in the second transformer, $\frac{-E \times \sqrt{3}}{2e}$ primary turns.

In the secondary circuit the first phase, $E' \sin \beta$, will give in the first transformer: $\frac{E'}{e}$ secondary turns; in the second transformer: 0 secondary turns.

The second phase: $E' \sin (\beta - 90)$ will give in the first transformer: 0 secondary turns; in the second transformer, $\frac{E'}{e}$ secondary turns.

Or, if

$$E = 5000, \ E' = 100, \ e = 10.$$

	PRIMARY			SECONDARY	
	1st.	**2d.**	**3d.**	**1st.**	**2d.**
First transformer	+500	−250	−250	10	0
Second transformer	0	+433	−433	0	10 turns.

Using autotransformer connection in the three-phase primaries of the first transformer, that is, using as coils of the second and the third phase the two halves of the coil of the first phase, this gives the well known T-connection of three-phase-quarter-phase transformation.

That means:

Any balanced polyphase system can be transformed by two transformers only, without storage of energy, into any other balanced polyphase system.

Or more generally stated:

Any polyphase system can be transformed by two transformers only, without storage of energy, into any other polyphase system of the same balance factor.

61. Some of the more common methods of transformation between polyphase systems are:

1. The *delta-Y connection* of transformers between three-phase systems, shown in Fig. 39. One side of the transformers is connected in delta, the other in *Y*. This arrangement becomes necessary for feeding four-wire three-phase secondary distributions. The Y connection of the secondary allows the bringing out of a neutral wire, while the delta connection of the primary maintains the balance, in regard to the voltage between the phases at unequal distribution of load.

The delta-Y connection of step-up transformers is frequently used in long-distance transmissions, to allow grounding of the high-potential neutral. Under certain conditions—which therefore have to be guarded against—it is liable to induce excessive voltages by resonance with the line capacity.

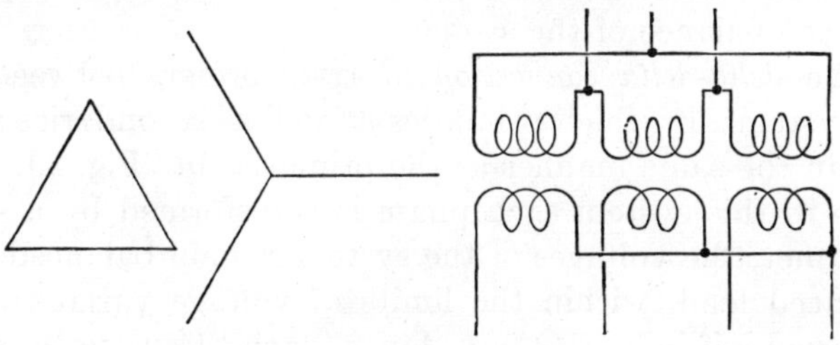

Fig. 39.

The reverse thereof, or the *Y-delta connection*, is undesirable on unbalanced load, since it gives what has been called a "floating neutral;" the three primary Y voltages do not remain even approximately constant, at unequal distribution of load on the secondary delta, but the primary voltage corresponding to the heavier loaded secondary, and, therefore, also the corresponding secondary voltage, collapses. Thereby the common connection of the primary shifts toward one corner of the e.m.f. triangle, away from the center of the triangle, and may even fall outside of the triangle. As result thereof the secondary triangle becomes very greatly distorted even at moderate inequality of load, and the system thus loses all ability to maintain constant voltage at unequal distribution of load, that is, becomes inoperative. In high-potential systems in this case excessive voltages may be induced by resonance with the line capacity.

For instance, if only one phase of the secondary triangle is loaded, the other two unloaded, the primary current of the loaded phase must return over the other two transformers, which, at open secondaries, act as very high reactances, thus limiting the current and consuming practically all the voltage, and the loaded primary, and thus its secondary, receive practically no voltage.

Y-delta connection is satisfactory if the secondary load is balanced, as induction—or synchronous motors, or if the primary neutral is connected with the generator neutral or the secondary neutral of step-up transformers in which the primaries are connected in delta, and the unbalanced current can return over the neutral. If with Y-delta connection, in addition to an unbalanced load, the secondary carries polyphase motors, the motors take different currents in the different phases, so that the total current is approximately the same in all three phases. That is, the motors act as phase converters, and so partially restore the balance of the system.

2. The *delta-delta connection* of transformers between three-phase systems, in which primaries as well as secondaries are connected in the same manner as the primaries in Fig. 39.

Since in this system each phase is transformed by a separate transformer, the voltages of the system remain balanced even at unbalanced load, within the limits of voltage variation due to the internal self-inductive impedance (or short-circuit impedance) of the transformers—which is small, while the exciting impedance (or open-circuit impedance) of the transformers, which causes the unbalancing in the Y-delta connection above discussed is enormous.

3. Y-*Y connection* of transformers between three-phase systems. Primaries and secondaries connected as the secondaries in Fig. 39.

In this case, if the neutral is not fixed by connection with a fixed neutral, either directly or by grounding it, the neutral also is floating, and so abnormal voltages may be produced between the lines and the neutral, without appearing in the voltages between the lines, and may lead to disruptive effects, or to overheating of the transformers, so that this connection is not an entirely safe one.

Where in transformer connections in polyphase systems, a neutral or common connection of the transformers exists, care must, therefore, be taken to have this neutral a fixed voltage point, irrespective of the variation of the load or its distribution, which may occur; otherwise harmful phenomena may result from a "floating" or "unstable" neutral.

In connections (2) and (3), the secondary-e.m.f. triangle is in phase with the primary-e.m.f. triangle, while in (1) it is displaced therefrom by 30°. Therefore, even if the voltages are equal, connection (1) cannot be operated in parallel with (2) or (3), but (2)

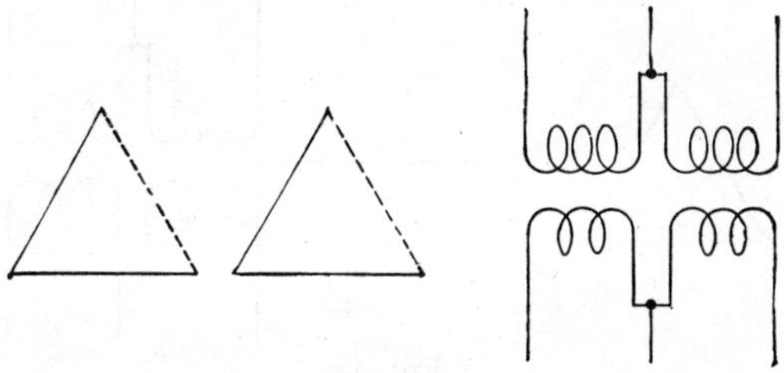

FIG. 40.

and (3) can be operated in parallel with each other, and with the connections (4) and (5), provided that the voltages are correct.

4. The *V connection* or *open delta connection* of transformers between three-phase systems, consists in using two sides of the triangle only, as shown in Fig. 40. This arrangement has the disadvantage of transforming one phase by two transformers in series, hence is less efficient, and is liable to unbalance the system

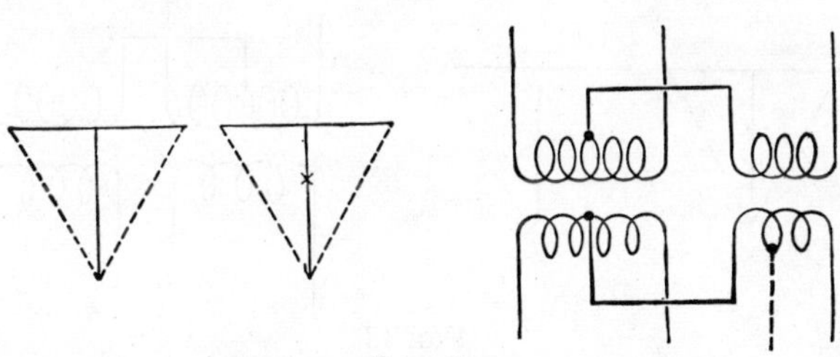

FIG. 41.

by the internal impedance of the transformers. It is convenient for small powers at moderate voltage, since it requires only two transformers, but is dangerous in high potential circuits, being liable to produce destructive voltages by its electrostatic unbalancing.

5. The *main and teaser*, or *T connection* of transformers between three-phase systems, is shown in Fig. 41. One of the

two transformers is wound for $\frac{\sqrt{3}}{2}$ times the voltage of the other (the altitude of the equilateral triangle), and connected with one of its ends to the center of the other transformer. From the point one-third inside of the teaser transformer, a neutral wire can be brought out in this connection.

6. The *monocyclic connection,* transforming between three-

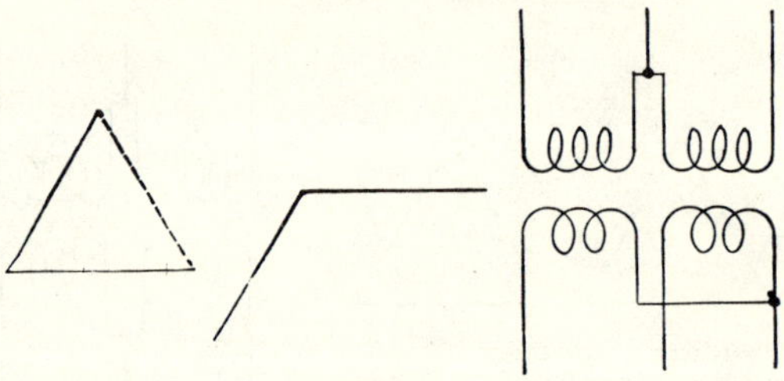

FIG. 42.

phase and inverted three-phase or polyphase monocyclic, by two transformers, the secondary of one being reversed regarding its primary, as shown in Fig. 42.

7. The *L connection* for transformation between quarter-phase and three-phase as described in the example, §291.

8. The *T connection* of transformation between quarter-phase and three-phase, as shown in Fig. 43. The quarter-phase sides

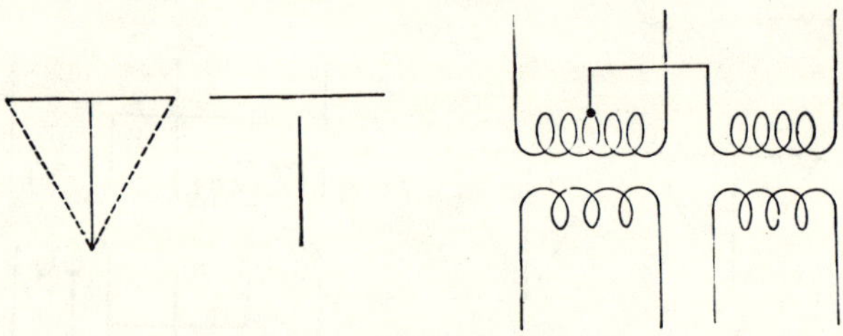

FIG. 43.

of the transformers contain two equal and independent (or inter-linked) coils, the three-phase sides two coils with the ratio of turns, $1 \div \frac{\sqrt{3}}{2}$, connected in T.

9. The *double delta connection* of transformation from three-phase to six-phase, shown in Fig. 44. Three transformers, with two secondary coils each, are used, one set of secondary coils connected in delta, the other set in delta also, but with reversed

terminals, so as to give a reversed e.m.f. triangle. These e.m.fs. thus give topographically a six-cornered star.

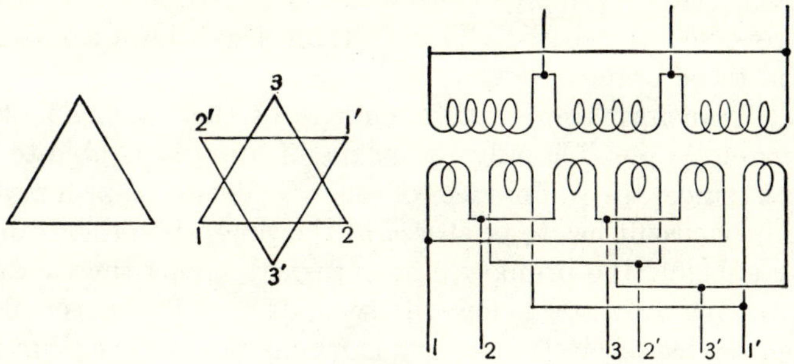

FIG. 44.

10. The *double Y connection* or *diametrical connection* of transformation from three-phase to six-phase, shown in Fig. 45. It

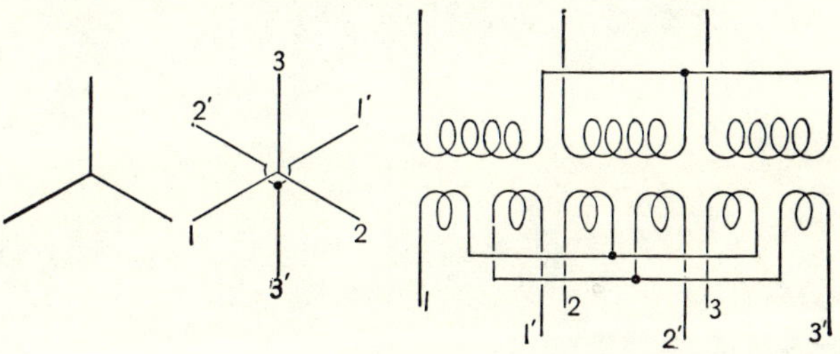

FIG. 45.

is analogous to (7), the delta connection merely being replaced by the Y connection. The neutrals of the two *Y*'s may be connected together and to an external neutral if desired.

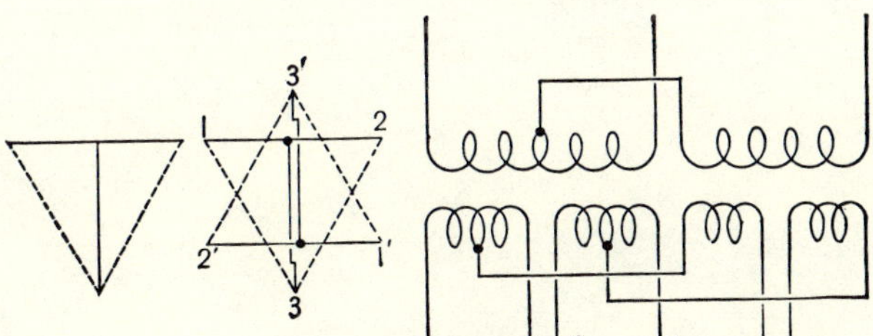

FIG. 46.

The primaries in 9 and 10 may be connected either delta or *Y*, and in the latter case a floating neutral must be guarded against.

11. The *double T connection* of transformation from three-phase to six-phase, shown in Fig. 45. Two transformers are used with two secondary coils which are T-connected, but one with reversed terminals. This method also allows a secondary neutral to be brought out.

62. Transformation with a change of the balance-factor of the system is possible only by means of apparatus able to store energy, since the difference of energy between primary and secondary circuit has to be stored at the time when the secondary power is below the primary, and returned during the time when the primary power is below the secondary. The most efficient storing device of electric energy is mechanical momentum in revolving machinery. It has, however, the disadvantage of requiring attendance; fairly efficient also are condensive and inductive reactances, but, as a rule, they have the disadvantage of not giving constant potential.

CHAPTER IX

COPPER EFFICIENCY OF SYSTEMS

63. In electric power transmission and distribution, wherever the place of consumption of the electric energy is distant from the place of production, the conductors which carry the current are a sufficiently large item to require consideration, when deciding which system and what potential is to be used.

In general, in transmitting a given amount of power at a given loss over a given distance, other things being equal, the amount of copper required in the conductors is inversely proportional to the square of the potential used. Since the total power transmitted is proportional to the product of current and e.m.f., at a given power, the current will vary inversely proportionally to the e.m.f., and therefore, since the loss is proportional to the product of current-square and resistance, to give the same loss the resistance must vary inversely proportional to the square of the current, that is, proportional to the square of the e.m.f.; and since the amount of copper is inversely proportional to the resistance, other things being equal, the amount of copper varies inversely proportional to the square of the e.m.f. used.

This holds for any system.

Therefore to compare the different systems, as two-wire single-phase, single-phase three-wire, three-phase and quarter-phase, equality of the potential must be assumed.

Some systems, however, as, for instance, the Edison three-wire system, or the inverted three-phase system, have different potentials in the different circuits constituting the system, and thus the comparison can be made either—

1st. On the basis of the maximum potential difference between any two conductors of the system; or

2nd. On the basis of the maximum potential difference between any conductor of the system and the ground; or

3rd. On the basis of the minimum potential difference in the system, or the potential difference per circuit or phase of the system.

In low-potential circuits, as secondary networks, where the potential is not limited by the insulation strain, but by the potential of the apparatus connected into the system, as incandescent lamps, the proper basis of comparison is equality of the potential per branch of the system, or per phase.

On the other hand, in long-distance transmissions where the potential is not restricted by any consideration of apparatus suitable for a certain maximum potential only, but where the limitation of potential depends upon the problem of insulating the conductors against disruptive discharge, the proper comparison is on the basis of equality of the maximum difference of potential; that is, equal maximum dielectric strain on the insulation.

In this case, the comparison voltage may be either the potential difference between any two conductors of the system, or it may be the potential difference between any conductor of the system and the ground, depending on the character of the circuit.

The dielectric stress is from conductor to conductor, or between any two conductors, in a system which is insulated from the ground, as is mostly the case in medium voltage overhead transmissions, and frequently in underground cables.

In an ungrounded cable system, in which all the conductors are enclosed in the same cable, the insulation stress is mainly from conductor to conductor, and this therefore is the basis of comparison. But even in an underground cable system with grounded neutral, as very commonly used, a direct path exists from conductor to conductor inside of the cables, for a disruptive voltage, and the comparison of systems, therefore, has to be made, in this case, on the basis of maximum potential difference between conductors as well as between conductor and ground.

In an ungrounded overhead system, the disruptive stress is from conductor to ground and back from ground to conductor. If the system is of considerable extent—as is the case where high voltages of serious disruptive strength have to be considered—the neutral of the system is maintained at approximate ground potential by the capacity of the system, and the normal voltage stress from conductor to ground therefore is that from conductor to neutral, that is, the same as in a system with grounded neutral, and the basis of comparison then is the voltage from line to ground, and not between lines. Since, however, one conductor of the system may temporarily ground, if it is required to maintain operation even with one conductor of the system grounded, the voltage between conductors must be the basis of comparison, since with one conductor grounded, the disruptive stress between the other conductors and ground is the potential difference between the conductors of the system.

In an overhead system with grounded neutral, frequently used for transmission systems of very high voltage, or in general in a grounded system, the disruptive stress is that due to the potential difference between conductor and ground or neutral, and this then is the basis of comparison.

In moderate-potential power circuits, in considering the danger to life from live wires entering buildings or otherwise accessible, the comparison on the basis of maximum potential also appears appropriate.

Thus the comparison of different systems of long-distance transmission at high potential or power distribution for motors is to be made on the basis of equality of the maximum difference of potential existing in the system; the comparison of low-potential distribution circuits for lighting on the basis of equality of the minimum difference of potential between any pair of wires connected to the receiving apparatus.

64. 1st. *Comparison on the basis of equality of the minimum difference of potential, in low-potential lighting circuits:*

In the single-phase, alternating-current circuit, if e = e.m.f., i = current, r = resistance per line, the total power is $= ei$, the loss of power, $2\ i^2r$.

Using, however, a three-wire system: the potential between outside wires and neutral being given equal to e, the potential between the outside wires is equal to $2\ e$, that is, the distribution takes place at twice the potential, or only one-fourth the copper is needed to transmit the same power at the same loss, if, as it is theoretically possible, the neutral wire has no cross-section. If, however, the neutral wire is made of the same cross-section as each of the outside wires, three-eighths as much copper as in the two-wire system is needed; if the neutral wire is one-half the cross-section of each of the outside wires, five-sixteenths as much copper is needed. Obviously, a single-phase, five-wire system will be a system of distribution at the potential, $4\ e$, and therefore require only one-sixteenth of the copper of the single-phase system in the outside wires; and if each of the three neutral wires is of one-half the cross-section of the outside wires, seven-sixty-fourths or 10.93 per cent. of the copper.

Coming now to the three-phase system with the potential, e, between the lines as delta potential, if i = the current per line or Y current, the current from line to line or delta current $= \frac{i_1}{\sqrt{3}}$; and since three branches are used, the total power is $\frac{3\,ei_1}{\sqrt{3}} = ei_1\sqrt{3}$. Hence if the same power has to be transmitted by the three-phase system as with the single-phase system, the three-phase line current must be $i_1 = \frac{i}{\sqrt{3}}$; where i = single-phase current, r = single-phase resistance per line, at equal power and loss: hence if r_1 = resistance of each of the three wires, the loss per wire is $i_1^2r_1 = \frac{i^2r_1}{3}$, and the total loss is i^2r_1, while in the single-phase system it is $2\,i^2r$. Hence, to get the same loss, it must be: $r_1 = 2\,r$, that is, each of the three three-phase lines has twice the resistance—that is, half the copper of each of the two single-phase lines; or in other words, the three-phase system requires three-fourths as much copper as the single-phase system of the same potential.

Introducing, however, a fourth or neutral wire into the three-phase system, and connecting the lamps between the neutral wire and the three outside wires—that is, in Y connection—the potential between the outside wires or delta potential will be $= e \times \sqrt{3}$, since the Y potential $= e$, and the potential of the system is raised thereby from e to $e\sqrt{3}$; that is, only one-third as much copper is required in the outside wires as before—that is one-fourth as much copper as in the single-phase two-wire system. Making the neutral of the same cross-section as the outside wires, requires one-third more copper, or $\frac{1}{3}$ = 33.3 per cent. of the copper of the single-sphase sytem; making the neutral of half cross-section, requires one-sixth more, or $\frac{7}{24}$ = 29.17 per cent. of the copper of the single-phase system. The system, however, now is a four-wire system.

The independent quarter-phase system with four wires is identical in efficiency to the two-wire, single-phase system, since it is nothing but two independent single-phase systems in quadrature.

The four-wire, quarter-phase system can be used as two independent Edison three-wire systems also, deriving therefrom the same saving by doubling the potential between the outside wires, and has in this case the advantage, that by interlinkage, the same neutral wire can be used for both phases, and thus one of the neutral wires saved.

In this case the quarter-phase system with common neutral of full cross-section requires $\frac{5}{16}$ or 31.25 per cent., the quarter-phase system with common neutral of one-half cross-section requires $\frac{9}{32}$ or 28.125 per cent. of the copper of the two-wire, single-phase system.

In this case, however, the system is a five-wire system, and as such far inferior in copper efficiency to the five-wire, single-phase system.

Coming now to the quarter-phase system with common return and potential e per branch, denoting the current in the outside wires by i_2, the current in the central wire is $i_2\sqrt{2}$; and if the same current density is chosen for all three wires, as the condition of maximum efficiency, and the resistance of each outside wire denoted by r_2, the resistance of the central wire $= \frac{r^2}{\sqrt{2}}$, and the loss of power per outside wire is $i_2^2r_2$, in the central wire $\frac{2\,i^2{}_2r^2}{\sqrt{2}}$ $= i_2^2r_2\sqrt{2}$; hence the total loss of power is $2\,i_2^2r_2 + i_2^2r_2\sqrt{2}$ $= i_2^2r_2(2+\sqrt{2})$. The power transmitted per branch is i_2e, hence the total power, $2\,i_2e$. To transmit the same power as by a single-phase system of power, ei, it must be $i_2 = \frac{i}{2}$; hence the loss, $\frac{i^2r_2(2+\sqrt{2})}{4}$. Since this loss shall be the same as the loss, $2\,i^2r$, in the single-phase system, it must be $2\,r = \frac{(2+\sqrt{2})}{4}\,r_2$, or $r_2 = \frac{8\,r}{2+\sqrt{2}}$. Therefore each of the outside wires must be $\frac{2+\sqrt{2}}{8}$ times as large as each single-phase wire, the central wire $\sqrt{2}$ times larger; hence the copper required for the quarter-phase system with common return bears to the copper required for the single-phase system the relation,

$$\frac{2\,(2+\sqrt{2})}{8} + \frac{(2+\sqrt{2})\sqrt{2}}{8} \div 2, \text{ or, } \frac{3+2\sqrt{2}}{8} \div 1 = 72.9$$

per cent. of the copper of the single-phase system.

Hence the quarter-phase system with common return saves 2 per cent. more copper than the three-phase system, but is inferior to the single-phase three-wire system.

The inverted three-phase system, consisting of two e.m.fs. e at 60° displacement, and three equal currents i_3 in the three lines of equal resistance r_3, gives the output $2\,ei_3$, that is, compared with the single-phase system, $i_3 = \frac{i}{2}$. The loss in the three lines is $3\,i_3^2r_3 = \frac{3}{4}\,i^2r_3$. Hence, to give the same loss, $2\,i^2r$, as the single-phase system, it must be $r_3 = \frac{8}{3}\,r$, that is, each of the three wires must have three-eighths of the copper cross-section of the wire in the two-wire single-phase system; or in other words, the inverted three-phase system requires nine-sixteenths of the copper of the two-wire single-phase system.

Thus if a given power has to be transmitted at a given loss, and a given *minimum* potential, as for instance 110 volts for lighting, the amount of copper necessary is:

2 Wires:	Single-phase system,	100.0
3 Wires:	Edison three-wire single-phase system, neutral full section,	37.5
	Edison three-wire single-phase system, neutral half-section,	31.25
	Inverted three-phase system,	56.25
	Quarter-phase system with common return,	72.9
	Three-phase system,	75.0
4 Wires:	Three-phase, with neutral-wire full section,	33.3
	Three-phase, with neutral-wire half-section,	29.17
	Independent quarter-phase system,	100.0
5 Wires:	Edison five-wire, single-phase system, full neutral,	15.625
	Edison five-wire, single-phase system, half-neutral,	10.93
	Four-wire, quarter-phase, with common-neutral full section,	31.25
	Four-wire, quarter-phase, with common-neutral half-section,	28.125

We see herefrom, that in distribution for lighting—that is, with the same minimum potential, and with the same number of wires—the single-phase system is superior to any polyphase system.

The continuous-current system is equivalent in this comparison to the single-phase alternating-current system of the same effective potential, since the comparison is made on the basis of effective potential, and the power depends upon the effective potential also.

65. *Comparison on the Basis of Equality of the Maximum Difference of Potential between any two Conductors of the System, in Long-distance Transmission, Power Distribution, etc.*

Wherever the potential is so high as to bring the question of the strain on the insulation into consideration, or in other cases, to approach the danger limit to life, the proper comparison of different systems is on the basis of equality of maximum potential in the system.

Hence in this case, since the maximum potential is fixed, nothing is gained by three- or five-wire, Edison systems. Thus, such systems do not come into consideration.

The comparison of the three-phase system with the single-phase system remains the same, since the three-phase system has the same maximum as minimum potential; that is:

The three-phase system requires three-fourths of the copper of the single-phase system to transmit the same power at the same loss over the same distance.

The four-wire, quarter-phase system requires the same amount of copper as the single-phase system, since it consists of two single-phase systems.

In a quarter-phase system with common return, the potential between the outside wires is $\sqrt{2}$ times the potential per branch, hence to get the same maximum strain on the insulation—that is, the same potential, e, between the outside wires as in the single-phase system—the potential per branch will be $\frac{e}{\sqrt{2}}$, hence the current $i_4 = \frac{i}{\sqrt{2}}$, if i equals the current of the single-phase system of equal power, and $i_4\sqrt{2} = i$ will be the current in the central wire.

Hence, if r_4 = resistance per outside wire, $\frac{r_4}{\sqrt{2}}$ = resistance of central wire, and the total loss in the system is

$$2\, i_4^2 r_4 + \frac{i_2^4\, 2\, r^4}{\sqrt{2}} = i_4^2 r_4 (2 + \sqrt{2}) = i^2 r_4 \frac{(2 + \sqrt{2})}{2}.$$

Since in the single-phase system, the loss $= 2\, i^2 r$, it is

$$r_4 = \frac{4\, r}{2 + \sqrt{2}}.$$

That is, each of the outside wires has to contain $\frac{2 + \sqrt{2}}{4}$ times as much copper as each of the single-phase wires. The central wires have to contain $\frac{2 + \sqrt{2}}{4} \sqrt{2}$ times as much copper; hence the total system contains $\frac{2\,(2 + \sqrt{2})}{4} + \frac{2 + \sqrt{2}}{4} \sqrt{2}$ times as much copper as each of the single-phase wires; that is, $\frac{3 + 2\sqrt{2}}{4}$ times the copper of the single-phase system.

Or, in other words,

A quarter-phase system with common return requires $\frac{3 + 2\sqrt{2}}{4}$ $= 1.457$ times as much copper as a single-phase system of the same maximum potential, same power, and same loss.

Since the comparison is made on the basis of equal maximum potential, and the maximum potential of an alternating system is $\sqrt{2}$ times that of a continuous-current circuit of equal effective potential, the alternating circuit of effective potential, e, compares with the continuous-current circuit of potential $e\sqrt{2}$, which latter requires only half the copper of the alternating system.

This comparison of the alternating with the continuous-current system is not proper, however, since the continuous-current voltage may introduce, besides the electrostatic strain, an electrolytic strain on the dielectric which does not exist in the alternating system, and thus may make the action of the continuous-current voltage on the insulation more severe than that of an equal alternating voltage. Besides, self-induction having no effect on a steady current, continuous-current circuits as a rule have a self-induction far in excess of any alternating circuit. During changes of current, as make and break, and changes of load, especially rapid changes, there may consequently be generated in these circuits e.m.fs. far exceeding their normal potentials. Inversely, however, with alternating voltages, dielectric hysteresis, etc., may cause heating and thereby lower the disruptive strength. At the voltages which came under consideration, the continuous current is usually excluded to begin with.

Thus we get:

If a given power is to be transmitted at a given loss, and a given *maximum* difference of potential in the system, that is, with the same strain on the insulation, the amount of copper required is:

2 Wires:	Single-phase system,	100.0
	[Continuous-current system,	50.0]
3 Wires:	Three-phase system,	75.0
	Quarter-phase system, with common return,	145.7
4 Wires:	Independent Quarter-phase system,	100.0

Hence the quarter-phase system with common return is practically excluded from long-distance transmission.

66. In a different way the same comparative results between single-phase, three-phase, and quarter-phase systems can be derived by resolving the systems into their single-phase branches.

The three-phase system of e.m.f., e, between the lines can be considered as consisting of three single-phase circuits of e.m.f., $\frac{e}{\sqrt{3}}$, and no return; the single-phase system of e.m.f., e, between lines as consisting of two single-phase circuits of e.m.f., $\frac{e}{2}$, and no return. Thus, the relative amount of copper in the two systems being inversely proportional to the square of e.m.f., bears the relation $\left(\frac{\sqrt{3}}{e}\right)^2 : \left(\frac{2}{e}\right)^2 = 3 : 4$; that is, the three-phase system requires 75 per cent. of the copper of the single-phase system.

The quarter-phase system with four equal wires requires the same copper as the single-phase system, since it consists of two single-phase circuits. Replacing two of the four quarter-phase wires by one wire of the same cross-section as each of the wires replaced thereby, the current in this wire is $\sqrt{2}$ times as large as in the other wires, hence, the loss is twice as large—that is, the same as in the two wires replaced by this common wire, or the total loss is not changed—while 25 per cent. of the copper is saved, and the system requires only 75 per cent. of the copper of the single-phase system, but produces $\sqrt{2}$ times as high a potential between the outside wires. Hence, to give the same maximum potential, the e.m.fs. of the system have to be reduced by $\sqrt{2}$, that is, the amount of copper doubled, and thus the quarter-phase system with common return of the same cross-section as the outside wires requires 150 per cent. of the copper of the single-phase system. In this case, however, the current density in the middle wire is higher, thus the copper not used most economically, and transferring a part of the copper from the outside wires to the middle wire, to bring all three wires to the same current density, reduces the loss, and thereby reduces the amount of copper at a given loss, to 145.7 per cent. of that of a single-phase system.

67. . *Comparison on the basis of equality of the maximum difference of potential between any conductor of the system and the ground, in long-distance, three-phase transmissions with grounded neutral, single-phase systems with ground return, etc.*

A system may be grounded by grounding its neutral point, for the purpose of maintaining constant-potential difference between the conductors and ground, without carrying any current through the ground, or the ground may be used as return conductor. In either case the system can be considered as consisting of and resolved into as many single-phase systems with ground return, as there are overhead conductors, and with zero resistance in the ground.

It immediately follows herefrom, that the copper efficiency of such a system is the same as that of a single-phase system with ground return, of the same voltage as exists between conductor and ground of the system under consideration. If then all the overhead conductors have the same potential difference against ground, as is the case in a three-phase or quarter-phase system with grounded neutral, a single-phase system with grounded neutral, or quarter-phase system with common ground return of both phases, the copper efficiency is the same. That is:

All grounded systems, whether with grounded neutral or with ground return, have the same copper efficiency, provided that all the overhead conductors have the same potential difference against ground.

Hence:

The three-phase system with grounded neutral has no superiority over the single-phase or the quarter-phase system with grounded neutral, in copper efficiency. The advantage of the three-phase system—which causes its practically universal use—

over the single-phase system is the greater usefulness of polyphase power, the advantage over the quarter-phase system is the use of three conductors, against four with the quarter-phase system.

No saving in copper results from the use of the ground (of zero resistance) as return circuit, but a single-phase or quarter-phase system with ground return, at equal dielectric strain on the insulation, requires the same amount of copper as a system with grounded neutral, but has a greater self-induction, due to the greater distance between conductor and return conductor or ground, and has the objection of establishing current through the ground and so disturbing neighboring circuits, by electromagnetic and electrostatic induction.

The apparent saving in copper, in the single-phase system, by replacing one of the conductors by the ground as return, therefore is a fallacy. By doing so, the potential difference of the other conductors against ground becomes twice what it would be with two conductors and grounded neutral, and at the same potential difference between conductors. That is, the single-phase system with ground return requires the same insulation as a single-phase system with grounded neutral, of twice the voltage, and then requires the same copper. A saving results only in the number of insulators required, etc. Only where the amount of power is so small that mechanical strength, and not power loss, determines the size of the conductor, a saving results by replacing one of the conductors by the ground.

The high-tension, direct-current system, whether insulated, or with grounded neutral, or with ground return, appears equal in copper efficiency to a single-phase system of the same character (insulated, or with grounded neutral, or with ground return) and of the same effective voltage, that is, with a sine wave of a maximum voltage $\sqrt{2}$ times that of the direct current. Due to the different character of unidirectional electric stress of the direct-current system, from the alternating stress, a general comparison of the system by a numerical factor appears hardly feasible. It is, however, claimed that usually the insulation stress with perfectly uniform continuous voltage is less than that of an alternating voltage of the same maximum value, so that continuous-current high-voltage transmission would offer advantages, if it were not for the difficulty of generating and utilizing very high continuous voltages, which with alternating voltages is overcome by the interposition of the stationary transformer.

CHAPTER X

METERING OF POLYPHASE CIRCUIT

68. . The power of a polyphase system or circuit is the sum of the powers of all the individual branch circuits, and the sum of the wattmeter readings of all the branch circuits thus gives the total power.

Let, then, in a general polyphase system, $e_1, e_2, e_3 \ . \ . \ . \ e_n =$ potentials at the n terminals or supply wires of the n-phase system.

These may be represented topographically by points in a plane, as shown in Fig. 47.

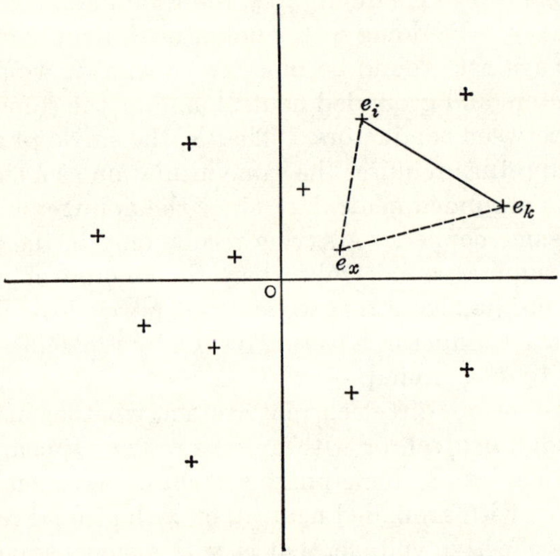

FIG. 47.

The voltage between any two terminals e_i and e_k then is:

$$e_{ik} = e_i - e_k \tag{1}$$

And this voltage, in any circuit connected between these two terminals, produces a current, i_{ik}, as the current, which flows from e_i to e_k through this circuit.

As there are $\frac{n(n-1)}{2}$ pairs of terminals e_i and e_k, there are existing in a general n-phase system $\frac{n(n-1)}{2}$ different phases, and there may thus be $\frac{n(n-1)}{2}$ different circuits, or rather sets of circuits—since a number of circuits may and usually are connected between the n terminals.

Consider one of these numerous circuits of the general n-phase system, that of the current i_{ik} passing from e_i to e_k. The power of this circuit is:

$$P_{ik} = [e_i - e_k, i_{ik}] \tag{2}$$

where the brackets denote the effective power, as discussed in Chapter XVI [Dover, Volume I, Part II].

Choosing any point e_x, which may be one of the terminals, or the neutral point of the system, if such exists, or any other point. Then the voltage $e_i - e_k$ can be resolved by the parallelogram (Fig. 218) into the voltages: $e_i - e_x$ and $e_x - e_i$, that is:

$$e_i - e_k = e_i - e_x + e_x - e_k \tag{3}$$

hence, substituted into (2):

$$\begin{aligned} P_{ik} &= [(e_i - {}_xe) + (e_x - e_k), i_{ik}] \\ &= [e_i - e_x, i_{ik}] + [e_x - e_k, i_{ik}] \end{aligned} \tag{4}$$

It is, however:

$$[e_x - e_k, i_{ik}] = [e_k - e_x, i_{ki}] \tag{5}$$

where i_{ki} is the current flowing from e_k to e_i, that is, the same current as i_{ik}, only considered in the reverse direction.

Thus it is, substituting (5) into (4):

$$P_{ik} = [e_i - e_x, i_{ik}] + [e_k - e_x, i_{ki}] \tag{6}$$

That is, the power of any branch circuit between two terminals, e_i and e_k, is the product of the powers giving by the two potential differences $e_i - e_x$ and $e_k - e_x$, of any arbitrarily chosen point e_x, with the current flowing into this branch circuit from the two terminals, e_i and e_k, that is, i_{ik} and i_{ki}, respectively.

69. The total power of the n-phase system, as the sum of the powers of all the branch circuits, then is:

$$\begin{aligned} P &= \sum_1^n{}^i \sum_1^n{}^k P_{ik} \\ &= \sum_1^n{}^i \sum_1^n{}^k [e_i - e_x, i_{ik}] \end{aligned} \tag{7}$$

where the double summation sign indicates that the summation is to be carried out for all values of k, from 1 to n, and for all values of i, from 1 to n.

As the term $e_i - e_x$ in (7) does not contain the index k, it is the same for all values of k, thus can be taken out from the second summation sign, that is:

$$P = \sum_1^n{}^i \left[e_i - e_x, \sum_1^n{}^k i_{ik} \right] \tag{8}$$

However:

$\sum_1^n{}^k i_{ik}$ is the sum of all the currents, flowing from the terminal e_i to all the other terminals e_k ($k = 1, 2 \ldots n$), that is, it is the total current issuing from the terminal e_i, or:

$$i_i = \sum_1^n{}^k i_{ik} \tag{9}$$

and, substituting this in 9, gives as the total power of the n-phase system:

$$P = \sum_1^n{}^i [e_i - e_x, i_i] \tag{10}$$

That is:

"The total power of a general n-phase system, is the sum of the n powers, given by the n currents i_i*, which issue from the n terminals* e_i*, with the n potential differences of these terminals* e_i *against any arbitrarily chosen point* e_x*."*

"The total power of the system, no matter how many branch circuits it contains, thus is measured by n wattmeters.

Choosing as the point, e_x one of the n-phase circuit terminals, that is one of the phase potentials (for instance, the neutral potential of the system, where such exists), as e_n, the number of terms in (10) reduces by one:

$$P = \sum_1^{n-1}{}^i [e_i - e_n, i_i] \tag{11}$$

That is:

"The total power of a general n-phase system is measured by n − 1 wattmeters, connected between one terminal e_n *and the n − 1 other terminals* e_1*."*

Thus for instance, a five-wire, four-phase system (Fig. 24), in which $\frac{5 \times 4}{2} = 10$ different sets of circuits are possible, is metered by $5 - 1 = 4$ meters.

A four-wire, three-phase system is metered by 3 meters.

A three-wire, three-phase system is metered by 2 meters.

70. In a three-phase system with ungrounded neutral, that is, a three-wire, three-phase system, the common method of measuring the total power thus is, by (11), as shown in Fig. 48. Often the two meters of Fig. 219 are arranged in one structure.

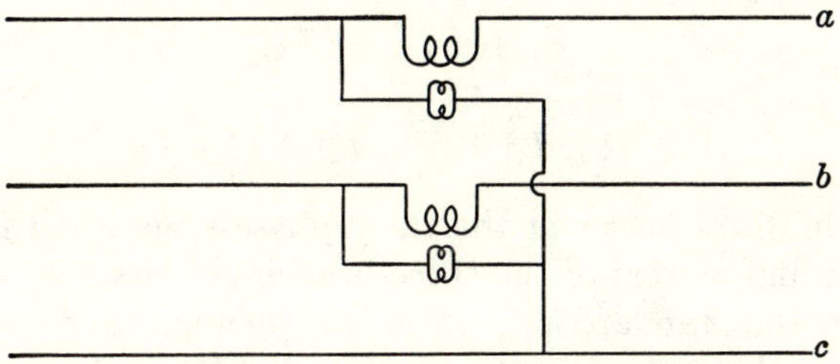

FIG. 48.

Thus, if Fig. 49 denotes a general three-wire, three-phase system, with the voltages and currents in the three phases:

$$\dot{E}_1, \dot{E}_2, \dot{E}_3 \text{ and } \dot{I}_1, \dot{I}_2, \dot{I}_3$$

counting voltages and currents in the direction indicated by the arrows in Fig. 49.

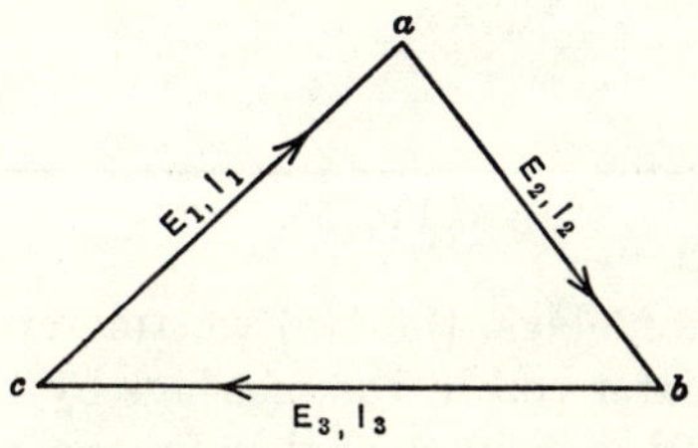

FIG. 49.

The voltages may be unequal in sizes and under unequal angles, by a distortion of the three-phase triangle, but it must be:

$$\dot{E}_1 + \dot{E}_2 + \dot{E}_3 = 0 \tag{12}$$

in a closed triangle.

Connecting then the current coils of the two wattmeters into the lines a and b, and the voltage coils between a respectively b, and c, the two wattmeter readings are:

$$\left.\begin{aligned} &[-\dot{E}_1, \dot{I}_2 - \dot{I}_1] = [\dot{E}_1, \dot{I}_1] - [\dot{E}_1, \dot{I}_2] \\ \text{and:} \quad & \\ &[\dot{E}_3, \dot{I}_3 - \dot{I}_2] = [\dot{E}_3, \dot{I}_3] - [\dot{E}_3, \dot{I}_2] \end{aligned}\right\} \tag{13}$$

and their sum is:

$$\left.\begin{aligned} P &= [\underset{\cdot}{E}_1, \underset{\cdot}{I}_1] - [\underset{\cdot}{E}_1, \underset{\cdot}{I}_2] - [\underset{\cdot}{E}_3, \underset{\cdot}{I}_2] + [\underset{\cdot}{E}_3, \underset{\cdot}{I}_3] \\ &= [\underset{\cdot}{E}_1, \underset{\cdot}{I}_1] - [\underset{\cdot}{E}_1 + \underset{\cdot}{E}_3, \underset{\cdot}{I}_2] + [\underset{\cdot}{E}_3, \underset{\cdot}{I}_3] \end{aligned}\right\} \quad (14)$$

and since by (12):

$$\underset{\cdot}{E}_1 + \underset{\cdot}{E}_3 = - \underset{\cdot}{E}_2,$$

it is:

$$\mathrm{P} = [\underset{\cdot}{E}_1, \underset{\cdot}{I}_1] + [\underset{\cdot}{E}_2, \underset{\cdot}{I}_2] + [\underset{\cdot}{E}_3, \underset{\cdot}{I}_3]$$

that is, the total power of the three-phase system is the sum of the individual powers of the three branch circuits.

71. In the standard polyphase wattmeter connection of the three-wire, three-phase system, **Fig. 48**, the voltage coils are out of phase with the current coils at non-inductive load, the one lagging, the other leading by 30°. Therefore, even in a balanced

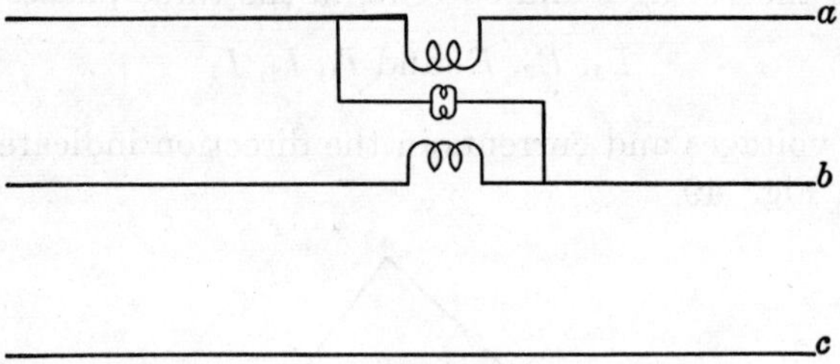

FIG. 50.

system, if the current lags, the two wattmeter coils do not read alike, as the voltmeter coil in the one lags by the angle of lag of the current plus 30°, and in the other by the angle of lag minus 30°. At 60° angle of lag, the voltage coil of the former lags $60 + 30 = 90°$, and the reading becomes zero, and at more than 60° lag, the one meter reads negative, but the algebraic sum of the two meter readings still remains the total power of the circuit, the one meter reading more than the total power, while the other meter reads negative.

In a balanced, or nearly balanced three-wire, three-phase system, instead of connecting the potential coils from *a* and *b* to *c*, Figs. 48 and 49, they are often connected from *a* to *b*. This interchanges the lagging and the leading coil, but on balanced loads leaves the same total. In this case, one voltage coil only may be used, acted upon by two current coils. That is, a single-phase wattmeter is constructed, similar to the Edison three-wire meter, with one current coil in the one, the other current coil in the other line, and the voltage coil connected between these two lines, as shown in Fig. 50.

If there is considerable unbalancing, this latter connection gives considerable error, and the double meter has to be used.

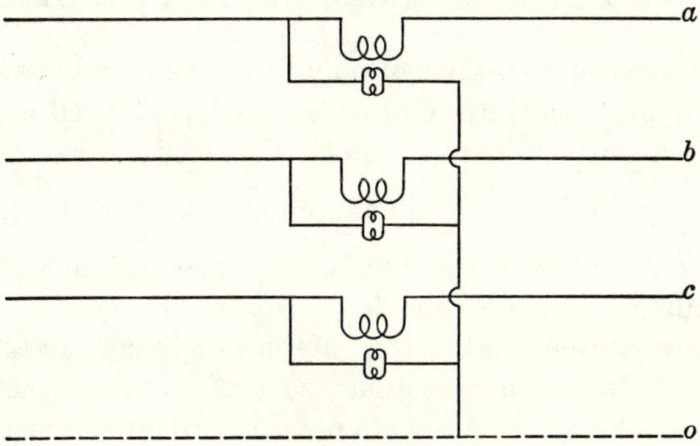

FIG. 51.

In a four-wire, three-phase system, the connection of the two meters obviously becomes wrong, if current flows in the neutral, and three meters must be used.

Most conveniently these are arranged with the three current coils in the three lines, and the voltage coils between these lines and the neutral, as shown in Fig. 51.

CHAPTER XI

BALANCED SYMMETRICAL POLYPHASE SYSTEMS

72. In most applications of polyphase systems the system is a balanced symmetrical system, or as nearly balanced as possible. That is, it consists of n equal e.m.fs. displaced in phase from each other by $\frac{1}{n}$ period, and producing equal currents of equal phase displacement against their e.m.fs. In such systems, each e.m.f. and its current can be considered separately as constituting a single-phase system, that is, the polyphase system can be resolved into n equal single-phase systems, each of which consists of one conductor of the polyphase system, with zero impedance as return circuit. Hereby the investigation of the polyphase system resolves itself into that of its constituent single-phase system.

So, for instance, the polyphase system shown in Fig. 37, at balanced load, can be considered as consisting of the equal single-phase systems : $0 - 1$; $0 - 2$; $0 - 3$; . . . $0 - n$, each of which consists of one conductor, 1, 2, 3, . . . n, and the return conductor, 0. Since the sum of all the currents equals 0, there is no current in conductor 0, that is, no voltage is consumed in this conductor; this is equivalent to assuming this conductor as of zero impedance. This common return conductor, 0, since it carries no current, can be omitted, as is usually the case. With star connection of an apparatus into a polyphase system, as in Fig. 29, the impedance of the equivalent single-phase system is the impedance of one conductor or circuit; if, however, the apparatus is ring connected, as shown diagrammatically in Fig. 30, the impedance of the ring-connected part of the circuit has to be reduced to star connection, in the usual manner of reducing a circuit to another circuit of different voltage, by the ratio

$$c = \frac{\text{ring voltage}}{\text{star voltage}};$$

or, as these voltages are usually called in a three-phase system,

$$c = \frac{\text{delta voltage}}{Y \text{ voltage}}.$$

That is, all ring voltages are divided, all ring currents multiplied with c; all ring impedances are divided, all ring admittances multiplied with the square of the ratio, c^2.

For instance, if in a three-phase induction motor with delta-connected circuits, the impedance of each circuit is

$$Z = r + jx,$$

and the voltage impressed upon the circuit terminals E, and the motor is supplied over a line of impedance, per line wire,

$$Z_0 = r_0 + jx_0,$$

the motor impedance, reduced to star connection, or Y impedance, is

$$Z' = r \frac{r + jx}{c^2} = \frac{1}{3}(r + jx),$$

and the impressed voltage, reduced to Y circuit,

$$E' = \frac{E}{\sqrt{3}},$$

and the total impedance of the equivalent single-phase circuit is therefore

$$Z_0 + Z' = (r_0 + jx_0) + \frac{1}{3}(r + jx).$$

Inversely, however, where this appears more convenient, all quantities may be reduced to ring or delta connection, or one of the ring connections considered as equivalent single-phase circuit, of impedance

$$Z + c^2 Z_0 = (r + jx) + 3(r_0 + jx_0).$$

Since the line impedances, line currents and the voltages consumed in the lines of a polyphase system are star, or (in a three-phase system) Y quantities, it usually is more convenient to reduce all quantities to Y connection, and use one of the Y-circuits as the equivalent single-phase circuit.

73. As an example may be considered the calculation of a long-distance transmission line, delivering 10,000 kw., three-phase power at 60 cycles, 80,000 volts and 90 per cent. power-factor at 100 miles from the generating station, with approximately 10 per cent. loss of power in the transmission line, and with the line conductors arranged in a triangle 6 ft. distant from each other.

10,000 kw. total power delivered gives 3,333 kw. per line or single-phase branch (Y power).

3,333 kw. at 90 per cent. power-factor gives 3,700 kv.-amp.

80,000 volts between the lines gives $80,000 \div \sqrt{3} = 46,100$ volts from line to neutral, or per single-phase circuit.

3,700 kv.-amp. per circuit, at 46,100 volts, gives 80 amp. per line.

10 per cent. loss gives 333 kw. loss per line, and at 80 amp., this gives a resistance per line,

$$333,000 \div 80^2 = 52 \text{ ohms},$$

or, 0.52 ohms per mile.

The nearest standard size of wire is No. 0 B. & S., which has a resistance of 0.52 ohms, and a weight of 1680 lb. per mile.

Choosing this size of wire so requires for the 300 miles of line conductor, $300 \times 1680 = 500,000$ lb. of copper.

At 0.52 ohms per mile, the resistance per transmission line or circuit of 100 miles length is,

$$r = 52 \text{ ohms}.$$

The inductance of wire No. 0, with $d = 0.325$ in. diameter, and 6 ft. = 72 in. distance from the return conductor, is calculated from the formula of line inductance[1] as, 2.3 mil-henrys per mile; hence, per circuit,

$$L = 0.23 \text{ henry},$$

and herefrom the reactance,

$$\begin{aligned} x &= 2\,\pi f L \\ &= 88 \text{ ohms}. \end{aligned}$$

The capacity of the transmission line may be calculated directly, or more conveniently it may be derived from the inductance. If C is the capacity of the circuit, of which the inductance is L, then

$$f_1 = \frac{1}{4\sqrt{LC}}$$

is the fundamental frequency of oscillation, or natural period, that is, the frequency which makes the length, l, of the line a quarter-wave length.

Since the velocity of propagation of the electric field is the ve-

[1] *Elements of Electrical Engineering*, Dover edition Volume I, Part I.

locity of light, v, with a wave-length, $4\,l$, the number of waves per second, or frequency of oscillation of the line, is

$$f_1 = \frac{v}{4\,l}$$

and herefrom then follows:

$$\frac{v}{l} = \frac{1}{\sqrt{LC}}.^{1}$$

hence, for

$$\begin{aligned} l &= 100 \text{ miles}, \\ v &= 186{,}000 \text{ miles per second}, \\ L &= 0.23 \text{ henry}, \\ C &= 1.26 \text{ mf}. \end{aligned}$$

and the capacity susceptance,

$$b = 2\,\pi f C = 475 \times 10^{-6}.$$

Representing, as approximation, the line capacity by a condenser shunted across the middle of the line

We have, impedance of half the line,

$$Z = \frac{r}{2} + j\frac{x}{2} = 26 + 44\,j \text{ ohms}.$$

Choosing the voltage at the receiving end as zero vector,

$$e = 46{,}100 \text{ volts},$$

at 90 per cent. power-factor and therefore 43.6 per cent. inductance factor, the current is represented by

$$\dot{I} = 80\,(0.9 - 0.436\,j) = 72 - 35\,j.$$

[1] Or, if μ = permeability, κ = dielectric constant of the medium surrounding the conductor, it is

$$f_1 = \frac{v}{4\,l\sqrt{\mu\kappa}}$$

hence,

$$\frac{v}{l} = \sqrt{\frac{\mu\kappa}{LC}},$$

or,

$$C = \frac{\left(\frac{l}{v}\right)^2}{L}.$$

This gives:

Voltage at receiver circuit, $e = 46{,}100$ volts;
current in receiver circuit, $\dot{I} = 72 - 35\,j$ amp.;
impedance voltage of half the line, $Z\dot{I} = 3410 + 2260\,j$ volts.

Hence, the condenser voltage, $\dot{E}_1 = e + Z\dot{I} = 49{,}510 + 2260\,j$ volts;

and the condenser current, $+ jb\dot{E}_1 = - 1.1 + 23.8\,j$ amp.;

hence, the total, or generator current, $\dot{I}_0 = \dot{I} + jb\dot{E}_1 = 70.9 - 11.2\,j$ amp.

The impedance voltage of the other half of the line, $Z\dot{I}_0 = 2330 - 2830\,j$ volts;

hence, the generator voltage, $\dot{E}_0 = \dot{E}_1 + Z\dot{I}_0 = 51{,}840 + 5090\,j$ volts;

and the phase angle of the generator current,

$$\tan \theta_1 = \frac{11.2}{70.9} = 0.158; \quad \theta_1 = 9.0°$$

The phase angle of the generator voltage,

$$\tan \theta_2 = - \frac{5090}{51{,}840} = - 0.098; \; \theta_2 = - 5.6°;$$

the lag of the generator current, $\theta_0 = \theta_1 - \theta_2 = 14.6°$;
hence the power-factor at the generator, $\cos \theta_0 = 96.7$ per cent.
And the power output, $3\,[\dot{I}, e]^1 = 10{,}000$ kw.;
the power input, $3\,[\dot{I}_0, \dot{E}_0]^1 = 11{,}190$ kw.;
the efficiency $= 89.35$ per cent.;
the volt-ampere output, $3\,ie = 11{,}110$ kv.-amp.;
the volt-ampere input, $3\,i_0e_0 = 11{,}220$ kv.-amp.;
ratio: $= 99.02$ per cent.

And the absolute values are:
receiver current, $i = 80$ amp.;
receiver voltage, $e = 46{,}100 \times \sqrt{3} = 80{,}000$ volts;
generator current, $i_0 = 71.8$ amp.;
generator voltage, $e_0 = 52{,}100 \times \sqrt{3} = 90{,}000$ volts;
voltage drop in line, $= 11.1$ per cent.

74. Balanced polyphase systems thus can be calculated as single-phase systems, and this has been done in many preceding chapters, as in those on the induction machines, synchronous machines, etc., that is, apparatus which is usually operated on polyphase circuits.

Only in dealing with those phenomena which are resultants of all the phases of the polyphase system, in the resolution of the polyphase system into its constituent single-phase systems the effective value of the constant has to be used, which corresponds to the resultant effect. This, for instance, is the case in calculating the magnetic field of the induction machine—which is energized by the combination of all phases—or the armature reaction of synchronous machines, etc.

For instance, in the induction machine, from the generated e.m.f., e the magnetic flux of the machine is calculated, and from the magnetic flux and the dimensions of the magnetic circuit: length and section of air-gap, and length and section of the iron part, follows the ampere-turns excitation, that is, the ampere turns, F_0, required to produce the magnetic flux.

The resultant m.m.f. of m equal magnetizing coils displaced in position by $\frac{1}{m}$ cycle, energized by m equal currents of an m-phase system, is given by §40, page 61 as

$$F_0 = \frac{nmI}{\sqrt{2}}$$

where

I = current per phase, or per magnetizing coil,
n = number of turns per coil,
m = number of phases.

The exciting current per phase required to produce the resulting m.m.f., F_0, therefore, is

$$I = \frac{F_0 \sqrt{2}}{nm};$$

hence, for a three-phase system,

$$I = \frac{F_0 \sqrt{2}}{3n},$$

and for a quarter-phase system, with two coils in quadrature,

$$I = \frac{F_0}{n\sqrt{2}}.$$

In the investigation of the armature reaction of synchronous machines, the armature reaction of an m-phase machine is, by §40, page 61,

$$F = \frac{n_0 m I}{\sqrt{2}};$$

where

m = number of phases,

n_0 = number of turns per phase, effective, that is, allowing for the spread of turns over an arc of the periphery in machines of distributed winding,

I = current per phase,

and when the armature reaction is given by nI, the number of effective turns, n, is, accordingly, for a polyphase alternator,

$$n = \frac{m}{\sqrt{2}} n_0;$$

hence, in a three-phase machine,

$$n = \frac{3 n_0}{\sqrt{2}} = 1.5 \, n_0 \sqrt{2};$$

in a quadrature-phase machine,

$$n = n_0 \sqrt{2}.$$

75. When replacing a balanced symmetrical polyphase system by its constituent single-phase systems, it must be considered, that the constants of the constituent single-phase circuit may not be the same which this circuit would have as independent single-phase circuit.

If the branches of the polyphase circuit, which constitute the equivalent single-phase circuits, are electrically or magnetically interlinked, the constants, as admittance, impedance, etc., of the equivalent single-phase circuit often are different from those of the same circuit on single-phase supply, and the polyphase values then must be used in the equivalent single-phase circuits which replace the polyphase system.

This is the case in induction machines, in the armatures of synchronous machines, etc., where the phases are in mutual induction with each other.

Let, in a star or Y-connected three-phase induction motor:

$$Y = g - jb$$

be the exciting admittance and e the impressed voltage per three-phase Y circuit or constituent single-phase circuit.

The exciting current per circuit then is:

$$\dot{I} = eY$$

or, absolute:

$$i = ey$$

if n = number of turns per circuit,
$f = ni$ = effective value of the m.m.f. per phase, and
$F = 1.5 \times \sqrt{2}\, ni$ = resultant m.m.f. of all three phases.

F then produces in the magnetic circuit the flux Φ which consumes the impressed voltage e.

Assuming now, that instead of impressing three three-phase voltage e on the three constituent single-phase circuits of the motor, we impress only a single-phase voltage e on one of the three circuits.

The current in this circuit then must produce the same flux Φ, and have the same maximum m.m.f. F, as was given by the resultant of all three phases.

With n turns, that means, the current i_1 under the single-phase e.m.f. e is given by:

$$F = \sqrt{2}\, ni_1$$

and since we had, under the same voltage e and flux Φ, three-phase:

$$F = 1.5\sqrt{2}\, ni$$

it follows:

$$i_1 = 1.5\, i$$

That is, with a single-phase voltage, e, the current, i_1, and thus the admittance, Y_1, of the circuit, is 1.5 times the current, i, and thus the admittance, Y, which is produced in the same circuit by the three-phase voltage:

$$Y_1 = 1.5\, Y$$

or:

$$Y = \tfrac{2}{3}\, Y_1$$

That is:

If we measure the admittance of one of the motor circuits by single-phase supply voltage, this is not the admittance of this circuit as constituent single-phase circuit of the three-phase motor, but

The admittance of the constituent or equivalent single-phase circuit of a three-phase induction motor is two-thirds of the admittance of this same circuit as independent single-phase circuit.

We can look at this in a different way:

As the three-phase circuits combine to a resultant which is 1.5 times the m.m.f. of each circuit, each circuit requires only two-thirds of the m.m.f., and thus two-thirds of the exciting admittance, as equivalent single-phase circuit of a three-phase motor, which it would require, if as independent single-phase circuit it had to produce the entire m.m.f.

76. The same applies to the self-inductive reactance: as the self-inductive or leakage flux, which consumes the reactance voltage, is produced by the resultant of the currents of all three phases, and this resultant is 1.5 times the maximum of one phase, each phase produces only two-thirds, that is, the impedance current of each phase of the motor on three-phase voltage supply is only two-thirds that of the same circuit at the same voltage of single-phase supply, and the impedance thus is $3/2 = 1.5$ times.

That is:

The effective admittance of the equivalent or constituent single-phase circuit of a three-phase induction machine is two-thirds of the admittance, and the effective impedance is 1.5 times the impedance of this circuit as independent single-phase circuit.

The same applies to synchronous machines:

The three-phase synchronous reactance per armature circuit, that is, the synchronous reactance of this armature circuit as equivalent single-phase circuit of the three-phase system, is 1.5 times the single-phase synchronous reactance of the same armature circuit, that is, synchronous reactance of this circuit as single-phase machine.

In dealing with the constituent single-phase circuits of a three-phase system, the proper "three-phase" values of the constants of the equivalent circuit must be used.

CHAPTER XII

THREE-PHASE SYSTEM

77. With equal load of the same phase displacement in all three branches, the symmetrical three-phase system offers no special features over those of three equally loaded single-phase systems, and can be treated as such; since the mutual reactions between the three phases balance at equal distribution of load, that is, since each phase is acted upon by the preceding phase in an equal but opposite manner as by the following phase.

With unequal distribution of load between the different branches, the voltages and phase differences become more or less unequal. These unbalancing effects are obviously maximum if some of the phases are fully loaded, others unloaded.

Let E = e.m.f. between branches 1 and 2 of a three-phaser. Then

$$\epsilon E = \text{e.m.f. between 2 and 3,}$$

$$\epsilon^2 E = \text{e.m.f. between 3 and 1;}$$

where

$$\epsilon = \sqrt[3]{1} = \frac{-1 + j\sqrt{3}}{2}.$$

Let

Z_1, Z_2, Z_3 = impedances of the lines issuing from generator terminals 1, 2, 3,

and Y_1, Y_2, Y_3 = admittances of the consumer circuits connected between lines 2 and 3, 3 and 1, 1 and 2.

If then,

I_1, I_2, I_3, are the currents issuing from the generator terminals into the lines, it is,

$$I_1 + I_2 + I_3 = 0. \qquad (1)$$

If, I'_1, I'_2, I'_3 = currents through the admittances, Y_1, Y_2, Y_3, from 2 to 3, 3 to 1, 1 to 2, it is,

$$\left.\begin{array}{l} I_1 = I'_3 - I'_2, \text{ or, } I_1 + I'_2 - I'_3 = 0 \\ I_2 = I'_1 - I'_3, \text{ or, } I_2 + I'_3 - I'_1 = 0 \\ I_3 = I'_2 - I'_1, \text{ or, } I_3 + I'_1 - I'_2 = 0 \end{array}\right\} \qquad (2)$$

These three equations (2) added, give (1) as dependent equation.

At the ends of the lines 1, 2, 3, it is:

$$\left.\begin{aligned} \dot{E}'_1 &= \dot{E}_1 - Z_2\dot{I}_2 + Z_3\dot{I}_3 \\ \dot{E}'_2 &= \dot{E}_2 - Z_3\dot{I}_3 + Z_1\dot{I}_1 \\ \dot{E}'_3 &= \dot{E}_3 - Z_1\dot{I}_1 + Z_2\dot{I}_2 \end{aligned}\right\} \qquad (3)$$

the differences of potential, and:

$$\left.\begin{aligned} \dot{I}'_1 &= \dot{E}'_1 Y_1 \\ \dot{I}'_2 &= \dot{E}'_2 Y_2 \\ \dot{I}'_3 &= \dot{E}'_3 Y_3 \end{aligned}\right\} \qquad (4)$$

the currents in the receiver circuits.

These nine equations (2), (3), (4), determine the nine quantities: $\dot{I}_1, \dot{I}_2, \dot{I}_3, \dot{I}_1', \dot{I}_2', \dot{I}_3', \dot{E}_1', \dot{E}_2', \dot{E}_3'$.

Equations (4) substituted in (2) give:

$$\left.\begin{aligned} \dot{I}_1 &= \dot{E}'_3 Y_3 - \dot{E}'_2 Y_2 \\ \dot{I}_2 &= \dot{E}'_1 Y_1 - \dot{E}'_3 Y_3 \\ \dot{I}_3 &= \dot{E}'_2 Y_2 - \dot{E}'_1 Y_1 \end{aligned}\right\} \qquad (5)$$

These equations (5) substituted in (3), and transposed, give:

since $\left.\begin{aligned} \dot{E}_1 &= \epsilon\,\dot{E} \\ \dot{E}_2 &= \epsilon^2\dot{E} \\ \dot{E}_3 &= \dot{E} \end{aligned}\right\}$ as e.m.fs. at the generator terminals.

$$\left.\begin{aligned} \epsilon\,\dot{E} - \dot{E}'_1(1 + Y_1Z_2 + Y_1Z_3) + \dot{E}'_2Y_2Z_3 + \dot{E}'_3Y_3Z_2 &= 0 \\ \epsilon^2\dot{E} - \dot{E}'_2(1 + Y_2Z_3 + Y_2Z_1) + \dot{E}'_3Y_3Z_1 + \dot{E}'_1Y_1Z_3 &= 0 \\ \dot{E} - \dot{E}'_3(1 + Y_3Z_1 + Y_3Z_2) + \dot{E}'_1Y_1Z_2 + \dot{E}'_2Y_2Z_1 &= 0 \end{aligned}\right\} \qquad (6)$$

as three linear equations with the three quantities, $\dot{E}'_1$, $\dot{E}'_2$, $\dot{E}'_3$.

Substituting the abbreviations:

$$\left.\begin{array}{l}
K = \begin{bmatrix} -(1 + Y_1Z_2 + Y_1Z_3), & Y_2Z_3, & Y_3Z_2 \\ Y_1Z_3, & -(1 + Y_2Z_3 + Y_2Z_1), & Y_3Z_1 \\ Y_1Z_2, & Y_2Z_1, & -(1 + Y_3Z_1 + Y_3Z_2) \end{bmatrix} \\[2ex]
K_1 = \begin{bmatrix} \epsilon, & Y_2Z_3, & Y_3Z_2 \\ \epsilon^2, & -(1 + Y_2Z_3 + Y_2Z_1), & Y_3Z_1 \\ 1, & Y_2Z_1, & -(1 + Y_3Z_1 + Y_3Z_2) \end{bmatrix} \\[2ex]
K_2 = \begin{bmatrix} -(1 + Y_1Z_2 + Y_1Z_3), & \epsilon, & Y_3Z_2 \\ Y_1Z_3, & \epsilon^2, & Y_3Z_1 \\ Y_1Z_2, & 1, & -(1 + Y_3Z_1 + Y_3Z_2) \end{bmatrix} \\[2ex]
K_3 = \begin{bmatrix} -(1 + Y_1Z_2 + Y_1Z_3), & Y_2Z_3, & \epsilon \\ Y_1Z_3, & -(1 + Y_2Z_3 + Y_2Z_1), & \epsilon^2 \\ Y_1Z_2, & Y_2Z_1, & 1 \end{bmatrix}
\end{array}\right\} \qquad (7)$$

we have:

$$\left.\begin{array}{l}
\underset{\cdot}{E}'_1 = \dfrac{\underset{\cdot}{E}K_1}{K} \\[2ex]
\underset{\cdot}{E}'_2 = \dfrac{\underset{\cdot}{E}K_2}{K} \\[2ex]
\underset{\cdot}{E}'_3 = \dfrac{\underset{\cdot}{E}K_3}{K}
\end{array}\right\} \qquad (8)$$

$$\left.\begin{array}{l}
\underset{\cdot}{I}'_1 = \dfrac{Y_1\underset{\cdot}{E}K_1}{K} \\[2ex]
\underset{\cdot}{I}'_2 = \dfrac{Y_2\underset{\cdot}{E}K_2}{K} \\[2ex]
\underset{\cdot}{I}'_3 = \dfrac{Y_3\underset{\cdot}{E}K_3}{K}
\end{array}\right\} \qquad (9)$$

$$\left.\begin{array}{l}
\underset{\cdot}{I}_1 = \dfrac{Y_3K_3 - Y_2K_2\underset{\cdot}{E}}{K} \\[2ex]
\underset{\cdot}{I}_2 = \dfrac{Y_1K_1 - Y_3K_3\underset{\cdot}{E}}{K} \\[2ex]
\underset{\cdot}{I}_3 = \dfrac{Y_2K_2 - Y_1K_1\underset{\cdot}{E}}{K}
\end{array}\right\} \qquad (10)$$

hence,

$$\left.\begin{array}{l}
\underset{\cdot}{E}'_1 + \underset{\cdot}{E}'_2 + \underset{\cdot}{E}'_3 = 0 \\
\underset{\cdot}{I}_1 + \underset{\cdot}{I}_2 + \underset{\cdot}{I}_3 = 0
\end{array}\right\} \qquad (11)$$

78. Special Cases.

A. Balanced System

$$Y_1 = Y_2 = Y_3 = Y$$
$$Z_1 = Z_2 = Z_3 = Z.$$

Substituting this in (6), and transposing:

$$\left.\begin{array}{ll} \left.\begin{array}{l} \dot{E}_1 = \epsilon \dot{E} \\ \dot{E}_2 = \epsilon^2 \dot{E} \\ \dot{E}_3 = \dot{E} \end{array}\right\} & \begin{array}{l} \dot{E}'_1 = \dfrac{\epsilon \dot{E}}{1 + 3\,YZ} \\ \dot{E}'_2 = \dfrac{\epsilon^2 \dot{E}}{1 + 3\,YZ} \\ \dot{E}'_3 = \dfrac{\dot{E}}{1 + 3\,YZ} \end{array} \\ \left.\begin{array}{l} \dot{I}'_1 = \dfrac{\epsilon \dot{E} Y}{1 + 3\,YZ} \\ \dot{I}'_2 = \dfrac{\epsilon^2 \dot{E} Y}{1 + 3\,YZ} \\ \dot{I}'_3 = \dfrac{\dot{E} Y}{1 + 3\,YZ} \end{array}\right\} & \begin{array}{l} \dot{I}_1 = \dfrac{\epsilon^2 (\epsilon - 1)\, \dot{E} Y}{1 + 3\,YZ} \\ \dot{I}_2 = \dfrac{(\epsilon - 1)\, \dot{E} Y}{1 + 3\,YZ} \\ \dot{I}_3 = \dfrac{\epsilon (\epsilon - 1)\, \dot{E} Y}{1 + 3\,YZ} \end{array} \end{array}\right\} \quad (12)$$

The equations of the symmetrical balanced three-phase system.

B. One Circuit Loaded, Two Unloaded

$$Y_1 = Y_2 = 0, \; Y_3 = Y,$$
$$Z_1 = Z_2 = Z_3 = Z.$$

Substituted in equations (6):

$$\left.\begin{array}{l} \epsilon\, \dot{E} - \dot{E}'_1 + \dot{E}_3'YZ = 0 \\ \epsilon^2 \dot{E} - \dot{E}'_2 + \dot{E}'_3 YZ = 0 \end{array}\right\} \text{unloaded branches.}$$

$$\dot{E} - \dot{E}'_3 (1 + 2\,YZ) = 0, \text{ loaded branch.}$$

hence:

$$\left.\begin{array}{ll} \left.\begin{array}{l} \dot{E}'_1 = \dfrac{\dot{E}\{\epsilon + (1 + 2\,\epsilon)\, YZ\}}{1 + 2\,YZ} \\ \dot{E}'_2 = \dfrac{\dot{E}\{\epsilon^2 + (1 + 2\,\epsilon^2)\, YZ\}}{1 + 2\,YZ} \end{array}\right\} & \text{unloaded;} \\ \dot{E}'_3 = \dfrac{\dot{E}}{1 + 2\,YZ} & \text{loaded;} \end{array}\right\} \begin{array}{c} \text{all three} \\ \text{e.m.fs.} \\ \text{unequal, and} \\ \text{of unequal} \\ \text{phase angles.} \end{array} \quad (13)$$

$$\left.\begin{aligned} \dot{I}'_1 &= \dot{I}'_2 = 0 \\ \dot{I}'_3 &= \frac{\dot{E}Y}{1 + 2\,YZ} \end{aligned}\right\} \qquad (13)$$

$$\left.\begin{aligned} \dot{I}_1 &= \frac{\dot{E}Y}{1 + 2\,YZ} \\ \dot{I}_2 &= -\frac{\dot{E}Y}{1 + 2\,YZ} \\ \dot{I}_3 &= 0 \end{aligned}\right\} \qquad (13)$$

C. Two Circuits Loaded, One Unloaded

$$Y_1 = Y_2 = Y, \; Y_3 = 0,$$
$$Z_1 = Z_2 = Z_3 = Z.$$

Substituting this in equations (6), it is

$$\left.\begin{aligned} \epsilon\,\dot{E} - \dot{E}'_1(1 + 2\,YZ) + \dot{E}'_2YZ &= 0 \\ \epsilon^2\dot{E} - \dot{E}'_2(1 + 2\,YZ) + \dot{E}'_1YZ &= 0 \end{aligned}\right\} \text{ loaded branches.}$$

$$\dot{E} - \dot{E}'_3 + (\dot{E}'_1 + \dot{E}'_2)YZ = 0 \quad \text{unloaded branch.}$$

or, since

$$(\dot{E}'_1 + \dot{E}'_2) = -\dot{E}'_3;$$
$$\dot{E} - \dot{E}'_3 - \dot{E}'_3YZ = 0,$$
$$\dot{E}'_3 = \frac{\dot{E}}{1 + YZ};$$

thus,

$$\left.\begin{aligned} &\left.\begin{aligned} \dot{E}'_1 &= \frac{\dot{E}\epsilon\,\{1 + (2 + \epsilon)\,YZ\}}{1 + 4\,YZ + 3\,Y^2Z^2} \\ \dot{E}'_2 &= \frac{\dot{E}\epsilon^2\,\{1 + (2 + \epsilon^1)\,YZ\}}{1 + 4\,YZ + 3\,Y^2Z^2} \end{aligned}\right\} \text{ loaded branches.} \\ &\dot{E}'_3 = \frac{\dot{E}}{1 + YZ} \qquad \text{unloaded branch.} \end{aligned}\right\} \qquad (14)$$

As seen, with unsymmetrical distribution of load, all three branches become more or less unequal, and the phase displacement between them unequal also.

CHAPTER XIII

QUARTER-PHASE SYSTEM

79. In a three-wire quarter-phase system, or quarter-phase system with common return-wire of both phases, let the two outside terminals and wires be denoted by 1 and 2, the middle wire or common return by 0.

It is then,

$E_1 = E$ = e.m.f. between 0 and 1 in the generator.

$E_2 = jE$ = e.m.f. between 0 and 2 in the generator.

Let I_1 and I_2 = currents in 1 and in 2,

I_0 = current in 0,

Z_1 and Z_2 = impedances of lines 1 and 2,

Z_0 = impedance of line 0,

Y_1 and Y_2 = admittances of circuits 0 to 1, and 0 to 2,

I'_1 and I'_2 = currents in circuits 0 to 1, and 0 to 2,

E'_1 and E'_2 = potential differences at circuit 0 to 1, and 0 to 2.

it is then,

$$\left.\begin{aligned} &I_1 + I_2 + I_0 = 0, \\ \text{or,}\quad &I_0 = -(I_1 + I_2); \end{aligned}\right\} \qquad (1)$$

that is, I_0 is common return of I_1 and I_2.

Further, we have:

$$\left.\begin{aligned} E'_1 &= E - I_1Z_1 + I_0Z_0 = E - I_1(Z_1 + Z_0) - I_2Z_0 \\ E'_2 &= jE - I_2Z_0 + I_0Z_0 = jE - I_2(Z_2 + Z_0) - I_1Z_1 \end{aligned}\right\} \qquad (2)$$

and

$$\left.\begin{aligned} I_1 &= Y_1E'_1 \\ I_2 &= Y_2E'_2 \\ I_0 &= -(Y_1E'_1 + Y_2E'_2) \end{aligned}\right\} \qquad (3)$$

Substituting (3) in (2), and expanding,

$$\left.\begin{aligned} E'_1 &= E\frac{1 + Y_2Z_2 + Y_2Z_0(1 - j)}{(1 + Y_1Z_0 + Y_1Z_1)(1 + Y_2Z_0 + Y_2Z_2) - Y_1Y_2Z_0^2} \\ E'_2 &= jE\frac{1 + Y_1Z_1 + Y_1Z_0(1 + j)}{(1 + Y_1Z_0 + Y_1Z_1)(1 + Y_2Z_0 + Y_2Z_2) - Y_1Y_2Z_0^2} \end{aligned}\right\} \qquad (4)$$

Hence, the two e.m.fs. at the end of the line are unequal in magnitude, and not in quadrature any more.

80. Special Cases:

A. Balanced System

$$Z_1 = Z_2 = Z$$

$$Z_0 = \frac{Z}{\sqrt{2}};$$

$$Y_1 = Y_2 = Y.$$

Substituting these values in (4), gives:

$$\left.\begin{aligned} \dot{E}'_1 &= \dot{E}\frac{1 + \frac{1 + \sqrt{2} - j}{\sqrt{2}} YZ}{1 + \sqrt{2}(1 + \sqrt{2})\, YZ + (1 + \sqrt{2})\, Y^2Z^2} \\ &= \dot{E}\frac{1 + (1.707 + 0.707\, j)\, YZ}{1 + 3.414\, YZ + 2.414\, Y^2Z^2} \\ \dot{E}'_2 &= j\dot{E}\frac{1 + \frac{1 + \sqrt{2} + j}{\sqrt{2}} YZ}{1 + \sqrt{2}(1 + \sqrt{2})\, YZ + (1 + \sqrt{2})\, Y^2Z^2} \\ &= j\dot{E}\frac{1 + (1.707 + 0.707\, j)\, YZ}{1 + 3.414\, YZ + 2.414\, Y^2Z^2} \end{aligned}\right\} \quad (5)$$

Hence, the balanced quarter-phase system with common return is unbalanced with regard to voltage and phase relation, or in other words, even if in a quarter-phase system with common return both branches or phases are loaded equally, with a load of the same phase displacement, nevertheless the system becomes unbalanced, and the two e.m.fs. at the end of the line are neither equal in magnitude, nor in quadrature with each other.

B. One Branch Loaded, One Unloaded

$$Z_1 = Z_2 = Z,$$

$$Z_0 = \frac{Z}{\sqrt{2}}.$$

(a) $$Y_1 = 0,\ Y_2 = Y,$$

(b) $$Y_1 = Y,\ Y_2 = 0.$$

Substituting these values in (4), gives:

$$
\left.
\begin{aligned}
(a)\qquad \dot{E}'_1 &= \dot{E}\,\frac{1 + YZ\dfrac{1+\sqrt{2}-j}{\sqrt{2}}}{1 + YZ\dfrac{1+\sqrt{2}}{\sqrt{2}}} \\
&= \dot{E}\left\{1 - \frac{j}{1+\sqrt{2}+\dfrac{\sqrt{2}}{YZ}}\right\} \\
&= \dot{E}\left\{1 - \frac{j}{2.414+\dfrac{1.414}{YZ}}\right\} \\
\dot{E}'_2 &= j\dot{E}\,\frac{1}{1 + YZ\dfrac{1+\sqrt{2}}{\sqrt{2}}} \\
&= j\dot{E}\,\frac{1}{1 + 1.707\,YZ}
\end{aligned}
\right\} \qquad (6)
$$

$$
\left.
\begin{aligned}
(b)\qquad \dot{E}'_1 &= \dot{E}\,\frac{1}{1 + YZ\dfrac{1+\sqrt{2}}{\sqrt{2}}} \\
&= \dot{E}\,\frac{1}{1 + 1.707\,YZ} \\
\dot{E}'_2 &= j\dot{E}\,\frac{1 + YZ\dfrac{1+\sqrt{2}+j}{\sqrt{2}}}{1 + YZ\dfrac{1+\sqrt{2}}{\sqrt{2}}} \\
&= j\dot{E}\left\{1 + \frac{j}{1+\sqrt{2}+\dfrac{\sqrt{2}}{YZ}}\right\} \\
&= j\dot{E}\left\{1 + \frac{j}{2.414+\dfrac{1.414}{YZ}}\right\}
\end{aligned}
\right\} \qquad (7)
$$

These two e.m.fs. are unequal, and not in quadrature with each other.

But the values in case (*a*) are different from the values in case (*b*).

That means:

The two phases of a three-wire, quarter-phase system are

unsymmetrical, and the leading phase, 1, reacts upon the lagging phase, 2, in a different manner than 2 reacts upon 1.

It is thus undesirable to use a three-wire, quarter-phase system, except in cases where the line impedances, Z, are negligible.

In all other cases, the four-wire, quarter-phase system is preferable, which essentially consists of two independent single-phase circuits, and is treated as such.

Obviously, even in such an independent quarter-phase system, at unequal distribution of load, unbalancing effects may take place.

If one of the branches or phases is loaded differently from the other, the drop of voltage and the shift of the phase will be different from that in the other branch; and thus the e.m.fs. at the end of the lines will be neither equal in magnitude, nor in quadrature with each other.

With both branches, however, loaded equally, the system remains balanced in voltage and phase, just like the three-phase system under the same conditions.

Thus the four-wire, quarter-phase system and the three-phase system are balanced with regard to voltage and phase at equal distribution of load, but are liable to become unbalanced at unequal distribution of load; the three-wire, quarter-phase system is unbalanced in voltage and phase, even at equal distribution of load.

PART II

ELEMENTARY LECTURES ON ELECTRIC DISCHARGES, WAVES AND IMPULSES

CHAPTER I

NATURE AND ORIGIN OF TRANSIENTS

1. Electrical engineering deals with electric energy and its flow, that is, electric power. Two classes of phenomena are met: permanent and transient phenomena. To illustrate: Let G in Fig. 1 be a direct-current generator, which over a circuit A connects to a load L, as a number of lamps, etc. In the generator G, the line A, and the load L, a current i flows, and voltages e

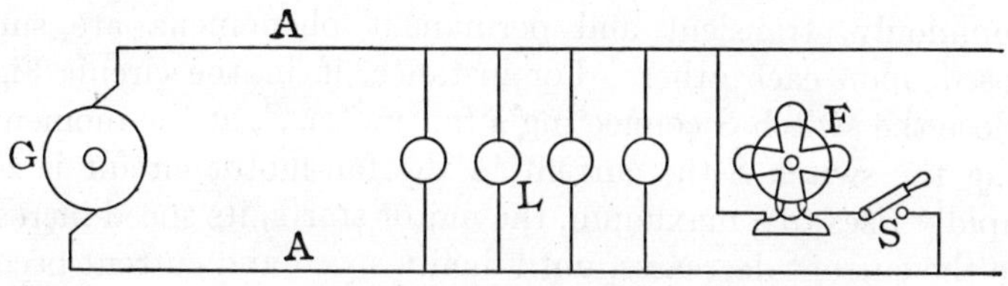

Fig. 1.

exist, which are constant, or permanent, as long as the conditions of the circuit remain the same. If we connect in some more lights, or disconnect some of the load, we get a different current i', and possibly different voltages e'; but again i' and e' are permanent, that is, remain the same as long as the circuit remains unchanged.

Let, however, in Fig. 2, a direct-current generator G be connected to an electrostatic condenser C. Before the switch S is closed, and therefore also in the moment of closing the switch, no current flows in the line A. Immediately after the switch S is closed, current begins to flow over line A into the condenser C, charging this condenser up to the voltage given by the generator. When the condenser C is charged, the current in the line A and the condenser C is zero again. That is, the permanent condition before closing the switch S, and also some time after the closing of the switch, is zero current in the line. Immediately after the closing of the switch, however, current flows for a more or less short time. With the condition of the circuit unchanged: the same generator voltage, the switch S closed on the same circuit, the current nevertheless changes, increasing from zero, at the moment of closing the switch S, to a maximum, and then decreasing again to zero, while the condenser charges from zero voltage to the generator voltage. We then here meet a transient phenomenon, in the charge of the condenser from a source of continuous voltage.

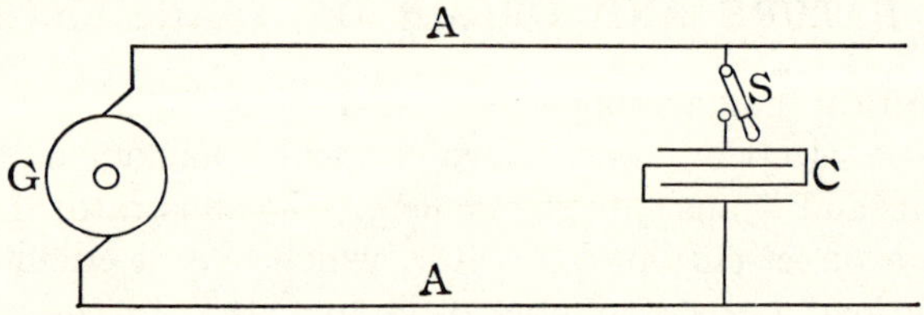

Fig. 2.

Commonly, transient and permanent phenomena are superimposed upon each other. For instance, if in the circuit Fig. 1 we close the switch S connecting a fan motor F, at the moment of closing the switch S the current in the fan-motor circuit is zero. It rapidly rises to a maximum, the motor starts, its speed increases while the current decreases, until finally speed and current become constant; that is, the permanent condition is reached.

The transient, therefore, appears as intermediate between two permanent conditions: in the above instance, the fan motor disconnected, and the fan motor running at full speed. The question then arises, why the effect of a change in the conditions of an electric circuit does not appear instantaneously, but only after a transition period, requiring a finite, though frequently very short, time.

2. Consider the simplest case: an electric power transmission (Fig. 3). In the generator G electric power is produced from mechanical power, and supplied to the line A. In the line A some of this power is dissipated, the rest transmitted into the load L, where the power is used. The consideration of the electric power in generator, line, and load does not represent the entire phenomenon. While electric power flows over the line A, there is a magnetic field surrounding the line conductors, and an electrostatic field issuing from the line conductors. The magnetic field and the electrostatic or "dielectric" field represent stored energy. Thus, during the permanent conditions of the flow of power through the circuit Fig. 3, there is electric energy stored in the space surrounding the line conductors. There is energy stored also in the generator and in the load; for instance, the mechanical momentum of the revolving fan in Fig. 1, and the heat energy of the incandescent lamp filaments. The permanent condition of the circuit Fig. 3 thus represents not only flow of power, but also storage of energy. When the switch S is open, and no power flows, no energy is stored in the system. If we now close the switch, before the permanent condition corresponding to the closed switch can occur,

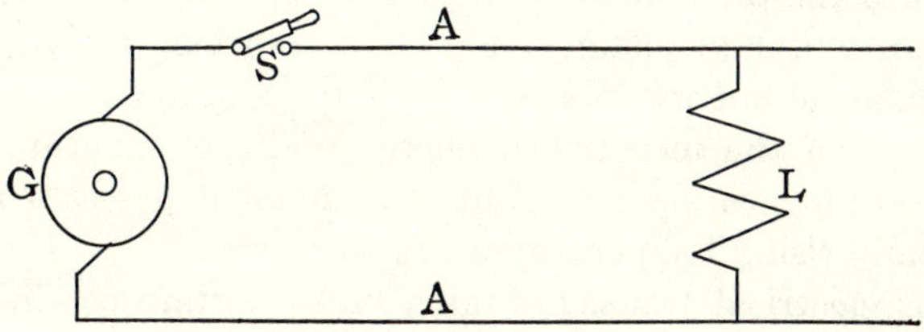

Fig. 3.

the stored energy has to be supplied from the source of power; that is, for a short time power, in supplying the stored energy, flows not only through the circuit, but also from the circuit into the space surrounding the conductors, etc. This flow of power, which supplies the energy stored in the permanent condition of the circuit, must cease as soon as the stored energy has been supplied, and thus is a transient.

Inversely, if we disconnect some of the load L in Fig. 3, and thereby reduce the flow of power, a smaller amount of stored energy would correspond to that lesser flow, and before the conditions of the circuit can become stationary, or permanent (corresponding to the lessened flow of power), some of the stored energy has to be returned to the circuit, or dissipated, by a transient.

Thus the transient is the result of the change of the amount of stored energy, required by the change of circuit conditions, and is the phenomenon by which the circuit readjusts itself to the change of stored energy. It may thus be said that the permanent phenomena are the phenomena of electric power, the transients the phenomena of electric energy.

3. It is obvious, then, that transients are not specifically electrical phenomena, but occur with all forms of energy, under all conditions where energy storage takes place.

Thus, when we start the motors propelling an electric car, a transient period, of acceleration, appears between the previous permanent condition of standstill and the final permanent condition of constant-speed running; when we shut off the motors, the permanent condition of standstill is not reached instantly, but a transient condition of deceleration intervenes. When we open the water gates leading to an empty canal, a transient condition of flow and water level intervenes while the canal is filling, until the permanent condition is reached. Thus in the case of the fan motor in instance Fig. 1, a transient period of speed and mechanical energy appeared while the motor was speeding up and gathering the mechanical energy of its momentum. When turning on an incandescent lamp, the filament passes a transient of gradually rising temperature.

Just as electrical transients may, under certain conditions, rise to destructive values; so transients of other forms of energy may become destructive, or may require serious consideration, as, for instance, is the case in governing high-head water powers. The column of water in the supply pipe represents a considerable amount of stored mechanical energy, when flowing at velocity, under load. If, then, full load is suddenly thrown off, it is not possible to suddenly stop the flow of water, since a rapid stopping would lead to a pressure transient of destructive value, that is, burst the pipe. Hence the use of surge tanks, relief valves, or deflecting nozzle governors. Inversely, if a heavy load comes on suddenly, opening the nozzle wide does not immediately take care of the load, but momentarily drops the water pressure at the nozzle, while gradually the water column acquires velocity, that is, stores energy.

The fundamental condition of the appearance of a transient thus is such a disposition of the stored energy in the system as differs from that required by the existing conditions of the system; and any change of the condition of a system, which requires a change of the stored energy, of whatever form this energy may be, leads to a transient.

Electrical transients have been studied more than transients of other forms of energy because:

(*a*) Electrical transients generally are simpler in nature, and therefore yield more easily to a theoretical and experimental investigation.

(*b*) The theoretical side of electrical engineering is further advanced than the theoretical side of most other sciences, and especially:

(*c*) The destructive or harmful effects of transients in electrical systems are far more common and more serious than with other forms of energy, and the engineers have therefore been driven by necessity to their careful and extensive study.

4. The simplest form of transient occurs where the effect is directly proportional to the cause. This is generally the case in electric circuits, since voltage, current, magnetic flux, etc., are proportional to each other, and the electrical transients therefore are usually of the simplest nature. In those cases, however, where this direct proportionality does not exist, as for instance in inductive circuits containing iron, or in electrostatic fields exceeding the corona voltage, the transients also are far more complex, and very little work has been done, and very little is known, on these more complex electrical transients.

Assume that in an electric circuit we have a transient current, as represented by curve i in Fig. 4; that is, some change of circuit condition requires a readjustment of the stored energy, which occurs by the flow of transient current i. This current starts at the value i_1, and gradually dies down to zero. Assume now that the law of proportionality between cause and effect applies; that is, if the transient current started with a different value, i_2, it would traverse a curve i', which is the same as curve i, except that all values are changed proportionally, by the ratio $\frac{i_2}{i_1}$; that is, $i' = i \times \frac{i_2}{i_1}$.

Starting with current i_1, the transient follows the curve i; starting with i_2, the transient follows the proportional curve i'. At some time, t, however, the current i has dropped to the value i_2, with which the curve i' started. At this moment t, the conditions in the first case, of current i, are the same as the conditions in the second case, of current i', at the moment t_1; that is, from t onward, curve i is the same as curve i' from time t_1 onward. Since

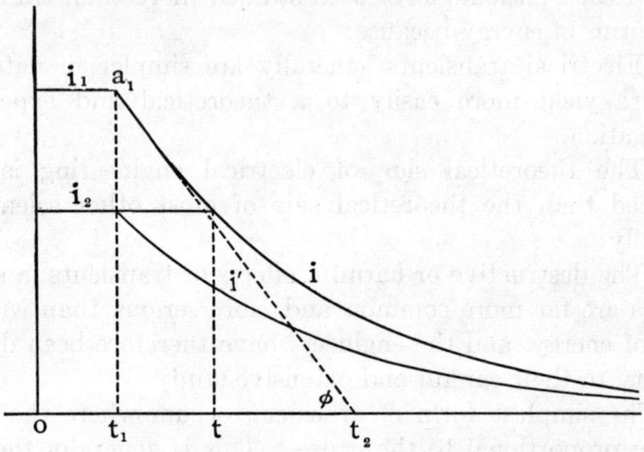

Fig. 4. — Curve of Simple Transient: Decay of Current.

i' is proportional to i, from any point t onward curve i is proportional to the same curve i from t_1 onward. At time t_1, it is

$$\frac{di_2}{dt_1} = \frac{di_1}{dt_1} \times \frac{i_2}{i_1}.$$

But since $\frac{di_2}{dt_1}$ and i_2 at t_1 are the same as $\frac{di}{dt}$ and i at time t, it follows:

$$\frac{di}{dt} = \frac{di_1}{dt_1}\frac{i}{i_1},$$

or,

$$\frac{di}{dt} = - ci,$$

where $c = -\frac{1}{i_1}\frac{di_1}{dt} =$ constant, and the minus sign is chosen, as $\frac{di}{dt}$ is negative.

As in Fig. 4:

$$\tan \phi = - \frac{di_1}{dt},$$

$$\overline{a_1 t_1} = i_1,$$

$$c = -\frac{1}{i_1}\frac{di_1}{dt} = \frac{\tan \phi}{\overline{a_1 t_1}} = \frac{1}{\overline{t_1 t_2}};$$

that is, c is the reciprocal of the projection $T = \overline{t_1 t_2}$ on the zero line of the tangent at the starting moment of the transient.

Since
$$c = \frac{1}{T},$$
$$\frac{di}{i} = -cdt;$$

that is, the percentual change of current is constant, or in other words, in the same time, the current always decreases by the same fraction of its value, no matter what this value is.

Integrated, this equation gives:

$$\log i = -ct + C,$$
$$i = A\epsilon^{-ct},$$

or,
$$i = A\epsilon^{-\frac{t}{T}};$$

that is, the curve is the exponential.

The exponential curve thus is the expression of the simplest form of transient. This explains its common occurrence in electrical and other transients. Consider, for instance, the decay of radioactive substances: the radiation, which represents the decay, is proportional to the amount of radiating material; it is $\frac{dm}{dt} = cm$, which leads to the same exponential function.

Not all transients, however, are of this simplest form. For instance, the deceleration of a ship does not follow the exponential, but at high velocities the decrease of speed is a greater fraction of the speed than during the same time interval at lower velocities, and the speed-time curves for different initial speeds are not proportional to each other, but are as shown in Fig. 5. The reason is, that the frictional resistance is not proportional to the speed, but to the square of the speed.

5. Two classes of transients may occur:

1. Energy may be stored in one form only, and the only energy change which can occur thus is an increase or a decrease of the stored energy.

2. Energy may be stored in two or more different forms, and the possible energy changes thus are an increase or decrease of the total stored energy, or a change of the stored energy from one form to another. Usually both occur simultaneously.

An instance of the first case is the acceleration or deceleration of a train, or a ship, etc.: here energy can be stored only as mechanical momentum, and the transient thus consists of an increase of the stored energy, during acceleration, or of a decrease, during

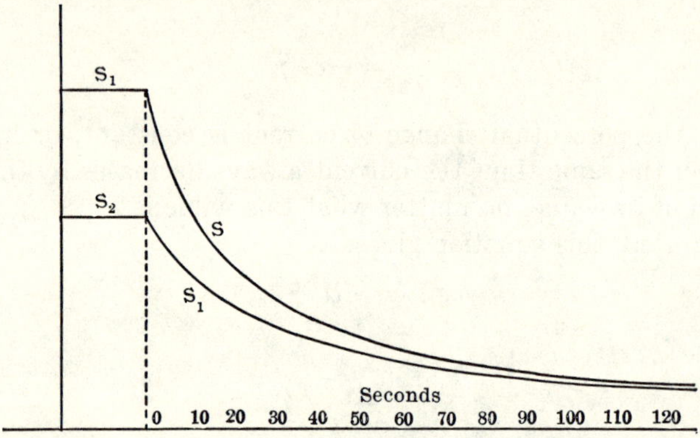

Fig. 5. — Deceleration of Ship.

deceleration. Thus also in a low-voltage electric circuit of negligible capacity, energy can be stored only in the magnetic field, and the transient represents an increase of the stored magnetic energy, during increase of current, or a decrease of the magnetic energy, during a decrease of current.

An instance of the second case is the pendulum, Fig. 6: with the weight at rest in maximum elevation, all the stored energy is potential energy of gravitation. This energy changes to kinetic mechanical energy until in the lowest position, *a*, when all the potential gravitational energy has been either converted to kinetic mechanical energy or dissipated. Then, during the rise of the weight, that part of the energy which is not dissipated again changes to potential gravitational energy, at *c*, then back again to kinetic energy, at *a*; and in this manner the total stored energy is gradually dissipated, by a series of successive oscillations or changes between potential gravitational and kinetic mechanical

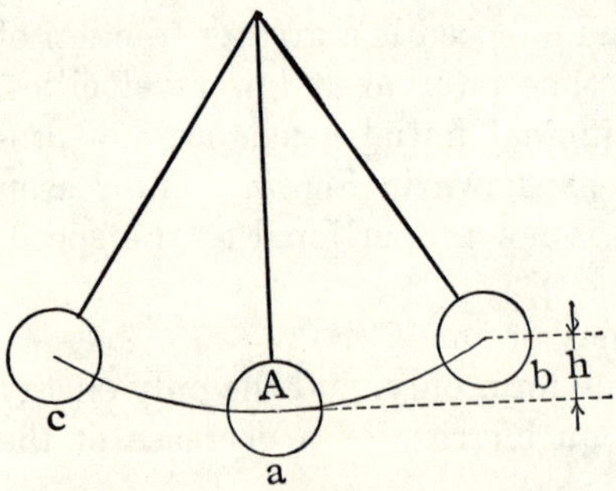

Fig. 6. — Double-energy Transient of Pendulum.

energy. Thus in electric circuits containing energy stored in the magnetic and in the dielectric field, the change of the amount of stored energy — decrease or increase — frequently occurs by a series of successive changes from magnetic to dielectric and back again from dielectric to magnetic stored energy. This for instance is the case in the charge or discharge of a condenser through an inductive circuit.

If energy can be stored in more than two different forms, still more complex phenomena may occur, as for instance in the hunting of synchronous machines at the end of long transmission lines, where energy can be stored as magnetic energy in the line and apparatus, as dielectric energy in the line, and as mechanical energy in the momentum of the motor.

6. The study and calculation of the permanent phenomena in electric circuits are usually far simpler than are the study and calculation of transient phenomena. However, only the phenomena of a continuous-current circuit are really permanent. The alternating-current phenomena are transient, as the e.m.f. continuously and periodically changes, and with it the current, the stored energy, etc. The theory of alternating-current phenomena, as periodic transients, thus has been more difficult than that of continuous-current phenomena, until methods were devised to treat the periodic transients of the alternating-current circuit as permanent phenomena, by the conception of the "effective values," and more completely by the introduction of the general number or complex quantity, which represents the periodic function of time by a constant numerical value. In this feature lies the advantage and the power of the symbolic method of dealing with alternating-current phenomena, — the reduction of a periodic transient to a permanent or constant quantity. For this reason, wherever periodic transients occur, as in rectification, commutation, etc., a considerable advantage is frequently gained by their reduction to permanent phenomena, by the introduction of the symbolic expression of the equivalent sine wave.

Hereby most of the periodic transients have been eliminated from consideration, and there remain mainly the nonperiodic transients, as occur at any change of circuit conditions. Since they are the phenomena of the readjustment of stored energy, a study of the energy storage of the electric circuit, that is, of its magnetic and dielectric field, is of first importance.

CHAPTER II

THE ELECTRIC FIELD

7. Let, in Fig. 7, a generator G transmit electric power over line A into a receiving circuit M.

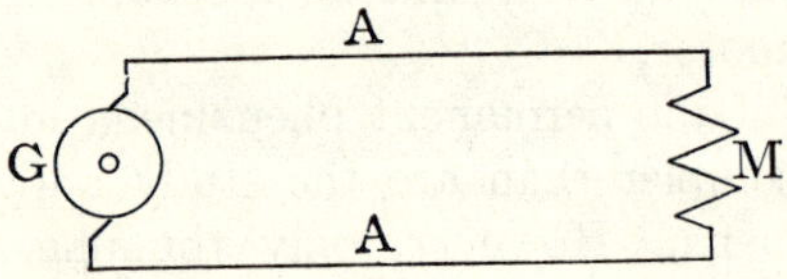

Fig. 7.

While power flows through the conductors A, power is consumed in these conductors by conversion into heat, represented by i^2r. This, however, is not all, but in the space surrounding the conductor certain phenomena occur: magnetic and electrostatic forces appear.

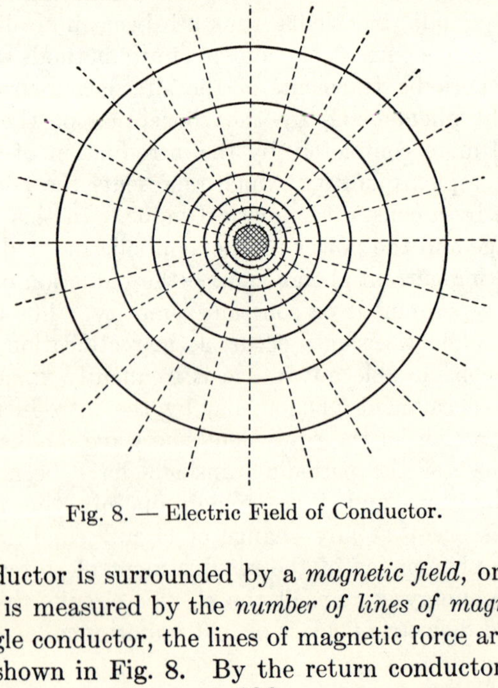

Fig. 8. — Electric Field of Conductor.

The conductor is surrounded by a *magnetic field*, or a magnetic flux, which is measured by the *number of lines of magnetic force* Φ. With a single conductor, the lines of magnetic force are concentric circles, as shown in Fig. 8. By the return conductor, the circles are crowded together between the conductors, and the magnetic field consists of eccentric circles surrounding the conductors, as shown by the drawn lines in Fig. 9.

An *electrostatic,* or, as more properly called, *dielectric field,* issues from the conductors, that is, a *dielectric flux* passes between the conductors, which is measured by the *number of lines of dielectric force* Ψ. With a single conductor, the lines of dielectric force are radial straight lines, as shown dotted in Fig. 8. By the return conductor, they are crowded together between the conductors, and form arcs of circles, passing from conductor to return conductor, as shown dotted in Fig. 9.

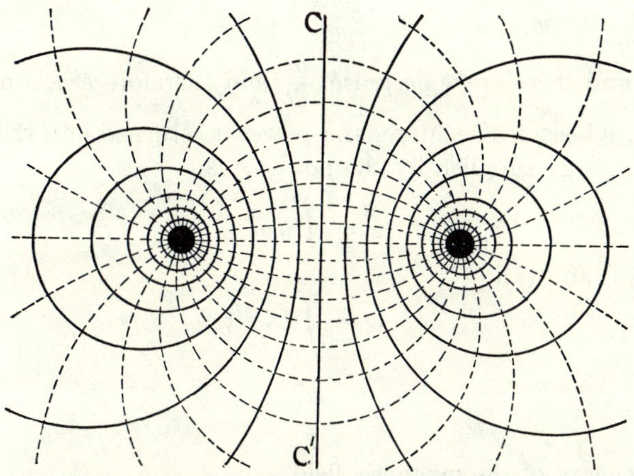

Fig. 9. — Electric Field of Circuit.

The magnetic and the dielectric field of the conductors both are included in the term *electric field,* and are the two components of the electric field of the conductor.

8. The magnetic field or *magnetic flux* of the circuit, Φ, is proportional to the current, i, with a proportionality factor, L, which is called the *inductance of the circuit.*

$$\Phi = Li.* \tag{1}$$

The magnetic field represents stored energy w. To produce it, power, p, must therefore be supplied by the circuit.

Since power is current times voltage:

$$p = e'i, \tag{2}$$

* $n\Phi$, if the flux Φ interlinks the circuit n fold.

to produce the magnetic field Φ of the current i, a voltage e' must be consumed in the circuit, which with the current i gives the power p, which supplies the stored energy w of the magnetic field Φ. This voltage e' is called the *inductance voltage*, or *voltage consumed by self-induction.*

Since no power is required to maintain the field, but power is required to produce it, the inductance voltage must be proportional to the rate of increase of the magnetic field:

$$e' = \frac{d\Phi}{dt}, \tag{3}$$

or by (1),

$$e' = L\frac{di}{dt}. \tag{4}$$

If i and therefore Φ decrease, $\frac{di}{dt}$ and therefore e' are negative; that is, p becomes negative, and power is returned into the circuit.

The energy supplied by the power p is

$$w = \int p\, dt,$$

or by (2) and (4),

$$w = \int Li\, di;$$

hence

$$w = \frac{Li^2}{2} \tag{5}$$

is the energy of the magnetic field

$$\Phi = Li$$

of the circuit.

9. Exactly analogous relations exist in the dielectric field.

The dielectric field, or *dielectric flux*, Ψ, is proportional to the voltage e, with a proportionality factor, C, which is called the *capacity of the circuit:*

$$\Psi = Ce. \tag{6}$$

The dielectric field represents stored energy, w. To produce it, power, p, must, therefore, be supplied by the circuit.

Since power is current times voltage:

$$p = i'e, \tag{7}$$

to produce the dielectric field Ψ of the voltage e, a current i' must be consumed in the circuit, which with the voltage e gives the power p, which supplies the stored energy w of the dielectric field Ψ. This current i' is called the *capacity current*, or, wrongly, *charging current* or *condenser current*.

Since no power is required to maintain the field, but power is required to produce it, the capacity current must be proportional to the rate of increase of the dielectric field:

$$i' = \frac{d\Psi}{dt}, \tag{8}$$

or by (6),

$$i' = C\frac{de}{dt}. \tag{9}$$

If e and therefore Ψ decrease, $\frac{de}{dt}$ and therefore i' are negative; that is, p becomes negative, and power is returned into the circuit.

The energy supplied by the power p is

$$w = \int p\, dt, \tag{10}$$

or by (7) and (9),

$$w = \int Ce\, de;$$

hence

$$w = \frac{Ce^2}{2} \tag{11}$$

is the energy of the dielectric field

$$\Psi = Ce$$

of the circuit.

As seen, the capacity current is the exact analogy, with regard to the dielectric field, of the inductance voltage with regard to the magnetic field; the representations in the electric circuit, of the energy storage in the field.

The dielectric field of the circuit thus is treated and represented in the same manner, and with the same simplicity and perspicuity, as the magnetic field, by using the same conception of lines of force.

Unfortunately, to a large extent in dealing with the dielectric fields the prehistoric conception of the electrostatic charge on the conductor still exists, and by its use destroys the analogy between the two components of the electric field, the magnetic and the dielectric, and makes the consideration of dielectric fields unnecessarily complicated.

There obviously is no more sense in thinking of the capacity current as current which charges the conductor with a quantity of electricity, than there is of speaking of the inductance voltage as charging the conductor with a quantity of magnetism. But while the latter conception, together with the notion of a quantity of magnetism, etc., has vanished since Faraday's representation of the magnetic field by the lines of magnetic force, the terminology of electrostatics of many textbooks still speaks of electric charges on the conductor, and the energy stored by them, without considering that the dielectric energy is not on the surface of the conductor, but in the space outside of the conductor, just as the magnetic energy.

10. All the lines of magnetic force are closed upon themselves, all the lines of dielectric force terminate at conductors, as seen in Fig. 8, and the magnetic field and the dielectric field thus can be considered as a *magnetic circuit* and a *dielectric circuit.*

To produce a *magnetic flux* Φ, a *magnetomotive force* F is required. Since the magnetic field is due to the current, and is proportional to the current, or, in a coiled circuit, to the current times the number of turns, magnetomotive force is expressed in *current turns* or *ampere turns.*

$$F = ni. \tag{12}$$

If F is the m.m.f., l the length of the magnetic circuit, energized by F,

$$f = \frac{F}{l} \tag{13}$$

is called the *magnetizing force,* or *magnetic gradient,* and is expressed in *ampere turns per cm.* (or industrially sometimes in ampere turns per inch).

In empty space, and therefore also, with very close approximation, in all nonmagnetic material, f ampere turns per cm. length of magnetic circuit produce $\mathcal{H} = 4\pi f\,10^{-1}$ lines of magnetic force per square cm. section of the magnetic circuit. (Here the factor 10^{-1} results from the ampere being 10^{-1} of the absolute or cgs. unit of current.)

$$\mathcal{H} = 4\pi f\,10^{-1}\,* \tag{14}$$

* The factor 4π is a survival of the original definition of the magnetic field intensity from the conception of the magnetic mass, since unit magnetic mass was defined as that quantity of magnetism which acts on an equal quantity at

is called the *magnetic-field intensity.* It is the *magnetic density,* that is, the number of lines of magnetic force per cm^2, produced by the magnetizing force of f ampere turns per cm. in empty space.

The *magnetic density,* in lines of magnetic force per cm^2, produced by the field intensity $\mathcal{H}$ in any material is

$$\mathcal{B} = \mu\mathcal{H}, \tag{15}$$

where μ is a constant of the material, a "magnetic conductivity," and is called the *permeability.* $\mu = 1$ or very nearly so for most materials, with the exception of very few, the so-called *magnetic materials:* iron, cobalt, nickel, and some alloys and oxides of these metals and of manganese and chromium.

If then A is the section of the magnetic circuit, the total magnetic flux is

$$\Phi = A\mathcal{B}. \tag{16}$$

Obviously, if the magnetic field is not uniform, equations (13) and (16) would be correspondingly modified; f in (13) would be the average magnetizing force, while the actual magnetizing force would vary, being higher at the denser, and lower at the less dense, parts of the magnetic circuit:

$$f = \frac{dF}{dl}. \tag{17}$$

In (16), the magnetic flux Φ would be derived by integrating the densities $\mathcal{B}$ over the total section of the magnetic circuit.

11. Entirely analogous relations exist in the dielectric circuit. To produce a *dielectric flux* Ψ, an *electromotive force* e is required, which is measured in volts. The e.m.f. per unit length of the dielectric circuit then is called the *electrifying force* or the *voltage gradient,* and is

$$G = \frac{e}{l}. \tag{18}$$

unit distance with unit force. The unit field intensity, then, was defined as the field intensity at unit distance from unit magnetic mass, and represented by one line (or rather "tube") of magnetic force. The magnetic flux of unit magnetic mass (or "unit magnet pole") hereby became 4π lines of force, and this introduced the factor 4π into many magnetic quantities. An attempt to drop this factor 4π has failed, as the magnetic units were already too well established.

The factor 10^{-1} also appears undesirable, but when the electrical units were introduced the absolute unit appeared as too large a value of current as practical unit, and one-tenth of it was chosen as unit, and called "ampere."

This gives the average voltage gradient, while the actual gradient in an ununiform field, as that between two conductors, varies, being higher at the denser, and lower at the less dense, portion of the field, and is

$$G = \frac{de}{dl}. \tag{19}$$

$$K = \frac{G}{4\pi}\text{*}$$

then is the *dielectric-field intensity*, and

$$D = \kappa K \tag{20}$$

would be the *dielectric density*, where κ is a constant of the material, the electrostatic or dielectric conductivity, and is called the *specific capacity* or *permittivity*.

For empty space, and thus with close approximation for air and other gases,

$$\kappa = \frac{1}{v^2},$$

where

$$v = 3 \times 10^{10}$$

is the velocity of light.

It is customary, however, and convenient, to use the permittivity of empty space as unity: $\kappa = 1$. This changes the unit of dielectric-field intensity by the factor $\frac{1}{v^2}$, and gives: dielectric-field intensity,

$$K = \frac{G}{4\pi v^2}; \tag{21}$$

dielectric density,

$$D = \kappa K, \tag{22}$$

where $\kappa = 1$ for empty space, and between 2 and 6 for most solids and liquids, rarely increasing beyond 6, except in materials of appreciable electric conductivity.

The dielectric flux then is

$$\Psi = AD. \tag{23}$$

12. As seen, the dielectric and the magnetic fields are entirely analogous, and the corresponding values are tabulated in the following Table I.

* The factor 4π appears here in the denominator as the result of the factor 4π in the magnetic-field intensity $\mathcal{H}$, due to the relations between these quantities.

TABLE I.

Magnetic Field.	Dielectric Field.
Magnetic flux: $\Phi = Li\,10^8$ lines of magnetic force.	Dielectric flux: $\Psi = Ce$ lines of dielectric force, or coulombs.
Inductance voltage: $e' = n\frac{d\Phi}{dt}10^{-8} = L\frac{di}{dt}$ volts.	Capacity current: $i' = \frac{d\psi}{dt} = C\frac{de}{dt}$ amperes.
Magnetic energy: $w = \frac{Li^2}{2}$ joules.	Dielectric energy: $w = \frac{Ce^2}{2}$ joules.
Magnetomotive force: $F = ni$ ampere turns.	Electromotive force: $e =$ volts.
Magnetizing force: $f = \frac{F}{l}$ ampere turns per cm.	Electrifying force or voltage gradient: $G = \frac{e}{l}$ volts per cm.
Magnetic-field intensity: $\mathcal{H} = 4\pi f\,10^{-1}$ lines of magnetic force per cm².	Dielectric-field intensity: $K = \frac{G}{4\pi v^2}10^9$ lines of dielectric force per cm², or coulombs per cm².
Magnetic density: $\mathcal{B} = \mu\mathcal{H}$ lines of magnetic force per cm².	Dielectric density: $D = \kappa K$ lines of dielectric force per cm², or coulombs per cm².
Permeability: μ	Permittivity or specific capacity: κ
Magnetic flux: $\Phi = A\mathcal{B}$ lines of magnetic force.	Dielectric flux: $\Psi = AD$ lines of dielectric force, or coulombs.
$v = 3 \times 10^{10} =$ velocity of light.	

The powers of 10, which appear in some expressions, are reduction factors between the absolute or cgs. units which are used for Φ, $\mathcal{H}$, $\mathcal{B}$, and the practical electrical units, used for other constants.

As the magnetic field and the dielectric field also can be considered as the magnetic circuit and the dielectric circuit, some analogy exists between them and the electric circuit, and in Table II the corresponding terms of the magnetic circuit, the dielectric circuit, and the electric circuit are given.

TABLE II.

Magnetic Circuit.	Dielectric Circuit.	Electric Circuit.
Magnetic flux (magnetic current): Φ = lines of magnetic force.	Dielectric flux (dielectric current): Ψ = lines of dielectric force.	Electric current: i = electric current.
Magnetomotive force: $F = ni$ ampere turns.	Electromotive force: e = volts.	Voltage: e = volts.
Permeance: $M = \frac{\Phi}{.4\pi F}$. Inductance: $L = \frac{n^2\Phi}{F} 10^{-8} = \frac{n\Phi}{i} 10^{-8}$ henry.	Permittance or capacity: $C = \frac{\Psi}{e}$ farads.	Conductance: $g = \frac{i}{e}$ mhos.
Reluctance: $R = \frac{F}{\Phi}$.	(Elastance): $\frac{1}{C} = \frac{e}{\Psi}$.	Resistance: $r = \frac{e}{i}$ ohms.
Magnetic energy: $w = \frac{Li^2}{2} = \frac{F\Phi}{2} 10^{-8}$ joules.	Dielectric energy: $w = \frac{Ce^2}{2} = \frac{e\Psi}{2}$ joules.	Electric power: $p = ri^2 = ge^2 = ei$ watts.
Magnetic density: $\mathcal{B} = \frac{\Phi}{A} = \mu\mathcal{H}$ lines per cm².	Dielectric density: $D = \frac{\Psi}{A} = \kappa K$ lines per cm².	Electric-current density: $I = \frac{i}{A} = \gamma G$ amperes per cm².
Magnetizing force: $f = \frac{F}{l}$ ampere turns per cm.	Dielectric gradient: $G = \frac{e}{l}$ volts per cm.	Electric gradient: $G = \frac{e}{l}$ volts per cm.
Magnetic-field intensity: $\mathcal{H} = .4\pi f$.	Dielectric-field intensity: $K = \frac{G}{4\pi v^2} 10^9$.	
Permeability: $\mu = \frac{\mathcal{B}}{\mathcal{H}}$.	Permittivity or specific capacity: $\kappa = \frac{D}{K}$.	Conductivity: $\gamma = \frac{I}{G}$ mho-cm.
Reluctivity: $\rho = \frac{f}{\mathcal{B}}$.	(Elastivity ?): $\frac{1}{\kappa} = \frac{K}{D}$.	Resistivity: $\rho = \frac{1}{\gamma} = \frac{G}{I}$ ohm-cm.
Specific magnetic energy: $w_0 = \frac{.4\pi\mu f^2}{2} = \frac{f\mathcal{B}}{2} 10^{-8} = \frac{\mathcal{H}\mathcal{B}}{8\pi} 10^{-7}$ joules per cm³.	Specific dielectric energy: $w_0 = \frac{\kappa G^2}{4\pi v^2} 10^9 = \frac{GD}{2} 10^9 = 2\pi v^2 KD$ joules per cm³.	Specific power: $p_0 = \rho I^2 = G^2 = GI$ watts per cm³.

CHAPTER III

SINGLE-ENERGY TRANSIENTS IN CONTINUOUS-CURRENT CIRCUITS

13. The simplest electrical transients are those in circuits in which energy can be stored in one form only, as in this case the change of stored energy can consist only of an increase or decrease; but no surge or oscillation between several forms of energy can exist. Such circuits are most of the low- and medium-voltage circuits, — 220 volts, 600 volts, and 2200 volts. In them the capacity is small, due to the limited extent of the circuit, resulting from the low voltage, and at the low voltage the dielectric energy thus is negligible, that is, the circuit stores appreciable energy only by the magnetic field.

A circuit of considerable capacity, but negligible inductance, if of high resistance, would also give one form of energy storage only, in the dielectric field. The usual high-voltage capacity circuit, as that of an electrostatic machine, while of very small inductance, also is of very small resistance, and the momentary discharge currents may be very considerable, so that in spite of the very small inductance, considerable magnetic-energy storage may occur; that is, the system is one storing energy in two forms, and oscillations appear, as in the discharge of the Leyden jar.

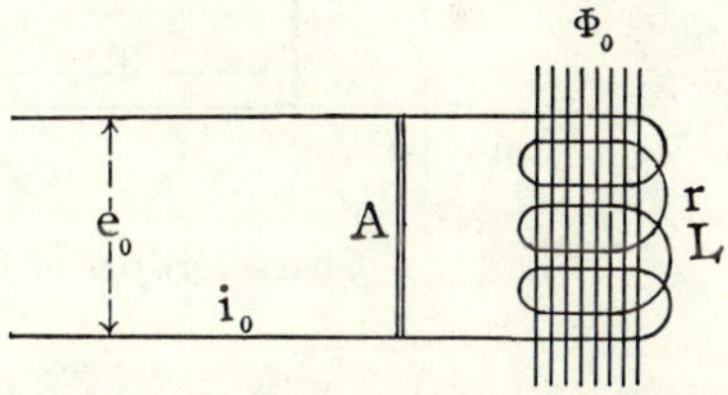

Fig. 10. — Magnetic Single-energy Transient.

Let, as represented in Fig. 10, a continuous voltage e_0 be impressed upon a wire coil of resistance r and inductance L (but negligible capacity). A current $i_0 = \frac{e_0}{r}$ flows through the coil and a magnetic field $\Phi_0\, 10^{-8} = \frac{Li_0}{n}$ interlinks with the coil. Assuming now that the voltage e_0 is suddenly withdrawn, without changing the constants of the coil circuit, as for instance by short-circuiting the terminals of the coil, as indicated at *A*. With no voltage impressed upon the coil, and thus no power supplied to it, current i and magnetic flux Φ of the coil must finally be zero. However, since the magnetic flux represents stored energy, it cannot instantly vanish, but the magnetic flux must gradually decrease from its initial value Φ_0, by the dissipation of its stored energy in the resistance of the coil circuit as i^2r. Plotting, therefore, the magnetic flux of the coil as function of the time, in Fig. 11*A*, the flux is constant and denoted by Φ_0 up to the moment of

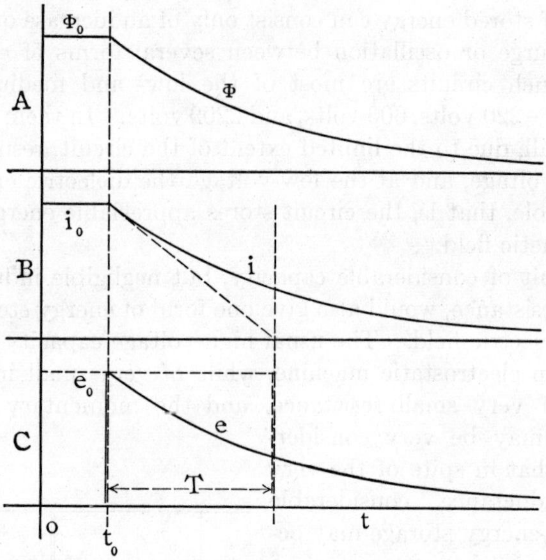

Fig. 11. — Characteristics of Magnetic Single-energy Transient.

time where the short circuit is applied, as indicated by the dotted line t_0. From t_0 on the magnetic flux decreases, as shown by curve Φ. Since the magnetic flux is proportional to the current, the latter must follow a curve proportional to Φ, as shown in Fig. 11*B*. The impressed voltage is shown in Fig. 11*C* as a dotted line; it is e_0 up to t_0, and drops to 0 at t_0. However, since after t_0 a current i flows, an e.m.f. must exist in the circuit, proportional to the current.

$$e = ri.$$

This is the e.m.f. induced by the decrease of magnetic flux Φ, and is therefore proportional to the rate of decrease of Φ, that is, to $\frac{d\Phi}{dt}$. In the first moment of short circuit, the magnetic flux Φ still has full value Φ_0, and the current i thus also full value i_0. Hence, at the first moment of short circuit, the induced e.m.f. e must be equal to e_0, that is, the magnetic flux Φ must begin to decrease at such rate as to induce full voltage e_0, as shown in Fig. 11*C*.

The three curves Φ, i, and e are proportional to each other, and as e is proportional to the rate of change of Φ, Φ must be proportional to its own rate of change, and thus also i and e. That is, the transients of magnetic flux, current, and voltage follow the law of proportionality, hence are simple exponential functions, as seen in Lecture I:

$$\left.\begin{aligned} \Phi &= \Phi_0 \epsilon^{-c(t-t_0)}, \\ i &= i_0 \epsilon^{-c(t-t_0)}, \\ e &= e_0 \epsilon^{-c(t-t_0)}. \end{aligned}\right\} \qquad (1)$$

Φ, i, and e decrease most rapidly at first, and then slower and slower, but can theoretically never become zero, though practically they become negligible in a finite time.

The voltage e is induced by the rate of change of the magnetism, and equals the decrease of the number of lines of magnetic force, divided by the time during which this decrease occurs, multiplied by the number of turns n of the coil. The induced voltage e times the time during which it is induced thus equals n times the decrease of the magnetic flux, and the total induced voltage, that is, the area of the induced-voltage curve, Fig. 11*C*, thus equals n times the total decrease of magnetic flux, that is, equals the initial current i_0 times the inductance L:

$$\Sigma et = n\Phi_0 10^{-8} = Li_0. \qquad (2)$$

Whatever, therefore, may be the rate of decrease, or the shape of the curves of Φ, i, and e, the total area of the voltage curve must be the same, and equal to $n\Phi_0 = Li_0$.

If then the current i would continue to decrease at its initial rate, as shown dotted in Fig. 11*B* (as could be caused, for instance, by a gradual increase of the resistance of the coil circuit), the induced voltage would retain its initial value e_0 up to the moment of time $t = t_0 + T$, where the current has fallen to zero, as shown dotted in Fig. 11C. The area of this new voltage curve would be e_0T, and since it is the same as that of the curve e, as seen above, it follows that the area of the voltage curve e is

$$\left.\begin{aligned}\Sigma et &= e_0T,\\ &= ri_0T,\end{aligned}\right\} \tag{3}$$

and, combining (2) and (3), i_0 cancels, and we get the value of T:

$$T = \frac{L}{r}. \tag{4}$$

That is, the initial decrease of current, and therefore of magnetic flux and of induced voltage, is such that if the decrease continued at the same rate, the current, flux, and voltage would become zero after the time $T = \frac{L}{r}$.

The total induced voltage, that is, voltage times time, and therefore also the total current and magnetic flux during the transient, are such that, when maintained at their initial value, they would last for the time $T = \frac{L}{r}$.

Since the curves of current and voltage theoretically never become zero, to get an estimate of the duration of the transient we may determine the time in which the transient decreases to half, or to one-tenth, etc., of its initial value. It is preferable, however, to estimate the duration of the transient by the time T, which it would last if maintained at its initial value. That is, the duration of a transient is considered as the time $T = \frac{L}{r}$.

This time T has frequently been called the "time constant" of the circuit.

The higher the inductance L, the longer the transient lasts, obviously, since the stored energy which the transient dissipates is proportional to L.

The higher the resistance r, the shorter is the duration of the transient, since in the higher resistance the stored energy is more rapidly dissipated.

Using the time constant $T = \frac{L}{r}$ as unit of length for the abscissa, and the initial value as unit of the ordinates, all exponential transients have the same shape, and can thereby be constructed by the numerical values of the exponential function, $y = \epsilon^{-x}$, given in Table III.

TABLE III.

Exponential Transient of Initial Value 1 and Duration 1.

$y = \epsilon^{-x}$. $\epsilon = 2.71828$.

x	y	x	y
0	1.000	1.0	0.368
0.05	0.951	1.2	0.301
0.1	0.905	1.4	0.247
0.15	0.860	1.6	0.202
0.2	0.819	1.8	0.165
0.25	0.779	2.0	0.135
0.3	0.741	2.5	0.082
0.35	0.705	3.0	0.050
0.4	0.670	3.5	0.030
0.45	0.638	4.0	0.018
0.5	0.607	4.5	0.011
0.6	0.549	5.0	0.007
0.7	0.497	6.0	0.002
0.8	0.449	7.0	0.001
0.9	0.407	8.0	0.000
1.0	0.368	....	....

As seen in Lecture I, the coefficient of the exponent of the single-energy transient, c, is equal to $\frac{1}{T}$, where T is the projection of the tangent at the starting moment of the transient, as shown in Fig. 11, and by equation (4) was found equal to $\frac{L}{r}$. That is,

$$c = \frac{1}{T} = \frac{r}{L},$$

and the equations of the single-energy magnetic transient, (1), thus may be written in the forms:

$$\left.\begin{aligned} \Phi &= \Phi_0 \epsilon^{-c(t-t_0)} = \Phi_0 \epsilon^{-\frac{t-t_0}{T}} = \Phi_0 \epsilon^{-\frac{r}{L}(t-t_0)}, \\ i &= i_0 \epsilon^{-c(t-t_0)} = i_0 \epsilon^{-\frac{t-t_0}{T}} = i_0 \epsilon^{-\frac{r}{L}(t-t_0)}, \\ e &= e_0 \epsilon^{-c(t-t_0)} = e_0 \epsilon^{-\frac{t-t_0}{T}} = e_0 \epsilon^{-\frac{r}{L}(t-t_0)}. \end{aligned}\right\} \quad (5)$$

Usually, the starting moment of the transient is chosen as the zero of time, $t_0 = 0$, and equations (5) then assume the simpler form:

$$\left.\begin{aligned}\Phi &= \Phi_0\epsilon^{-ct} = \Phi_0\epsilon^{-\frac{t}{T}} = \Phi_0\epsilon^{-\frac{rt}{L}},\\ i &= i_0\epsilon^{-ct} = i_0\epsilon^{-\frac{t}{T}} = i_0\epsilon^{-\frac{rt}{L}},\\ e &= e_0\epsilon^{-ct} = e_0\epsilon^{-\frac{t}{T}} = e_0\epsilon^{-\frac{rt}{L}}.\end{aligned}\right\} \tag{6}$$

The same equations may be derived directly by the integration of the differential equation:

$$L\frac{di}{dt} + ri = 0, \tag{7}$$

where $L\frac{di}{dt}$ is the inductance voltage, ri the resistance voltage, and their sum equals zero, as the coil is short-circuited.

Equation (7) transposed gives

$$\frac{di}{i} = -\frac{r}{L}\,dt,$$

hence

$$logi = -\frac{r}{L}t + logC,$$

$$i = C\epsilon^{-\frac{r}{L}t},$$

and, as for $t = 0$: $i = i_0$, it is: $C = i_0$; hence

$$i = i_0\epsilon^{-\frac{r}{L}t}.$$

14. Usually single-energy transients last an appreciable time, and thereby become of engineering importance, only in highly inductive circuits, as motor fields, magnets, etc.

To get an idea on the duration of such magnetic transients, consider a motor field:

A 4-polar motor has 8 ml. (megalines) of magnetic flux per pole, produced by 6000 ampere turns m.m.f. per pole, and dissipates normally 500 watts in the field excitation.

That is, if i_0 = field-exciting current, n = number of field turns per pole, r = resistance, and L = inductance of the field-exciting circuit, it is

$$i_0^2 r = 500,$$

hence

$$r = \frac{500}{i_0^2}.$$

The magnetic flux is $\Phi_0 = 8 \times 10^6$, and with $4n$ total turns the total number of magnetic interlinkages thus is

$$4 n\Phi_0 = 32 n \times 10^6,$$

hence the inductance

$$L = \frac{4 n\Phi_0 10^{-8}}{i_0} = \frac{.32 n}{i_0} \text{ henrys.}$$

The field excitation is

$$ni_0 = 6000 \text{ ampere turns,}$$

hence

$$n = \frac{6000}{i_0}$$

hence

$$L = \frac{.32 \times 6000}{i_0^2} \text{ henrys,}$$

and

$$T = \frac{L}{r} = \frac{1920}{500} = 3.84 \text{ sec.}$$

That is, the stored magnetic energy could maintain full field excitation for nearly 4 seconds.

It is interesting to note that the duration of the field discharge does not depend on the voltage, current, or size of the machine, but merely on, first, the magnetic flux and m.m.f., — which determine the stored magnetic energy, — and, second, on the excitation power, which determines the rate of energy dissipation.

15. Assume now that in the moment where the transient begins the resistance of the coil in Fig. 10 is increased, that is, the

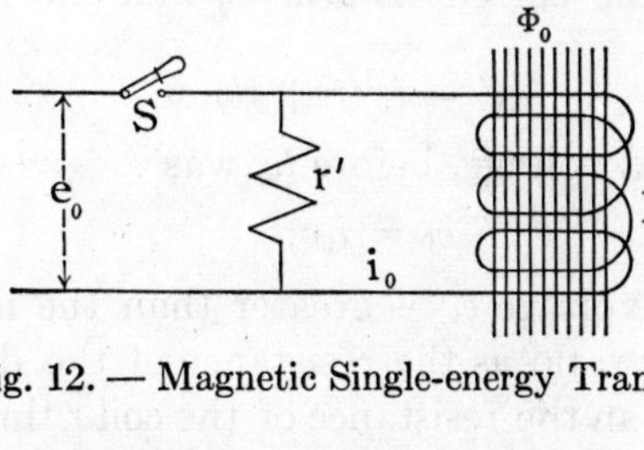

Fig. 12. — Magnetic Single-energy Transient.

coil is not short-circuited upon itself, but its circuit closed by a resistance r'. Such would, for instance, be the case in Fig. 12, when opening the switch S.

The transients of magnetic flux, current, and voltage are shown as *A*, *B*, and *C* in Fig. 13.

The magnetic flux and therewith the current decrease from the initial values Φ_0 and i_0 at the moment t_0 of opening the switch *S*, on curves which must be steeper than those in Fig. 11, since the current passes through a greater resistance, $r + r'$, and thereby dissipates the stored magnetic energy at a greater rate.

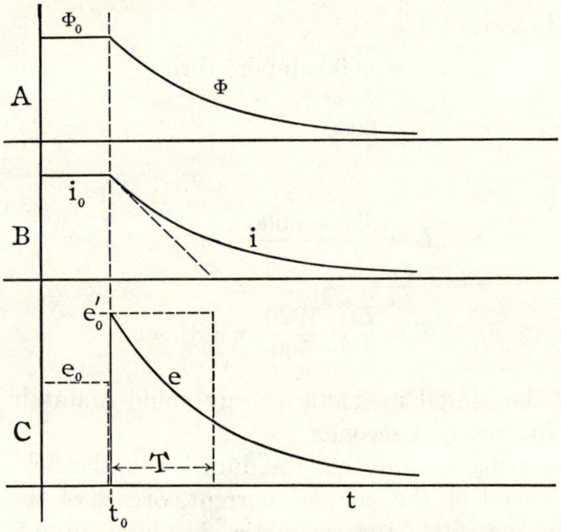

Fig. 13. — Characteristics of Magnetic Single-energy Transient.

The impressed voltage e_0 is withdrawn at the moment t_0, and a voltage thus induced from this moment onward, of such value as to produce the current i through the resistance $r + r'$. In the first moment, t_0, the current is still i_0, and the induced voltage thus must be

$$e_0' = i_0 (r + r'),$$

while the impressed voltage, before t_0, was

$$e_0 = i_0 r;$$

hence the induced voltage e_0' is greater than the impressed voltage e_0, in the same ratio as the resistance of the discharge circuit $r + r'$ is greater than the resistance of the coil r through which the impressed voltage sends the current

$$\frac{e_0'}{e_0} = \frac{r + r'}{r}.$$

The duration of the transient now is

$$T = \frac{L}{r + r'},$$

that is, shorter in the same proportion as the resistance, and thereby the induced voltage is higher.

If $r' = \infty$, that is, no resistance is in shunt to the coil, but the circuit is simply opened, if the opening were instantaneous, it would be: $e_0' = \infty$; that is, an infinite voltage would be induced. That is, the insulation of the coil would be punctured and the circuit closed in this manner.

The more rapid, thus, the opening of an inductive circuit, the higher is the induced voltage, and the greater the danger of breakdown. Hence it is not safe to have too rapid circuit-opening devices on inductive circuits.

To some extent the circuit protects itself by an arc following the blades of the circuit-opening switch, and thereby retarding the circuit opening. The more rapid the mechanical opening of the switch, the higher the induced voltage, and further, therefore, the arc follows the switch blades and maintains the circuit.

16. Similar transients as discussed above occur when closing a circuit upon an impressed voltage, or changing the voltage, or the current, or the resistance or inductance of the circuit. A discussion of the infinite variety of possible combinations obviously would be impossible. However, they can all be reduced to the same simple case discussed above, by considering that several currents, voltages, magnetic fluxes, etc., in the same circuit add algebraically, without interfering with each other (assuming, as done here, that magnetic saturation is not approached).

If an e.m.f. e_1 produces a current i_1 in a circuit, and an e.m.f. e_2 produces in the same circuit a current i_2, then the e.m.f. $e_1 + e_2$ produces the current $i_1 + i_2$, as is obvious.

If now the voltage $e_1 + e_2$, and thus also the current $i_1 + i_2$, consists of a permanent term, e_1 and i_1, and a transient term, e_2 and i_2, the transient terms e_2, i_2 follow the same curves, when combined with the permanent terms e_1, i_1, as they would when alone in the circuit (the case above discussed). Thus, the preceding discussion applies to all magnetic transients, by separating the transient from the permanent term, investigating it separately, and then adding it to the permanent term.

The same reasoning also applies to the transient resulting from several forms of energy storage (provided that the law of proportionality of i, e, Φ, etc., applies), and makes it possible, in investigating the phenomena during the transition period of energy readjustment, to separate the permanent and the transient term, and discuss them separately.

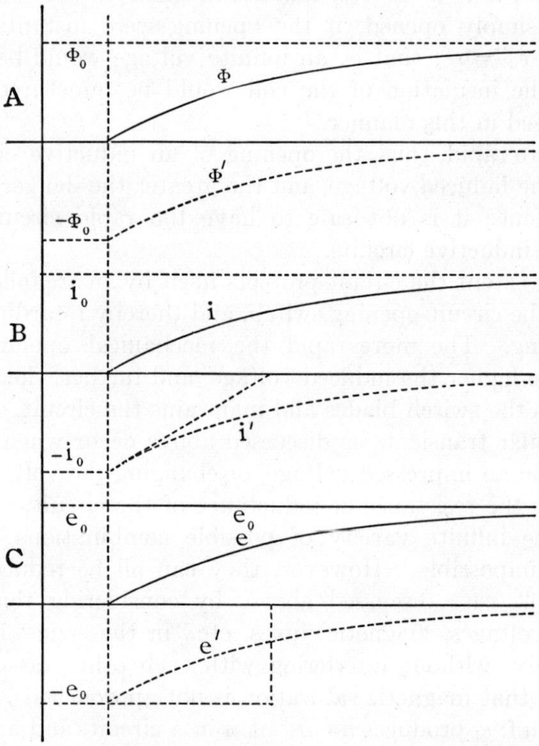

Fig. 14. — Single-energy Starting Transient of Magnetic Circuit.

For instance, in the coil shown in Fig. 10, let the short circuit A be opened, that is, the voltage e_0 be impressed upon the coil. At the moment of time, t_0, when this is done, current i, magnetic flux Φ, and voltage e on the coil are zero. In final condition, after the transient has passed, the values i_0, Φ_0, e_0 are reached. We may then, as discussed above, separate the transient from the permanent term, and consider that at the time t_0 the coil has a permanent current i_0, permanent flux Φ_0, permanent voltage e_0, and in addition thereto a transient current $-i_0$, a transient flux $-\Phi_0$, and a transient voltage $-e_0$. These transients are the same as in Fig. 11 (only with reversed direction). Thus the same curves result, and to them are added the permanent values i_0, Φ_0, e_0. This is shown in Fig. 14.

A shows the permanent flux Φ_0, and the transient flux $-\Phi_0$, which are assumed, up to the time t_0, to give the resultant zero flux. The transient flux dies out by the curve Φ', in accordance with Fig. 11. Φ' added to Φ_0 gives the curve Φ, which is the transient from zero flux to the permanent flux Φ_0.

In the same manner *B* shows the construction of the actual current change i by the addition of the permanent current i_0 and the transient current i', which starts from $-i_0$ at t_0.

C then shows the voltage relation: e_0 the permanent voltage, e' the transient voltage which starts from $-e_0$ at t_0, and e the resultant or effective voltage in the coil, derived by adding e_0 and e'.

CHAPTER IV

SINGLE-ENERGY TRANSIENTS IN ALTERNATING-CURRENT CIRCUITS

17. Whenever the conditions of an electric circuit are changed in such a manner as to require a change of stored energy, a transition period appears, during which the stored energy adjusts itself from the condition existing before the change to the condition after the change. The currents in the circuit during the transition period can be considered as consisting of the superposition of the permanent current, corresponding to the conditions after the change, and a transient current, which connects the current value before the change with that brought about by the change. That is, if i_1 = current existing in the circuit immediately before, and thus at the moment of the change of circuit condition, and i_2 = current which should exist at the moment of change in accordance with the circuit condition after the change, then the actual current i_1 can be considered as consisting of a part or component i_2, and a component $i_1 - i_2 - i_0$. The former, i_2, is permanent, as resulting from the established circuit condition. The current component i_0, however, is not produced by any power supply, but is a remnant of the previous circuit condition, that is, a transient, and therefore gradually decreases in the manner as discussed in paragraph 13, that is, with a duration $T = \frac{L}{r}\cdot$

The permanent current i_2 may be continuous, or alternating, or may be a changing current, as a transient of long duration, etc.

The same reasoning applies to the voltage, magnetic flux, etc.

Thus, let, in an alternating-current circuit traversed by current i_1, in Fig. 15A, the conditions be changed, at the moment $t = 0$, so as to produce the current i_2. The instantaneous value of the current i_1 at the moment $t = 0$ can be considered as consisting of the instantaneous value of the permanent current i_2, shown dotted, and the transient $i_0 = i_1 - i_2$. The latter gradually dies down, with the duration $T = \frac{L}{r}$, on the usual exponential transient, shown dotted in Fig. 15. Adding the transient current i_0 to the permanent current i_2 gives the total current during the transition period, which is shown in drawn line in Fig. 15.

As seen, the transient is due to the difference between the instantaneous value of the current i_1 which exists, and that of the current i_2 which should exist at the moment of change, and

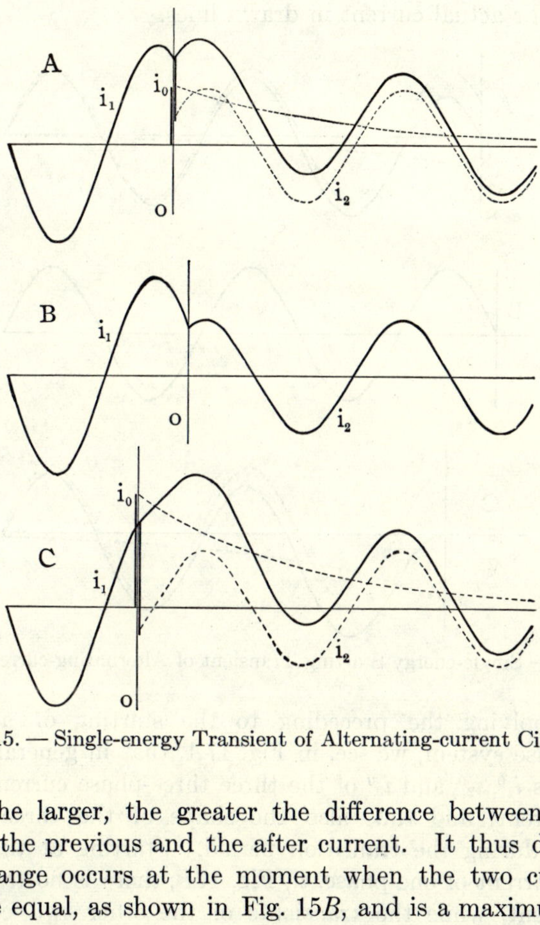

Fig. 15. — Single-energy Transient of Alternating-current Circuit.

thus is the larger, the greater the difference between the two currents, the previous and the after current. It thus disappears if the change occurs at the moment when the two currents i_1 and i_2 are equal, as shown in Fig. 15*B*, and is a maximum, if the change occurs at the moment when the two currents i_1 and i_2 have the greatest difference, as shown in Fig. 15*C*, that is, at a point one-quarter period or 90 degrees distant from the intersection of i_1 and i_2.

If the current i_1 is zero, we get the starting of the alternating current in an inductive circuit, as shown in Figs. 16, *A*, *B*, *C*. The starting transient is zero, if the circuit is closed at the moment when the permanent current would be zero (Fig. 16*B*), and is a maximum when closing the circuit at the maximum point of the permanent-current wave (Fig. 16*C*). The permanent current and the transient components are shown dotted in Fig. 16, and the resultant or actual current in drawn lines.

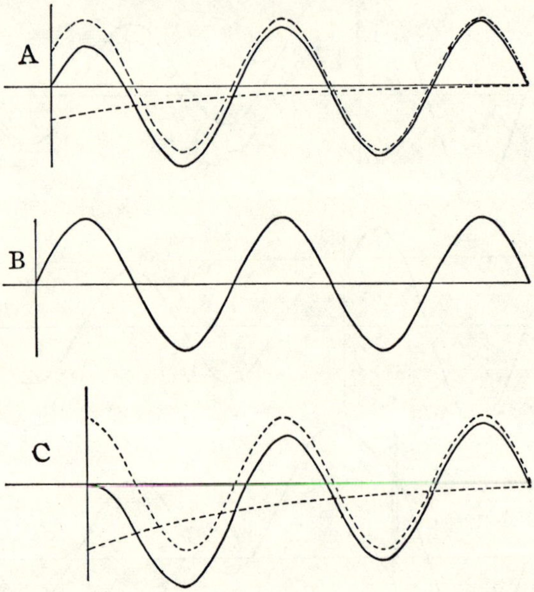

Fig. 16. — Single-energy Starting Transient of Alternating-current Circuit.

18. Applying the preceding to the starting of a balanced three-phase system, we see, in Fig. 17*A*, that in general the three transients i_1^0, i_2^0, and i_3^0 of the three three-phase currents i_1, i_2, i_3 are different, and thus also the shape of the three resultant currents during the transition period. Starting at the moment of zero current of one phase, i_1, Fig. 17*B*, there is no transient for this current, while the transients of the other two currents, i_2 and i_3, are equal and opposite, and near their maximum value. Starting, in Fig. 17*C*, at the maximum value of one current i_3, we have the maximum value of transient for this current i_3^0, while the transients of the two other currents, i_1 and i_2, are equal, have half the value of $i_3{}^0$, and are opposite in direction thereto. In any case, the three transients must be distributed on both sides of the zero line. This is obvious: if i_1', i_2', and i_3' are the instantaneous values of the permanent three-phase currents, in Fig. 17, the initial values of their transients are: $-i_1'$, $-i_2'$, $-i_3'$.

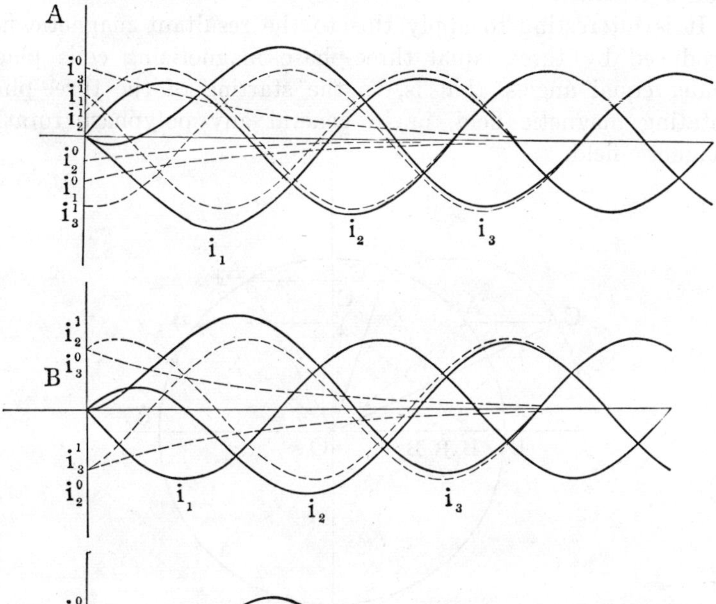

Fig. 17. — Single-energy Starting Transient of Three-phase Circuit.

Since the sum of the three three-phase currents at every moment is zero, the sum of the initial values of the three transient currents also is zero. Since the three transient curves $i_1{}^0$, $i_2{}^0$, $i_3{}^0$ are proportional to each other $\left(\text{as exponential curves of the same duration } T = \frac{L}{r}\right)$, and the sum of their initial values is zero, it follows that the sum of their instantaneous values must be zero at any moment, and therefore the sum of the instantaneous values of the resultant currents (shown in drawn line) must be zero at any moment, not only during the permanent condition, but also during the transition period existing before the permanent condition is reached.

It is interesting to apply this to the resultant magnetic field produced by three equal three-phase magnetizing coils placed under equal angles, that is, to the starting of the three-phase rotating magnetic field, or in general any polyphase rotating magnetic field.

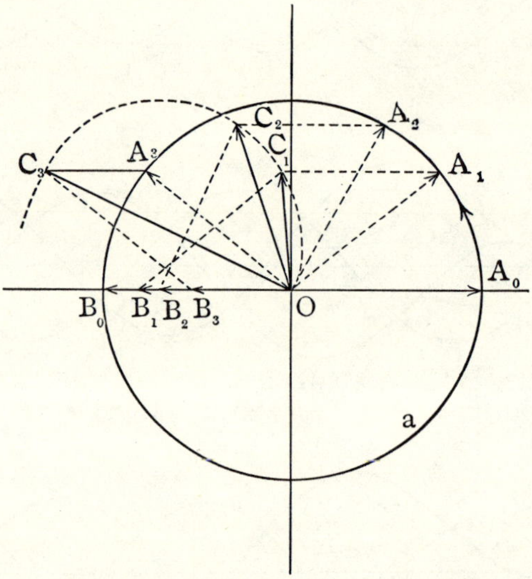

Fig. 18. — Construction of Starting Transient of Rotating Field.

As is well known, three equal magnetizing coils, placed under equal angles and excited by three-phase currents, produce a resultant magnetic field which is constant in intensity, but revolves synchronously in space, and thus can be represented by a concentric circle *a*, Fig. 18.

This, however, applies only to the permanent condition. In the moment of start, all the three currents are zero, and their resultant magnetic field thus also zero, as shown above. Since the magnetic field represents stored energy and thus cannot be produced instantly, a transient must appear in the building up of the rotating field. This can be studied by considering separately the permanent and the transient components of the three currents, as is done in the preceding. Let i_1', i_2', i_3' be the instantaneous values of the permanent currents at the moment of closing the circuit, $t = 0$. Combined, these would give the resultant field $\overline{OA_0}$ in Fig. 18. The three transient currents in this moment are $i_1^0 = -i_1'$, $i_2^0 = -i_2'$, $i_3^0 = -i_3'$, and combined these give a resultant field $\overline{OB_0}$, equal and opposite to $\overline{OA_0}$ in Fig. 18. The permanent field rotates synchronously on the concentric circle a; the transient field $\overline{OB}$ remains constant in the direction $\overline{OB_0}$, since all three transient components of current decrease in proportion to each other. It decreases, however, with the decrease of the transient current, that is, shrinks together on the line $\overline{B_0O}$. The resultant or actual field thus is the combination of the permanent fields, shown as $\overline{OA_1}$ $\overline{OA_2}$, . . . , and the transient fields, shown as $\overline{OB_1}$, $\overline{OB_2}$, etc., and derived thereby by the parallelogram law, as shown in Fig. 18, as $\overline{OC_1}$, $\overline{OC_2}$, etc. In this diagram, $\overline{B_1C_1}$, $\overline{B_2C_2}$, etc., are equal to $\overline{OA_1}$, $\overline{OA_2}$, etc., that is, to the radius of the permanent circle a. That is, while the rotating field in permanent condition is represented by the concentric circle a, the resultant field during the transient or starting period is represented by a succession of arcs of circles c, the centers of which move from B_0 in the moment of start, on the line $\overline{B_0O}$ toward O, and can be constructed hereby by drawing from the successive points B_0, B_1, B_2, which correspond to successive moments of time 0, t_1, t_2 . . . , radii $\overline{B_1C_1}$, $\overline{B_2C_2}$, etc., under the angles, that is, in the direction corresponding to the time 0, t_1, t_2, etc. This is done in Fig. 19, and thereby the transient of the rotating field is constructed.

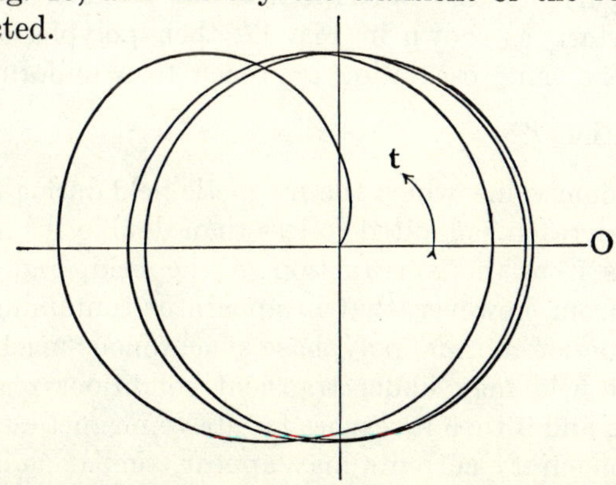

Fig. 19. — Starting Transient of Rotating Field: Polar Form.

From this polar diagram of the rotating field, in Fig. 19, values $\overline{OC}$ can now be taken, corresponding to successive moments of time, and plotted in rectangular coördinates, as done in Fig. 20. As seen, the rotating field builds up from zero at the moment of closing the circuit, and reaches the final value by a series of oscillations; that is, it first reaches beyond the permanent value, then drops below it, rises again beyond it, etc.

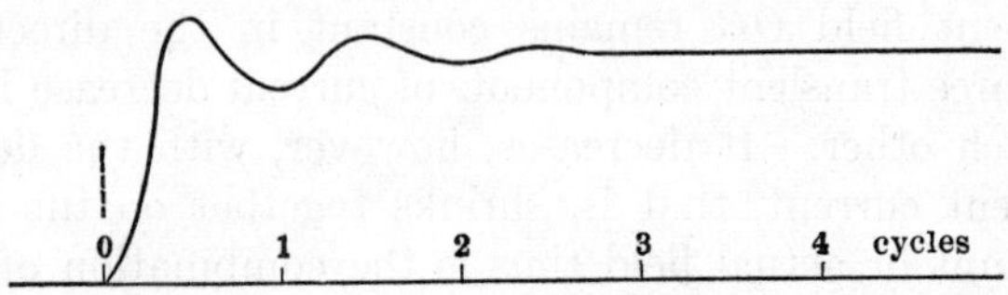

Fig. 20. — Starting Transient of Rotating Field: Rectangular Form.

We have here an oscillatory transient, produced in a system with only one form of stored energy (magnetic energy), by the combination of several simple exponential transients. However, it must be considered that, while energy can be stored in one form only, as magnetic energy, it can be stored in three electric circuits, and a transfer of stored magnetic energy between the three electric circuits, and therewith a surge, thus can occur.

It is interesting to note that the rotating-field transient is independent of the point of the wave at which the circuit is closed. That is, while the individual transients of the three three-phase currents vary in shape with the point of the wave at which they start, as shown in Fig. 17, their polyphase resultant always has the same oscillating approach to a uniform rotating field, of duration $T = \frac{L}{r}$.

The maximum value, which the magnetic field during the transition period can reach, is limited to less than double the final value, as is obvious from the construction of the field, Fig. 19. It is evident herefrom, however, that in apparatus containing rotating fields, as induction motors, polyphase synchronous machines, etc., the resultant field may under transient conditions reach nearly double value, and if then it reaches far above magnetic saturation, excessive momentary currents may appear, similar as in starting transformers of high magnetic density. In polyphase rotary apparatus, however, these momentary starting currents usually are far more limited than in transformers, by the higher stray field (self-inductive reactance), etc., of the apparatus, resulting from the air gap in the magnetic circuit.

19. As instance of the use of the single-energy transient in engineering calculations may be considered the investigation of the momentary short-circuit phenomena of synchronous alternators. In alternators, especially high-speed high-power machines as turboalternators, the momentary short-circuit current may be many times greater than the final or permanent short-circuit current, and this excess current usually decreases fairly slowly, lasting for many cycles. At the same time, a big current rush occurs in the field. This excess field current shows curious pulsations, of single and of double frequency, and in the beginning the armature currents also show unsymmetrical shapes. Some oscillograms of three-phase, quarter-phase, and single-phase short circuits of turboalternators are shown in Figs. 25 to 28.

By considering the transients of energy storage, these rather complex-appearing phenomena can be easily understood, and predetermined from the constants of the machine with reasonable exactness.

In an alternator, the voltage under load is affected by armature reaction and armature self-induction. Under permanent condition, both usually act in the same way, reducing the voltage at noninductive and still much more at inductive load, and increasing it at antiinductive load; and both are usually combined in one quantity, the synchronous reactance x_0. In the transients resulting from circuit changes, as short circuits, the self-inductive armature reactance and the magnetic armature reaction act very differently:* the former is instantaneous in its effect, while the latter requires time. The self-inductive armature reactance x_1 consumes a voltage $x_1 i$ by the magnetic flux surrounding the armature conductors, which results from the m.m.f. of the armature current, and therefore requires a component of the magnetic-field flux for its production. As the armature magnetic flux and the current which produces it must be simultaneous (the former being an integral part of the phenomenon of current flow, as seen in Lecture II), it thus follows that the armature reactance appears together

* So also in their effect on synchronous operation, in hunting, etc.

with the armature current, that is, is instantaneous. The armature reaction, however, is the m.m.f. of the armature current in its reaction on the m.m.f. of the field-exciting current. That is, that part $x_2 = x_0 - x_1$ of the synchronous reactance which corresponds to the armature reaction is not a true reactance at all, consumes no voltage, but represents the consumption of field ampere turns by the m.m.f. of the armature current, and the corresponding change of field flux. Since, however, the field flux represents stored magnetic energy, it cannot change instantly, and the armature reaction thus does not appear instantaneously with the armature current, but shows a transient which is determined essentially by the constants of the field circuit, that is, is the counterpart of the field transient of the machine.

If then an alternator is short-circuited, in the first moment only the true self-inductive part x_1 of the synchronous reactance exists, and the armature current thus is $i_1 = \frac{e_0}{x_1}$, where e_0 is the induced e.m.f., that is, the voltage corresponding to the magnetic-field excitation flux existing before the short circuit. Gradually the armature reaction lowers the field flux, in the manner as represented by the synchronous reactance x_0, and the short-circuit current decreases to the value $i_0 = \frac{e_0}{x_0}$.

The ratio of the momentary short-circuit current to the permanent short-circuit current thus is, approximately, the ratio $\frac{i_1}{i_0} = \frac{x_0}{x_1}$, that is, synchronous reactance to self-inductive reactance, or armature reaction plus armature self-induction, to armature self-induction. In machines of relatively low self-induction and high armature reaction, the momentary short-circuit current thus may be many times the permanent short-circuit current.

The field flux remaining at short circuit is that giving the voltage consumed by the armature self-induction, while the decrease of field flux between open circuit and short circuit corresponds to the armature reaction. The ratio of the open-circuit field flux to the short-circuit field flux thus is the ratio of armature reaction plus self-induction, to the self-induction; or of the synchronous reactance to the self-inductive reactance: $\frac{x_0}{x_1}$.

Thus it is:

$$\frac{\text{momentary short-circuit current}}{\text{permanent short-circuit current}} = \frac{\text{open-circuit field flux}^{*}}{\text{short-circuit field flux}} =$$

$$\frac{\text{armature reaction plus self-induction}}{\text{self-induction}} = \frac{\text{synchronous reactance}}{\text{self-inductive reactance}} = \frac{x_0}{x_1}.$$

20. Let Φ_1 = field flux of a three-phase alternator (or, in general, polyphase alternator) at open circuit, and this alternator be short-circuited at the time $t = 0$. The field flux then gradually dies down, by the dissipation of its energy in the field circuit, to the short-circuit field flux Φ_0, as indicated by the curve Φ in Fig. 21*A*. If m = ratio

$$\frac{\text{armature reaction plus self-induction}}{\text{armature self-induction}} = \frac{x_0}{x_1},$$

it is $\Phi_1 = m\Phi_0$, and the initial value of the field flux consists of the permanent part Φ_0, and the transient part $\Phi' = \Phi_1 - \Phi_0 = (m-1)\Phi_0$. This is a rather slow transient, frequently of a duration of a second or more.

The armature currents i_1, i_2, i_3 are proportional to the field flux Φ which produces them, and thus gradually decrease, from initial values, which are as many times higher than the final values as Φ_1 is higher than Φ_0, or m times, and are represented in Fig. 21*B*.

The resultant m.m.f. of the armature currents, or the armature reaction, is proportional to the currents, and thus follows the same field transient, as shown by F in Fig. 21*C*.

The field-exciting current is i_0 at open circuit as well as in the permanent condition of short circuit. In the permanent condition of short circuit, the field current i_0 combines with the armature reaction F_0, which is demagnetizing, to a resultant m.m.f., which produces the short-circuit flux Φ_0. During the transition period the field flux Φ is higher than Φ_0, and the resultant m.m.f. must therefore be higher in the same proportion. Since it is the difference between the field current and the armature reaction F, and the latter is proportional to Φ, the field current thus must also be

* If the machine were open-circuited before the short circuit, otherwise the field flux existing before the short circuit. It herefrom follows that the momentary short-circuit current essentially depends on the field flux, and thereby the voltage of the machine, before the short circuit, but is practically independent of the load on the machine before the short circuit and the field excitation corresponding to this load.

proportional to Φ. Thus, as it is $i = i_0$ at Φ_0, during the transition period it is $i = \frac{\Phi}{\Phi_0} i_0$. Hence, the field-exciting current traverses the same transient, from an initial value i_0' to the normal value i_0, as the field flux Φ and the armature currents.

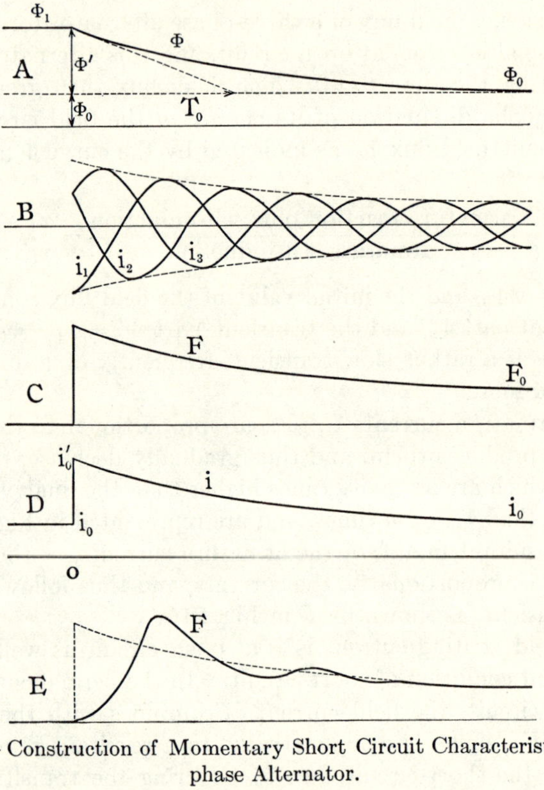

Fig. 21. — Construction of Momentary Short Circuit Characteristic of Polyphase Alternator.

Thus, at the moment of short circuit a sudden rise of field current must occur, to maintain the field flux at the initial value Φ_1 against the demagnetizing armature reaction. In other words, the field flux Φ decreases at such a rate as to induce in the field circuit the e.m.f. required to raise the field current in the proportion m, from i_0 to i_0', and maintain it at the values corresponding to the transient i, Fig. 21*D*.

As seen, the transients Φ; i_1, i_2, i_3; F; i are proportional to each other, and are a field transient. If the field, excited by current i_0

at impressed voltage e_0, were short-circuited upon itself, in the first moment the current in the field would still be i_0, and therefore the voltage e_0 would have to be induced by the decrease of magnetic flux; and the duration of the field transient, as discussed in Lecture III, would be $T_0 = \frac{L_0}{r_0}$.

The field current in Fig. 21*D*, of the alternator short-circuit transient, starts with the value $i_0' = mi_0$, and if e_0 is the e.m.f. supplied in the field-exciting circuit from a source of constant voltage supply, as the exciter, to produce the current i_0', the voltage $e_0' = me_0$ must be acting in the field-exciting circuit; that is, in addition to the constant exciter voltage e_0, a voltage $(m-1)e_0$ must be induced in the field circuit by the transient of the magnetic flux. As a transient of duration $\frac{L_0}{r_0}$ induces the voltage e_0, to induce the voltage $(m-1)e_0$ the duration of the transient must be

$$T_0 = \frac{L_0}{(m-1)\,r_0},$$

where L_0 = inductance, r_0 = total resistance of field-exciting circuit (inclusive of external resistance).

The short-circuit transient of an alternator thus usually is of shorter duration than the short-circuit transient of its field, the more so, the greater m, that is, the larger the ratio of momentary to permanent short-circuit current.

In Fig. 21 the decrease of the transient is shown greatly exaggerated compared with the frequency of the armature currents, and Fig. 22 shows the curves more nearly in their actual proportions.

The preceding would represent the short-circuit phenomena, if there were no armature transient. However, the armature circuit contains inductance also, that is, stores magnetic energy, and thereby gives rise to a transient, of duration $T = \frac{L}{r}$, where L = inductance, r = resistance of armature circuit. The armature transient usually is very much shorter in duration than the field transient.

The armature currents thus do not instantly assume their symmetrical alternating values, but if in Fig. 21*B*, i_1', i_2', i_3' are the instantaneous values of the armature currents in the moment of start, $t = 0$, three transients are superposed upon these, and start with the values $-i_1'$, $-i_2'$, $-i_3'$. The resultant armature currents are derived by the addition of these armature transients upon the permanent armature currents, in the manner as discussed in paragraph 18, except that in the present case even the permanent armature currents i_1, i_2, i_3 are slow transients.

In Fig. 22*B* are shown the three armature short-circuit currents, in their actual shape as resultant from the armature transient and the field transient. The field transient (or rather its beginning) is shown as Fig. 22*A*. Fig. 22*B* gives the three armature

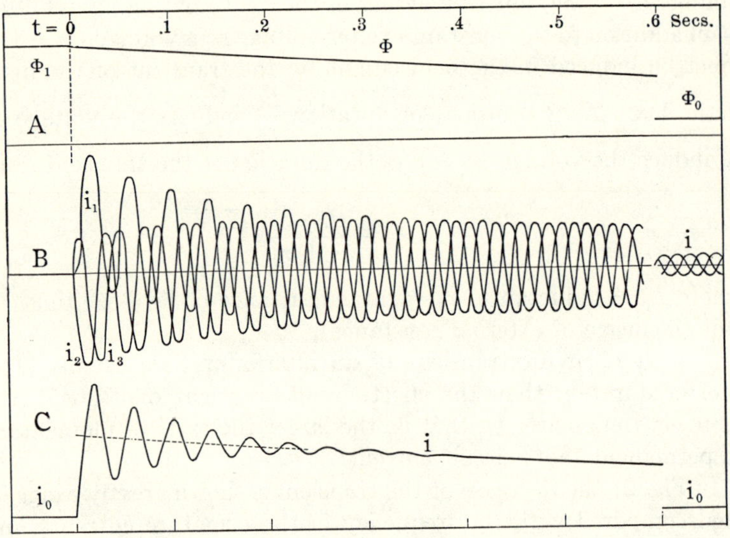

Fig. 22. — Momentary Short Circuit Characteristic of Three-phase Alternator.

currents for the case where the circuit is closed at the moment when i_1 should be maximum; i_1 then shows the maximum transient, and i_2 and i_3 transients in opposite direction, of half amplitude. These armature transients rapidly disappear, and the three currents become symmetrical, and gradually decrease with the field transient to the final value indicated in the figure.

The resultant m.m.f. of three three-phase currents, or the armature reaction, is constant if the currents are constant, and as the currents decrease with the field transient, the resultant armature reaction decreases in the same proportion as the field, as is shown in Fig. 21*C* by *F*. During the initial part of the short circuit, however, while the armature transient is appreciable and the armature currents thus unsymmetrical, as seen in Fig. 22*B*, their resultant polyphase m.m.f. also shows a transient, the transient of the rotating magnetic field discussed in paragraph 18. That is, it approaches the curve *F* of Fig. 21*C* by a series of oscillations, as indicated in Fig. 21*E*.

Since the resultant m.m.f. of the machine, which produces the flux, is the difference of the field excitation, Fig. 21*D* and the armature reaction, then if the armature reaction shows an initial oscillation, in Fig. 21*E*, the field-exciting current must give the same oscillation, since its m.m.f. minus the armature reaction gives the resultant field excitation corresponding to flux Φ. The starting transient of the polyphase armature reaction thus appears in the field current, as shown in Fig. 22*C*, as an oscillation of full machine frequency. As the mutual induction between armature and field circuit is not perfect, the transient pulsation of armature reaction appears with reduced amplitude in the field current, and this reduction is the greater, the poorer the mutual inductance, that is, the more distant the field winding is from the armature winding. In Fig. 22*C* a damping of 20 per cent is assumed, which corresponds to fairly good mutual inductance between field and armature, as met in turboalternators.

If the field-exciting circuit contains inductance outside of the alternator field, as is always the case to a slight extent, the pulsations of the field current, Fig. 22*C*, are slightly reduced and delayed in phase; and with considerable inductance intentionally inserted into the field circuit, the effect of this inductance would require consideration.

From the constants of the alternator, the momentary short-circuit characteristics can now be constructed.

Assuming that the duration of the field transient is

$$T_0 = \frac{L_0}{(m-1)\,r_0} = 1 \text{ sec.},$$

the duration of the armature transient is

$$T = \frac{L}{r} = .1 \text{ sec.}$$

And assuming that the armature reaction is 5 times the armature self-induction, that is, the synchronous reactance is 6 times the self-inductive reactance, $\frac{x_0}{x_1} = m = 6$. The frequency is 25 cycles.

If Φ_1 is the initial or open-circuit flux of the machine, the short-circuit flux is $\Phi_0 = \frac{\Phi_1}{m} = \frac{1}{6}\Phi_1$, and the field transient Φ is a transient of duration 1 sec., connecting Φ_1 and Φ_0, Fig. 22*A*, represented by the expression

$$\Phi = \Phi_0 + (\Phi_1 - \Phi_0)\epsilon^{-\frac{t}{T_0}}.$$

The permanent armature currents i_1, i_2, i_3 then are currents starting with the values $m\frac{e_0}{x_0}$, and decreasing to the final short-circuit current $\frac{e_0}{x_0}$, on the field transient of duration T_0. To these currents are added the armature transients, of duration T, which start with initial values equal but opposite in sign to the initial values of the permanent (or rather slowly transient) armature currents, as discussed in paragraph 18, and thereby give the asymmetrical resultant currents, Fig. 22*B*.

The field current i gives the same slow transient as the flux Φ, starting with $i_0' = mi_0$, and tapering to the final value i_0. Upon this is superimposed the initial full-frequency pulsation of the armature reaction. The transient of the rotating field, of duration $T = .1$ sec., is constructed as in paragraph 18, and for its instantaneous values the percentage deviation of the resultant field from its permanent value is calculated. Assuming 20 per cent damping in the reaction on the field excitation, the instantaneous values of the slow field transient (that is, of the current $(i - i_0)$, since i_0 is the permanent component) then are increased or decreased by 80 per cent of the percentage variation of the transient field of armature reaction from uniformity, and thereby the field curve, Fig. 22*C*, is derived. Here the correction for the external field inductance is to be applied, if considerable.

Since the transient of the armature reaction does not depend on the point of the wave where the short circuit occurs, it follows that the phenomena at the short circuit of a polyphase alternator are always the same, that is, independent of the point of the wave at which the short circuit occurs, with the exception of the initial wave shape of the armature currents, which individually depend on the point of the wave at which the phenomenon begins, but not so in their resultant effect.

21. The conditions with a single-phase short circuit are different, since the single-phase armature reaction is pulsating, varying between zero and double its average value, with double the machine frequency.

The slow field transient and its effects are the same as shown in Fig. 21, *A* to *D*.

However, the pulsating armature reaction produces a corresponding pulsation in the field circuit. This pulsation is of double

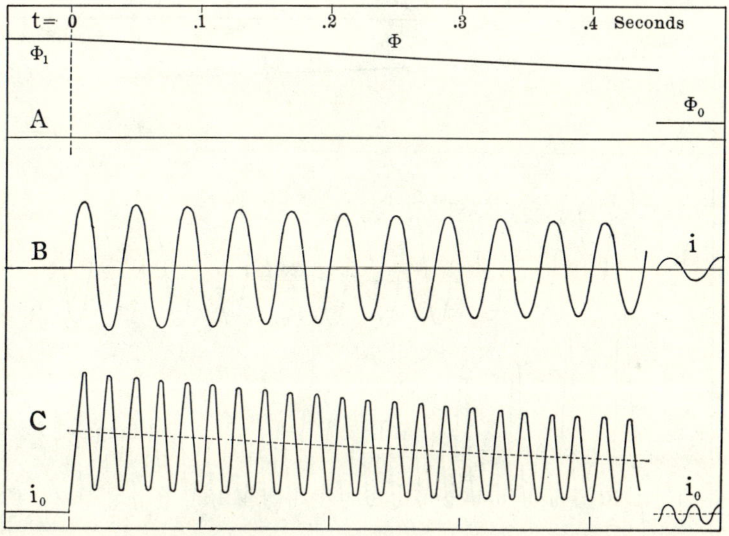

Fig. 23. — Symmetrical Momentary Single-phase Short Circuit of Alternator.

frequency, and is not transient, but equally exists in the final short-circuit current.

Furthermore, the armature transient is not constant in its reaction on the field, but varies with the point of the wave at which the short circuit starts.

Assume that the short circuit starts at that point of the wave where the permanent (or rather slowly transient) armature current should be zero: then no armature transient exists, and the armature current is symmetrical from the beginning, and shows the slow transient of the field, as shown in Fig. 23, where *A* is the field transient Φ (the same as in Fig. 22*A*) and *B* the armature current, decreasing from an initial value, which is *m* times the final value, on the field transient.

Assume then that the mutual induction between field and armature is such that 60 per cent of the pulsation of armature reaction appears in the field current. Forty per cent damping for the double-frequency reaction would about correspond to the 20 per cent damping assumed for the transient full-frequency pulsation of the polyphase machine. The transient field current thus pulsates by 60 per cent around the slow field transient, as shown by Fig. 23*C*; passing a maximum for every maximum of armature

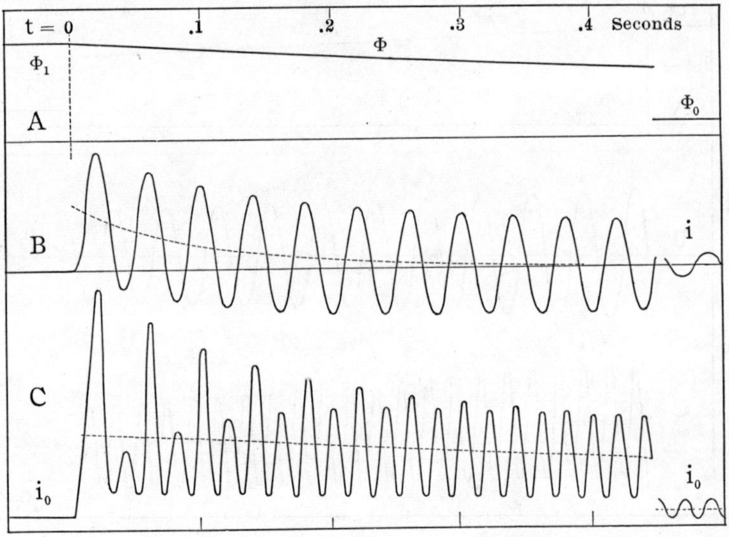

Fig. 24.—Asymmetrical Momentary Single-phase Short Circuit of Alternator.

current, and thus maximum of armature reaction, and a minimum for every zero value of armature current, and thus armature reaction.

Such single-phase short-circuit transients have occasionally been recorded by the oscillograph, as shown in Fig. 27. Usually, however, the circuit is closed at a point of the wave where the permanent armature current would not be zero, and an armature transient appears, with an initial value equal, but opposite to, the initial value of the permanent armature current. This is shown in Fig. 24 for the case of closing the circuit at the moment where the armature current should be a maximum, and its transient thus a maximum. The field transient Φ is the same as before. The armature current shows the initial asymmetry resulting from the armature transient, and superimposed on the slow field transient.

On the field current, which, due to the single-phase armature reaction, shows a permanent double-frequency pulsation, is now superimposed the transient full-frequency pulsation resultant from the transient armature reaction, as discussed in paragraph 20. Every second peak of the permanent double-frequency pulsation then coincides with a peak of the transient full-frequency pulsation, and is thereby increased, while the intermediate peak of the double-frequency pulsation coincides with a minimum of the full-frequency pulsation, and is thereby reduced. The result is that successive waves of the double-frequency pulsation of the field current are unequal in amplitude, and high and low peaks alternate. The difference between successive double-frequency waves is a maximum in the beginning, and gradually decreases, due to the decrease of the transient full-frequency pulsation, and finally the double-frequency pulsation becomes symmetrical, as shown in Fig. 24*C*.

In the particular instance of Fig. 24, the double-frequency and the full-frequency peaks coincide, and the minima of the field-current curve thus are symmetrical. If the circuit were closed at another point of the wave, the double-frequency minima would become unequal, and the maxima more nearly equal, as is easily seen.

While the field-exciting current is pulsating in a manner determined by the full-frequency transient and double-frequency permanent armature reaction, the potential difference across the field winding may pulsate less, if little or no external resistance or inductance is present, or may pulsate so as to be nearly alternating and many times higher than the exciter voltage, if considerable external resistance or inductance is present; and therefore it is not characteristic of the phenomenon, but may become important by its disruptive effects, if reaching very high values of voltage.

With a single-phase short circuit on a polyphase machine, the double-frequency pulsation of the field resulting from the single-phase armature reaction induces in the machine phase, which is in quadrature to the short-circuited phase, an e.m.f. which contains the frequencies $f(2 \pm 1)$, that is, full frequency and triple

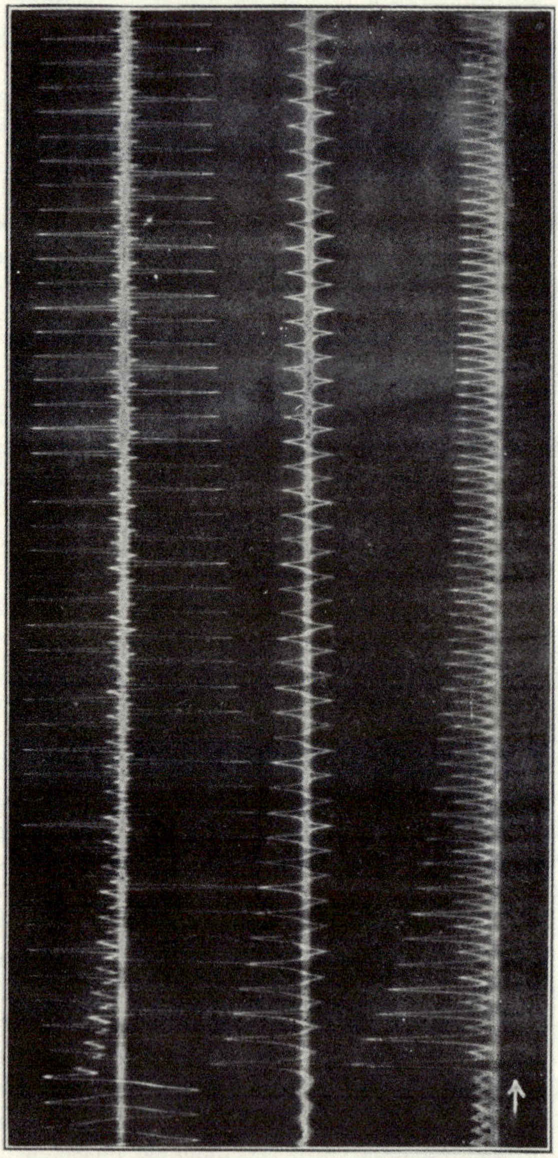

Fig. 25.—CD9762.—Momentary Single-phase Short Circuit of 1875-Kw. 2300-Volt Two-phase Alternator (AQB-4-1875M-1800). Oscillogram of Armature Voltage of Open Phase, Armature Current of Short-circuited Phase, and Field Current.

frequency, and as the result an increase of voltage and a distortion of the quadrature phase occurs, as shown in the oscillogram Fig. 25.

Various momentary short-circuit phenomena are illustrated by the oscillograms Figs. 26 to 28.

Figs. 26*A* and 26*B* show the momentary three-phase short circuit of a 4-polar 25-cycle 1500-kw. steam turbine alternator. The

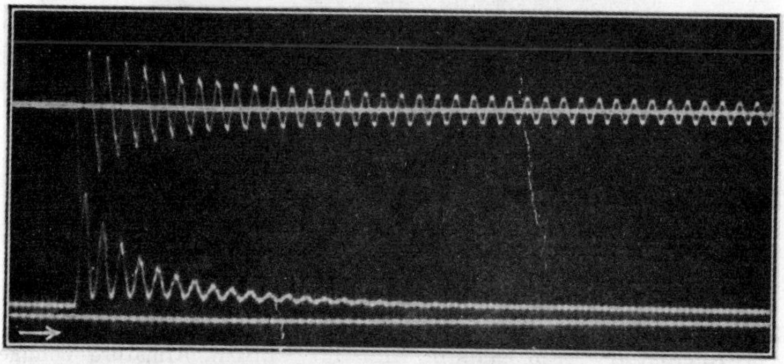

Fig. 26*A*. — CD9399. — Symmetrical.

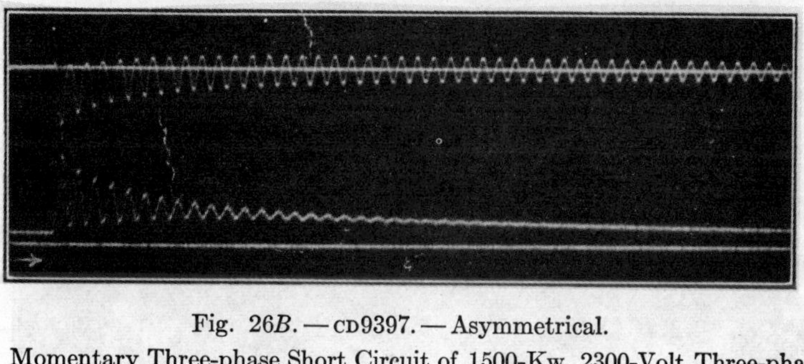

Fig. 26*B*. — CD9397. — Asymmetrical.

Momentary Three-phase Short Circuit of 1500-Kw. 2300-Volt Three-phase Alternator (ATB–4–1500–1800). Oscillograms of Armature Current and Field Current.

lower curve gives the transient of the field-exciting current, the upper curve that of one of the armature currents, — in Fig. 26*A* that current which should be near zero, in Fig. 26*B* that which should be near its maximum value at the moment where the short circuit starts.

Fig. 27 shows the single-phase short circuit of a pair of machines in which the short circuit occurred at the moment in which the armature short-circuit current should be zero; the armature current wave, therefore, is symmetrical, and the field current shows only the double-frequency pulsation. Only a few half-waves were recorded before the circuit breaker opened the short circuit.

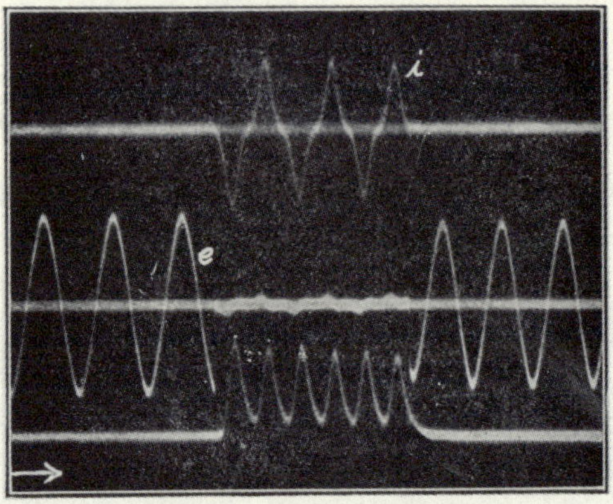

Fig. 27. — CD5128. — Symmetrical. Momentary Single-phase Short Circuit of Alternator. Oscillogram of Armature Current, Armature Voltage, and Field Current. (Circuit breaker opens.)

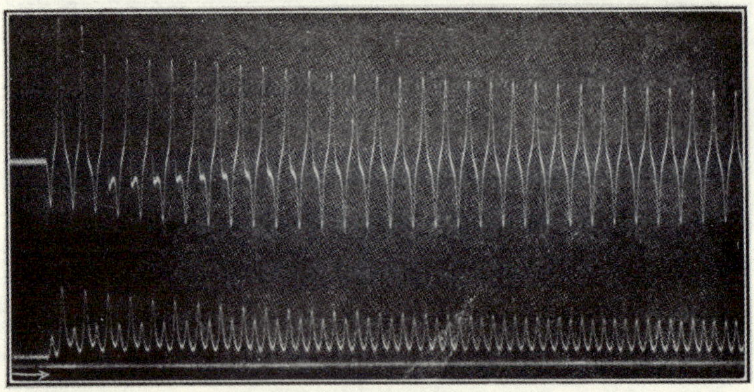

Fig. 28. — CD6565. — Asymmetrical. Momentary Single-phase Short Circuit of 5000-Kw. 11,000-Volt Three-phase Alternator (ATB-6-5000-500). Oscillogram of Armature Current and Field Current.

Fig. 28 shows the single-phase short circuit of a 6-polar 5000-kw. 11,000-volt steam turbine alternator, which occurred at a point of the wave where the armature current should be not far from its maximum. The transient armature current, therefore, starts unsymmetrical, and the double-frequency pulsation of the field current shows during the first few cycles the alternate high and low peaks resulting from the full-frequency transient pulsation of the rotating magnetic field of armature reaction. The irregular initial decrease of the armature current and the sudden change of its wave shape are due to the transient of the current transformer, through which the armature current was recorded.

Fig. 25 shows a single-phase short circuit of a quarter-phase alternator; the upper wave is the voltage of the phase which is not short-circuited, and shows the increase and distortion resulting from the double-frequency pulsation of the armature reaction.

While the synchronous reactance x_0 can be predetermined with fair accuracy, the self-inductive x_1 is not such a definite quantity. It includes a transient component. The armature magnetic circuit is in mutual inductive relation with the field-exciting circuit. At constant alternating current in the armature, the resultant of the armature m.m.f's. and e.m.f's. is constant with regard to the field, and the mutual inductance thus does not come into play. During a transient, however, the armature conditions change, and the self-inductance of the exciting circuit is partly transformed into the armature circuit by the ratio of field turns to armature turns, giving rise to a transient effective component of armature self-induction, which depends on the relative rate of change of the armature and the field, and thereby is a maximum in the beginning, and gradually decreases to zero in stationary conditions. This tends to lower the maximum values of the field transients and to increase the duration of the armature transients. This effect is materially affected by the amount of resistance and reactance in the exciting circuit outside of the field winding.

There also exists a mutual inductance between the armature circuits of the three-phase machine, which results in an energy transfer between the phases, during the armature transient.

The instantaneous power of the momentary short-circuit current, and with it the forces acting on driving shaft and prime mover, are proportional to the short-circuit current, being short-circuit current times magnetic field flux. The forces exerted between the armature conductors — which tend to tear and strip the end windings, etc. — are proportional to the square of the short-circuit current.

CHAPTER V

SINGLE-ENERGY TRANSIENT OF IRONCLAD CIRCUIT

22. Usually in electric circuits, current, voltage, the magnetic field and the dielectric field are proportional to each other, and the transient thus is a simple exponential, if resulting from one form of stored energy, as discussed in the preceding lectures. This, however, is no longer the case if the magnetic field contains iron or other magnetic materials, or if the dielectric field reaches densities beyond the dielectric strength of the carrier of the field, etc.; and the proportionality between current or voltage and their respective fields, the magnetic and the dielectric, thus ceases, or, as it may be expressed, the inductance L is not constant, but varies with the current, or the capacity is not constant, but varies with the voltage.

The most important case is that of the ironclad magnetic circuit, as it exists in one of the most important electrical apparatus, the alternating-current transformer. If the iron magnetic circuit contains an air gap of sufficient length, the magnetizing force consumed in the iron, below magnetic saturation, is small compared with that consumed in the air gap, and the magnetic flux, therefore, is proportional to the current up to the values where magnetic saturation begins. Below saturation values of current, the transient thus is the simple exponential discussed before.

If the magnetic circuit is closed entirely by iron, the magnetic flux is not proportional to the current, and the inductance thus not constant, but varies over the entire range of currents, following the permeability curve of the iron. Furthermore, the transient due to a decrease of the stored magnetic energy differs in shape and in value from that due to an increase of magnetic energy, since the rising and decreasing magnetization curves differ, as shown by the hysteresis cycle.

Since no satisfactory mathematical expression has yet been found for the cyclic curve of hysteresis, a mathematical calculation is not feasible, but the transient has to be calculated by an approximate step-by-step method, as illustrated for the starting transient of an alternating-current transformer in Part III, Chapter XII, pp. 451 ff. Such methods are very cumbersome and applicable only to numerical instances.

An approximate calculation, giving an idea of the shape of the transient of the ironclad magnetic circuit, can be made by neglecting the difference between the rising and decreasing magnetic characteristic, and using the approximation of the magnetic characteristic given by Fröhlich's formula:

$$\mathcal{B} = \frac{\mathcal{H}}{\alpha + \sigma\mathcal{H}}, \tag{1}$$

which is usually represented in the form given by Kennelly:

$$\rho = \frac{\mathcal{H}}{\mathcal{B}} = \alpha + \sigma\mathcal{H}; \tag{2}$$

that is, the reluctivity is a linear function of the field intensity. It gives a fair approximation for higher magnetic densities.

This formula is based on the fairly rational assumption that the permeability of the iron is proportional to its remaining magnetizability. That is, the magnetic-flux density $\mathcal{B}$ consists of a component $\mathcal{H}$, the field intensity, which is the flux density in space, and a component $\mathcal{B}' = \mathcal{B} - \mathcal{H}$, which is the additional flux density carried by the iron. $\mathcal{B}'$ is frequently called the "metallic-flux density." With increasing $\mathcal{H}$, $\mathcal{B}'$ reaches a finite limiting value, which in iron is about

$$\mathcal{B}_\infty' = 20{,}000 \text{ lines per cm}^2.\ *$$

At any density $\mathcal{B}'$, the remaining magnetizability then is $\mathcal{B}_\infty' - \mathcal{B}'$, and, assuming the (metallic) permeability as proportional hereto, gives

$$\mu = c(\mathcal{B}_\infty' - \mathcal{B}'),$$

and, substituting

$$\mu = \frac{\mathcal{B}'}{\mathcal{H}'},$$

gives

$$\mathcal{B}' = \frac{c\mathcal{B}_\infty'\mathcal{H}'}{1 + c\mathcal{H}'},$$

* See "On the Law of Hysteresis," Part II, A.I.E.E. Transactions, 1892, page 621.

or, substituting

$$\frac{1}{c\mathfrak{B}_\infty'} = \alpha, \ \frac{1}{\mathfrak{B}_\infty'} = \sigma,$$

gives equation (1).

For $\mathfrak{H} = 0$ in equation (1), $\frac{\mathfrak{B}}{\mathfrak{H}} = \frac{1}{\alpha}$; for $\mathfrak{H} = \infty$, $\mathfrak{B} = \frac{1}{\sigma}$; that is, in equation (1), $\frac{1}{\alpha}$ = initial permeability, $\frac{1}{\sigma}$ = saturation value of magnetic density.

If the magnetic circuit contains an air gap, the reluctance of the iron part is given by equation (2), that of the air part is constant, and the total reluctance thus is

$$\rho = \beta + \sigma\mathfrak{H},$$

where $\beta = \alpha$ plus the reluctance of the air gap. Equation (1), therefore, remains applicable, except that the value of α is increased.

In addition to the metallic flux given by equation (1), a greater or smaller part of the flux always passes through the air or through space in general, and then has constant permeance, that is, is given by

$$\mathfrak{B} = c\mathfrak{H}.$$

23. In general, the flux in an ironclad magnetic circuit can, therefore, be represented as function of the current by an expression of the form

$$\Phi = \frac{ai}{1 + bi} + ci, \tag{3}$$

where $\frac{ai}{1 + bi} = \Phi'$ is that part of the flux which passes through the iron and whatever air space may be in series with the iron, and ci is the part of the flux passing through nonmagnetic material.

Denoting now

$$\left. \begin{aligned} L_1 &= na\,10^{-8}, \\ L_2 &= nc\,10^{-8}, \end{aligned} \right\} \tag{4}$$

where n = number of turns of the electric circuit, which is interlinked with the magnetic circuit, L_2 is the inductance of the air part of the magnetic circuit, L_1 the (virtual) initial inductance, that is, inductance at very small currents, of the iron part of the magnetic circuit, and $\frac{a}{b}$ the saturation value of the flux in the iron. That is, for $i = 0, \frac{n\Phi'}{i} = L_1$; and for $i = \infty, \Phi' = \frac{a}{b}$.

If r = resistance, the duration of the component of the transient resulting from the air flux would be

$$T_2 = \frac{L_2}{r} = \frac{nc\,10^{-8}}{r}, \tag{5}$$

and the duration of the transient which would result from the initial inductance of the iron flux would be

$$T_1 = \frac{L_1}{r} = \frac{na\,10^{-8}}{r}. \tag{6}$$

The differential equation of the transient is: induced voltage plus resistance drop equal zero; that is,

$$n\frac{d\Phi}{dt}10^{-8} + ri = 0.$$

Substituting (3) and differentiating gives

$$\frac{na\,10^{-8}}{(1+bi)^2}\frac{di}{dt} + nc\,10^{-8}\frac{di}{dt} + ri = 0,$$

and, substituting (5) and (6),

$$\left\{\frac{T_1}{(1+bi)^2} + T_2\right\}\frac{di}{dt} + i = 0;$$

hence, separating the variables,

$$\frac{T_1\,di}{i(1+bi)^2} + \frac{T_2\,di}{i} + dt = 0. \tag{7}$$

The first term is integrated by resolving into partial fractions:

$$\frac{1}{i(1+bi)^2} = \frac{1}{i} - \frac{b}{1+bi} - \frac{b}{(1+bi)^2},$$

and the integration of differential equation (7) then gives

$$T_1\log\frac{i}{1+bi} + T_2\log i + \frac{T_1}{1+bi} + t + C = 0. \tag{8}$$

If then, for the time $t = t_0$, the current is $i = i_0$, these values substituted in (8) give the integration constant C:

$$T_1\log\frac{i_0}{1+bi_0} + T_2\log i_0 + \frac{T_1}{1+bi_0} + t_0 + C = 0, \tag{9}$$

and, subtracting (8) from (9), gives

$$t - t_0 = T_1 \log \frac{i_0 (1 + bi)}{i (1 + bi_0)} + T_2 \log \frac{i_0}{i} + T_1 \left\{ \frac{1}{1 + bi_0} - \frac{1}{1 + bi} \right\}. \quad (10)$$

This equation is so complex in i that it is not possible to calculate from the different values of t the corresponding values of i; but inversely, for different values of i the corresponding values of t can be calculated, and the corresponding values of i and t, derived in this manner, can be plotted as a curve, which gives the single-energy transient of the ironclad magnetic circuit.

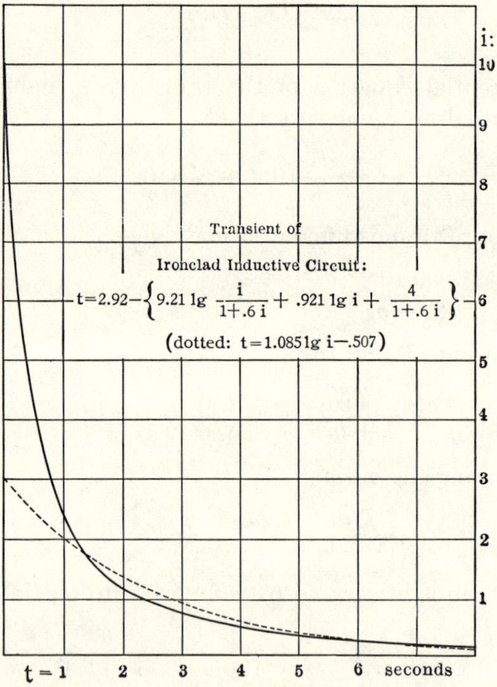

Fig. 29.

Such is done in Fig. 29, for the values of the constants:

$$r = .3,$$
$$a = 4 \times 10^5,$$
$$c = 4 \times 10^4,$$
$$b = .6,$$
$$n = 300.$$

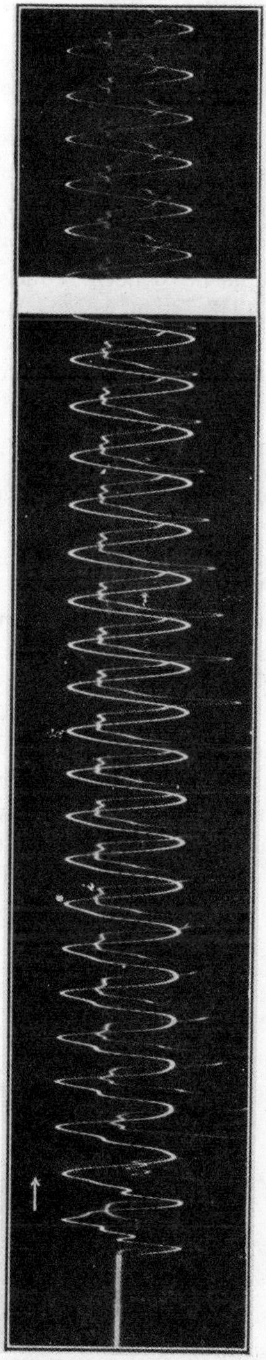

20 Cycles Later.

Fig. 30. — CD10075. — Transient Starting Current of Alternating-current Transformer.

This gives

$$T_1 = 4,$$
$$T_2 = .4.$$

Assuming $i_0 = 10$ amperes for $t_0 = 0$, gives from (10) the equation:

$$T = 2.92 - \left\{ 9.21 \, log^{10} \frac{i}{1 + .6\,i} + .921 \, log^{10} \, i + \frac{4}{1 + .6\,i} \right\}.$$

Herein, the logarithms have been reduced to the base 10 by division with $log^{10}\epsilon = .4343$.

For comparison is shown, in dotted line, in Fig. 29, the transient of a circuit containing no iron, and of such constants as to give about the same duration:

$$t = 1.085 \, log^{10} \, i - .507.$$

As seen, in the ironclad transient the current curve is very much steeper in the range of high currents, where magnetic saturation is reached, but the current change is slower in the range of medium magnetic densities.

Thus, in ironclad transients very high-current values of short duration may occur, and such transients, as those of the starting current of alternating-current transformers, may therefore be of serious importance by their excessive current values.

An oscillogram of the voltage and current waves in an 11,000-kw. high-voltage 60-cycle three-phase transformer, when switching onto the generating station near the most unfavorable point of the wave, is reproduced in Fig. 30. As seen, an excessive current rush persists for a number of cycles, causing a distortion of the voltage wave, and the current waves remain unsymmetrical for many cycles.

CHAPTER VI

DOUBLE-ENERGY TRANSIENTS

24. In a circuit in which energy can be stored in one form only, the change in the stored energy which can take place as the result of a change of the circuit conditions is an increase or decrease. The transient can be separated from the permanent condition, and then always is the representation of a gradual decrease of energy. Even if the stored energy after the change of circuit conditions is greater than before, and during the transition period an increase of energy occurs, the representation still is by a decrease of the transient. This transient then is the difference between the energy storage in the permanent condition and the energy storage during the transition period.

If the law of proportionality between current, voltage, magnetic flux, etc., applies, the single-energy transient is a simple exponential function:

$$y = y_0 \epsilon^{-\frac{t}{T_0}}, \tag{1}$$

where

y_0 = initial value of the transient, and
T_0 = duration of the transient,

that is, the time which the transient voltage, current, etc., would last if maintained at its initial value.

The duration T_0 is the ratio of the energy-storage coefficient to the power-dissipation coefficient. Thus, if energy is stored by the current i, as magnetic field,

$$T_0 = \frac{L}{r}, \tag{2}$$

where L = inductance = coefficient of energy storage by the current, r = resistance = coefficient of power dissipation by the current.

If the energy is stored by the voltage e, as dielectric field, the duration of the transient would be

$$T_0' = \frac{C}{g}, \tag{3}$$

where C = capacity = coefficient of energy storage by the voltage, in the dielectric field, and g = conductance = coefficient of power consumption by the voltage, as leakage conductance by the voltage, corona, dielectric hysteresis, etc.

Thus the transient of the spontaneous discharge of a condenser would be represented by

$$e = e_0 \epsilon^{-\frac{g}{C}t}. \tag{4}$$

Similar single-energy transients may occur in other systems. For instance, the transient by which a water jet approaches constant velocity when falling under gravitation through a resisting medium would have the duration

$$T = \frac{v_0}{g}, \tag{5}$$

where v_0 = limiting velocity, g = acceleration of gravity, and would be given by

$$v = v_0 \left(1 - \epsilon^{-\frac{t}{T}}\right). \tag{6}$$

In a system in which energy can be stored in two different forms, as for instance as magnetic and as dielectric energy in a circuit containing inductance and capacity, in addition to the gradual decrease of stored energy similar to that represented by the single-energy transient, a transfer of energy can occur between its two different forms.

Thus, if i = transient current, e = transient voltage (that is, the difference between the respective currents and voltages existing in the circuit as result of the previous circuit condition, and the values which should exist as result of the change of circuit conditions), then the total stored energy is

$$\left.\begin{aligned} W &= \frac{Li^2}{2} + \frac{Ce^2}{2}, \\ &= W_m + W_d. \end{aligned}\right\} \tag{7}$$

While the total energy W decreases by dissipation, W_m may be converted into W_d, or inversely.

Such an energy transfer may be periodic, that is, magnetic energy may change to dielectric and then back again; or unidirectional, that is, magnetic energy may change to dielectric (or inversely, dielectric to magnetic), but never change back again; but the energy is dissipated before this. This latter case occurs when the dissipation of energy is very rapid, the resistance (or conductance) high, and therefore gives transients, which rarely are of industrial importance, as they are of short duration and of low power. It therefore is sufficient to consider the oscillating double-energy transient, that is, the case in which the energy changes periodically between its two forms, during its gradual dissipation.

This may be done by considering separately the periodic transfer, or pulsation of the energy between its two forms, and the gradual dissipation of energy.

A. Pulsation of energy.

25. The magnetic energy is a maximum at the moment when the dielectric energy is zero, and when all the energy, therefore, is magnetic; and the magnetic energy is then

$$\frac{Li_0^2}{2},$$

where i_0 = maximum value of transient current.

The dielectric energy is a maximum at the moment when the magnetic energy is zero, and all the energy therefore dielectric, and is then

$$\frac{Ce_0^2}{2},$$

where e_0 = maximum value of transient voltage.

As it is the same stored energy which alternately appears as magnetic and as dielectric energy, it obviously is

$$\frac{Li_0^2}{2} = \frac{Ce_0^2}{2}. \qquad (8)$$

This gives a relation between the maximum value of transient current and the maximum value of transient voltage:

$$\frac{e_0}{i_0} = \sqrt{\frac{L}{C}}. \qquad (9)$$

$\sqrt{\frac{L}{C}}$ therefore is of the nature of an impedance z_0, and is called the *natural impedance*, or the *surge impedance*, of the circuit; and its reciprocal, $\sqrt{\frac{C}{L}} = y_0$, is the *natural admittance*, or the *surge admittance*, of the circuit.

The maximum transient voltage can thus be calculated from the maximum transient current:

$$e_0 = i_0 \sqrt{\frac{L}{C}} = i_0 z_0, \tag{10}$$

and inversely,

$$i_0 = e_0 \sqrt{\frac{C}{L}} = e_0 y_0. \tag{11}$$

This relation is very important, as frequently in double-energy transients one of the quantities e_0 or i_0 is given, and it is important to determine the other.

For instance, if a line is short-circuited, and the short-circuit current i_0 suddenly broken, the maximum voltage which can be induced by the dissipation of the stored magnetic energy of the short-circuit current is $e_0 = i_0 z_0$.

If one conductor of an ungrounded cable system is grounded, the maximum momentary current which may flow to ground is $i_0 = e_0 y_0$, where e_0 = voltage between cable conductor and ground.

If lightning strikes a line, and the maximum voltage which it may produce on the line, as limited by the disruptive strength of the line insulation against momentary voltages, is e_0, the maximum discharge current in the line is limited to $i_0 = e_0 y_0$.

If L is high but C low, as in the high-potential winding of a high-voltage transformer (which winding can be considered as a circuit of distributed capacity, inductance, and resistance), z_0 is high and y_0 low. That is, a high transient voltage can produce only moderate transient currents, but even a small transient current produces high voltages. Thus reactances, and other reactive apparatus, as transformers, stop the passage of large oscillating currents, but do so by the production of high oscillating voltages.

Inversely, if L is low and C high, as in an underground cable, z_0 is low but y_0 high, and even moderate oscillating voltages produce large oscillating currents, but even large oscillating currents produce only moderate voltages. Thus underground cables are little liable to the production of high oscillating voltages. This is fortunate, as the dielectric strength of a cable is necessarily relatively much lower than that of a transmission line, due to the close proximity of the conductors in the former. A cable, therefore, when receiving the moderate or small oscillating currents which may originate in a transformer, gives only very low oscillating voltages, that is, acts as a short circuit for the transformer oscillation, and therefore protects the latter. Inversely, if the large oscillating current of a cable enters a reactive device, as a current transformer, it produces enormous voltages therein. Thus, cable oscillations are more liable to be destructive to the reactive apparatus, transformers, etc., connected with the cable, than to the cable itself.

A transmission line is intermediate in the values of z_0 and y_0 between the cable and the reactive apparatus, thus acting like a reactive apparatus to the former, like a cable toward the latter. Thus, the transformer is protected by the transmission line in oscillations originating in the transformer, but endangered by the transmission line in oscillations originating in the transmission line.

The simple consideration of the relative values of $z_0 = \sqrt{\frac{L}{C}}$ in the different parts of an electric system thus gives considerable information on the relative danger and protective action of the parts on each other, and shows the reason why some elements, as current transformers, are far more liable to destruction than others; but also shows that disruptive effects of transient voltages, observed in one apparatus, may not and very frequently do not originate in the damaged apparatus, but originate in another part of the system, in which they were relatively harmless, and become dangerous only when entering the former apparatus.

26. If there is a periodic transfer between magnetic and dielectric energy, the transient current i and the transient voltage e successively increase, decrease, and become zero.

The current thus may be represented by

$$i = i_0 \cos (\phi - \gamma), \tag{12}$$

where i_0 is the maximum value of current, discussed above, and

$$\phi = 2\pi ft, \tag{13}$$

where f = the frequency of this transfer (which is still undetermined), and γ the phase angle at the starting moment of the transient; that is,

$$i_1 = i_0 \cos \gamma = \text{initial transient current}. \tag{14}$$

As the current i is a maximum at the moment when the magnetic energy is a maximum and the dielectric energy zero, the voltage e

must be zero when the current is a maximum, and inversely; and if the current is represented by the cosine function, the voltage thus is represented by the sine function, that is,

$$e = e_0 \sin (\phi - \gamma), \tag{15}$$

where

$$e_1 = -e_0 \sin \gamma = \text{initial value of transient voltage.} \tag{16}$$

The frequency f is still unknown, but from the law of proportionality it follows that there must be a frequency, that is, the successive conversions between the two forms of energy must occur in equal time intervals, for this reason: If magnetic energy converts to dielectric and back again, at some moment the proportion between the two forms of energy must be the same again as at the starting moment, but both reduced in the same proportion by the power dissipation. From this moment on, the same cycle then must repeat with proportional, but proportionately lowered values.

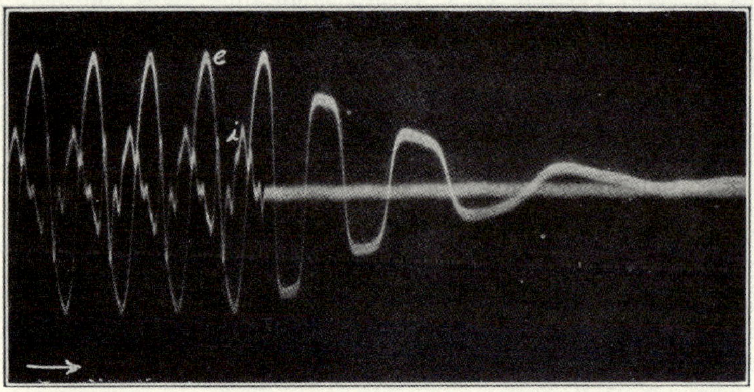

Fig. 31.—CD10017.—Oscillogram of Stationary Oscillation of Varying Frequency: Compound Circuit of Step-up Transformer and 28 Miles of 100,000-volt Transmission Line.

If, however, the law of proportionality does not exist, the oscillation may not be of constant frequency. Thus in Fig. 31 is shown an oscillogram of the voltage oscillation of the compound circuit consisting of 28 miles of 100,000-volt transmission line and the 2500-kw. high-potential step-up transformer winding, caused by switching transformer and 28-mile line by low-tension switches off a substation at the end of a 153-mile transmission line, at 88 kv. With decreasing voltage, the magnetic density in the transformer decreases, and as at lower magnetic densities the permeability of the iron is higher, with the decrease of voltage the permeability of the iron and thereby the inductance of the electric circuit interlinked with it increases, and, resulting from this increased magnetic energy storage coefficient L, there follows a slower period of oscillation, that is, a decrease of frequency, as seen on the oscillogram, from 55 cycles to 20 cycles per second.

If the energy transfer is not a simple sine wave, it can be represented by a series of sine waves, and in this case the above equations (12) and (15) would still apply, but the calculation of the frequency f would give a number of values which represent the different component sine waves.

The dielectric field of a condenser, or its "charge," is capacity times voltage: Ce. It is, however, the product of the current flowing into the condenser, and the time during which this current flows into it, that is, it equals $i\,t$.

Applying the law

$$Ce = it \tag{17}$$

to the oscillating energy transfer: the voltage at the condenser changes during a half-cycle from $-e_0$ to $+e_0$, and the condenser charge thus is

$$2\,e_0C;$$

the current has a maximum value i_0, thus an average value $\frac{2}{\pi}i_0$, and as it flows into the condenser during one-half cycle of the frequency f, that is, during the time $\frac{1}{2f}$, it is

$$2\,e_0C = \frac{2}{\pi}\,i_0\,\frac{1}{2f},$$

which is the expression of the condenser equation (17) applied to the oscillating energy transfer.

Transposed, this equation gives

$$f = \frac{i_0}{2\,\pi e_0 C}, \tag{18}$$

and substituting equation (10) into (18), and canceling with i_0, gives

$$f = \frac{1}{2\,\pi\,\sqrt{LC}} = \frac{1}{2\,\pi\sigma} \tag{19}$$

as the expression of the frequency of the oscillation, where

$$\sigma = \sqrt{LC} \tag{20}$$

is a convenient abbreviation of the square root.

The transfer of energy between magnetic and dielectric thus occurs with a definite frequency $f = \frac{1}{2\pi\sigma}$, and the oscillation thus is a sine wave without distortion, as long as the law of proportionality applies. When this fails, the wave may be distorted, as seen on the oscillogram Fig. 31.

The equations of the *periodic part of the transient* can now be written down by substituting (13), (19), (14), and (16) into (12) and (15):

$$\begin{aligned} i = i_0 \cos(\phi - \gamma) &= i_0 \cos\gamma \cos\phi + i_0 \sin\gamma \sin\phi \\ &= i_1 \cos\frac{t}{\sigma} - e_1 \frac{i_0}{e_0} \sin\frac{t}{\sigma}, \end{aligned}$$

and by (11):

$$i = i_1 \cos\frac{t}{\sigma} - y_0 e_1 \sin\frac{t}{\sigma}, \tag{21}$$

and in the same manner:

$$e = e_1 \cos\frac{t}{\sigma} + z_0 i_1 \sin\frac{t}{\sigma}, \tag{22}$$

where e_1 is the initial value of transient voltage, i_1 the initial value of transient current.

B. Power dissipation.

27. In Fig. 32 are plotted as A the periodic component of the oscillating current i, and as B the voltage e, as C the stored magnetic energy $\frac{Li^2}{2}$, and as D the stored dielectric energy $\frac{Ce^2}{2}$.

As seen, the stored magnetic energy pulsates, with double frequency, $2f$, between zero and a maximum, equal to the total stored energy. The average value of the stored magnetic energy thus is one-half of the total stored energy, and the dissipation of magnetic energy thus occurs at half the rate at which it would occur if all the energy were magnetic energy; that is, the transient resulting from the power dissipation of the magnetic energy lasts twice as long as it would if all the stored energy were magnetic, or in other words, if the transient were a single (magnetic) energy transient. In the latter case, the duration of the transient would be

$$T_0 = \frac{L}{r},$$

and with only half the energy magnetic, the duration thus is twice as long, or

$$T_1 = 2\,T_0 = \frac{2\,L}{r}, \tag{23}$$

and hereby the factor

$$h = \epsilon^{-\frac{t}{T_1}}$$

multiplies with the values of current and voltage (21) and (22).

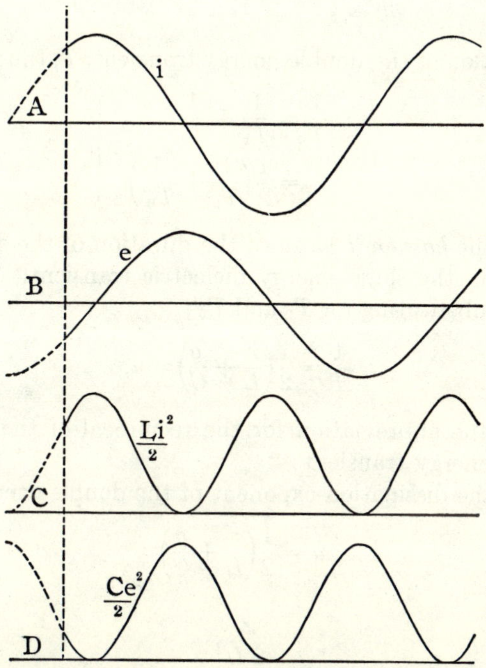

Fig. 32. — Relation of Magnetic and Dielectric Energy of Transient.

The same applies to the dielectric energy. If all the energy were dielectric, it would be dissipated by a transient of the duration:

$$T_0' = \frac{C}{g};$$

as only half the energy is dielectric, the dissipation is half as rapid, that is, the dielectric transient has the duration

$$T_2 = 2\,T_0' = \frac{2\,C}{g}, \tag{24}$$

and therefore adds the factor

$$k = \epsilon^{-\frac{t}{T_2}}$$

to the equations (21) and (22).

While these equations (21) and (22) constitute the periodic part of the phenomenon, the part which represents the dissipation of power is given by the factor

$$hk = \epsilon^{-\frac{t}{T_1}}\epsilon^{-\frac{t}{T_2}} = \epsilon^{-t\left(\frac{1}{T_1}+\frac{1}{T_2}\right)}. \tag{25}$$

The duration of the double-energy transient, T, thus is given by

$$\left.\begin{aligned} \frac{1}{T} &= \frac{1}{T_1} + \frac{1}{T_2}, \\ &= \frac{1}{2}\left(\frac{1}{T_0} + \frac{1}{T_0'}\right), \end{aligned}\right\} \tag{26}$$

and this is the *harmonic mean* of the duration of the single-energy magnetic and the single-energy dielectric transient.

It is, by substituting for T_0 and T_0',

$$\frac{1}{T} = \frac{1}{2}\left(\frac{r}{L} + \frac{g}{C}\right) = u, \tag{27}$$

where u is the abbreviation for the reciprocal of the duration of the double-energy transient.

Usually, the dissipation exponent of the double-energy transient

$$u = \frac{1}{2}\left(\frac{r}{L} + \frac{g}{C}\right)$$

is given as

$$\frac{r}{2\,L}.$$

This is correct only if $g = 0$, that is, the conductance, which represents the power dissipation resultant from the voltage (by leakage, dielectric induction and dielectric hysteresis, corona, etc.), is negligible. Such is the case in most power circuits and transmission lines, except at the highest voltages, where corona appears. It is not always the case in underground cables, high-potential transformers, etc., and is not the case in telegraph or telephone lines, etc. It is very nearly the case if the capacity is due to electrostatic condensers, but not if the capacity is that of electrolytic condensers, aluminum cells, etc.

Combining now the power-dissipation equation (25) as factor with the equations of periodic energy transfer, (21) and (22), gives the complete equations of the double-energy transient of the circuit containing inductance and capacity:

$$\left.\begin{aligned} i &= \epsilon^{-ut}\left\{i_1 \cos\frac{t}{\sigma} - y_0 e_1 \sin\frac{t}{\sigma}\right\}, \\ e &= \epsilon^{-ut}\left\{e_1 \cos\frac{t}{\sigma} + z_0 i_1 \sin\frac{t}{\sigma}\right\}, \end{aligned}\right\} \quad (28)$$

where

$$\left.\begin{aligned} z_0 &= \sqrt{\frac{L}{C}} = \frac{1}{y_0}, \\ u &= \frac{1}{2}\left\{\frac{r}{L} + \frac{g}{C}\right\}, \end{aligned}\right\} \quad (29)$$

$$\sigma = \sqrt{LC}, \quad (30)$$

and i_1 and e_1 are the initial values of the transient current and voltage respectively.

As instance are constructed, in Fig. 33, the transients of current and of voltage of a circuit having the constants:

Inductance,	$L = 1.25\, mh = 1.25 \times 10^{-3}$ henrys;
Capacity,	$C = 2\, mf = 2 \times 10^{-6}$ farads;
Resistance,	$r = 2.5$ ohms;
Conductance,	$g = 0.008$ mho,

in the case, that

The initial transient current,	$i_1 = 140$ amperes;
The initial transient voltage,	$e_1 = 2000$ volts.

It is, by the preceding equations:

$$\sigma = \sqrt{LC} = 5 \times 10^{-5},$$

$$f = \frac{1}{2\pi\sigma} = 3180 \text{ cycles per second},$$

$$z_0 = \sqrt{\frac{L}{C}} = 25 \text{ ohms},$$

$$y_0 = \sqrt{\frac{C}{L}} = 0.04 \text{ mho},$$

$$T_1 = \frac{2\,L}{r} = 0.001 \text{ sec.} = 1 \text{ millisecond},$$

$$T_2 = \frac{2\,C}{g} = 0.0005 \text{ sec.} = 0.5 \text{ millisecond},$$

$$T = \frac{1}{\frac{1}{T_1} + \frac{1}{T_2}} = 0.000333 \text{ sec.} = 0.33 \text{ millisecond};$$

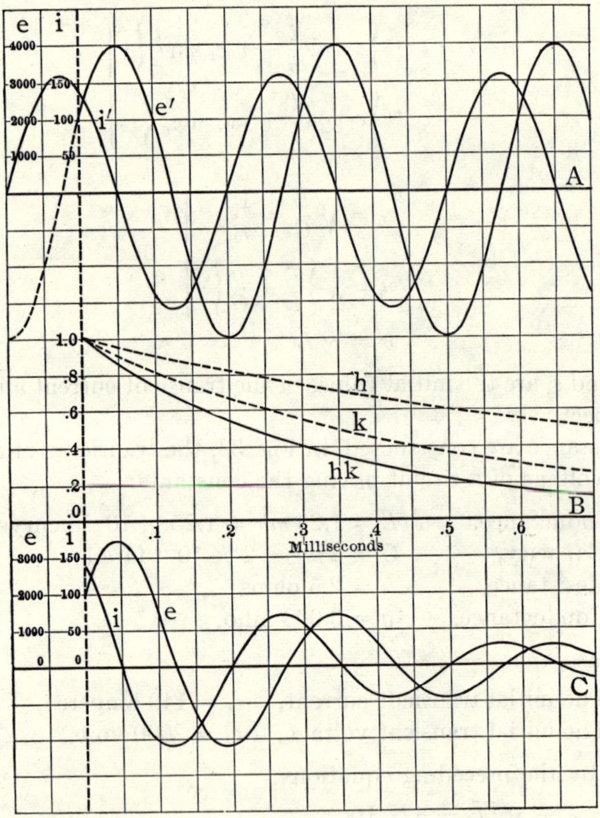

Fig. 33.

hence, substituted in equation (28),

$$\left.\begin{aligned} i &= \epsilon^{-3t}\{140 \cos 0.2\,t - 80 \sin 0.2\,t\}, \\ e &= \epsilon^{-3t}\{2000 \cos 0.2\,t + 3500 \sin 0.2\,t\}, \end{aligned}\right\}$$

where the time t is given in milliseconds.

Fig. 33A gives the periodic components of current and voltage:

$$i' = 140 \cos 0.2\,t - 80 \sin 0.2\,t,$$
$$e' = 2000 \cos 0.2\,t + 3500 \sin 0.2\,t.$$

Fig. 33B gives

The magnetic-energy transient, $h = \epsilon^{-t}$,
The dielectric-energy transient, $k = \epsilon^{-2t}$,
And the resultant transient, $hk = \epsilon^{-3t}$.

And Fig. 33C gives the transient current, $i = hki'$, and the transient voltage, $e = hke'$.

CHAPTER VII

LINE OSCILLATIONS

28. In a circuit containing inductance and capacity, the transient consists of a periodic component, by which the stored energy surges between magnetic $\frac{Li^2}{2}$ and dielectric $\frac{Ce^2}{2}$, and a transient component, by which the total stored energy decreases.

Considering only the periodic component, the maximum value of magnetic energy must equal the maximum value of dielectric energy,

$$\frac{Li_0^2}{2} = \frac{Ce_0^2}{2}, \tag{1}$$

where i_0 = maximum value of transient current, e_0 = maximum value of transient voltage.

This gives the relation between e_0 and i_0,

$$\frac{e_0}{i_0} = \sqrt{\frac{L}{C}} = z_0 = \frac{1}{y_0}, \tag{2}$$

where z_0 is called the natural impedance or surge impedance, y_0 the natural or surge admittance of the circuit.

As the maximum of current must coincide with the zero of voltage, and inversely, if the one is represented by the cosine function, the other is the sine function; hence the periodic components of the transient are

$$\left.\begin{aligned} i_1 &= i_0 \cos(\phi - \gamma) \\ e_1 &= e_0 \sin(\phi - \gamma) \end{aligned}\right\}, \tag{3}$$

where

$$\phi = 2\pi f t, \tag{4}$$

and

$$f = \frac{1}{2\pi\sqrt{LC}} \tag{5}$$

is the frequency of oscillation.

The dissipative or "transient" component is

$$hk = \epsilon^{-ut}, \tag{6}$$

where

$$u = \frac{1}{2}\left(\frac{r}{L} + \frac{g}{C}\right); \tag{7}$$

hence the total expression of transient current and voltage is

$$\left.\begin{aligned} i &= i_0\epsilon^{-ut}\cos(\phi - \gamma) \\ e &= e_0\epsilon^{-ut}\sin(\phi - \gamma) \end{aligned}\right\}. \tag{8}$$

γ, e_0, and i_0 follow from the initial values e' and i' of the transient, at $t = 0$ or $\phi = 0$:

$$\left.\begin{aligned} i' &= i_0 \cos\gamma \\ e' &= -e_0 \sin\gamma \end{aligned}\right\}; \tag{9}$$

hence

$$\tan\gamma = -\frac{e'}{i'}\frac{i_0}{e_0} = -y_0\frac{e'}{i'}. \tag{10}$$

The preceding equations of the double-energy transient apply to the circuit in which capacity and inductance are massed, as, for instance, the discharge or charge of a condenser through an inductive circuit.

Obviously, no material difference can exist, whether the capacity and the inductance are separately massed, or whether they are intermixed, a piece of inductance and piece of capacity alternating, or uniformly distributed, as in the transmission line, cable, etc.

Thus, the same equations apply to any point of the transmission line.

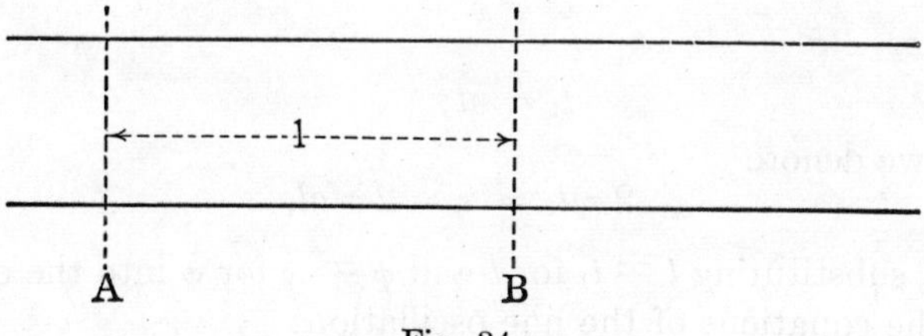

Fig. 34.

However, if (8) are the equations of current and voltage at a point A of a line, shown diagrammatically in Fig. 34, at any other point B, at distance l from the point A, the same equations will apply, but the phase angle γ, and the maximum values e_0 and i_0, may be different.

Thus, if

$$\left.\begin{aligned} i &= c_0\epsilon^{-ut}\cos(\phi - \gamma) \\ e &= z_0c_0\epsilon^{-ut}\sin(\phi - \gamma) \end{aligned}\right\} \tag{11}$$

are the current and voltage at the point A, this oscillation will appear at a point B, at distance l from A, at a moment of time later than at A by the time of propagation t_1 from A to B, if the oscillation is traveling from A to B; that is, in the equation (11), instead of t the time $(t - t_1)$ enters.

Or, if the oscillation travels from B to A, it is earlier at B by the time t_1; that is, instead of the time t, the value $(t + t_1)$ enters the equation (11). In general, the oscillation at A will appear at B, and the oscillation at B will appear at A, after the time t_1; that is, both expressions of (11), with $(t - t_1)$ and with $(t + t_1)$, will occur.

The general form of the line oscillation thus is given by substituting $(t \mp t_1)$ instead of t into the equations (11), where t_1 is the time of propagation over the distance l.

If v = velocity of propagation of the electric field, which in air, as with a transmission line, is approximately

$$v = 3 \times 10^{10}, \tag{12}$$

and in a medium of permeability μ and permittivity (specific capacity) κ is

$$v = \frac{3 \times 10^{10}}{\sqrt{\mu\kappa}}, \tag{13}$$

and we denote

$$a = \frac{1}{v}, \tag{14}$$

then

$$t_1 = al; \tag{15}$$

and if we denote

$$2\pi f t_1 = \omega = 2\pi f a l, \tag{16}$$

we get, substituting $t \mp t_1$ for t and $\phi \mp \omega$ for ϕ into the equation (11), the equations of the line oscillation:

$$\left.\begin{aligned} i &= c\epsilon^{-ut} \cos (\phi \mp \omega - \gamma) \\ e &= z_0 c\epsilon^{-ut} \sin (\phi \mp \omega - \gamma) \end{aligned}\right\}. \tag{17}$$

In these equations,

$$\phi = 2\pi f t$$

is the time angle, and (18)

$$\omega = 2\pi f a l$$

is the space angle, and $c = c_0\epsilon^{\pm ut_1}$ is the maximum value of current, $z_0 c$ the maximum value of voltage at the point l.

Resolving the trigonometric expressions of equation (17) into functions of single angles, we get as equations of current and of voltage products of the transient ϵ^{-ut}, and of a combination of the trigonometric expressions:

$$\left.\begin{array}{l}\cos\phi\cos\omega,\\ \sin\phi\cos\omega,\\ \cos\phi\sin\omega,\\ \sin\phi\sin\omega.\end{array}\right\}\qquad(19)$$

Line oscillations thus can be expressed in two different forms, either as functions of the sum and difference of time angle ϕ and distance angle ω: $(\phi \pm \omega)$, as in (17); or as products of functions of ϕ and functions of ω, as in (19). The latter expression usually is more convenient to introduce the terminal conditions in stationary waves, as oscillations and surges; the former is often more convenient to show the relation to traveling waves.

In Figs. 35 and 36 are shown oscillograms of such line oscillations. Fig. 35 gives the oscillation produced by switching 28 miles of 100-kv. line by high-tension switches onto a 2500-kw. step-up transformer in a substation at the end of a 153-mile three-phase line; Fig. 36 the oscillation of the same system caused by switching on the low-tension side of the step-up transformer.

29. As seen, the phase of current i and voltage e changes progressively along the line l, so that at some distance l_0 current and voltage are 360 degrees displaced from their values at the starting point, that is, are again in the same phase. This distance l_0 is called the wave length, and is the distance which the electric field travels during one period $t_0 = \frac{1}{f}$ of the frequency of oscillation.

As current and voltage vary in phase progressively along the line, the effect of inductance and of capacity, as represented by the inductance voltage and capacity current, varies progressively, and the resultant effect of inductance and capacity, that is, the effective inductance and the effective capacity of the circuit, thus are not the sum of the inductances and capacities of all the line elements, but the resultant of the inductances and capacities of all the line elements combined in all phases. That is, the effective inductance and capacity are derived by multiplying the total inductance and total capacity by avg/cos/, that is, by $\frac{2}{\pi}$.

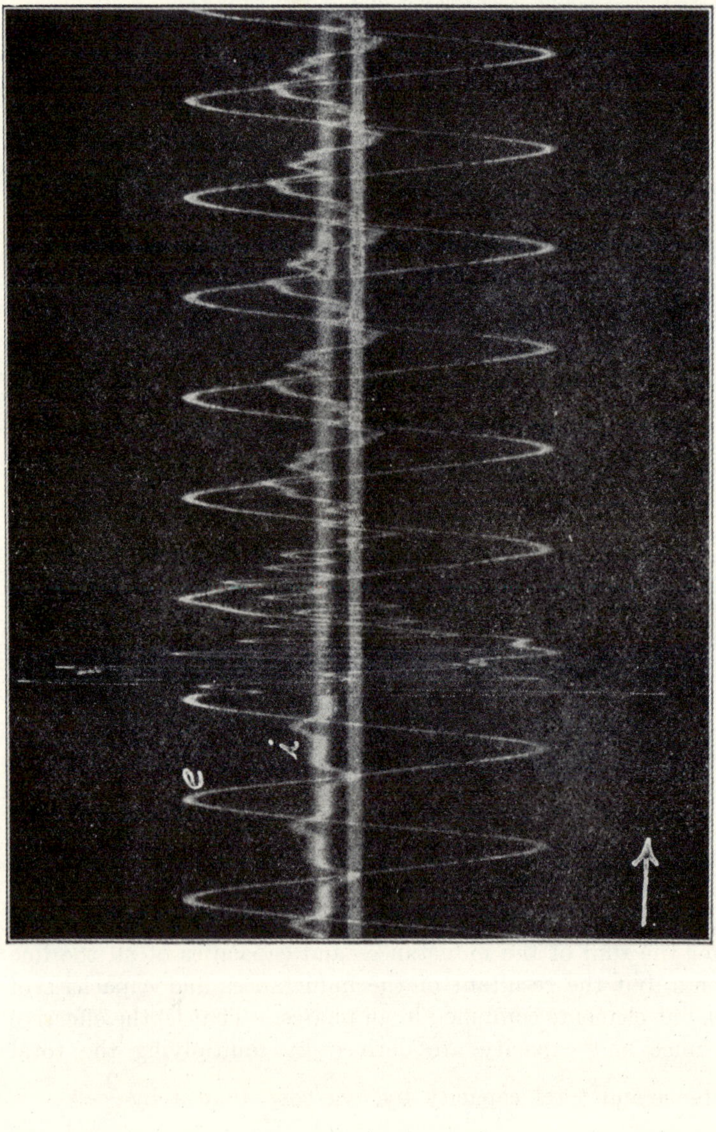

Fig. 35.—CD10116.—Oscillogram of Starting Oscillation of 28 Miles of 100,000-volt Transmission Line: High-tension Switching.

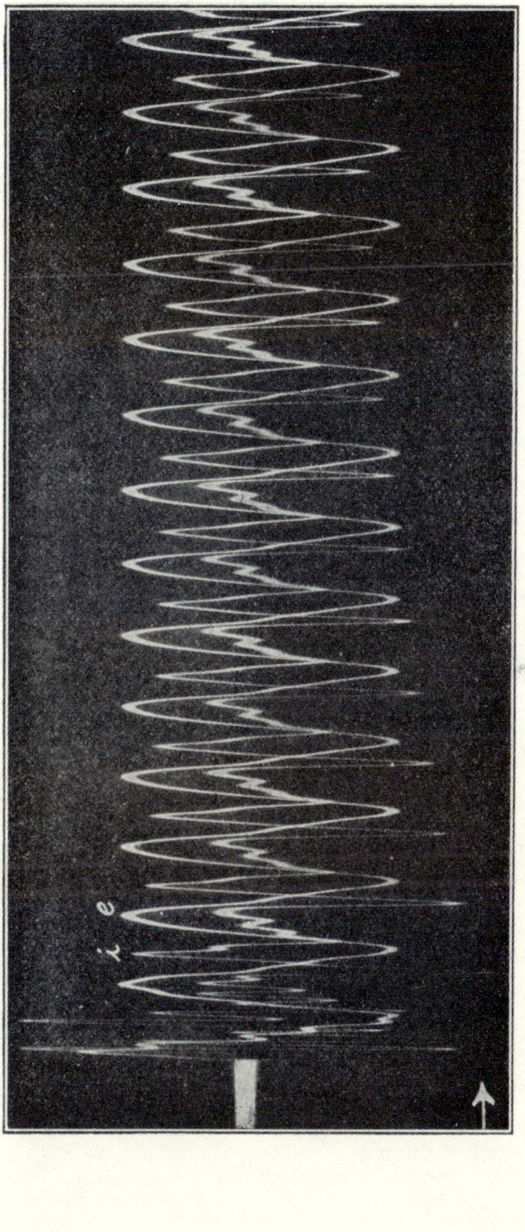

Fig. 36.—CD10002.—Oscillogram of Starting Oscillation of 28 Miles of 100,000-volt Transmission Line: Low-tension Switching.

Instead of L and C, thus enter into the equation of the double-energy oscillation of the line the values $\frac{2L}{\pi}$ and $\frac{2C}{\pi}$.

In the same manner, instead of the total resistance r and the total conductance g, the values $\frac{2r}{\pi}$ and $\frac{2g}{\pi}$ appear.

The values of z_0, y_0, u, ϕ, and ω are not changed hereby.

The frequency f, however, changes from the value corresponding to the circuit of massed capacity, $f = \frac{1}{2\pi\sqrt{LC}}$, to the value

$$f = \frac{1}{4\sqrt{LC}}.$$

Thus the frequency of oscillation of a transmission line is

$$f = \frac{1}{4\sqrt{LC}} = \frac{1}{4\sigma}, \tag{20}$$

where

$$\sigma = \sqrt{LC}. \tag{21}$$

If l_1 is the length of the line, or of that piece of the line over which the oscillation extends, and we denote by

$$L_0, C_0, r_0, g_0 \tag{22}$$

the inductance, capacity, resistance, and conductance per unit length of line, then

$$u = \frac{1}{2}\left(\frac{r_0}{L_0} + \frac{g_0}{C_0}\right); \tag{23}$$

that is, the rate of decrease of the transient is independent of the length of the line, and merely depends on the line constants per unit length.

It then is

$$\sigma = l_1\sigma_0, \tag{24}$$

where

$$\sigma_0 = \sqrt{L_0C_0} \tag{25}$$

is a constant of the line construction, but independent of the length of the line.

The frequency then is

$$f = \frac{1}{4\,l_1\sigma_0}. \tag{26}$$

The frequency f depends upon the length l_1 of the section of line in which the oscillation occurs. That is, the oscillations occurring in a transmission line or other circuit of distributed capacity have no definite frequency, but any frequency may occur, depending on the length of the circuit section which oscillates (provided that this circuit section is short compared with the entire length of the circuit, that is, the frequency high compared with the frequency which the oscillation would have if the entire line oscillates as a whole).

If l_1 is the oscillating line section, the wave length of this oscillation is four times the length

$$l_0 = 4\,l_1. \tag{27}$$

This can be seen as follows:

At any point l of the oscillating line section l_1, the effective power

$$p_0 = \text{avg } ei = 0 \tag{28}$$

is always zero, since voltage and current are 90 degrees apart.

The instantaneous power

$$p = ei, \tag{29}$$

however, is not zero, but alternately equal amounts of energy flow first one way, then the other way.

Across the ends of the oscillating section, however, no energy can flow, otherwise the oscillation would not be limited to this section. Thus at the two ends of the section, the instantaneous power, and thus either e or i, must continuously be zero.

Three cases thus are possible:

1. $e = 0$ at both ends of l_1;
2. $i = 0$ at both ends of l_1;
3. $e = 0$ at one end, $i = 0$ at the other end of l_1.

In the third case, $i = 0$ at one end, $e = 0$ at the other end of the line section l_1, the potential and current distribution in the line section l_1 must be as shown in Fig. 37, A, B, C, etc. That is, l_1 must be a quarter-wave or an odd multiple thereof.

If l_1 is a three-quarters wave, in Fig. 37B, at the two points C and D the power is also zero, that is, l_1 consists of three separate and independent oscillating sections, each of the length $\frac{l_1}{3}$; that is, the unit of oscillation is $\frac{l_1}{3}$, or also a quarter-wave. The same is the case in Fig. 37*C*, etc.

In the case 2, $i = 0$ at both ends of the line, the current and voltage distribution are as sketched in Fig. 38, *A*, *B*, *C*, etc.

That is, in *A*, the section l_1 is a half-wave, but the middle, *C*, of l_1 is a node or point of zero power, and the oscillating unit again is a quarter-wave. In the same way, in Fig. 38*B*, the section l_1 consists of 4 quarter-wave units, etc.

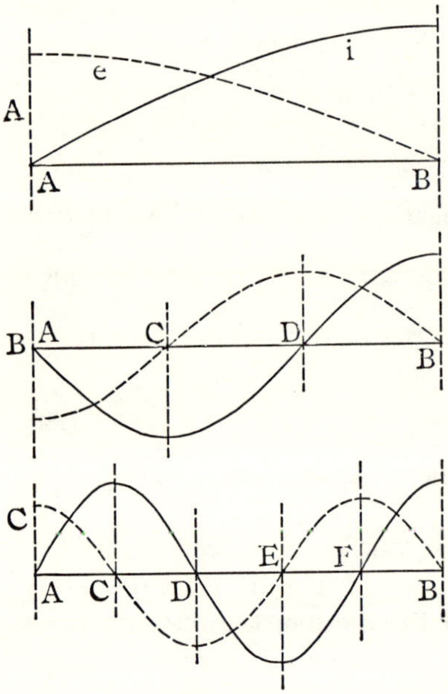

Fig. 37.

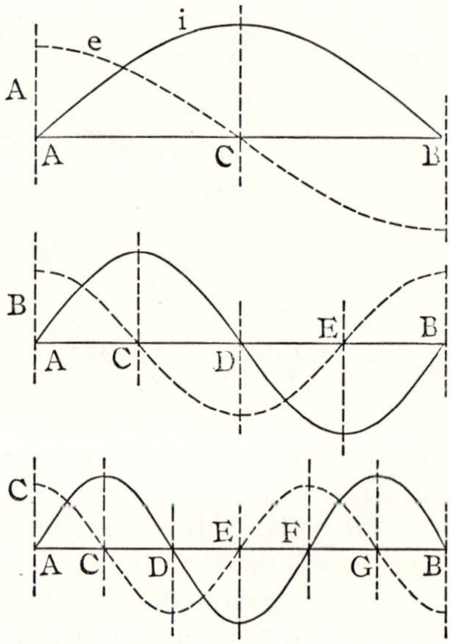

Fig. 38.

The same applies to case 1, and it thus follows that the wave length l_0 is four times the length of the oscillation l_1.

30. Substituting $l_0 = 4\, l_1$ into (26) gives as the frequency of oscillation

$$f = \frac{1}{l_0 \sigma_0}. \tag{30}$$

However, if f = frequency, and $v = \frac{1}{a}$, velocity of propagation, the wave length l_0 is the distance traveled during one period:

$$t_0 = \frac{1}{f} = \text{period}, \tag{31}$$

thus is

$$l_0 = vt_0 = \frac{1}{af}, \tag{32}$$

and, substituting (32) into (31), gives

$$a = \sigma_0, \tag{33}$$

or

$$v = \frac{1}{\sigma_0} = \frac{1}{\sqrt{L_0 C_0}}. \tag{34}$$

This gives a very important relation between inductance L_0 and capacity C_0 per unit length, and the velocity of propagation.

It allows the calculation of the capacity from the inductance,

$$C_0 = \frac{1}{v^2 L_0}, \tag{35}$$

and inversely. As in complex overhead structures the capacity usually is difficult to calculate, while the inductance is easily derived, equation (35) is useful in calculating the capacity by means of the inductance.

This equation (35) also allows the calculation of the mutual capacity, and thereby the static induction between circuits, from the mutual magnetic inductance.

The reverse equation,

$$L_0 = \frac{1}{v^2 C_0}, \tag{36}$$

is useful in calculating the inductance of cables from their measured capacity, and the velocity of propagation equation (13).

31. If l_1 is the length of a line, and its two ends are of different electrical character, as the one open, the other short-circuited, and thereby $i = 0$ at one end, $e = 0$ at the other end, the oscillation of this line is a quarter-wave or an odd multiple thereof.

The longest wave which may exist in this circuit has the wave length $l_0 = 4\,l_1$, and therefore the period $t_0 = \sigma_0 l_0 = 4\,\sigma_0 l_1$, that is, the frequency $f_0 = \frac{1}{4\,\sigma_0 l_1}$. This is called the *fundamental* wave of oscillation. In addition thereto, all its odd multiples can exist as higher harmonics, of the respective wave lengths $\frac{l_0}{2\,k - 1}$ and the frequencies $(2\,k - 1)f_0$, where $k = 1, 2, 3 \ldots$

If then ϕ denotes the time angle and ω the distance angle of the fundamental wave, that is, $\phi = 2\pi$ represents a complete cycle and $\omega = 2\pi$ a complete wave length of the fundamental wave, the time and distance angles of the higher harmonics are

$$3\phi,\ 3\omega,$$
$$5\phi,\ 5\omega,$$
$$7\phi,\ 7\omega,\ \text{etc.}$$

A complex oscillation, comprising waves of all possible frequencies, thus would have the form

$$a_1 \cos(\phi \mp \omega - \gamma_1) + a_3 \cos 3(\phi \mp \omega - \gamma_3) + a_5 \cos 5(\phi \mp \omega - \gamma_5) + \ldots, \tag{37}$$

and the length l_1 of the line then is represented by the angle $\omega = \frac{\pi}{2}$, and the oscillation called a *quarter-wave oscillation.*

If the two ends of the line l_1 have the same electrical characteristics, that is, $e = 0$ at both ends, or $i = 0$, the longest possible wave has the length $l_0 = 2\,l_1$, and the frequency

$$f_0 = \frac{1}{\sigma_0 l_0} = \frac{1}{2\,\sigma_0 l_1},$$

or any multiple (odd or even) thereof.

If then ϕ and ω again represent the time and the distance angles of the fundamental wave, its harmonics have the respective time and distance angles

$$2\phi,\ 2\omega,$$
$$3\phi,\ 3\omega,$$
$$4\phi,\ 4\omega,\ \text{etc.}$$

A complex oscillation then has the form

$$a_1 \cos(\phi \mp \omega - \gamma_1) + a_2 \cos 2(\phi \mp \omega - \gamma_2) + a_3 \cos 3(\phi \mp \omega - \gamma_3) + \ldots, \tag{38}$$

and the length l_1 of the line is represented by angle $\omega_1 = \pi$, and the oscillation is called a *half-wave oscillation.*

The half-wave oscillation thus contains even as well as odd harmonics, and thereby may have a wave shape, in which one half wave differs from the other.

Equations (37) and (38) are of the form of equation (17), but usually are more conveniently resolved into the form of equation (19).

At extremely high frequencies $(2k-1)f$, that is, for very large values of k, the successive harmonics are so close together that a very small variation of the line constants causes them to overlap, and as the line constants are not perfectly constant, but may vary slightly with the voltage, current, etc., it follows that at very high frequencies the line responds to any frequency, has no definite frequency of oscillation, but oscillations can exist of any frequency, provided this frequency is sufficiently high. Thus in long transmission lines, resonance phenomena can occur only with moderate frequencies, but not with frequencies of hundred thousands or millions of cycles.

32. The line constants r_0, g_0, L_0, C_0 are given per unit length, as per cm., mile, 1000 feet, etc.

The most convenient unit of length, when dealing with transients in circuits of distributed capacity, is the velocity unit v.

That is, choosing as unit of length the distance of propagation in unit time, or 3×10^{10} cm. in overhead circuits, this gives $v = 1$, and therefore

$$\left.\begin{aligned} \sigma_0 = \sqrt{L_0 C_0} = 1, \\ L_0 C_0 = 1, \end{aligned}\right\} \qquad (39)$$

or

$$C_0 = \frac{1}{L_0}; \; L_0 = \frac{1}{C_0}.$$

That is, the capacity per unit of length, in velocity measure, is inversely proportional to the inductance. In this velocity unit of length, distances will be represented by λ.

Using this unit of length, σ_0 disappears from the equations of the transient.

This velocity unit of length becomes specially useful if the transient extends over different circuit sections, of different constants and therefore different wave lengths, as for instance an overhead line, the underground cable, in which the wave length is about one-half what it is in the overhead line ($\kappa = 4$) and coiled windings, as the high-potential winding of a transformer, in which the wave length usually is relatively short. In the velocity measure of length, the wave length becomes the same throughout all these circuit sections, and the investigation is thereby greatly simplified.

Substituting $\sigma_0 = 1$ in equations (30) and (31) gives

$$\left.\begin{aligned} t_0 &= \lambda_0, \\ f &= \frac{1}{\lambda_0}, \\ \phi &= 2\,\pi f t = \frac{2\,\pi t}{\lambda_0}, \\ \omega &= 2\,\pi f \lambda = \frac{2\,\pi\lambda}{\lambda_0}, \end{aligned}\right\} \tag{40}$$

and the natural impedance of the line then becomes, in velocity measure,

$$z_0 = \sqrt{\frac{L_0}{C_0}} = L_0 = \frac{1}{C_0} = \frac{1}{y_0} = \frac{e_0}{i_0} \tag{41}$$

where e_0 = maximum voltage, i_0 = maximum current.

That is, the natural impedance is the inductance, and the natural admittance is the capacity, per velocity unit of length, and is the main characteristic constant of the line.

The equations of the current and voltage of the line oscillation then consist, by (19), of trigonometric terms

$$\begin{aligned} &\cos\phi\cos\omega, \\ &\sin\phi\cos\omega, \\ &\cos\phi\sin\omega, \\ &\sin\phi\sin\omega, \end{aligned}$$

multiplied with the transient, ϵ^{-ut}, and would thus, in the most general case, be given by an expression of the form

$$\left.\begin{aligned} i &= \epsilon^{-ut}\{a_1\cos\phi\cos\omega + b_1\sin\phi\cos\omega + c_1\cos\phi\sin\omega \\ &\qquad + d_1\sin\phi\sin\omega\}, \\ e &= \epsilon^{-ut}\{a_1'\cos\phi\cos\omega + b_1'\sin\phi\cos\omega + c_1'\cos\phi\sin\omega \\ &\qquad + d_1'\sin\phi\sin\omega\}, \end{aligned}\right\} \tag{42}$$

and its higher harmonics, that is, terms, with

$$\begin{aligned} &2\,\phi,\ 2\,\omega, \\ &3\,\phi,\ 3\,\omega, \\ &4\,\phi,\ 4\,\omega, \text{ etc.} \end{aligned}$$

In these equations (42), the coefficients a, b, c, d, a', b', c', d' are determined by the *terminal conditions* of the problem, that is, by the values of i and e at all points of the circuit ω, at the beginning of time, that is, for $\phi = 0$, and by the values of i and e at all times t (or ϕ respectively) at the ends of the circuit, that is, for $\omega = 0$ and $\omega = \frac{\pi}{2}$.

For instance, if:

(*a*) The circuit is open at one end $\omega = 0$, that is, the current is zero at all times at this end. That is,

$$i = 0 \text{ for } \omega = 0;$$

the equations of i then must not contain the terms with $\cos \omega$, $\cos 2\omega$, etc., as these would not be zero for $\omega = 0$. That is, it must be

$$\left.\begin{aligned} a_1 &= 0, \quad b_1 = 0, \\ a_2 &= 0, \quad b_2 = 0, \\ a_3 &= 0, \quad b_3 = 0, \text{ etc.} \end{aligned}\right\} \tag{43}$$

The equation of i contains only the terms with $\sin \omega$, $\sin 2\omega$, etc. Since, however, the voltage e is a maximum where the current i is zero, and inversely, at the point where the current is zero, the voltage must be a maximum; that is, the equations of the voltage must contain only the terms with $\cos \omega$, $\cos 2\omega$, etc. Thus it must be

$$\left.\begin{aligned} c_1' &= 0, \quad d_1' = 0, \\ c_2' &= 0, \quad d_2' = 0, \\ c_3' &= 0, \quad d_3' = 0, \text{ etc.} \end{aligned}\right\} \tag{44}$$

Substituting (43) and (44) into (42) gives

$$\left.\begin{aligned} i &= \epsilon^{-ut}\{c_1 \cos \phi + d_1 \sin \phi\} \sin \omega, \\ e &= \epsilon^{-ut}\{a_1' \cos \phi + b_1' \sin \phi\} \cos \omega \end{aligned}\right\} \tag{45}$$

and the higher harmonics hereof.

(*b*) If in addition to (*a*), the open circuit at one end $\omega = 0$, the line is short-circuited at the other end $\omega = \frac{\pi}{2}$, the voltage e must be zero at this latter end. Cos ω, $\cos 3\omega$, $\cos 5\omega$, etc., become zero for $\omega = \frac{\pi}{2}$, but $\cos 2\omega$, $\cos 4\omega$, etc., are not zero for $\omega = \frac{\pi}{2}$, and the latter functions thus cannot appear in the expression of e.

That is, the voltage e can contain no even harmonics. If, however, the voltage contains no even harmonics, the current produced by this voltage also can contain no even harmonics. That is, it must be

$$\left.\begin{array}{llll} c_2 = 0, & d_2 = 0, & a_2' = 0, & b_2' = 0, \\ c_4 = 0, & d_4 = 0, & a_4' = 0, & b_4' = 0, \\ c_6 = 0, & d_6 = 0, & a_6' = 0, & b_6' = 0, \text{ etc.} \end{array}\right\} \tag{46}$$

The complete expression of the stationary oscillation in a circuit open at the end $\omega = 0$ and short-circuited at the end $\omega = \frac{\pi}{2}$ thus would be

$$\left.\begin{array}{l} i = \epsilon^{-ut} \{(c_1 \cos \phi + d_1 \sin \phi) \sin \omega + (c_3 \cos 3\,\phi + d_3 \sin 3\,\phi) \\ \qquad \sin 3\,\omega + \ldots \}, \\ e = \epsilon^{-ut} \{(a_1' \cos \phi + b_1' \sin \phi) \cos \omega + (a_3' \cos 3\,\phi + b_3' \sin 3\,\phi) \\ \qquad \cos 3\,\omega + \ldots \}. \end{array}\right\} \tag{47}$$

(*c*) Assuming now as instance that, in such a stationary oscillation as given by equation (47), the current in the circuit is zero at the starting moment of the transient for $\phi = 0$. Then the equation of the current can contain no terms with $\cos \phi$, as these would not be zero for $\phi = 0$.

That is, it must be

$$\left.\begin{array}{l} c_1 = 0, \\ c_3 = 0, \\ c_5 = 0, \text{ etc.} \end{array}\right\} \tag{48}$$

At the moment, however, when the current is zero, the voltage of the stationary oscillation must be a maximum. As $i = 0$ for $\phi = 0$, at this moment the voltage e must be a maximum, that is, the voltage wave can contain no terms with $\sin \phi$, $\sin 3\,\phi$, etc. This means

$$\left.\begin{array}{l} b_1' = 0, \\ b_3' = 0, \\ b_5' = 0, \text{ etc.} \end{array}\right\} \tag{49}$$

Substituting (48) and (49) into equation (47) gives

$$\left.\begin{array}{l} i = \epsilon^{-ut} \{d_1 \sin \phi \sin \omega + d_3 \sin 3\,\phi \sin 3\,\omega + d_5 \sin 5\,\phi \sin 5\,\omega \\ \qquad + \ldots \}, \\ e = \epsilon^{-ut} \{a_1' \cos \phi \cos \omega + a_3' \cos 3\,\phi \cos 3\,\omega + a_5' \cos 5\,\phi \cos 5\,\omega \\ \qquad + \ldots \}. \end{array}\right\} \tag{50}$$

In these equations (50), d and a' are the maximum values of current and of voltage respectively, of the different harmonic waves. Between the maximum values of current, i_0, and of voltage, e_0, of a stationary oscillation exists, however, the relation

$$\frac{e_0}{i_0} = z_0 \sqrt{\frac{L}{C}},$$

where z_0 is the natural impedance or surge impedance. That is

$$a' = dz_0, \tag{51}$$

and substituting (51) into (50) gives

$$\left.\begin{aligned} i &= \epsilon^{-ut} \{ d_1 \sin \phi \sin \omega + d_3 \sin 3 \phi \sin 3 \omega + d_5 \sin 5 \phi \sin 5 \omega \\ &\quad + \dots \}, \\ e &= z_0 \epsilon^{-ut} \{ d_1 \cos \phi \cos \omega + d_3 \cos 3 \phi \cos 3 \omega + d_5 \cos 5 \phi \cos 5 \omega \\ &\quad + \dots \}. \end{aligned}\right\} \tag{52}$$

(*d*) If then the distribution of voltage e along the circuit is given at the moment of start of the transient, for instance, the voltage is constant and equals e_1 throughout the entire circuit at the starting moment $\phi = 0$ of the transient, this gives the relation, by substituting in (52),

$$e_1 = z_0 \epsilon^{-ut} \{ d_1 \cos \omega + d_3 \cos 3 \omega + d_5 \cos 5 \omega + \dots \}, \tag{53}$$

for all values of ω.

Herefrom then calculate the values of d_1, d_3, d_5, etc., in the manner as discussed in *Engineering Mathematics*, Chapter III.*

* See chapter X, Philip Alger, *Mathematics for Science and Engineering*, (Based on *Engineering Mathematics* by C. P. Steinmetz), Second Edition; McGraw-Hill, 1969.

CHAPTER VIII

TRAVELING WAVES

33. In a stationary oscillation of a circuit having uniformly distributed capacity and inductance, that is, the transient of a circuit storing energy in the dielectric and magnetic field, current and voltage are given by the expression

$$\left.\begin{aligned} i &= i_0 \epsilon^{-ut} \cos (\phi \mp \omega - \gamma), \\ e &= e_0 \epsilon^{-ut} \sin (\phi \mp \omega - \gamma), \end{aligned}\right\} \tag{1}$$

where ϕ is the time angle, ω the distance angle, u the exponential decrement, or the "power-dissipation constant," and i_0 and e_0 the maximum current and voltage respectively.

The power flow at any point of the circuit, that is, at any distance angle ω, and at any time t, that is, time angle ϕ, then is

$$\begin{aligned} p &= ei, \\ &= e_0 i_0 \epsilon^{-2ut} \cos (\phi \mp \omega - \gamma) \sin (\phi \mp \omega - \gamma), \\ &= \frac{e_0 i_0}{2} \epsilon^{-2ut} \sin 2 (\phi \mp \omega - \gamma), \end{aligned} \tag{2}$$

and the average power flow is

$$\begin{aligned} P_0 &= \text{avg } p, \\ &= 0. \end{aligned} \tag{3}$$

Hence, in a stationary oscillation, or standing wave of a uniform circuit, the average flow of power, p_0, is zero, and no power flows along the circuit, but there is a surge of power, of double frequency. That is, power flows first one way, during one-quarter cycle, and then in the opposite direction, during the next quarter-cycle, etc.

Such a transient wave thus is analogous to the permanent wave of reactive power.

As in a stationary wave, current and voltage are in quadrature with each other, the question then arises, whether, and what physical meaning a wave has, in which current and voltage are in phase with each other:

$$\left.\begin{aligned} i &= i_0\epsilon^{-ut}\cos(\phi \mp \omega - \gamma), \\ e &= e_0\epsilon^{-ut}\cos(\phi \mp \omega - \gamma). \end{aligned}\right\} \tag{4}$$

In this case the flow of power is

$$\begin{aligned} p &= ei, \\ &= e_0 i_0 \epsilon^{-2ut}\cos^2(\phi \mp \omega - \gamma), \\ &= \frac{e_0 i_0}{2}\epsilon^{-2ut}[1 + \cos 2(\phi \mp \omega - \gamma)], \end{aligned} \tag{5}$$

and the average flow of power is

$$\begin{aligned} p_0 &= \text{avg } p, \\ &= \frac{e_0 i_0}{2}\epsilon^{-2ut}. \end{aligned} \tag{6}$$

Such a wave thus consists of a combination of a steady flow of power along the circuit, p_0, and a pulsation or surge, p_1, of the same nature as that of the standing wave (2):

$$p_1 = \frac{e_0 i_0}{2}\epsilon^{-2ut}\cos 2(\phi \mp \omega - \gamma). \tag{7}$$

Such a flow of power along the circuit is called a *traveling wave.* It occurs very frequently. For instance, it may be caused if by a lightning stroke, etc., a quantity of dielectric energy is impressed

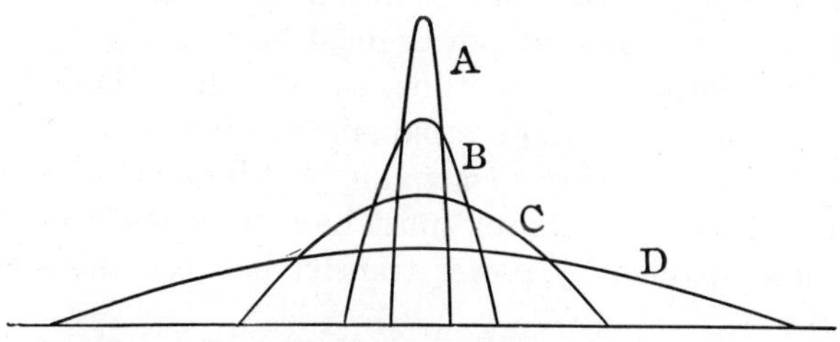

Fig. 39.—Starting of Impulse, or Traveling Wave.

upon a part of the circuit, as shown by curve A in Fig. 39, or if by a local short circuit a quantity of magnetic energy is impressed upon a part of the circuit. This energy then gradually distributes over the circuit, as indicated by the curves B, C, etc., of Fig. 39, that is, moves along the circuit, and the dissipation of the stored energy thus occurs by a flow of power along the circuit.

Such a flow of power must occur in a circuit containing sections of different dissipation constants u. For instance, if in a circuit consisting of an unloaded transformer and a transmission line, as indicated in Fig. 40, at no load on the step-down trans-

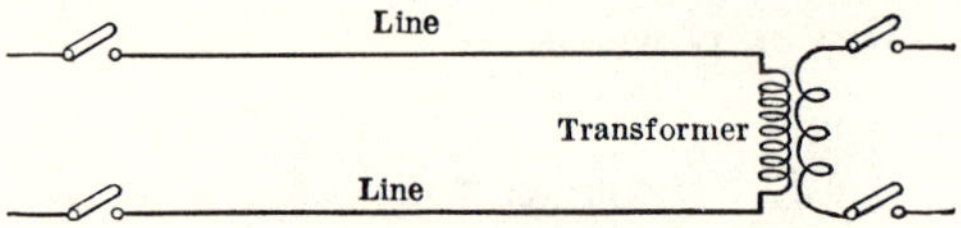

Fig. 40.

former, the high-tension switches are opened at the generator end of the transmission line. The energy stored magnetically and dielectrically in line and transformer then dissipates by a transient, as shown in the oscillogram Fig. 41. This gives the oscillation of a circuit consisting of 28 miles of line and 2500-kw. 100-kv. step-up and step-down transformers, and is produced by disconnecting this circuit by low-tension switches. In the transformer, the duration of the transient would be very great, possibly several seconds, as the stored magnetic energy (L) is very large, the dissipation of power (r and g) relatively small; in the line, the transient is of fairly short duration, as r (and g) are considerable. Left to themselves, the line oscillations thus would die out much more rapidly, by the dissipation of their stored energy, than the transformer oscillations. Since line and transformer are connected together, both must die down simultaneously by the same transient. It then follows that power must flow during the transient from the transformer into the line, so as to have both die down together, in spite of the more rapid energy dissipation in the line. Thus a transient in a *compound circuit*, that is, a circuit comprising sections of different constants, must be a traveling wave, that is, must be accompanied by power transfer between the sections of the circuit.*

A traveling wave, equation (4), would correspond to the case of effective power in a permanent alternating-current circuit, while the stationary wave of the uniform circuit corresponds to the case of reactive power.

Since one of the most important applications of the traveling wave is the investigation of the compound circuit, it is desirable

* In oscillogram Fig. 41, the current wave is shown reversed with regard to the voltage wave for greater clearness.

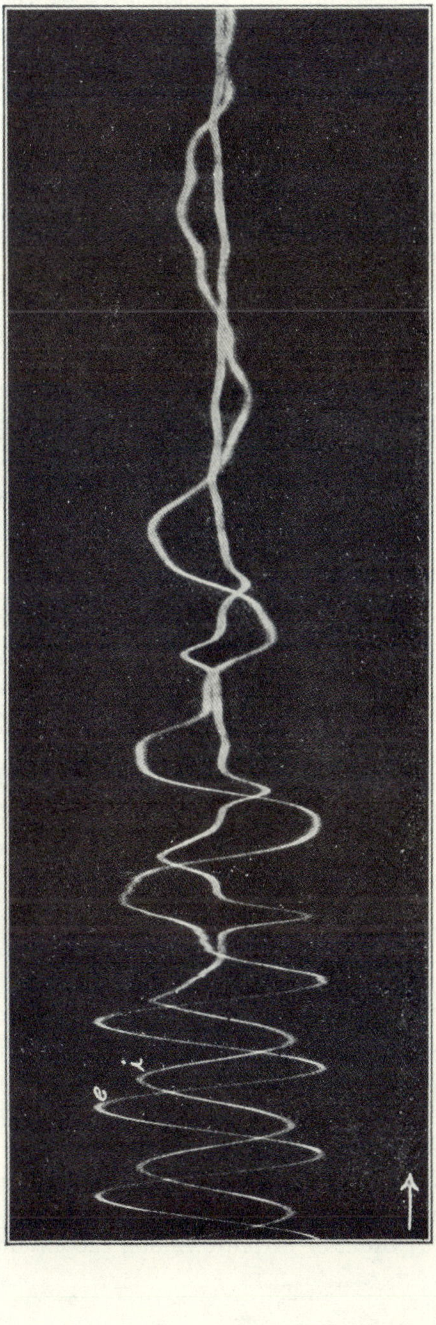

Fig. 41.—CD10045.—Oscillogram of Compound Circuit of 154 Miles 100,000-volt Transmission Line and High-tension Coils of 10,000-kw. Step-up Transformers, Switching off by Low-tension Switches. High-tension Current and Low-tension Voltage.

to introduce, when dealing with traveling waves, the velocity unit as unit of length, that is, measure the length with the distance of propagation during unit time (3×10^{10} cm. with a straight conductor in air) as unit of length. This allows the use of the same distance unit through all sections of the circuit, and expresses the wave length λ_0 and the period T_0 by the same numerical values, $\lambda_0 = T_0 = \frac{1}{f}$, and makes the time angle ϕ and the distance angle ω directly comparable:

$$\left.\begin{aligned} \phi &= 2\,\pi f t = 2\,\pi \frac{t}{\lambda_0}, \\ \omega &= 2\,\pi \frac{\lambda}{\lambda_0} = 2\,\pi f \lambda. \end{aligned}\right\} \qquad (8)$$

34. If power flows along the circuit, three cases may occur:

(*a*) The flow of power is uniform, that is, the power remains constant in the direction of propagation, as indicated by *A* in Fig. 42.

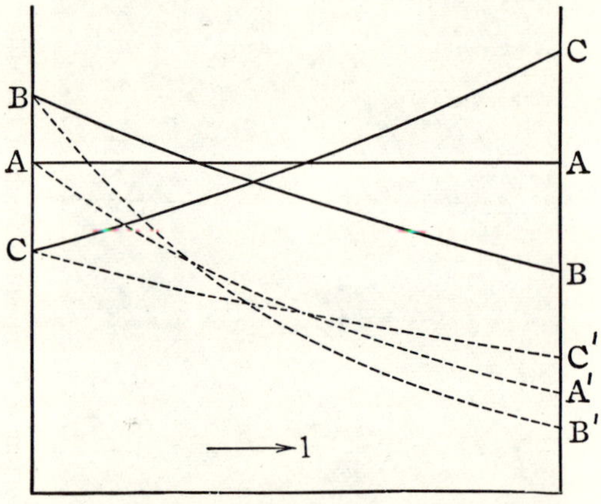

Fig. 42.—Energy Transfer by Traveling Wave.

(*b*) The flow of power is decreasing in the direction of propagation, as illustrated by *B* in Fig. 42.

(*c*) The flow of power is increasing in the direction of propagation, as illustrated by *C* in Fig. 42.

Obviously, in all three cases the flow of power decreases, due to the energy dissipation by *r* and *g*, that is, by the decrement ϵ^{-ut}. Thus, in case (*a*) the flow of power along the circuit decreases at the rate ϵ^{-ut}, corresponding to the dissipation of the stored energy by ϵ^{-ut}, as indicated by A' in Fig. 42; while in the case (b) the power flow decreases faster, in case (c) slower, than corresponds to the energy dissipation, and is illustrated by B' and C' in Fig. 42.

(a) If the flow of power is constant in the direction of propagation, the equation would be

$$\begin{aligned} i &= i_0\epsilon^{-ut}\cos(\phi - \omega - \gamma), \\ e &= e_0\epsilon^{-ut}\cos(\phi - \omega - \gamma), \\ p_0 &= \frac{e_0 i_0}{2}\epsilon^{-2ut}. \end{aligned} \tag{9}$$

In this case there must be a continuous power supply at the one end, and power abstraction at the other end, of the circuit or circuit section in which the flow of power is constant. This could occur approximately only in special cases, as in a circuit section of medium rate of power dissipation, u, connected between a section of low- and a section of high-power dissipation. For instance, if as illustrated in Fig. 43 we have a transmission line

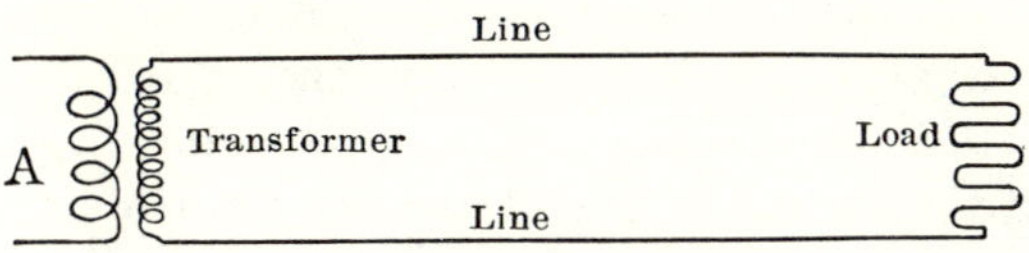

Fig. 43.—Compound Circuit.

connecting the step-up transformer with a load on the step-down end, and the step-up transformer is disconnected from the generating system, leaving the system of step-up transformer, line, and load to die down together in a stationary oscillation of a compound circuit, the rate of power dissipation in the transformer then is much lower, and that in the load may be greater, than the average rate of power dissipation of the system, and the transformer will supply power to the rest of the oscillating system, the load receive power. If then the rate of power dissipation of the line u should happen to be exactly the average, u_0, of the entire system, power would flow from the transformer over the line into the load, but in the line the flow of power would be uniform, as the line neither receives energy from nor gives off energy to the rest of the system, but its stored energy corresponds to its rate of power dissipation.

(*b*) If the flow of power decreases along the line, every line element receives more power at one end than it gives off at the other end. That is, energy is supplied to the line elements by the flow of power, and the stored energy of the line element thus decreases at a slower rate than corresponds to its power dissipation by r and g. Or, in other words, a part of the power dissipated in the line element is supplied by the flow of power along the line, and only a part supplied by the stored energy.

Since the current and voltage would decrease by the term ϵ^{-ut}, if the line element had only its own stored energy available, when receiving energy from the power flow the decrease of current and voltage would be slower, that is, by a term

$$\epsilon^{-(u-s)t}; \tag{10}$$

hence the exponential decrement u is decreased to $(u - s)$, and s then is the exponential coefficient corresponding to the energy supply by the flow of power.

Thus, while u is called the *dissipation constant* of the circuit, s may be called the *power-transfer constant* of the circuit.

Inversely, however, in its propagation along the circuit, λ, such a traveling wave must decrease in intensity more rapidly than corresponds to its power dissipation, by the same factor by which it increased the energy supply of the line elements over which it passed. That is, as function of the distance, the factor $\epsilon^{-s\lambda}$ must enter.* In other words, such a traveling wave, in passing along the line, leaves energy behind in the line elements, at the rate ϵ^{+st}, and therefore decreases faster in the direction of progress by $\epsilon^{-s\lambda}$. That is, it scatters a part of its energy along its path of travel, and thus dies down more rapidly with the distance of travel.

Thus, in a traveling wave of decreasing power flow, the time decrement is changed to $\epsilon^{-(u-s)t}$, and the distance decrement $\epsilon^{+s\lambda}$ added, and the equation of a traveling wave of decreasing power flow thus is

$$\left.\begin{aligned} i &= i_0\epsilon^{-(u-s)t}\epsilon^{-s\lambda}\cos(\phi-\omega-\gamma) = i_0\epsilon^{-ut}\epsilon^{+s(t-\lambda)}\cos(\phi-\omega-\gamma), \\ e &= e_0\epsilon^{-(u-s)t}\epsilon^{-s\lambda}\cos(\phi-\omega-\gamma) = e_0\epsilon^{-ut}\epsilon^{+s(t-\lambda)}\cos(\phi-\omega-\gamma); \end{aligned}\right\} \tag{11}$$

* Due to the use of the velocity unit of length λ, distance and time are given the same units, $t_0 = \lambda_0$; and the time decrement, ϵ^{+st}, and the distance decrement, $\epsilon^{-s\lambda}$, give the same coefficient s in the exponent. Otherwise, the velocity of propagation would enter as factor in the exponent.

the average power then is

$$p_0 = \text{avg } ei,$$
$$= \frac{e_0 i_0}{2} \epsilon^{-2(u-s)t} \epsilon^{-2s\lambda} = \frac{e_0 i_0}{2} \epsilon^{-2ut} \epsilon^{+2s(t-\lambda)}. \tag{12}$$

Both forms of the expressions of i, e, and p_0 of equations (11) and (12) are of use. The first form shows that the wave decreases slower with the time t, but decreases with the distance λ. The second form shows that the distance λ enters the equation only in the form $t - \lambda$ and $\phi - \omega$ respectively, and that thus for a constant value of $t - \lambda$ the decrement is ϵ^{-2ut}, that is, in the direction of propagation the energy dies out by the power dissipation constant u.

Equations (10) to (12) apply to the case, when the direction of propagation, that is, of wave travel, is toward increasing λ. For a wave traveling in opposite direction, the sign of λ and thus of ω is reversed.

(*c*) If the flow of power increases along the line, more power leaves every line element than enters it; that is, the line element is drained of its stored energy by the passage of the wave, and thus the transient dies down with the time at a greater rate than corresponds to the power dissipation by r and g. That is, not all the stored energy of the line elements supplies the power which is being dissipated in the line element, but a part of the energy leaves the line element in increasing the power which flows along the line. The rate of dissipation thus is increased, and instead of u, $(u + s)$ enters the equation. That is, the exponential time decrement is

$$\epsilon^{-(u+s)t}, \tag{13}$$

but inversely, along the line λ the power flow increases, that is, the intensity of the wave increases, by the same factor $\epsilon^{+s\lambda}$, or rather, the wave decreases along the line at a slower rate than corresponds to the power dissipation.

The equations then become:

$$\left.\begin{aligned} i &= i_0 \epsilon^{-(u+s)t} \epsilon^{+s\lambda} \cos(\phi - \omega - \gamma) = i_0 \epsilon^{-ut} \epsilon^{-s(t-\lambda)} \cos(\phi - \omega - \gamma), \\ e &= e_0 \epsilon^{-(u+s)t} \epsilon^{+s\lambda} \cos(\phi - \omega - \gamma) = e_0 \epsilon^{-ut} \epsilon^{-s(t-\lambda)} \cos(\phi - \omega - \gamma), \end{aligned}\right\} \tag{14}$$

and the average power is

$$p_0 = \frac{e_0 i_0}{2} \epsilon^{-2(u+s)t} \epsilon^{+2s\lambda} = \frac{e_0 i_0}{2} \epsilon^{-2ut} \epsilon^{-2s(t-\lambda)}, \tag{15}$$

that is, the power decreases with the time at a greater, but with the distance at a slower, rate than corresponds to the power dissipation.

For a wave moving in opposite direction, again the sign of λ and thus of ω would be reversed.

35. In the equations (10) to (15), the power-transfer constant s is assumed as positive. In general, it is more convenient to assume that s may be positive or negative; positive for an increasing, negative for a decreasing, flow of power. The equations (13) to (15) then apply also to the case (b) of decreasing power flow, but in the latter case s is negative. They also apply to the case (a) for $s = 0$.

The equation of current, voltage, and power of a traveling wave then can be combined in one expression:

$$\left.\begin{aligned} i &= i_0\epsilon^{-(u+s)t}\epsilon^{\pm s\lambda}\cos(\phi\mp\omega-\gamma) = i_0\epsilon^{-ut}\epsilon^{-s(t\mp\lambda)}\cos(\phi\mp\omega-\gamma), \\ e &= e_0\epsilon^{-(u+s)t}\epsilon^{\pm s\lambda}\cos(\phi\mp\omega-\gamma) = e_0\epsilon^{-ut}\epsilon^{-s(t\mp\lambda)}\cos(\phi\mp\omega-\gamma), \end{aligned}\right\} \quad (16)$$

$$p_0 = \frac{e_0 i_0}{2}\epsilon^{-2(u+s)t}\epsilon^{\pm 2s\lambda} = \frac{e_0 i_0}{2}\epsilon^{-2ut}\epsilon^{-2s(t\mp\lambda)}, \quad (17)$$

where the upper sign applies to a wave traveling in the direction toward rising values of λ, the lower sign to a wave traveling in opposite direction, toward decreasing λ. Usually, waves of both directions of travel exist simultaneously (and in proportions depending on the terminal conditions of the oscillating system, as the values of i and e at its ends, etc.).

$s = 0$ corresponds to a traveling wave of constant power flow (case (a)).

$s > 0$ corresponds to a traveling wave of increasing power flow, that is, a wave which drains the circuit over which it travels of some of its stored energy, and thereby increases the time rate of dying out (case (c)).

$s < 0$ corresponds to a traveling wave of decreasing power flow, that is, a wave which supplies energy to the circuit over which it travels, and thereby decreases the time rate of dying out of the transient.

If s is negative, for a transient wave, it always must be

$$-s \leqq u,$$

since, if $-s > u$, $u + s$ would be negative, and $\epsilon^{-(u+s)t}$ would increase with the time; that is, the intensity of the transient would increase with the time, which in general is not possible, as the transient must decrease with the time, by the power dissipation in r and g.

Standing waves and traveling waves, in which the coefficient in the exponent of the time exponential is positive, that is, the wave increases with the time, may, however, occur in electric circuits in which the wave is supplied with energy from some outside source, as by a generating system flexibly connected (electrically) through an arc. Such waves then are "cumulative oscillations." They may either increase in intensity indefinitely, that is, up to destruction of the circuit insulation, or limit themselves by the power dissipation increasing with the increasing intensity of the oscillation, until it becomes equal to the power supply. Such oscillations, which frequently are most destructive ones, are met in electric systems as "arcing grounds," "grounded phase," etc. They are frequently called "undamped oscillations," and as such find a use in wireless telegraphy and telephony. Thus far, the only source of cumulative oscillation seems to be an energy supply over an arc, especially an unstable arc. In the self-limiting cumulative oscillation, the so-called damped oscillation, the transient becomes a permanent phenomenon. Our theoretical knowledge of the cumulative oscillations thus far is rather limited, however.

An oscillogram of a "grounded phase" on a 154-mile three-phase line, at 82 kilovolts, is given in Figs. 44 and 45. Fig. 44 shows current and voltage at the moment of formation of the ground; Fig. 45 the same one minute later, when the ground was fully developed.

An oscillogram of a cumulative oscillation in a 2500-kw. 100,000-volt power transformer (60-cycle system) is given in Fig. 46. It is caused by switching off 28 miles of line by high-tension switches, at 88 kilovolts. As seen, the oscillation rapidly increases in intensity, until it stops by the arc extinguishing, or by the destruction of the transformer.

Of special interest is the limiting case,

$$- s = u;$$

in this case, $u + s = 0$, and the exponential function of time vanishes, and current and voltage become

$$\left.\begin{aligned} i &= i_0 \epsilon^{\pm s\lambda} \cos (\phi \mp \omega - \gamma), \\ e &= e_0 \epsilon^{\pm s\lambda} \cos (\phi \mp \omega - \gamma), \end{aligned}\right\} \tag{18}$$

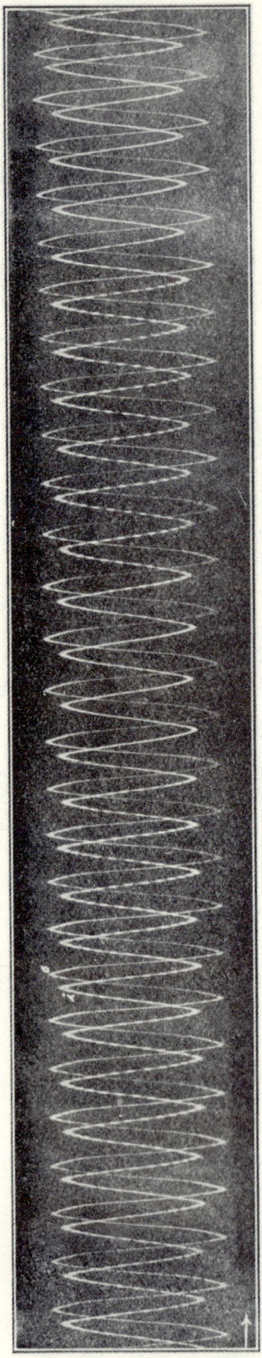

Fig. 44.—CD10057.—Oscillogram of Beginning of Arcing Ground on 154 Miles of 100,000-volt Transmission Line.

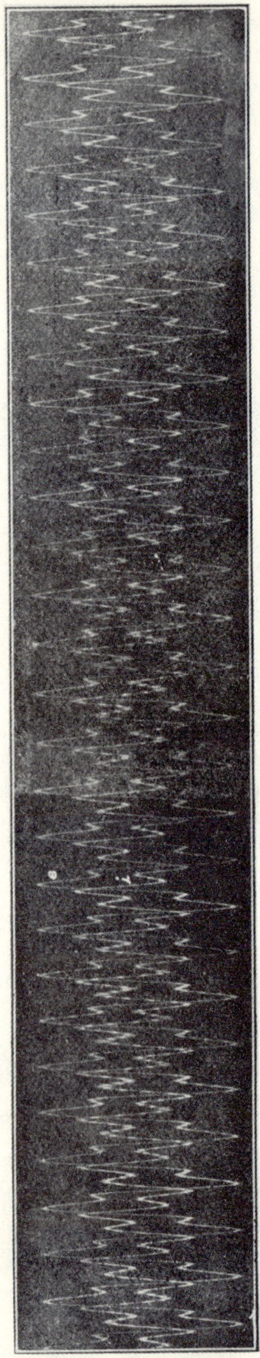

Fig. 45.—CD10058.—Oscillogram of Arcing Ground on 154 Miles of 100,000-volt Transmission Line.

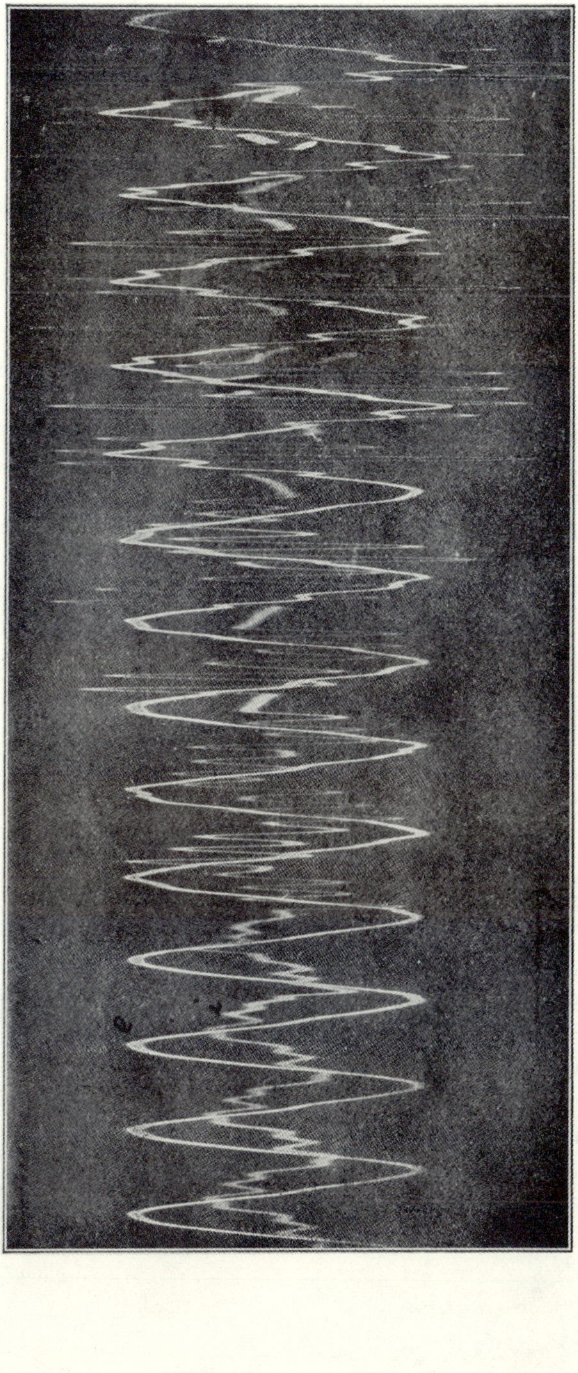

Fig. 46.—cd10001.—Oscillogram of Cumulative Oscillation in High-potential Coil of 2500-kw. Step-up Transformer Caused by Disconnecting 28 Miles of 100,000-volt Transmission Line; High-tension Switching.

that is, are not transient, but permanent or alternating currents and voltages.

Writing the two waves in (18) separately gives

$$\left.\begin{aligned} i &= i_0\epsilon^{+s\lambda}\cos(\phi - \omega - \gamma_1) - i_0'\epsilon^{-s\lambda}(\phi + \omega - \gamma_2), \\ e &= e_0\epsilon^{+s\lambda}\cos(\phi - \omega - \gamma_1) + e_0'\epsilon^{-s\lambda}(\phi + \omega - \gamma_2), \end{aligned}\right\} \quad (19)$$

and these are the equations of the alternating-current transmission line, and reduce, by the substitution of the complex quantity for the function of the time angle ϕ, to the standard form given in "Transients in Space" [Dover, Volume III, pp. 59–221].

36. Obviously, traveling waves and standing waves may occur simultaneously in the same circuit, and usually do so, just as in alternating-current circuits effective and reactive waves occur simultaneously. In an alternating-current circuit, that is, in permanent condition, the wave of effective power (current in phase with the voltage) and the wave of reactive power (current in quadrature with the voltage) are combined into a single wave, in which the current is displaced from the voltage by more than 0 but less than 90 degrees. This cannot be done with transient waves. The transient wave of effective power, that is, the traveling wave,

$$\begin{aligned} i &= i_0\epsilon^{-ut}\epsilon^{-s(t\pm\lambda)}\cos(\phi \mp \omega - \gamma), \\ e &= e_0\epsilon^{-ut}\epsilon^{-s(t\pm\lambda)}\cos(\phi \mp \omega - \gamma), \end{aligned}$$

cannot be combined with the transient wave of reactive power, that is, the stationary wave,

$$\begin{aligned} i &= i_0'\epsilon^{-ut}\cos(\phi \mp \omega - \gamma'), \\ e &= e_0'\epsilon^{-ut}\sin(\phi \mp \omega - \gamma'), \end{aligned}$$

to form a transient wave, in which current and voltage differ in phase by more than 0 but less than 90 degrees, since the traveling wave contains the factor $\epsilon^{-s(t\mp\lambda)}$, resulting from its progression along the circuit, while the stationary wave does not contain this factor, as it does not progress.

This makes the theory of transient waves more complex than that of alternating waves.

Thus traveling waves and standing waves can be combined only locally, that is, the resultant gives a wave in which the phase angle between current and voltage changes with the distance λ and with the time t.

When traveling waves and stationary waves occur simultaneously, very often the traveling wave precedes the stationary wave.

The phenomenon may start with a traveling wave or impulse, and this, by reflection at the ends of the circuit, and combination of the reflected waves and the main waves, gradually changes to a stationary wave. In this case, the traveling wave has the same frequency as the stationary wave resulting from it. In Fig. 47 is shown the reproduction of an oscillogram of the formation of a stationary oscillation in a transmission line by the repeated re-

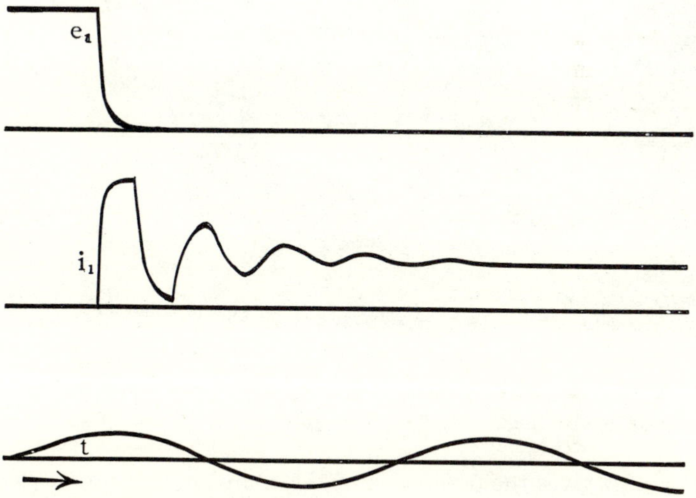

Fig. 47.—CD11168.—Reproduction of an Oscillogram of Stationary Line Oscillation by Reflection of Impulse from Ends of Line. (The lowest curve gives a 60-cycle current as time measure.)

flection from the ends of the line of the single impulse caused by short circuiting the energized line at one end. In the beginning of a stationary oscillation of a compound circuit, that is, a circuit comprising sections of different constants, traveling waves frequently occur, by which the energy stored magnetically or dielectrically in the different circuit sections adjusts itself to the proportion corresponding to the stationary oscillation of the complete circuit. Such traveling waves then are local, and therefore of much higher frequency than the final oscillation of the complete circuit, and thus die out at a faster rate. Occasionally they are shown by the oscillograph as high-frequency oscillations intervening between the alternating waves before the beginning of the transient and the low-frequency stationary oscillation of the complete circuit. Such oscillograms are given in Figs. 48 to 49.

Fig. 48*A* gives the oscillation of the compound circuit consisting of 28 miles of line and the high-tension winding of the 2500-kw. step-up transformer, caused by switching off, by low-tension switches, from a substation at the end of a 153-mile three-phase transmission line, at 88 kilovolts.

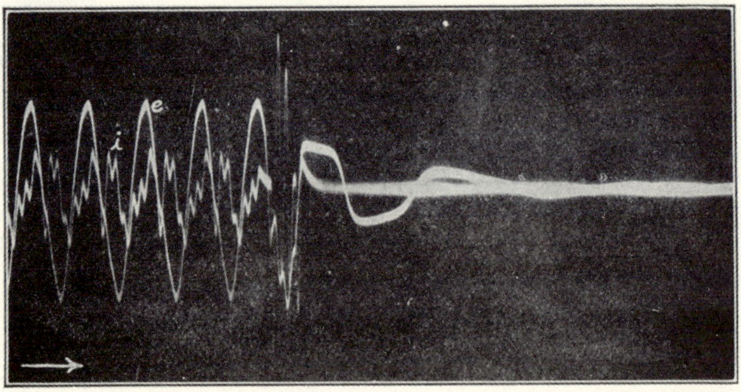

Fig. 48*A*. — CD10002. — Oscillogram of High-frequency Oscillation Preceding Low-frequency Oscillation of Compound Circuit of 28 Miles of 100,000-volt Line and Step-up Transformer; Low-tension Switching.

Fig. 48*B* gives the oscillation of the compound circuit consisting of 154 miles of three-phase line and 10,000-kw. step-down transformer, when switching this line, by high-tension switches, off the end of another 154 miles of three-phase line, at 107 kilovolts. The voltage at the end of the supply line is given as e_1, at the beginning of the oscillating circuit as e_2.

Fig. 49 shows the oscillations and traveling waves appearing in a compound circuit consisting of 154 miles of three-phase line and 10,000-kw. step-down transformer, by switching it on and off the generating system, by high-tension switches, at 89 kilovolts.

Frequently traveling waves are of such high frequency — reaching into the millions of cycles — that the oscillograph does not record them, and their existence and approximate magnitude are determined by inserting a very small inductance into the

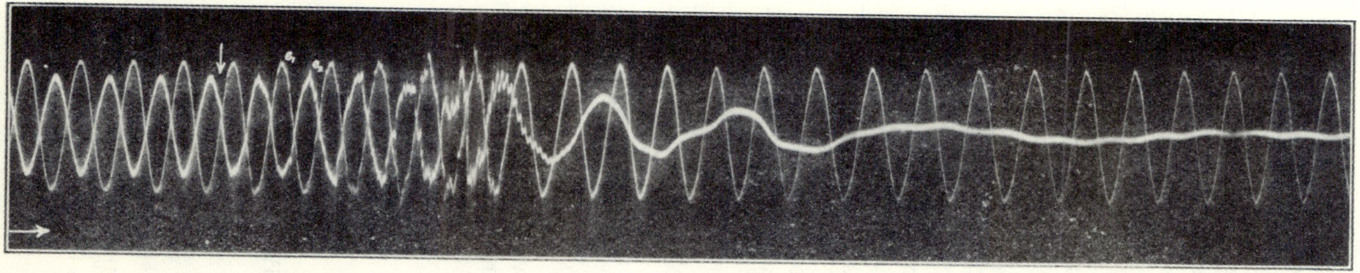

Fig. 48*B*. — CD10049. — Oscillogram of High-frequency Oscillation Preceding Low-frequency Oscillation of Compound Circuit Caused by Switching 154 miles of 100,000 Volts Transmission Line and Step-down Transformer off another 154 Miles of 100,000 Volts Line; High-tension Switching.

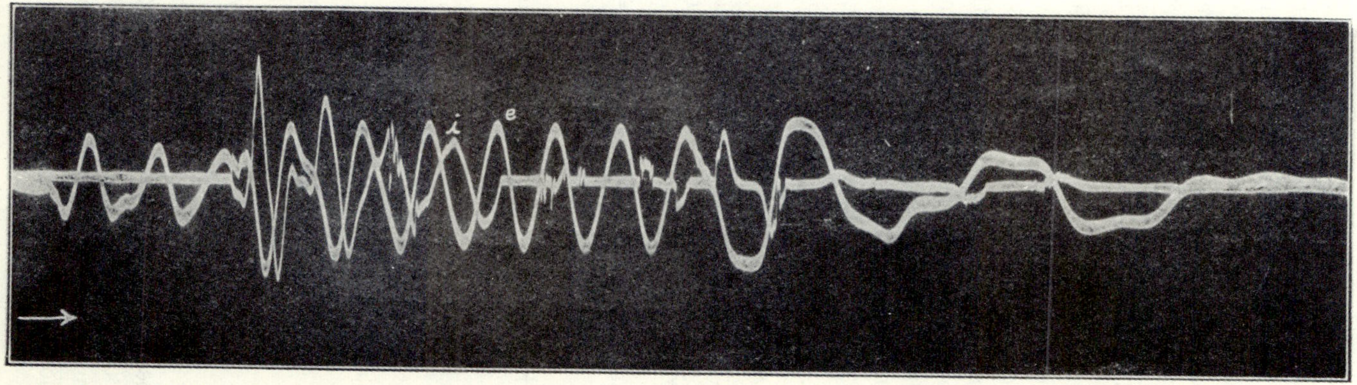

Fig. 49. — CD10036. — Oscillogram of Oscillation of Compound Circuit Consisting of 154 Miles of 100,000 Volts Line and Step-up Transformer; Connecting and Disconnecting by Low-tension Switches. High-tension Current and Low-tension Voltage.

circuit and measuring the voltage across the inductance by spark gap. These traveling waves of very high frequency are extremely local, often extending over a few hundred feet only.

An approximate estimate of the effective frequency of these very high frequency local traveling waves can often be made from their striking distance across a small inductance, by means of the relation $\frac{e_0}{i_0} = \sqrt{\frac{L_0}{C_0}} = z_0$, discussed in Chapter VI (page 187).

For instance, in the 100,000-volt transmission line of Fig. 48*A*, the closing of the high-tension oil switch produces a high-frequency oscillation which at a point near its origin, that is, near the switch, jumps a spark gap of 3.3 cm. length, corresponding to $e_1 = 35{,}000$ volts, across the terminals of a small inductance consisting of 34 turns of 1.3 cm. copper rod, of 15 cm. mean diameter and 80 cm. length. The inductance of this coil is calculated as approximately 13 microhenrys. The line constants, from line to neutral, are $L = 0.323$ henry, $C = 2.2 \times 10^{-6}$ farad; hence $z_0 = \sqrt{\frac{L}{C}} = \sqrt{0.1465} \times 10^3 = 383$ ohms.

The sudden change of voltage at the line terminals, produced by closing the switch, is $\frac{100{,}000}{\sqrt{3}} = 57{,}700$ volts effective, or a maximum of $e_0 = 57{,}700 \times \sqrt{2} = 81{,}500$ volts, and thus gives a maximum transient current in the impulse, of $i_0 = \frac{e_0}{z_0} = 212$ amperes. $i_0 = 212$ amperes maximum, traversing the inductance of 13 microhenrys, thus give the voltage, recorded by the spark gap, of $e_1 = 35{,}000$. If then f = frequency of impulse, it is

$$e_1 = 2\pi f L i_0.$$

Or,

$$f = \frac{e_1}{2\pi L i_0},$$

$$= \frac{35{,}000}{2\pi \times 13 \times 10^{-6} \times 212}$$

$$= 2{,}000{,}000 \text{ cycles.}$$

37. A common form of traveling wave is the discharge of a local accumulation of stored energy, as produced for instance by a direct or induced lightning stroke, or by the local disturbance caused by a change of circuit conditions, as by switching, the blowing of fuses, etc.

Such simple traveling waves frequently are called "*impulses.*"

When such an impulse passes along the line, at any point of the line, the wave energy is zero up to the moment where the wave front of the impulse arrives. The energy then rises, more or less rapidly, depending on the steepness of the wave front, reaches a maximum, and then decreases again, about as shown in Fig. 50. The impulse thus is the combination of two waves,

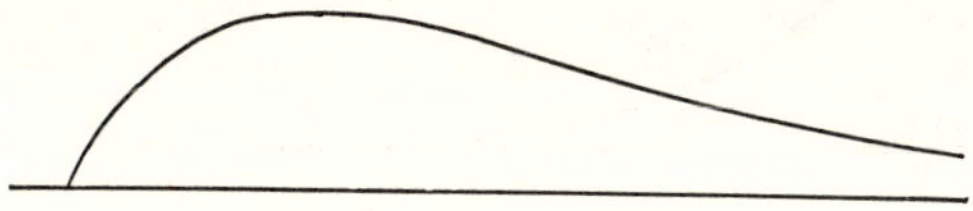

Fig. 50. — Traveling Wave.

one, which decreases very rapidly, $\epsilon^{-(u+s)t}$, and thus preponderates in the beginning of the phenomenon, and one, which decreases slowly, $\epsilon^{-(u-s)t}$. Hence it may be expressed in the form:

$$p_0 = a_1\epsilon^{-2(u+s)t}\epsilon^{+2s\lambda} + a_2\epsilon^{-2(u-s)t}\epsilon^{-2s\lambda}, \tag{20}$$

where the value of the power-transfer constant s determines the " steepness of wave front."

Figs. 51 to 53 show oscillograms of the propagation of such an impulse over an (artificial) transmission line of 130 miles,* of the constants:

$$r = 93.6 \text{ ohms},$$
$$L = 0.3944 \text{ henrys},$$
$$C = 1.135 \text{ microfarads},$$

thus of surge impedance $z_0 = \sqrt{\frac{L}{C}} = 590$ ohms.

The impulse is produced by a transformer charge.†

Its duration, as measured from the oscillograms, is $T_0 = 0.0036$ second.

In Fig. 51, the end of the transmission line was connected to a noninductive resistance equal to the surge impedance, so as to

* For description of the line see "Design, Construction, and Test of an Artificial Transmission Line," by J. H. Cunningham, Proceedings A.I.E.E., January, 1911.

† In the manner as described in "Disruptive Strength of Air and Oil with Transient Voltages," by J. L. R. Hayden and C. P. Steinmetz, Transactions A.I.E.E., 1910, page 1125. The magnetic energy of the transformer is, however, larger, about 4 joules, and the transformer contained an air gap, to give constant inductance.

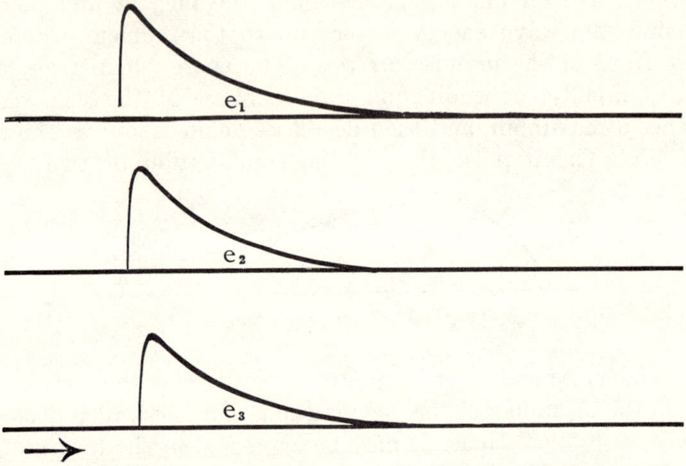

Fig. 51. — CD11145. — Reproduction of Oscillogram of Propagation of Impulse Over Transmission Line; no Reflection. Voltage.

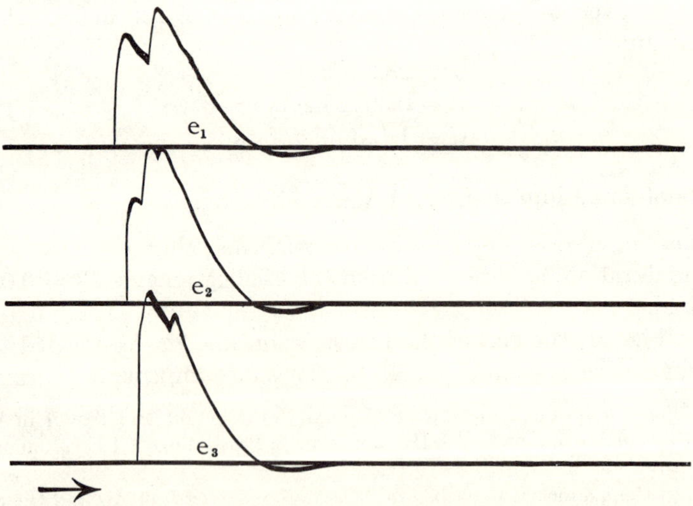

Fig. 52. — CD11152. — Reproduction of Oscillogram of Propagation of Impulse Over Transmission Line; Reflection from Open End of Line. Voltage.

give no reflection. The upper curve shows the voltage of the impulse at the beginning, the middle curve in the middle, and the lower curve at the end of the line.

Fig. 52 gives the same three voltages, with the line open at the end. This oscillogram shows the repeated reflections of the voltage impulse from the ends of the line,—the open end and the transformer inductance at the beginning. It also shows the increase of voltage by reflection.

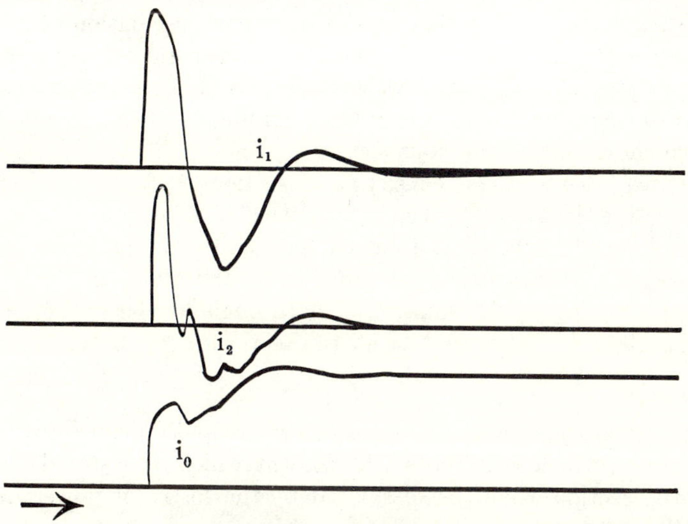

Fig. 53.—CD11153.—Reproduction of Oscillogram of Propagation of Impulse Over Transmission Line; Reflection from Open End of Line. Current.

Fig. 53 gives the current impulses at the beginning and the middle of the line, corresponding to the voltage impulses in Fig. 52, together with the primary current of the transformer, i_0. This oscillogram shows the reversals of current by reflection, and the formation of a stationary oscillation by the successive reflections of the traveling wave from the ends of the line.

CHAPTER IX

OSCILLATIONS OF THE COMPOUND CIRCUIT

38. The most interesting and most important application of the traveling wave is that of the stationary oscillation of a compound circuit, as industrial circuits are never uniform, but consist of sections of different characteristics, as the generating system, transformer, line, load, etc. Oscillograms of such circuits have been shown in the previous lecture.

If we have a circuit consisting of sections 1, 2, 3 . . . , of the respective lengths (in velocity measure) $\lambda_1, \lambda_2, \lambda_3 \ldots$, this entire circuit, when left to itself, gradually dissipates its stored energy by a transient. As function of the time, this transient must decrease at the same rate u_0 throughout the entire circuit. Thus the time decrement of all the sections must be

$$\epsilon^{-u_0 t}.$$

Every section, however, has a power-dissipation constant, $u_1, u_2, u_3 \ldots$, which represents the rate at which the stored energy of the section would be dissipated by the losses of power in the section,

$$\epsilon^{-u_1 t}, \epsilon^{-u_2 t}, \epsilon^{-u_3 t} \ldots$$

But since as part of the whole circuit each section must die down at the same rate $\epsilon^{-u_0 t}$, in addition to its power-dissipation decrement $\epsilon^{-u_1 t}, \epsilon^{-u_2 t} \ldots$, each section must still have a second time decrement, $\epsilon^{-(u_0 - u_1)t}, \epsilon^{-(u_0 - u_2)t} \ldots$ This latter does not represent power dissipation, and thus represents power transfer. That is,

$$\begin{aligned} s_1 &= u_0 - u_1, \\ s_2 &= u_0 - u_2, \\ &\cdot \\ &\cdot \\ &\cdot \end{aligned} \tag{1}$$

It thus follows that in a compound circuit, if u_0 is the average exponential time decrement of the complete circuit, or the average power-dissipation constant of the circuit, and u that of any section, this section must have a second exponential time decrement,

$$s = u_0 - u, \tag{2}$$

which represents power transfer from the section to other sections, or, if s is negative, power received from other sections. The oscillation of every individual section thus is a traveling wave, with a power-transfer constant s.

As u_0 is the average dissipation constant, that is, an average of the power-dissipation constants u of all the sections, and $s = u_0 - u$ the power-transfer constant, some of the s must be positive, some negative.

In any section in which the power-dissipation constant u is less than the average u_0 of the entire circuit, the power-transfer constant s is positive; that is, the wave, passing over this section, increases in intensity, builds up, or in other words, gathers energy, which it carries away from this section into other sections. In any section in which the power-dissipation constant u is greater than the average u_0 of the entire circuit, the power-transfer constant s is negative; that is, the wave, passing over this section, decreases in intensity and thus in energy, or in other words, leaves some of its energy in this section, that is, supplies energy to the section, which energy it brought from the other sections.

By the power-transfer constant s, sections of low energy dissipation supply power to sections of high energy dissipation.

39. Let for instance in Fig. 43 be represented a circuit consisting of step-up transformer, transmission line, and load. (The load, consisting of step-down transformer and its secondary circuit, may for convenience be considered as one circuit section.) Assume now that the circuit is disconnected from the power supply by low-tension switches, at A. This leaves transformer, line, and load as a compound oscillating circuit, consisting of four sections: the high-tension coil of the step-up transformer, the two lines, and the load.

Let then λ_1 = length of line, λ_2 = length of transformer circuit, and λ_3 = length of load circuit, in velocity measure.* If then

* If l_1 = length of circuit section in any measure, and L_0 = inductance, C_0 = capacity per unit of length l_1, then the length of the circuit in velocity measure is $\lambda_1 = \sigma_0 l_1$, where $\sigma_0 = \sqrt{L_0 C_0}$.

Thus, if L = inductance, C = capacity per transformer coil, n = number of transformer coils, for the transformer the unit of length is the coil; hence the

$u_1 = 900$ = power-dissipation constant of the line, $u_2 = 100$ = power-dissipation constant of transformer, and $u_3 = 1600$ = power-dissipation constant of the load, and the respective lengths of the circuit sections are

$$\lambda_1 = 1.5 \times 10^{-3}; \quad \lambda_2 = 1 \times 10^{-3}; \quad \lambda_3 = 0.5 \times 10^{-3},$$

it is:

		Line.	Transformer.	Line.	Load.	Sum.
Length:	$\lambda =$	1.5×10^{-3}	1×10^{-3}	1.5×10^{-3}	$.5\times10^{-3}$	4.5×10^{-3}
Power-dissipation constant:	$u =$	900	100	900	1600	
	$u\lambda =$	1.35	.1	1.35	.8	3.6

hence, $u_0 =$ average $u = \frac{\Sigma u\lambda}{\Sigma\lambda} = 800$, and:

	Line.	Transformer.	Line.	Load.
Power-transfer constant: $s = u_0 - u =$	-100	$+700$	-100	-800

The transformer thus dissipates power at the rate $u_2 = 100$, while it sends out power into the other sections at the rate of $s_2 = 700$, or seven times as much as it dissipates. That is, it supplies seven-eighths of its stored energy to other sections. The load dissipates power at the rate $u_3 = 1600$, and receives power at the rate $-s = 800$; that is, half of the power which it dissipates is supplied from the other sections, in this case the transformer.

The transmission line dissipates power at the rate $u_1 = 900$, that is, only a little faster than the average power dissipation of the entire circuit, $u_0 = 800$; and the line thus receives power only at the rate $-s = 100$, that is, receives only one-ninth of its power from the transformer; the other eight-ninths come from its stored energy.

The traveling wave passing along the circuit section thus increases or decreases in its power at the rate $\epsilon^{+2s\lambda}$; that is, in the line:

$p = p_1\epsilon^{-200\lambda}$, the energy of the wave decreases slowly;

in the transformer:

$p = p_2\epsilon^{+1400\lambda}$, the energy of the wave increases rapidly;

length $l_1 = n$, and the length in velocity measure, $\lambda = \sigma_0 n = n\sqrt{LC}$. Or, if L = inductance, C = capacity of the entire transformer, its length in velocity measure is $\lambda = \sqrt{LC}$.

Thus, the reduction to velocity measure of distance is very simple.

in the load:

$$p = p_3 \epsilon^{-1600\lambda}, \text{ the energy of the wave decreases rapidly.}$$

Here the coefficients of p_1, p_2, p_3 must be such that the wave at the beginning of one section has the same value as at the end of the preceding section.

In general, two traveling waves run around the circuit in opposite direction.

Each of the two waves reaches its maximum intensity in this circuit at the point where it leaves the transformer and enters the line, since in the transformer it increases, while in the line it again decreases, in intensity.

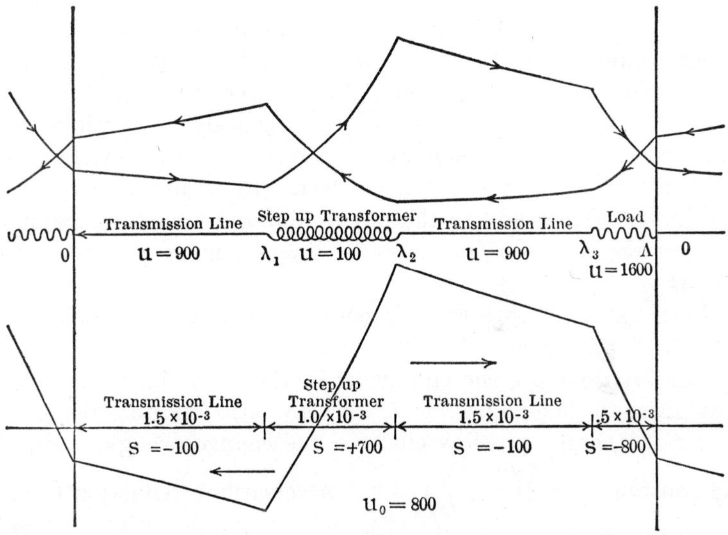

Fig. 54. — Energy Distribution in Compound Oscillation of Closed Circuit; High Line Loss.

Assuming that the maximum value of the one wave is 6, that of the opposite wave 4 megawatts, the power values of the two waves then are plotted in the upper part of Fig. 54, and their difference, that is, the resultant flow of power, in the lower part of Fig. 54. As seen from the latter, there are two power nodes, or points over which no power flows, one in the transformer and one in the load, and the power flows from the transformer over the line into the load; the transformer acts as generator of the power, and of this power a fraction is consumed in the line, the rest supplied to the load.

40. The diagram of this transient power transfer of the system thus is very similar to that of the permanent power transmission by alternating currents: a source of power, a partial consumption in the line, and the rest of the power consumed in the load.

However, this transient power-transfer diagram does not represent the entire power which is being consumed in the circuit, as power is also supplied from the stored energy of the circuit; and the case may thus arise — which cannot exist in a permanent power transmission — that the power dissipation of the line is less than corresponds to its stored energy, and the line also supplies power to the load, that is, acts as generator, and in this case the power would not be a maximum at the transformer terminals, but would still further increase in the line, reaching its maximum at the load terminals. This obviously is possible only with transient power, where the line has a store of energy from which it can draw in supplying power. In permanent condition the line could not add to the power, but must consume, that is, the permanent power-transmission diagram must always be like Fig. 54.

Not so, as seen, with the transient of the stationary oscillation. Assume, for instance, that we reduce the power dissipation in the line by doubling the conductor section, that is, reducing the resistance to one-half. As L thereby also slightly decreases, C increases, and g possibly changes, the change brought about in the constant $u = \frac{1}{2}\left(\frac{r}{L} + \frac{g}{C}\right)$ is not necessarily a reduction to one-half, but depends upon the dimensions of the line. Assuming therefore, that the power-dissipation constant of the line is by the doubling of the line section reduced from $u_1 = 900$ to $u_1 = 500$, this gives the constants:

	Line.	Transformer.	Line.	Load.	Sum.
$\lambda =$	1.5×10^{-3}	1×10^{-3}	1.5×10^{-3}	$.5 \times 10^{-3}$	4.5×10^{-3}
$u =$	500	100	500	1600	
$u\lambda =$	.75	.1	.75	.8	2.4

hence, u_0 = average $u = \frac{\Sigma u\lambda}{\Sigma\lambda} = 533$, and:

$s =$	+33	+433	+33	−1067

That is, the power-transfer constant of the line has become positive, $s_1 = 33$, and the line now assists the transformer in supplying power to the load. Assuming again the values of the two traveling waves, where they leave the transformer (which now are not the maximum values, since the waves still further increase in intensity in passing over the lines), as 6 and 4 megawatts respectively, the power diagram of the two waves, and the power diagram of their resultant, are given in Fig. 55.

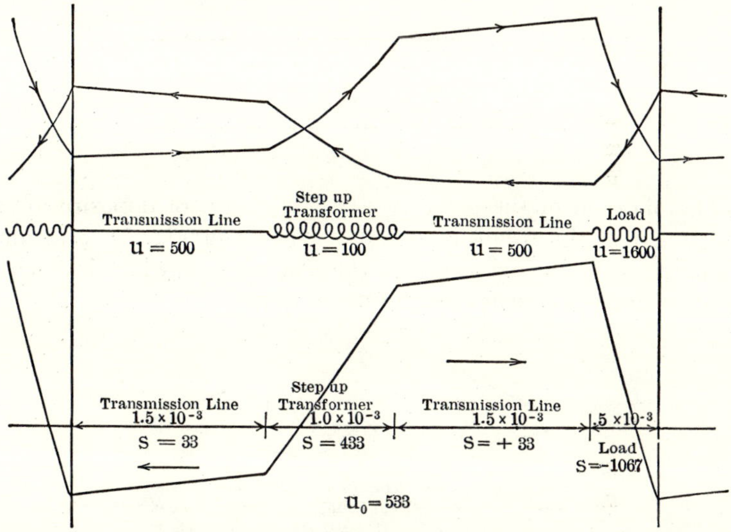

Fig. 55. — Energy Distribution in Compound Oscillation of Closed Circuit; Low Line Loss.

In a closed circuit, as here discussed, the relative intensity of the two component waves of opposite direction is not definite, but depends on the circuit condition at the starting moment of the transient.

In an oscillation of an open compound circuit, the relative intensities of the two component waves are fixed by the condition that at the open ends of the circuit the power transfer must be zero.

As illustration may be considered a circuit comprising the high-potential coil of the step-up transformer, and the two lines, which are assumed as open at the step-down end, as illustrated diagrammatically in Fig. 56.

Choosing the same lengths and the same power-dissipation constants as in the previous illustrations, this gives:

	Line.	Transformer.	Line.	Sum.
$\lambda =$	1.5×10^{-3}	1×10^{-3}	1.5×10^{-3}	4×10^{-3}
$u =$	900	100	900	
$u\lambda =$	1.35	.1	1.35	2.8

hence, $u_0 = \text{average } u = \frac{\Sigma u\lambda}{\Sigma\lambda} = 700$, and:

$s =$	-200	$+600$	-200

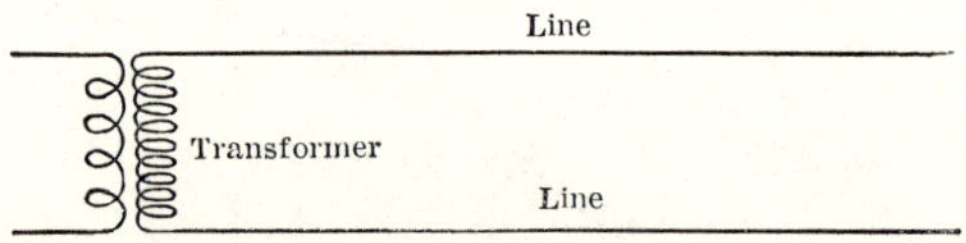

Fig. 56.

The diagram of the power of the two waves of opposite directions, and of the resultant power, is shown in Fig. 57, assuming 6 megawatts as the maximum power of each wave, which is reached at the point where it leaves the transformer.

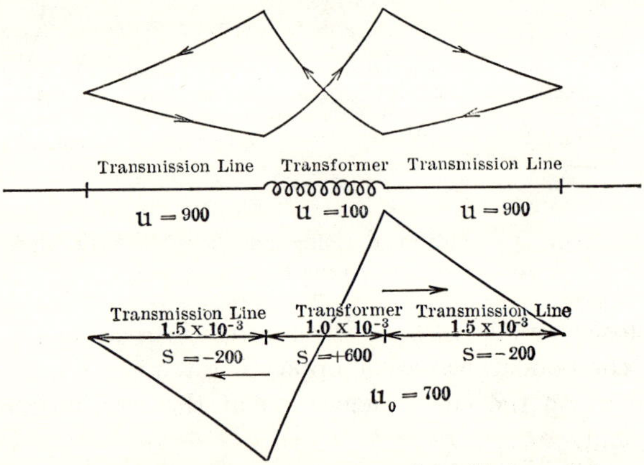

Fig. 57. — Energy Distribution in Compound Oscillation of Open Circuit.

In this case the two waves must be of the same intensity, so as to give 0 as resultant at the open ends of the line. A power node then appears in the center of the transformer.

41. A stationary oscillation of a compound circuit consists of two traveling waves, traversing the circuit in opposite direction, and transferring power between the circuit sections in such a manner as to give the same rate of energy dissipation in all circuit sections. As the result of this power transfer, the stored energy of the system must be uniformly distributed throughout the entire circuit, and if it is not so in the beginning of the transient, local traveling waves redistribute the energy throughout the oscillating circuit, as stated before. Such local oscillations are usually of very high frequency, but sometimes come within the range of the oscillograph, as in Fig. 47.

During the oscillation of the complex circuit, every circuit element $d\lambda$ (in velocity measure), or every wave length or equal part of the wave length, therefore contains the same amount of stored energy. That is, if e_0 = maximum voltage, i_0 = maximum current, and λ_0 = wave length, the average energy $\frac{e_0 i_0 \lambda_0}{2}$ must be constant throughout the entire circuit. Since, however, in velocity measure, λ_0 is constant and equal to the period T_0 throughout all the sections of the circuit, the product of maximum voltage and of maximum current, $e_0 i_0$, thus must be constant throughout the entire circuit.

The same applies to an ordinary traveling wave or impulse. Since it is the same energy which moves along the circuit at a constant rate, the energy contents for equal sections of the circuit must be the same except for the factor ϵ^{-2ut}, by which the energy decreases with the time, and thus with the distance traversed during this time.

Maximum voltage e_0 and maximum current i_0, however, are related to each other by the condition

$$\frac{e_0}{i_0} = z_0 = \sqrt{\frac{L_0}{C_0}}, \tag{3}$$

and as the relation of L_0 and C_0 is different in the different sections, and that very much so, z_0, and with it the ratio of maximum voltage to maximum current, differ for the different sections of the circuit.

If then e_1 and i_1 are maximum voltage and maximum current respectively of one section, and $z_1 = \sqrt{\frac{L_1}{C_1}}$ is the "natural impedance" of this section, and e_2, i_2, and $z_2 = \sqrt{\frac{L_2}{C_2}}$ are the corresponding values for another section, it is

$$e_2 i_2 = e_1 i_1, \tag{4}$$

and since

$$\frac{e_2}{i_2} = z_2; \quad \frac{e_1}{i_1} = z_1, \tag{5}$$

substituting

$$\left.\begin{aligned} e_2 &= i_2 z_2, \\ e_1 &= i_1 z_1, \end{aligned}\right\} \tag{6}$$

into (4) gives

$$i_2^2 z_2 = i_1^2 z_1,$$

or

$$\frac{i_2}{i_1} = \sqrt{\frac{z_1}{z_2}} = \sqrt[4]{\frac{L_1}{C_1}\frac{C_2}{L_2}}, \tag{7}$$

and

$$\frac{e_2^2}{z_2} = \frac{e_1^2}{z_1},$$

or

$$\frac{e_2}{e_1} = \sqrt{\frac{z_2}{z_1}} = \sqrt[4]{\frac{L_2}{C_2}\frac{C_1}{L_1}}. \tag{8}$$

That is, in the same oscillating circuit, the maximum voltages e_0 in the different sections are proportional to, and the maximum currents i_0 inversely proportional to, the square root of the natural impedances z_0 of the sections, that is, to the fourth root of the ratios of inductance to capacity $\frac{L_0}{C_0}$.

At every transition point between successive sections traversed by a traveling wave, as those of an oscillating system, a transformation of voltage and of current occurs, by a transformation ratio which is the square root of the ratio of the natural impedances, $z_0 = \sqrt{\frac{L_0}{C_0}}$, of the two respective sections.

When passing from a section of high capacity and low inductance, that is, low impedance z_0, to a section of low capacity and high inductance, that is, high impedance z_0, as when passing from a transmission line into a transformer, or from a cable into a transmission line, the voltage thus is transformed up, and the current transformed down, and inversely, with a wave passing in opposite direction.

A low-voltage high-current wave in a transmission line thus becomes a high-voltage low-current wave in a transformer, and inversely, and thus, while it may be harmless in the line, may become destructive in the transformer, etc.

42. At the transition point between two successive sections, the current and voltage respectively must be the same in the two sections. Since the maximum values of current and voltage respectively are different in the two sections, the phase angles of the waves of the two sections must be different at the transition point; that is, a change of phase angle occurs at the transition point.

This is illustrated in Fig. 58. Let $z_0 = 200$ in the first section (transmission line), $z_0 = 800$ in the second section (transformer). The transformation ratio between the sections then is $\sqrt{\frac{800}{200}} = 2$; that is, the maximum voltage of the second section is twice, and the maximum current half, that of the first section, and the waves of current and of voltage in the two sections thus may be as illustrated for the voltage in Fig. 58, by $e_1 e_2$.

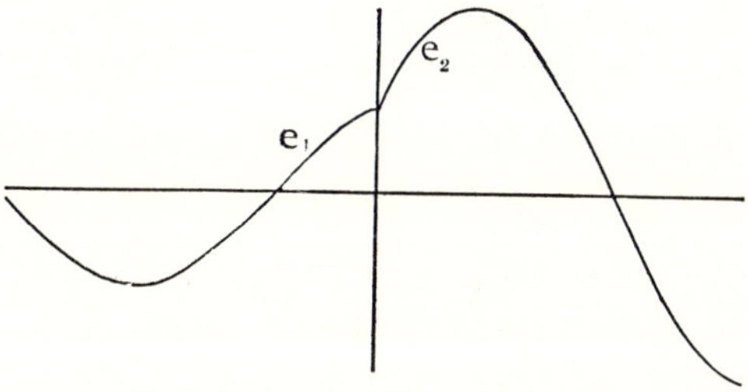

Fig. 58. — Effect of Transition Point on Traveling Wave.

If then e' and i' are the values of voltage and current respectively at the transition point between two sections 1 and 2, and e_1 and i_1 the maximum voltage and maximum current respectively of the first, e_2 and i_2 of the second, section, the voltage phase and current phase at the transition point are, respectively:

$$\left.\begin{array}{l} \text{For the wave of the first section:} \\ \qquad \dfrac{e'}{e_1} = \cos \gamma_1 \quad \text{and} \quad \dfrac{i'}{i_1} = \cos \delta_1. \\ \text{For the wave of the second section:} \\ \qquad \dfrac{e'}{e_2} = \cos \gamma_2 \quad \text{and} \quad \dfrac{i'}{i_2} = \cos \delta_2. \end{array}\right\} \tag{9}$$

Dividing the two pairs of equations of (9) gives

$$\left.\begin{aligned} \frac{\cos\gamma_2}{\cos\gamma_1} &= \frac{e_1}{e_2} = \sqrt{\frac{z_1}{z_2}}, \\ \frac{\cos\delta_2}{\cos\delta_1} &= \frac{i_1}{i_2} = \sqrt{\frac{z_2}{z_1}}, \end{aligned}\right\} \tag{10}$$

hence, multiplied,

$$\frac{\cos\gamma_2}{\cos\gamma_1} \times \frac{\cos\delta_2}{\cos\delta_1} = 1,$$

or

$$\frac{\cos\gamma_2}{\cos\gamma_1} = \frac{\cos\delta_1}{\cos\delta_2}, \tag{11}$$

or

$$\cos\gamma_1 \cos\delta_1 = \cos\gamma_2 \cos\delta_2;$$

that is, the ratio of the cosines of the current phases at the transition point is the reciprocal of the ratio of the cosines of the voltage phases at this point.

Since at the transition point between two sections the voltage and current change, from e_1, i_1 to e_2, i_2, by the transformation ratio $\sqrt{\frac{z_2}{z_1}}$, this change can also be represented as a partial reflection. That is, the current i_1 can be considered as consisting of a component i_2, which passes over the transition point, is " transmitted " current, and a component $i_1' = i_1 - i_2$, which is " reflected " current, etc. The greater then the change of circuit constants at the transition point, the greater is the difference between the currents and voltages of the two sections; that is, the more of current and voltage are reflected, the less transmitted, and if the change of constants is very great, as when entering from a transmission line a reactance of very low capacity, almost all the current is reflected, and very little passes into and through the reactance, but a high voltage is produced in the reactance.

CHAPTER X

CONTINUAL AND CUMULATIVE OSCILLATIONS

43. A transient is the phenomenon by which the stored energy readjusts itself to a change of circuit conditions. In an oscillatory transient, the difference of stored energy of the previous and the after condition of the circuit, at a circuit change, oscillates between magnetic and dielectric energy. As there always must be some energy dissipation in the circuit, the oscillating energy of the transient must steadily decline, that is, the transient must die out, at a rate depending on the energy dissipation in the circuit.

Thus, the oscillation resulting from a change of circuit conditions can become continual, that is, of constant amplitude, or cumulative, that is, of increasing amplitude, only if a steady supply of oscillating energy occurs.

Continual and cumulative oscillations thus involve a continual energy supply to the oscillating system, therefore cannot be mere readjustments of circuit conditions by the dissipation of stored energy.

If the continual energy supply is less than the energy dissipation in the circuit, the oscillation dies out, that is, is transient, but with a lowered attenuation constant. This for instance is the case with the transient in those sections of a compound circuit, in which the energy transfer constant is negative. If the continual energy supply equals the energy dissipation, the oscillation is continual, and if the energy supply is greater than the energy dissipation, the oscillation becomes cumulative, that is, increasing in amplitude, until either the system breaks down or, by the increase of the energy dissipation, it becomes equal to the energy supply, and the oscillation becomes continual.

A continual or cumulative oscillation thus involves an energy and frequency transformation, from the low-frequency or continuous-current energy of the power supply of the system to the high-frequency energy of the oscillation.

This energy transformation may be brought about by the transient of energy readjustment, resulting from a change of circuit conditions, producing again a change of circuit conditions and thereby an energy readjustment by transient, etc.

For instance, if in an isolated high-potential transmission line, the ground is brought within striking distance of one of the line conductors — as by the puncture of an insulator. A spark discharge then occurs to the ground, and the arc following the spark discharges the line by a transient oscillation, that is, brings it down to ground potential (and the other two lines, in a three-phase system, then correspondingly rise in voltage to ground, from the *Y* to the delta voltage). As soon as the line is discharged the arc ceases, that is, the spark gap to ground opens, and the line then charges again, from the power supply of the system, and its voltage to ground rises, until sufficient to jump to ground again and start a second transient oscillation, and so on continual transient oscillations follow each other, as a "continual transient," or "arcing ground." Oscillograms, Figs. 59 and 60, show such a series of successive or "recurrent" oscillations, resulting from an arcing ground on one phase of an isolated system consisting of 10 and 21.5 miles respectively of 22,000-volt three-phase 40-cycle circuit. In this case, three transients occur during each half wave of voltage. Oscillogram, Fig. 61, shows an arcing ground oscillations on 32 miles of the same system. Here each transient oscillation persists up to the beginning of the next transient of the same half wave of circuit voltage, and the recurrent oscillations thus tend to run into each other and approach a continuous oscillation.

When successive transients run into each other, they naturally synchronize, as the first oscillation of the second transient would start at a maximum voltage point, that is, at an oscillation maximum of the preceding transient.

Oscillograms CD15073 and CD15087, Figs. 62 and 63, show the formation of a continuous oscillation by the overlapping of successive recurrent oscillations, and finally oscillograms CD15091 and 15034, Figs. 64 and 65, show a continuous oscillation. In Figs. 62 and 63, the component recurrent oscillations are still noticeable by a periodical rise and fall of the amplitude of the wave, while in Figs. 64 and 65, the amplitude has become constant. These oscillo-

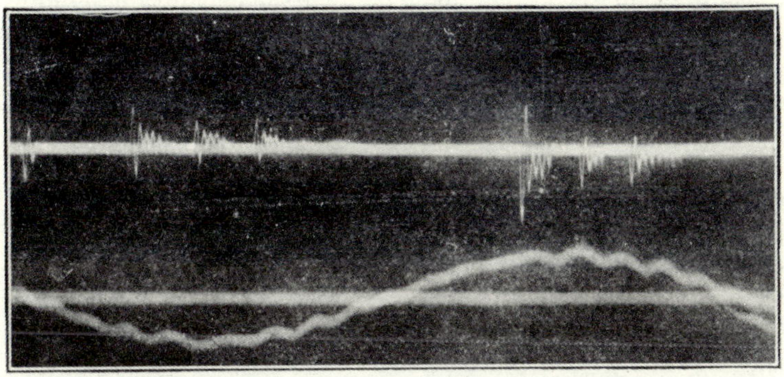

Fig. 59.—Recurrent Oscillation of Arcing Ground, in 10 Miles of 22,000-volt Three-phase 40-cycle Transmission Line.

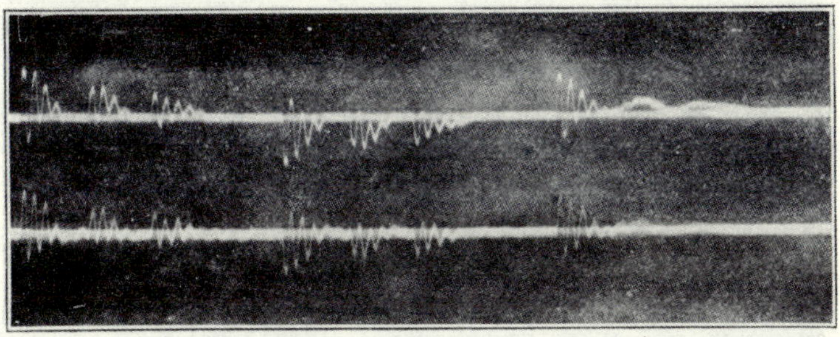

Fig. 60.—Recurrent Oscillation of Arcing Ground, in 21.5 Miles of 22,000-volt Three-phase 40-cycle Transmission Line.

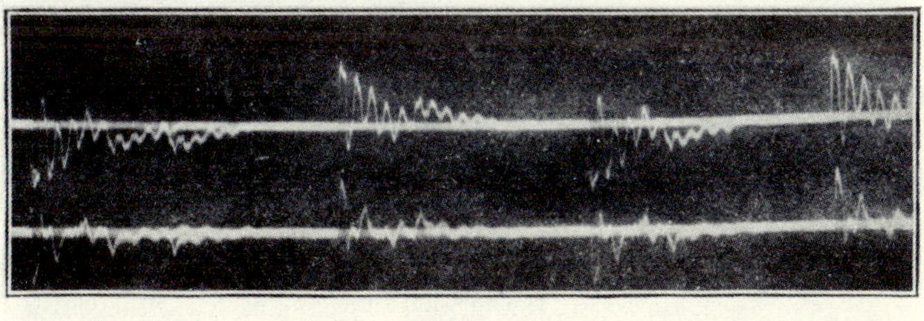

Fig. 61.—Change from Recurrent to Continuous Oscillation of Arcing Ground, in 32 Miles of 22,000-volt Three-phase 40-cycle Transmission Line.

grams, Figs. 62 to 65, were taken on an artificial transmission line.* Oscillations of the type 64 and 65 are industrially used, as "singing arc," in wireless telegraphy, and are produced by shunting a suitable arc by a circuit containing capacity and inductance in series with each other.

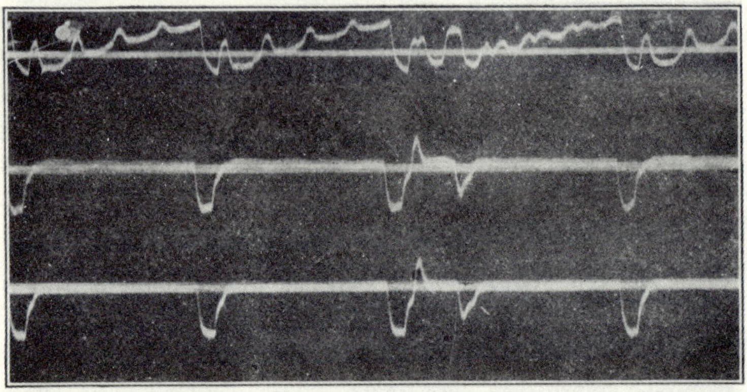

Fig. 62.—Semi-continuous Recurrent Oscillation of Arcing Ground in Transmission Line.

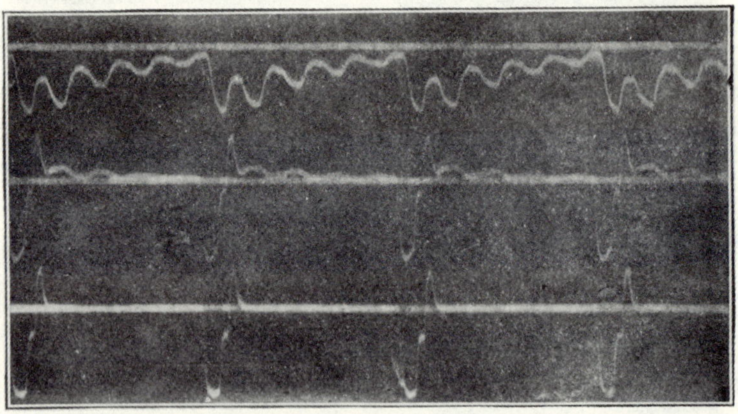

Fig. 63.—Semi-continuous Recurrent Oscillation of Arcing Ground in Transmission Line.

* "Design, Construction and Test of an Artificial Transmission Line," J. H. Cunningham, A. I. E. E. Transact., 1911.

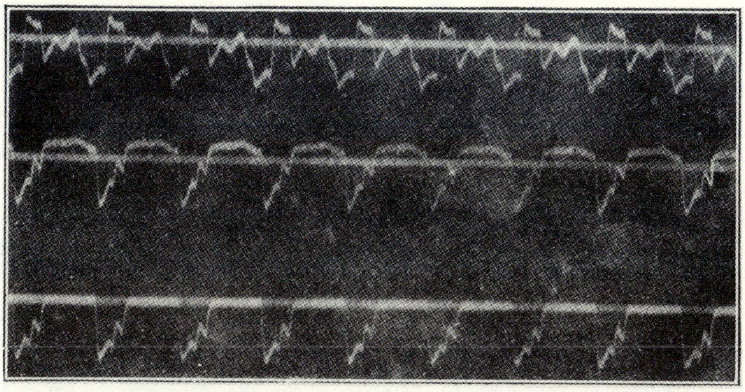

Fig. 64.—Continuous Oscillation of Arcing Ground in Transmission Line.

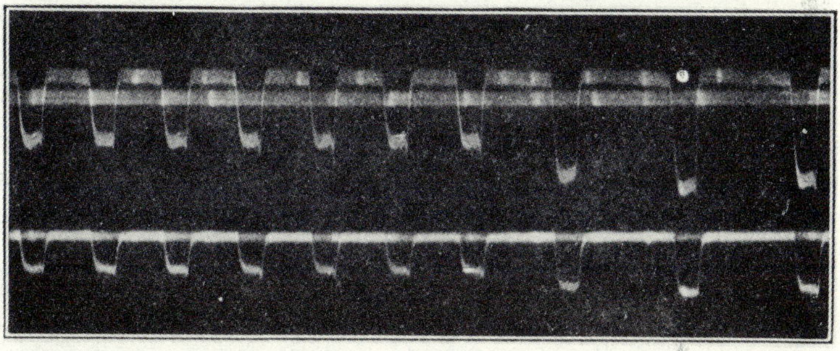

Fig. 65.—Continuous Oscillation of Arcing Ground in Transmission Line.

44. However, the formation of continuous oscillations, Figs. 61 to 65, from the recurrent oscillations, Figs. 59 and 60, is not a mere running together and overlapping of successive wave trains. In Fig. 59, the succeeding oscillation cannot start, until the preceding oscillation has died out and a sufficient time elapsed, for the line to charge again to a voltage which is high enough to discharge to ground and so start the next oscillation, that is, to store the energy for the next oscillation. If then, with an overlap of successive oscillations, no dead period occurs, during which the energy, which oscillates during the next wave train, is supplied to the line, this energy must be supplied during the oscillation, that is, there must be such a phase displacement or lag within the oscillation, which gives a negative energy cycle, or reversed hysteresis loop. Thus, essential for such a continual oscillation is the

existence of a hysteresis loop, formed by the lag of the effect behind the cause. Such a hysteresis loop exists in the transient arc, as illustrated by Fig. 66: the transient volt-ampere characteristic of a short high-temperature metal arc, between titanium and carbon. In this figure, the stationary arc characteristic, that is, the relation between arc voltage and arc current in stationary conditions, is shown in dotted lines, and the drawn line shows the cycle existing between arc current and arc voltage during a cyclic change of current, from zero to 4.1 amperes and back to zero, within $\frac{1}{120}$ of a second, with the current varying approximately as a sine function of the time. As seen, for rising current, the arc voltage is materially higher than for decreasing current. Close to zero current, the arc has ceased, and Geissler tube conduction passes the current through the residual vapor stream.

Other hysteresis cycles than those of the arc are instrumental in the energy supply to other systems of continual oscillation. Thus, for instance, the hysteresis cycle between synchronizing force and position displacement supplies the energy of the continual or cumulative oscillation, called hunting, in synchronous machines, as alternators, synchronous motors and converters. The mechanism, by which the hysteresis cycle supplies the energy of continual oscillations, has been investigated in the case of the hunting of synchronous machines,* but is still practically unknown in the case of continual oscillations between magnetic and dielectric energy in electric circuits.

Recurrent oscillations, as in Fig. 59, must be or very soon become continual, that is, the successive wave trains are of approximately constant amplitude, since each starts with the same energy, the stored energy of the supply system. Continual oscillations, however, in which the energy supply is through a hysteresis cycle, may be cumulative: the area of the hysteresis cycle, that is, the energy supply, depends on and increases with the voltage and current of the oscillation, and the voltage and current, that is, the intensity of the oscillation, depends on and increases with the energy supply, that is, the area of the hysteresis cycle, thus both increase together.

Such cumulative oscillations are represented for instance by Fig. 46, page 99.

* "Instability of Electric Circuits," A. I. E. E. Proceedings, Jan., 1914.

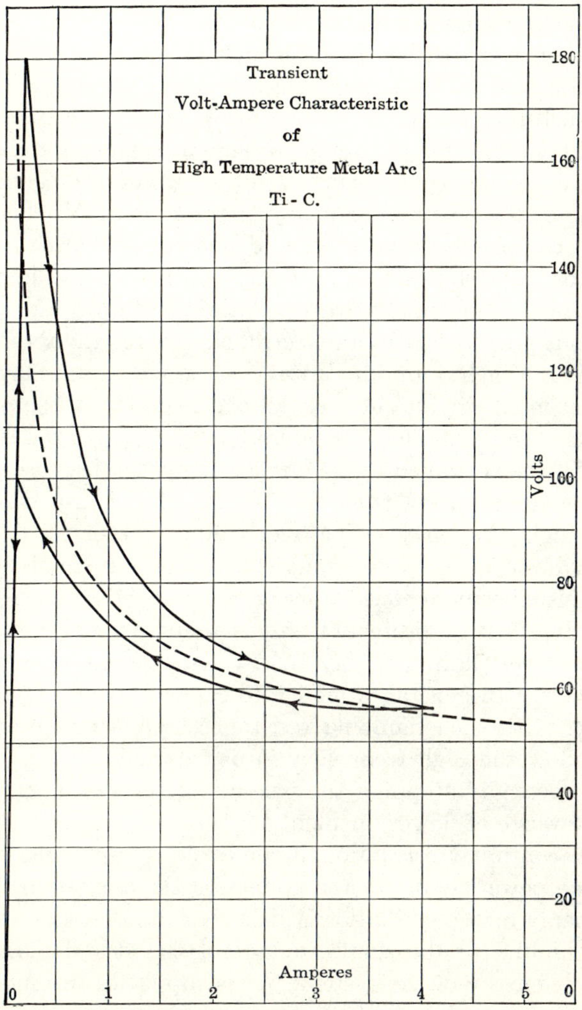
Transient
Volt-Ampere Characteristic
of
High Temperature Metal Arc
Ti - C.
180
160
140
120
Volts
100
80
60
40
20
0
Amperes
0
1
2
3
4
5

Fig. 66.

45. The frequency of the oscillations usually is the natural frequency of the oscillating circuit or section of circuit, but it may be some of the higher harmonics of the generator wave, where such harmonic is near the natural frequency of the system. The latter seems to be the case in the continual arcing ground oscillation in Figs. 44 and 45, page 224. In Fig. 44, the beginning of the disturbance, apparently a harmonic of the generator wave builds up by the energy supply through a beginning arc, and then builds down again, by being slightly out of resonance with a multiple of the natural frequency of the circuit. In Fig. 45, the arc has completely developed, and one of the harmonics of the generator wave appears as a steady continuous oscillation.

Continual and cumulative oscillations naturally are the most dangerous phenomena in an electric circuit, for the reasons:

1. While each individual wave or wave train may not be sufficient in energy and in over-voltage to do harm, the recurrence through seconds, minutes or hours finally destroys the insulation, either by the sustained over-voltage, or by the decrease of dielectric strength resulting from the heating of the insulation caused by the high frequency. The over-voltage protective devices frequently do not offer protection, as the over-voltage of the oscillation is insufficient to cause a discharge over the lightning arrester. The only effective protection seems to be a continuous dissipation of the oscillating energy by a resistance closing the oscillating circuit. In general, a moderate capacity would be connected in series with such damping resistance, and would be chosen so as to allow the high frequency to pass practically unobstructed, while practically stopping the passage of the machine frequency, and the waste of power, incident thereto.

2. A continual oscillation involves an energy transformation from the power supply of the system to the oscillation frequency. The energy of the oscillation which gives its destructiveness thus is not limited to the small amount of the stored magnetic and dielectric energy of the system, but is supplied continuously from the engine or turbine power.

3. The continual oscillation is not a resonance phenomenon which depends on the frequency of the exciting disturbance just coinciding with one of the natural frequencies of the oscillating system, and which thus can occur only very rarely. The dis-

turbance in the system, as lightning, change of load, etc., is only the exciting cause which starts the energy transformation to the oscillating frequency by the arc, etc., and the frequency with which the oscillation occurs then is determined by the circuit constants. Or, as is often stated: the electric arc has no frequency of its own, but oscillates with whatever frequency the circuit is able to oscillate. Thus such oscillations are not uncommon, and have in the last years been observed, measured and recorded in numerous instances, and experimentally produced in lines and high-potential transformer windings. The continual oscillations in transmission lines usually seem to be recurrent oscillations, as in Figs. 59 and 60, while in high-potential transformer windings, due to their much lesser damping, continuous oscillations seem to be more common, as in Fig. 46. Our knowledge of these phenomena is however still extremely incomplete.

CHAPTER XI

INDUCTANCE AND CAPACITY OF ROUND PARALLEL CONDUCTORS

A. Inductance and capacity.

46. As inductance and capacity are the two circuit constants which represent the energy storage, and which therefore are of fundamental importance in the study of transients, their calculation is discussed in the following.

The inductance is the ratio of the interlinkages of the magnetic flux to the current,

$$L = \frac{n\Phi}{i}, \tag{1}$$

where Φ = magnetic flux or number of lines of magnetic force, and n the number of times which each line of magnetic force interlinks with the current i.

The capacity is the ratio of the dielectric flux to the voltage,

$$C = \frac{\Psi}{e}, \tag{2}$$

where Ψ is the dielectric flux, or number of lines of dielectric force, and e the voltage which produces it.

With a single round conductor without return conductor (as wireless antennæ) or with the return conductor at infinite distance, the lines of magnetic force are concentric circles, shown by drawn lines in Fig. 8, page 136, and the lines of dielectric force are straight lines radiating from the conductor, shown dotted in Fig. 8.

Due to the return conductor, in a two-wire circuit, the lines of magnetic and dielectric force are crowded together between the conductors, and the former become eccentric circles, the latter circles intersecting in two points (the foci) inside of the conductors, as shown in Fig. 9, page 137. With more than one return conductor, and with phase displacement between the return currents, as in a three-phase three-wire circuit, the path of the lines of force is still more complicated, and varies during the cyclic change of current.

The calculation of such more complex magnetic and dielectric fields becomes simple, however, by the method of *superposition of fields*. As long as the magnetic and the dielectric flux are proportional respectively to the current and the voltage, — which is the case with the former in nonmagnetic materials, with the latter for all densities below the dielectric strength of the material, — the resultant field of any number of conductors at any point in space is the combination of the component fields of the individual conductors.

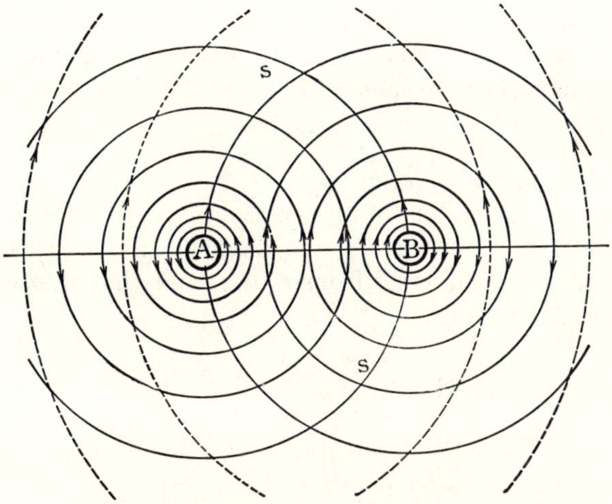

Fig. 67. — Magnetic Field of Circuit.

Thus the field of conductor A and return conductor B is the combination of the field of A, of the shape Fig. 8, and the field of B, of the same shape, but in opposite direction, as shown for the magnetic fields in Fig. 67.

All the lines of magnetic force of the resultant magnetic field must pass between the two conductors, since a line of magnetic force, which surrounds both conductors, would have no m.m.f., and thus could not exist. That is, the lines of magnetic force of A beyond B, and those of B beyond A, shown dotted in Fig. 67, neutralize each other and thereby vanish; thus, in determining the resultant magnetic flux of conductor and return conductor (whether the latter is a single conductor, or divided into two conductors out of phase with each other, as in a three-phase circuit), only the lines of magnetic force within the space from conductor to return conductor need to be considered. Thus, the resultant magnetic flux of a circuit consisting of conductor A and return conductor B, at distance s from each other, consists of the lines of magnetic force surrounding A up to the distance s, and the lines of magnetic force surrounding B up to the distance s. The former is attributed to the inductance of conductor A, the latter to the inductance of conductor B. If both conductors have the same size, they give equal inductances; if of unequal size, the smaller conductor has the higher inductance. In the same manner in a three-phase circuit, the inductance of each of the three conductors is that corresponding to the lines of magnetic force surrounding the respective conductor, up to the distance of the return conductor.

B. Calculation of inductance.

47. If r is the radius of the conductor, s the distance of the return conductor, in Fig. 68, the magnetic flux consists of that external to the conductor, from r to s, and that internal to the conductor, from 0 to r.

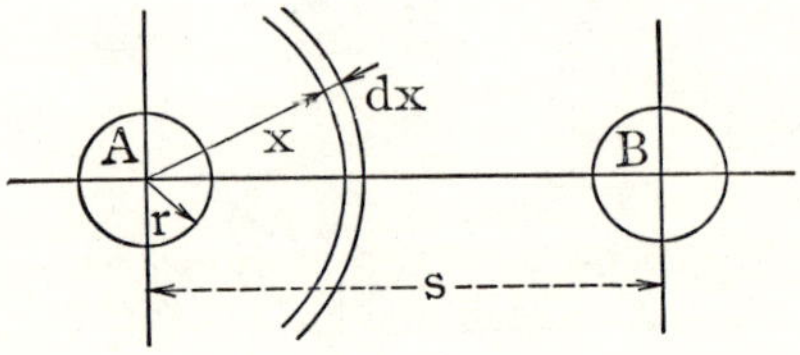

Fig. 68. — Inductance Calculation of Circuit.

At distance x from the conductor center, the length of the magnetic circuit is $2\pi x$, and if F = m.m.f. of the conductor, the magnetizing force is

$$f = \frac{F}{2\pi x}, \tag{3}$$

and the field intensity

$$\mathcal{H} = 4\pi f = \frac{2F}{x}, \tag{4}$$

hence the magnetic density

$$\mathcal{B} = \mu\mathcal{H} = \frac{2\mu F}{x}; \tag{5}$$

and the magnetic flux in the zone dx thus is

$$d\Phi = \frac{2\,\mu F}{x}\,dx, \tag{6}$$

and the magnetic flux interlinked with the conductor thus is

$$n d\Phi = \frac{2\,\mu n F}{x}\,dx, \tag{7}$$

hence the total magnetic flux between the distances x_1 and x_2 is

$$n\Phi]_1^2 = \int_{x_1}^{x_2} \frac{2\,\mu n F\,dx}{x};$$

thus the inductance

$$L]_1^2 = \frac{n\Phi]_1^2}{i} = \int_{x_1}^{x_2} 2\,\mu \frac{nF}{i}\frac{dx}{x}. \tag{8}$$

1. External magnetic flux. $x_1 = r$; $x_2 = s$; $F = i$, as this flux surrounds the total current; and $n = 1$, as each line of magnetic force surrounds the conductor once. $\mu = 1$ in air, thus:

$$L_1 = \int_r^s \frac{2\,dx}{x} = 2\log\frac{s}{r}. \tag{9}$$

2. Internal magnetic flux. Assuming uniform current density throughout the conductor section, it is

$$x_1 = 0;\ x_2 = r;\ \frac{F}{i} = \left(\frac{x}{r}\right)^2,$$

as the flux is produced by a part of the current only; and $n = \left(\frac{x}{r}\right)^2$, as each line of magnetic force surrounds only a part of the conductor

$$L_2 = \int_0^r \frac{2\,\mu x^3\,dx}{r^4} = \frac{\mu}{2}, \tag{10}$$

and the total inductance of the conductor thus is

$$L = L_1 + L_2 = 2\left\{\log\frac{s}{r} + \frac{\mu}{4}\right\} \text{ per cm. length of conductor,} \tag{11}$$

or, if the conductor consists of nonmagnetic material, $\mu = 1$:

$$L = 2\left\{\log\frac{s}{r} + \frac{1}{4}\right\}. \tag{12}$$

This is in absolute units, and, reduced to henrys, = 10^9 absolute units:

$$L = 2\left\{\log\frac{s}{r} + \frac{\mu}{4}\right\}10^{-9}\ h \text{ per cm.} \tag{13}$$

$$= 2\left\{\log\frac{s}{r} + \frac{1}{4}\right\}10^{-9}\ h \text{ per cm.} \tag{14}$$

In these equations the logarithm is the natural logarithm, which is most conveniently derived by dividing the common or 10 logarithm by 0.4343.*

C. Discussion of inductance.

48. In equations (11) to (14) s is the distance between the conductors. If s is large compared with r, it is immaterial whether as s is considered the distance between the conductor centers, or between the insides, or outsides, etc.; and, in calculating the inductance of transmission-line conductors, this is the case, and it therefore is immaterial which distance is chosen as s; and usually, in speaking of the "distance between the line conductors," no attention is paid to the meaning of s.

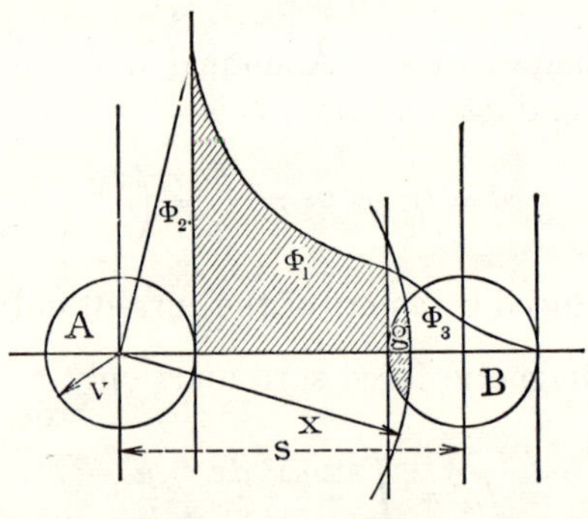

Fig. 69.—Inductance Calculation of Cable.

However, if s is of the same magnitude as r, as with the conductors of cables, the meaning of s has to be specified.

Let then in Fig. 69 r = radius of conductors, and s = distance between conductor centers. Assuming uniform current density in the conductors, the flux distribution of conductor A then is as indicated diagrammatically in Fig. 69.

* $0.4343 = \log_{10}\epsilon$.

The flux then consists of three parts:

Φ_1, between the conductors, giving the inductance

$$L_1 = 2 \log \frac{s - r}{r},$$

and shown shaded in Fig. 69.

Φ_2, inside of conductor A, giving the inductance

$$L_2 = \frac{\mu}{2}.$$

Φ_3, the flux external to A, which passes through conductor B and thereby incloses the conductor A and part of the conductor B, and thus has a m.m.f. less than i, that is, gives $\frac{F}{i} < 1$.

That is, a line of magnetic force at distance $s - r < x < s + r$ incloses the part q of the conductor B, thus incloses the fraction $\frac{q}{r^2\pi}$ of the return current, and thus has the m.m.f.

$$\frac{F}{i} = 1 - \frac{q}{r^2\pi}.$$

An exact calculation of the flux Φ_3, and the component inductance L_3 resulting from it, is complicated, and, due to the nature of the phenomenon, the result could not be accurate; and an approximation is sufficient in giving an accuracy as great as the variability of the phenomenon permits.

The magnetic flux Φ_3 does not merely give an inductance, but, if alternating, produces a potential difference between the two sides of conductor B, and thereby a higher current density on the side of B toward A; and as this effect depends on the conductivity of the conductor material, and on the frequency of the current, it cannot be determined without having the frequency, etc., given. The same applies for the flux Φ_1, which is reduced by unequal current density due to its screening effect, so that in the limiting case, for conductors of perfect conductivity, that is, zero resistance, or for infinite, that is, very high frequency, only the magnetic flux Φ_1 exists, which is shown shaded in Fig. 5; but Φ_2 and Φ_3 are zero, and the inductance is

$$L = 2 \log \frac{s - r}{r} 10^{-9} h. \tag{15}$$

That is, in other words, with small conductors and moderate currents, the total inductance in Fig. 69 is so small compared with the inductances in the other parts of the electric circuit that no very great accuracy of its calculation is required; with large conductors and large currents, however, the unequal current distribution and resultant increase of resistance become so considerable, with round conductors, as to make their use uneconomical, and leads to the use of flat conductors. With flat conductors, however, conductivity and frequency enter into the value of inductance as determining factors.

The exact determination of the inductance of round parallel conductors at short distances from each other thus is only of theoretical, but rarely of practical, importance.

An approximate estimate of the inductance L_3 is given by considering two extreme cases:

(*a*) The return conductor is of the shape Fig. 70, that is, from $s - r$ to $s + r$ the m.m.f. varies uniformly.

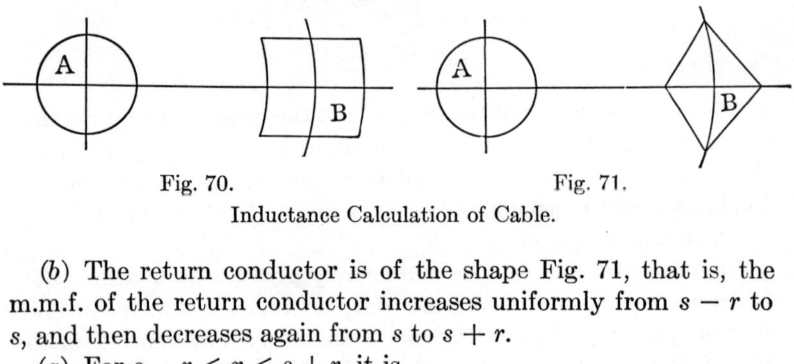

Fig. 70. Fig. 71.

Inductance Calculation of Cable.

(*b*) The return conductor is of the shape Fig. 71, that is, the m.m.f. of the return conductor increases uniformly from $s - r$ to s, and then decreases again from s to $s + r$.

(*a*) For $s - r < x < s + r$, it is

$$\frac{F}{i} = \frac{s + r - x}{2r} = \frac{s + r}{2r} - \frac{x}{2r}, \tag{16}$$

hence by (8),

$$L_3 = \int_{s-r}^{s+r} \frac{s + r}{r} \frac{dx}{x} - \int_{s-r}^{s+r} \frac{dx}{r}$$

$$= \frac{s + r}{r} \log \frac{s + r}{s - r} - 2, \tag{17}$$

by the approximation

$$\log (1 \pm x) = \pm x + \frac{x^2}{2} \pm \ . \ . \ . \tag{18}$$

it is

$$\log\frac{s+r}{s-r} = \log\frac{s+r}{s} - \log\frac{s-r}{s} = \log\left(1+\frac{r}{s}\right) - \log\left(1-\frac{r}{s}\right) = 2\frac{r}{s},$$

hence

$$L_3 = \frac{s+r}{r}\times\frac{2\,r}{s} - 2 = \frac{2\,r}{s}. \tag{19}$$

(*b*) For $s - r < x < s$, it is

$$\frac{F}{i} = 1 - \frac{1}{2}\left(\frac{x-s+r}{r}\right)^2, \tag{20}$$

and for $s < x < s + r$, it is

$$\frac{F}{i} = \frac{1}{2}\left(\frac{s+r-x}{r}\right)^2, \tag{21}$$

hence,

$$L_3 = \int_{s-r}^{s}\left[2 - \left(\frac{x-s+r}{r}\right)^2\right]\frac{dx}{x} + \int_{s}^{s+r}\left(\frac{s+r-x}{r}\right)^2\frac{dx}{x}, \tag{22}$$

and integrated this gives

$$L_3 = 2\log\frac{s}{s-r} + \frac{(s+r)^2}{r^2}\log\frac{s+r}{s} - \frac{(s-r)^2}{r^2}\log\frac{s}{s-r} - 3, \tag{23}$$

and by the approximation (18) this reduces to

$$L_3 = \frac{2\,r}{s}, \tag{24}$$

that is, the same value as (19); and as the actual case, Fig. 69, should lie between Figs. 70 and 71, the common approximation of the latter two cases should be a close approximation of case 4.

That is, for conductors close together it is

$$\begin{aligned} L &= L_1 + L_2 + L_3 \\ &= 2\left\{\log\frac{s-r}{r} + \frac{\mu}{4} + \frac{r}{s}\right\} 10^{-9}\,h. \end{aligned} \tag{25}$$

However, $\frac{r}{s}$ can be considered as the approximation of $-\log\left(1-\frac{r}{s}\right) = \log\frac{s}{s-r}$, and substituting this in (25) gives, by combining $\log\frac{s-r}{r} + \log\frac{s}{s-r} = \log\frac{s}{r}$:

$$L = 2\left\{\log\frac{s}{r} + \frac{\mu}{4}\right\} 10^{-9}\,h, \tag{26}$$

where s = distance between conductor centers, as the closest approximation in the case where the distance between the conductors is small. This is the same expression as (13).

In view of the secondary phenomena unavoidable in the conductors, equation (26) appears sufficiently accurate for all practical purposes, except when taking into consideration the secondary phenomena, as unequal current distribution, etc., in which case the frequency, conductivity, etc., are required.

D. Calculation of capacity.

49. The lines of dielectric force of the conductor A are straight radial lines, shown dotted in Fig. 72, and the dielectric equipotential lines are concentric circles, shown drawn in Fig. 72.

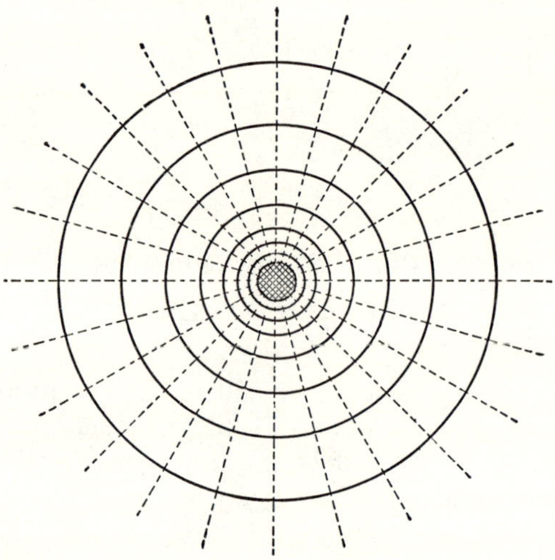

Fig. 72.—Electric Field of Conductor.

If e = voltage between conductor A and return conductor B, and s the distance between the conductors, the potential difference between the equipotential line at the surface of A, and the equipotential line which traverses B, must be e.

If e = potential difference or voltage, and l = distance, over which this potential difference acts,

$$G = \frac{e}{l} = \text{potential gradient, or electrifying force,} \qquad (27)$$

and $$K = \frac{G}{4\,\pi v^2} = \frac{e}{4\,\pi v^2 l} = \text{dielectric field intensity,} \tag{28}$$

where v^2 is the reduction factor from the electrostatic to the electromagnetic system of units, and

$$v = 3 \times 10^{10} \text{ cm. sec.} = \text{velocity of light;} \tag{29}$$

the dielectric density then is

$$D = \kappa K = \frac{\kappa e}{4\,\pi v^2 l}, \tag{30}$$

where κ = specific capacity of medium, = 1 in air. The dielectric flux then is

$$\Psi = AD = \frac{\kappa e A}{4\,\pi v^2 l}, \tag{31}$$

where A = section of dielectric flux. Or inversely:

$$e = \frac{4\,\pi v^2 l}{\kappa A}\Psi. \tag{32}$$

If then Ψ = dielectric flux, in Fig. 68, at a distance x from the conductor A, in a zone of thickness dx, and section $2\,\pi x$, the voltage is, by (32),

$$\begin{aligned} de &= \frac{4\,\pi v^2\,dx}{2\,\pi x\kappa}\Psi \\ &= \frac{2\,v^2}{\kappa}\Psi\frac{dx}{x}, \end{aligned} \tag{33}$$

and the voltage consumed between distances x_1 and x_2 thus is

$$e\Big] = \int_{x_1}^{x_2} de = \frac{2\,v^2\Psi}{\kappa}\log\frac{x_2}{x_1}, \tag{34}$$

hence the capacity of this space:

$$C\Big]_1^2 - \frac{\Psi}{e\Big]_1^2} = \frac{\kappa}{2\,v^2\log\frac{x_2}{x_1}}. \tag{35}$$

The capacity of the conductor A with the return conductor at distance s then is the capacity of the space from the distance $x_1 = r$ to the distance $x_2 = s$, hence is, by (35),

$$C = \frac{\kappa}{2\,v^2\log\frac{s}{r}} \text{ per cm.} \tag{36}$$

in absolute units, hence, reduced to farads,

$$C = \frac{\kappa\, 10^9}{2\, v^2 \log \frac{s}{r}} f \text{ per cm.,} \tag{37}$$

and in air, for $\kappa = 1$:

$$C = \frac{10^9}{2\, v^2 \log \frac{s}{r}} f \text{ per cm.} \tag{38}$$

is the capacity, per conductor, or " capacity to neutral," as often stated.

Immediately it follows: the external inductance was, by (9),

$$L_1 = 2 \log \frac{s}{r}\, 10^{-9}\, h \text{ per cm.,}$$

and multiplying this with (38) gives

$$\left.\begin{aligned} CL_1 &= \frac{1}{v^2}, \\ \text{or} \qquad C &= \frac{1}{v^2 L_1}; \end{aligned}\right\} \tag{39}$$

that is, the capacity equals the reciprocal of the external inductance L_1 times the velocity square of light. The external inductance L_1 would be the inductance of a conductor which had perfect conductivity, or zero losses of power. It is

$$v' = \frac{1}{\sqrt{LC}}$$

= velocity of propagation of the electric field, and this velocity is less than the velocity of light, due to the retardation by the power dissipation in the conductor, and becomes equal to the velocity of light v if there is no power dissipation, and, in the latter case, L would be equal to L_1, the external inductance.

The equation (39) is the most convenient to calculate capacities in complex systems of circuits from the inductances, or inversely, to determine the inductance of cables from the measured capacity, etc., More complete, this equation is

$$CL_1 = \frac{\kappa\mu}{v^2}, \tag{40}$$

where κ = specific capacity or permittivity, μ = permeability of the medium.

E. Conductor with ground return.

50. As seen in the preceding, in the electric field of conductor A and return conductor B, at distance s from each other, Fig. 9, the lines of magnetic force from conductor A to the center line CC' are equal in number and in magnetic energy to the lines of magnetic force which surround the conductor in Fig. 67, in concentric circles up to the distance s, and give the inductance L of conductor A. The lines of dielectric force which radiate from conductor A up to the center line CC', Fig. 9, are equal in number and in dielectric energy to the lines of dielectric force which issue as straight lines from the conductor, Fig. 8, up to the distance s, and represent the capacity C of the conductor A. The center line CC' is a dielectric equipotential line, and a line of magnetic force, and therefore, if it were replaced by a conducting plane of perfect conductivity, this would exert no effect on the magnetic or the dielectric field between the conductors A and B.

If then, in the electric field between overhead conductor and ground, we consider the ground as a plane of perfect conductivity, we get the same electric field as between conductor A and central plane CC' in Fig. 9. That is, the equations of inductance and capacity of a conductor with return conductor at distance s can be immediately applied to the inductance and capacity of a conductor with ground return, by using as distance s twice the distance of the conductor from the ground return. That is, the inductance and capacity of a conductor with ground return are the same as the inductance and capacity of the conductor against its *image conductor*, that is, against a conductor at the same distance below the ground as the conductor is above ground.

As the distance s between conductor and image conductor in the case of ground return is very much larger — usually 10 and more times — than the distance between conductor and overhead return conductor, the inductance of a conductor with ground return is much larger, and the capacity smaller, than that of the same conductor with overhead return. In the former case, however, this inductance and capacity are those of the entire circuit, since the ground return, as conducting plane, has no inductance and capacity; while in the case of overhead return, the inductance of the entire circuit of conductor and return conductor is twice, the capacity half, that of a single conductor, and therefore the total inductance of a circuit of two overhead conductors is greater, the capacity less, than that of a single conductor with ground return.

The conception of the image conductor is based on that of the ground as a conducting plane of perfect conductivity, and assumes that the return is by a current sheet at the ground surface. As regards the capacity, this is probably almost always the case, as even dry sandy soil or rock has sufficient conductivity to carry, distributed over its wide surface, a current equal to the capacity current of the overhead conductor. With the magnetic field, and thus with the inductance, this is not always the case, but the conductivity of the soil may be much below that required to conduct the return current as a surface current sheet. If the return current penetrates to a considerable depth into the ground, it may be represented approximately as a current sheet at some distance below the ground, and the "image conductor" then is not the image of the overhead conductor below ground, but much lower; that is, the distance s in the equation of the inductance is more, and often much more, than twice the distance of the overhead conductor above ground. However, even if the ground is of relatively low conductivity, and the return current thus has to penetrate to a considerable distance into the ground, the inductance of the overhead conductor usually is not very much increased, as it varies only little with the distance s. For instance, if the overhead conductor is $\frac{1}{2}$ inch diameter and 25 feet above ground, then, assuming perfect conductivity of the ground surface, the inductance would be

$$r = \tfrac{1}{4}''; \; s = 2 \times 25' = 600'', \text{ hence } \frac{s}{r} = 2400,$$

and

$$L = 2\left\{\log \frac{s}{r} + \frac{1}{4}\right\} 10^{-9} = 16.066 \times 10^{-9}\, h.$$

If, however, the ground were of such high resistance that the current would have to penetrate to a depth of over a hundred feet, and the mean depth of the ground current were at 50 feet, this would give $s = 2 \times 75' = 1800''$, hence $\frac{s}{r} = 7200$, and

$$L = 18.264 \times 10^{-9}\, h,$$

or only 13.7 per cent higher. In this case, however, the ground section available for the return current, assuming its effective width as 800 feet, would be 80,000 square feet, or 60 million times greater than the section of the overhead conductor.

Thus only with very high resistance soil, as very dry sandy soil, or rock, can a considerable increase of the inductance of the overhead conductor be expected over that calculated by the assumption of the ground as perfect conductor.

F. Mutual induction between circuits.

51. The mutual inductance between two circuits is the ratio of the current in one circuit into the magnetic flux produced by this current and interlinked with the second circuit. That is,

$$L_m = \frac{\Phi_2}{i_1} = \frac{\Phi_1}{i_2},$$

where Φ_2 is the magnetic flux interlinked with the second circuit, which is produced by current i_1 in the first circuit.

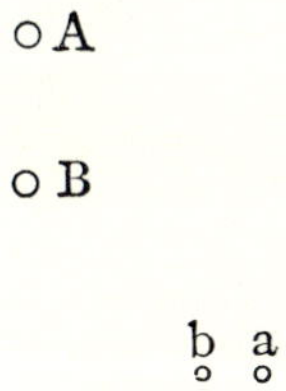

Fig. 73.

In the same manner as the self-inductance L, the mutual inductance L_m between two circuits is calculated; while the (external) self-inductance corresponds to the magnetic flux between the distances r and s, the mutual inductance of a conductor A upon a circuit ab corresponds to the magnetic flux produced by the conductor A and passing between the distances Aa and Ab, Fig. 73.

Thus the mutual inductance between a circuit AB and a circuit ab is mutual inductance of A upon ab,

$$L_m' = 2 \log \frac{Aa}{Ab} \times 10^{-9}\, h,$$

mutual inductance of B upon ab,

$$L_m'' = 2 \log \frac{Ba}{Bb} \times 10^{-9}\, h,$$

hence mutual inductance between circuits AB and ab,

$$\begin{aligned} L_m &= L_m'' - L_m', \\ &= 2 \log \frac{Ba \times Ab}{Aa \times Bb} 10^{-9}\, h, \end{aligned} \tag{41}$$

where Aa, Ab, Ba, Bb are the distances between the respective conductors, as shown in Fig. 74.

If one or both circuits have ground return, they are replaced by the circuit of the overhead conductor and its image conductor below ground, as discussed before.

If the distance D between the circuits AB and ab is great compared to the distance S between the conductors of circuit AB, and the distance s between the conductors of circuit ab, and ϕ = angle which the plane of circuit AB makes with the distance D, ψ the corresponding angle of circuit ab, as shown in Fig. 74, it is approximately

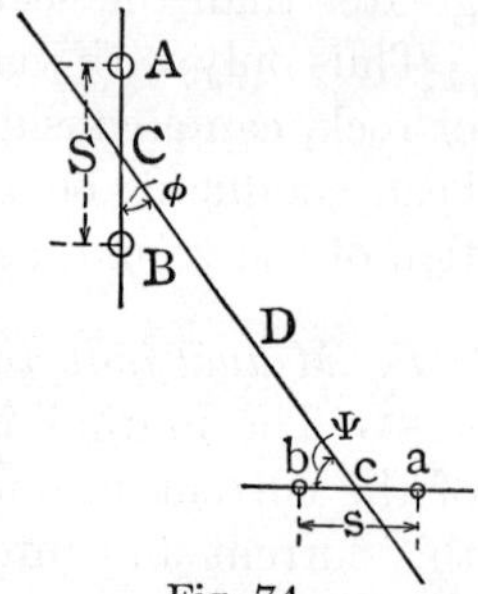

Fig. 74.

$$\left.\begin{aligned} Aa &= D + \frac{S}{2}\cos\phi + \frac{s}{2}\cos\psi, \\ Ab &= D + \frac{S}{2}\cos\phi - \frac{s}{2}\cos\psi, \\ Ba &= D - \frac{S}{2}\cos\phi + \frac{s}{2}\cos\psi, \\ Bb &= D - \frac{S}{2}\cos\phi - \frac{s}{2}\cos\psi, \end{aligned}\right\} \quad (42)$$

hence

$$L_m = 2\log\frac{\left(D+\frac{S}{2}\cos\phi-\frac{s}{2}\cos\psi\right)\left(D-\frac{S}{2}\cos\phi+\frac{s}{2}\cos\psi\right)}{\left(D+\frac{S}{2}\cos\phi+\frac{s}{2}\cos\psi\right)\left(D-\frac{S}{2}\cos\phi-\frac{s}{2}\cos\psi\right)}\,10^{-9}h$$

$$= 2\log\frac{D^2-\left(\frac{S}{2}\cos\phi-\frac{s}{2}\cos\psi\right)^2}{D^2-\left(\frac{S}{2}\cos\phi+\frac{s}{2}\cos\psi\right)^2}\,10^{-9}\,h$$

$$= 2\left\{\log\left[1-\left(\frac{\frac{S}{2}\cos\phi-\frac{s}{2}\cos\psi}{D}\right)^2\right] - \log\left[1-\left(\frac{\frac{S}{2}\cos\phi+\frac{s}{2}\cos\psi}{D}\right)^2\right]\right\}\times 10^{-9}\,h,$$

hence by (18)

$$L_m = 2\left\{\frac{\left(\frac{S}{2}\cos\phi+\frac{s}{2}\cos\psi\right)^2-\left(\frac{S}{2}\cos\phi-\frac{s}{2}\cos\psi\right)^2}{D^2}\right\}10^{-9}\,h;$$

thus

$$L_m = \frac{2\,Ss \cos\phi \cos\psi}{D^2} 10^{-9}\,h. \tag{43}$$

For $\phi = 90$ degrees or $\psi = 90$ degrees, L_m is a minimum, and the approximation (43) vanishes.

G. Mutual capacity between circuits.

52. The mutual capacity between two circuits is the ratio of the voltage between the conductors of one circuit into the dielectric flux produced by this voltage between the conductors of the other circuit. That is

$$C_m = \frac{\Psi_2}{e_1} = \frac{\Psi_1}{e_2},$$

where Ψ_2 is the dielectric flux produced between the conductors of the second circuit by the voltage e_1 between the conductors of the first circuit.

If e = voltage between conductors A and B, the dielectric flux of conductor A is, by (36),

$$\Psi = Ce = \frac{\kappa e}{2\,v^2 \log \frac{S}{R}}, \tag{44}$$

where R is the radius of these conductors and S their distance from each other.

This dielectric flux produces, by (32), between the distances Aa and Ab, the potential difference

$$e' = \frac{2\,v^2\Psi}{\kappa} \log \frac{Aa}{Ab}, \tag{45}$$

and the dielectric flux of conductor B produces the potential difference

$$e'' = \frac{2\,v^2\Psi}{\kappa} \log \frac{Ba}{Bb}; \tag{46}$$

hence the total potential difference between a and b is

$$e'' - e' = \frac{2\,v^2\Psi}{\kappa} \log \frac{Ab\,Ba}{Aa\,Bb}; \tag{47}$$

substituting (44) into (47),

$$e'' - e' = \frac{e}{\log \frac{S}{R}} \log \frac{Ab\,Ba}{Aa\,Bb},$$

and the dielectric flux produced by the potential difference $e'' - e'$ between the conductors a and b is

$$\Psi_{\prime\prime} = \frac{\kappa e}{2\, v^2 \log \frac{s}{r} \log \frac{S}{R}} \log \frac{Ab}{Aa} \frac{Ba}{Bb},$$

hence the mutual capacity

$$C_m = \frac{\kappa}{2\, v^2 \log \frac{s}{r} \log \frac{S}{r}} \log \frac{Ab}{Aa} \frac{Ba}{Bb} 10^9 f, \tag{48}$$

or, by approximation (18), as in (43),

$$C_m = \frac{\kappa Ss \cos \phi \cos \psi}{2\, D^2 v^2 \log \frac{s}{r} \log \frac{S}{R}} 10^9 f. \tag{49}$$

This value applies only if conductors A and B have the same voltage against ground, in opposite direction, as is the case if their neutral is grounded.

If the voltages are different, e_1 and e_2, where $e_1 + e_2 = 2\, e$, as for instance one conductor grounded:

$$e_1 = 0,\ e_2 = e, \tag{50}$$

the dielectric fluxes of the two conductors are different, and that of A is: $c_1 \Psi$; that of B is: $c_2 \Psi$, where

$$\left.\begin{aligned} c_1 &= \frac{e_1}{e}, \\ c_2 &= \frac{e_2}{e}, \end{aligned}\right\} \tag{51}$$

and

$$c_1 + c_2 = 2,$$

the equations (45) to (49) assume the forms:

$$e' = \frac{2\, v^2 c_1 \Psi}{\kappa} \log \frac{Aa}{Ab}, \tag{52}$$

$$e'' = \frac{2\, v^2 c_2 \Psi}{\kappa} \log \frac{Ba}{Bb}, \tag{53}$$

$$e'' - e' = \frac{2\, v^2 \Psi}{\kappa} \left\{ c_2 \log \frac{Ba}{Bb} - c_1 \log \frac{Aa}{Ab} \right\}, \tag{54}$$

$$\left.\begin{aligned} C_m &= \frac{\kappa}{2\, v^2 \log\frac{s}{r} \log\frac{S}{r}} \left\{ c_2 \log \frac{Ba}{Bb} - c_1 \log \frac{Aa}{Ab} \right\} 10^{-9} f \\ &= \frac{\kappa}{2\, v^2 \log\frac{s}{r} \log\frac{S}{R}} \log \left(\frac{Ab}{Aa}\right)^{c_1} \left(\frac{Ba}{Bb}\right)^{c_2} 10^{-9} f, \end{aligned}\right\} \quad (55)$$

and by (42):

$$\left\{ c_2 \log \frac{Ba}{Bb} - c_1 \log \frac{Aa}{Ab} \right\}$$

$$= + c_2 \left\{ \log\left(1 - \frac{S\cos\phi - s\cos\psi}{2\, D}\right) - \log\left(1 - \frac{S\cos\phi + s\cos\psi}{2\, D}\right) \right\}$$

$$- c_1 \left\{ \log\left(1 + \frac{S\cos\phi + s\cos\psi}{2\, D}\right) - \log\left(1 + \frac{S\cos\phi - s\cos\psi}{2\, D}\right) \right\}$$

and this gives:

$$= \frac{(c_2 - c_1)\, s\cos\psi}{D} + \frac{(c_1 + c_2)\, Ss\cos\phi\cos\psi}{2\, D^2}, \quad (56)$$

hence

$$C_m = \frac{\kappa}{2\, v^2 \log\frac{s}{r} \log\frac{S}{R}} \left\{ (c_2 - c_1) \frac{s\cos\psi}{D} + (c_2 + c_1) \frac{Ss\cos\phi\cos\psi}{2\, D^2} \right\} 10^9 f \quad (57)$$

and for $e_1 = 0$, and thus $c_1 = 0$, $c_2 = 2$:

$$C_m = \frac{\kappa s \cos\psi}{D v^2 \log\frac{s}{r} \log\frac{S}{R}} \left\{ 1 + \frac{S\cos\phi}{2\, D} \right\} 10^9 f, \quad (58)$$

hence very much larger than (49). However, equation (58) applies only, if the ground is at a distance very large compared with D, as it does not consider the ground as the static return of the conductor B.

H. The three-phase circuit.

53. The equations of the inductance and the capacity of a conductor

$$L = 2 \left\{ \log\frac{s}{r} + \frac{\mu}{4} \right\} 10^{-9}\, h, \quad (26)$$

$$C = \frac{\kappa}{2\, v^2 \log\frac{s}{r}}\, 10^9\, f \quad (37)$$

apply equally to the two-wire single-phase circuit, the single wire circuit with ground return, or the three-phase circuit.

In the expression of the energy per conductor:

$$\left.\begin{aligned} w_m &= \frac{Li^2}{2} \\ w_d &= \frac{Ce^2}{2}, \end{aligned}\right\} \qquad (59)$$

and of the inductance voltage e' and capacity current i', per conductor:

$$\left.\begin{aligned} e' &= 2\,\pi f L i, \\ i' &= 2\,\pi f C e, \end{aligned}\right\} \qquad (60)$$

i is the current in the conductor, thus in a three-phase system the Y or star current, and e is the voltage per conductor, that is, the voltage from conductor to ground, which is one-half the voltage between the conductors of a single-phase two-wire circuit, $\frac{1}{\sqrt{3}}$ the voltage between the conductors of a three-phase circuit (that is, it is the Y or star voltage), and is the voltage of the circuit in a conductor to ground, s is the distance between the conductors, and is twice the distance from conductor to ground in a single conductor with ground return.*

If the conductors of a three-phase system are arranged in a triangle, s is the same for all three conductors; otherwise the different conductors have different values of s, and the same conductor may have two different values of s, for its two return conductors or phases.

For instance, in the common arrangement of the three-phase conductors above each other, or beside each other, as shown in Fig. 75, if s is the distance between middle conductor and outside conductors, the distance between the two outside conductors is $2\,s$.

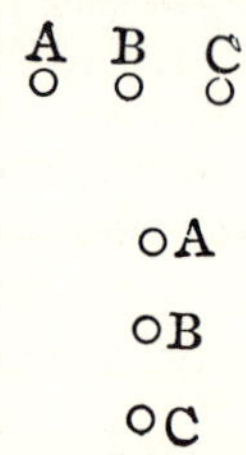

Fig. 75.

The inductance of the middle conductor then is:

$$L = 2\left\{\log\frac{s}{r} + \frac{\mu}{4}\right\} 10^{-9}\,h. \qquad (61)$$

The inductance of each of the outside conductors is, with respect to the middle conductor:

* See discussion in paragraph 50.

$$L = 2\left\{\log\frac{s}{r} + \frac{\mu}{4}\right\} 10^{-9}\,h. \tag{62}$$

With respect to the other outside conductor:

$$L = 2\left\{\log\frac{2\,s}{r} + \frac{\mu}{4}\right\} 10^{-9}\,h. \tag{63}$$

The inductance (62) applies to the component of current, which returns over the middle conductor, the inductance (63), which is larger, to the component of current which returns over the other outside conductor. These two currents are 60 degrees displaced in phase from each other. The inductance voltages, which are 90 degrees ahead of the current, thus also are 60 degrees displaced from each other. As they are unequal, their resultant is not 90 degrees ahead of the resultant current, but more in the one, less in the other outside conductor. The inductance voltage of the two outside conductors thus contains an energy component, which is positive in the one, negative in the other outside conductor. That is, a power transfer by mutual inductance occurs between the outside conductors of the three-phase circuit arranged as in Fig. **75**. The investigation of this phenomenon is given by C. M. Davis in the *Electrical Review and Western Electrician* for April 1, 1911.

If the line conductors are transposed sufficiently often to average their inductances, the inductances of all three conductors, and also their capacities, become equal, and can be calculated by using the average of the three distances s, s, $2\,s$ between the conductors, that is, $\frac{4}{3}\,s$, or more accurately, by using the average of the $\log\frac{s}{r}$, $\log\frac{s}{r}$ and $\log\frac{2\,s}{r}$, that is:

$$\frac{2\log\frac{s}{r} + \log\frac{2\,s}{r}}{3}. \tag{64}$$

In the same manner, with any other configuration of the line conductors, in case of transposition the inductance and capacity can be calculated by using the average value of the $\log\frac{s}{r}$ between the three conductors.

The calculation of the mutual inductance and mutual capacity between the three-phase circuit and a two-wire circuit is made in the same manner as in equation (41), except that three terms appear, and the phases of the three currents have to be considered.

Thus, if A, B, C are the three three-phase conductors, and a and b the conductors of the second circuit, as shown in Fig. 76, and if i_1, i_2, i_3 are the three currents, with their respective phase angles γ_1, γ_2, γ_3, and i the average current, denoting:

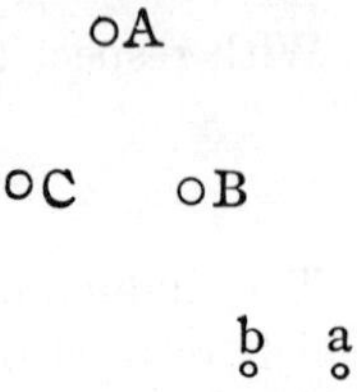

Fig. 76.

$$c_1 = \frac{i_1}{i}; \; c_2 = \frac{i_2}{i}; \; c_3 = \frac{i_3}{i};$$

conductor A gives:

$$L_m' = 2\, c_1 \cos (\beta - \gamma_1) \log \frac{Aa}{Ab},$$

conductor B:

$$L_m'' = 2\, c_2 \cos (\beta - 120° - \gamma_2) \log \frac{Ba}{Bb},$$

conductor C:

$$L_m''' = 2\, c_3 \cos (\beta - 240° - \gamma_3) \log \frac{Ca}{Cb};$$

hence,

$$L_m = 2 \left\{ c_1 \cos (\beta - \gamma_1) \log \frac{Aa}{Ab} + c_2 \cos (\beta - 120° - \gamma_2) \log \frac{Ba}{Bb}, \right.$$

$$\left. + \, c_3 \cos (\beta - 240° - \gamma_3) \log \frac{Ca}{Cb} \right\} 10^{-9}\, h,$$

and in analogous manner the capacity C_m is derived.

In these expressions, the trigonometric functions represent a rotation of the inductance combined with a pulsation.

PART III

TRANSIENTS IN TIME

CHAPTER I.

THE CONSTANTS OF THE ELECTRIC CIRCUIT

1. To transmit electric energy from one place where it is generated to another place where it is used, an electric circuit is required, consisting of conductors which connect the point of generation with the point of utilization.

When electric energy flows through a circuit, phenomena take place inside of the conductor as well as in the space outside of the conductor.

In the conductor, during the flow of electric energy through the circuit, electric energy is consumed continuously by being converted into heat. Along the circuit, from the generator to the receiver circuit, the flow of energy steadily decreases by the amount consumed in the conductor, and a power gradient exists in the circuit along or parallel with the conductor.

(Thus, while the voltage may decrease from generator to receiver circuit, as is usually the case, or may increase, as in an alternating-current circuit with leading current, and while the current may remain constant throughout the circuit, or decrease, as in a transmission line of considerable capacity with a leading or non-inductive receiver circuit, the flow of energy always decreases from generating to receiving circuit, and the power gradient therefore is characteristic of the direction of the flow of energy.)

In the space outside of the conductor, during the flow of energy through the circuit, a condition of stress exists which is called the *electric field* of the conductor. That is, the surrounding space is not uniform, but has different electric and magnetic properties in different directions.

No power is required to maintain the electric field, but energy is required to produce the electric field, and this energy is returned, more or less completely, when the electric field disappears by the stoppage of the flow of energy.

Thus, in starting the flow of electric energy, before a permanent condition is reached, a finite time must elapse during which the energy of the electric field is stored, and the generator therefore gives more power than consumed in the conductor and delivered at the receiving end; again, the flow of electric energy cannot be stopped instantly, but first the energy stored in the electric field has to be expended. As result hereof, where the flow of electric energy pulsates, as in an alternating-current circuit, continuously electric energy is stored in the field during a rise of the power, and returned to the circuit again during a decrease of the power.

The electric field of the conductor exerts magnetic and electrostatic actions.

The magnetic action is a maximum in the direction concentric, or approximately so, to the conductor. That is, a needle-shaped magnetizable body, as an iron needle, tends to set itself in a direction concentric to the conductor.

The electrostatic action has a maximum in a direction radial, or approximately so, to the conductor. That is, a light needle-shaped conducting body, if the electrostatic component of the field is powerful enough, tends to set itself in a direction radial to the conductor, and light bodies are attracted or repelled radially to the conductor.

Thus, the electric field of a circuit over which energy flows has three main axes which are at right angles with each other:

The electromagnetic axis, concentric with the conductor.

The electrostatic axis, radial to the conductor.

The power gradient, parallel to the conductor.

This is frequently expressed pictorially by saying that the lines of magnetic force of the circuit are concentric, the lines of electrostatic force radial to the conductor.

Where, as is usually the case, the electric circuit consists of several conductors, the electric fields of the conductors superimpose upon each other, and the resultant lines of magnetic and of electrostatic forces are not concentric and radial respectively, except approximately in the immediate neighborhood of the conductor.

In the electric field between parallel conductors the magnetic and the electrostatic lines of force are conjugate pencils of circles.

2. Neither the power consumption in the conductor, nor the electromagnetic field, nor the electrostatic field, are proportional to the flow of energy through the circuit.

The product, however, of the intensity of the magnetic field, Φ, and the intensity of the electrostatic field, Ψ, is proportional to the flow of energy or the power, P, and the power P is therefore resolved into a product of two components, i and e, which are chosen proportional respectively to the intensity of the magnetic field Φ and of the electrostatic field Ψ.

That is, putting

$$P = ie \tag{1}$$

we have

$$\Phi = Li = \text{the intensity of the electromagnetic field.} \tag{2}$$

$$\Psi = Ce = \text{the intensity of the electrostatic field.} \tag{3}$$

The component i, called the *current*, is defined as that factor of the electric power P which is proportional to the magnetic field, and the other component e, called the voltage, is defined as that factor of the electric power P which is proportional to the electrostatic field.

Current i and voltage e, therefore, are mathematical fictions, factors of the power P, introduced to represent respectively the magnetic and the electrostatic or "dielectric" phenomena.

The current i is measured by the magnetic action of a circuit, as in the ammeter; the voltage e, by the electrostatic action of a circuit, as in the electrostatic voltmeter, or by producing a current i by the voltage e and measuring this current i by its magnetic action, in the usual voltmeter.

The coefficients L and C, which are the proportionality factors of the magnetic and of the dielectric component of the electric field, are called the *inductance* and the *capacity* of the circuit, respectively.

As electric power P is resolved into the product of current i and voltage e, the power loss in the conductor, P_l, therefore can also be resolved into a product of current i and voltage e_l which is consumed in the conductor. That is,

$$P_l = ie_l.$$

It is found that the voltage consumed in the conductor, e_l, is proportional to the factor i of the power P, that is,

$$e_l = ri, \tag{4}$$

where r is the proportionality factor of the voltage consumed by the loss of power in the conductor, or by the power gradient, and is called the *resistance* of the circuit.

Any electric circuit therefore must have three constants, r, L, and C, where

r = circuit constant representing the power gradient, or the loss of power in the conductor, called *resistance.*

L = circuit constant representing the intensity of the electromagnetic component of the electric field of the circuit, called *inductance.*

C = circuit constant representing the intensity of the electrostatic component of the electric field of the circuit, called *capacity.*

In most circuits, there is no current consumed in the conductor, i_l, and proportional to the voltage factor e of the power P, that is:

$$i_l = ge$$

where g is the proportionality factor of the current consumed by the loss of power in the conductor, which depends on the voltage, such as dielectric losses, etc. Where such exist, a fourth circuit constant appears, the *conductance* g, regarding which see sections III and IV.

3. A change of the magnetic field of the conductor, that is, lf the number of lines of magnetic force Φ surrounding the conductor, generates an e.m.f.

$$e' = \frac{d\Phi}{dt} \tag{5}$$

in the conductor and thus absorbs a power

$$P' = ie' = i\frac{d\Phi}{dt} \tag{6}$$

or, by equation (2): $\Phi = Li$ by definition, thus:

$$\frac{d\Phi}{dt} = L\frac{di}{dt}, \text{ and: } P' = Li\frac{di}{dt}, \tag{7}$$

and the total energy absorbed by the magnetic field during the rise of current from zero to i is

$$W_M = \int P'dt \tag{8}$$

$$= L\int idi,$$

that is,

$$W_M = \frac{i^2L}{2}. \tag{9}$$

A change of the dielectric field of the conductor, Ψ, absorbs a current proportional to the change of the dielectric field:

$$i' = \frac{d\Psi}{dt}, \tag{10}$$

and absorbs the power

$$P'' = ei' = e\frac{d\Psi}{dt}, \tag{11}$$

or, by equation (3),

$$P'' = Ce\frac{de}{dt}, \tag{12}$$

and the total energy absorbed by the dielectric field during a rise of voltage from 0 to e is

$$\begin{aligned} W_K &= \int P''dt \qquad (13) \\ &= C\int ede, \end{aligned}$$

that is

$$W_K = \frac{e^2C}{2}. \tag{14}$$

The power consumed in the conductor by its resistance r is

$$P_r = ie_1, \tag{15}$$

and thus, by equation (4),

$$P_r = i^2r. \tag{16}$$

That is, when the electric power

$$P = ei \quad (1)$$

exists in a circuit, it is

$P_r = i^2r =$ power lost in the conductor, (16)

$W_M = \frac{i^2L}{2} =$ energy stored in the magnetic field of the circuit, (9)

$W_K = \frac{e^2C}{2} =$ energy stored in the dielectric field of the circuit, (14) and the three circuit constants r, L, C therefore appear as the components of the energy conversion into heat, magnetism, and electric stress, respectively, in the circuit.

4. The circuit constant, resistance r, depends only on the size and material of the conductor, but not on the position of the conductor in space, nor on the material filling the space surrounding the conductor, nor on the shape of the conductor section.

The circuit constants, inductance L and capacity C, almost entirely depend on the position of the conductor in space, on the material filling the space surrounding the conductor, and on the shape of the conductor section, but do not depend on the material of the conductor, except to that small extent as represented by the electric field inside of the conductor section.

5. The resistance r is proportional to the length and inversely proportional to the section of the conductor,

$$r = \rho \frac{l}{A}, \tag{17}$$

where ρ is a constant of the material, called the *resistivity* or *specific resistance.*

For different materials, ρ varies probably over a far greater range than almost any other physical quantity. Given in ohms per centimeter cube,* it is, approximately, at ordinary temperatures:

Metals:	Cu	1.6×10^{-6}
	Al	2.8×10^{-6}
	Fe	10×10^{-6}
	Hg	94×10^{-6}
	Gray cast iron	up to 100×10^{-6}
	High-resistance alloys	up to 150×10^{-6}
Electrolytes:	NO_3H	down to 1.3 at 30 per cent
	KOH	down to 1.9 at 25 per cent
	NaCl	down to 4.7 at 25 per cent
up to		
	Pure river water	10^4

and over alcohols, oils, etc., to practically infinity.

* Meaning a conductor of one centimeter length and one square centimeter section.

So-called "*insulators*":

Fiber	about 10^{12}
Paraffin oil	about 10^{13}
Paraffin	about 10^{14} to 10^{16}
Mica	about 10^{14}
Glass	about 10^{14} to 10^{16}
Rubber	about 10^{16}
Air	practically ∞

In the wide gap between the highest resistivity of metal alloys, about $\rho = 150 \times 10^{-6}$, and the lowest resistivity of electrolytes, about $\rho = 1$, are

Carbon:	metallic	down to 100×10^{-6}
	amorphous (dense)	0.04 and higher
	anthracite	very high

Silicon and *Silicon Alloys:*

Cast silicon	1 down to 0.04
Ferro silicon	0.04 down to 50×10^{-6}

The resistivity of *arcs* and of *Geissler tube discharges* is of about the same magnitude as electrolytic resistivity.

The resistivity, ρ, is usually a function of the temperature, rising slightly with increase of temperature in metallic conductors and decreasing in electrolytic conductors. Only with few materials, as silicon, the temperature variation of ρ is so enormous that ρ can no longer be considered as even approximately constant for all currents i which give a considerable temperature rise in the conductor. Such materials are commonly called pyroelectrolytes.

6. The inductance L is proportional to the section and inversely proportional to the length of the magnetic circuit surrounding the conductor, and so can be represented by

$$L = \frac{\mu A}{l}, \tag{18}$$

where μ is a constant of the material filling the space surrounding the conductor, which is called the magnetic *permeability*.

As in general neither section nor length is constant in different parts of the magnetic circuit surrounding an electric conductor, the magnetic circuit has as a rule to be calculated piecemeal, or by integration over the space occupied by it.

The permeability, μ, is constant and equals unity or very closely $\mu = 1$ for all substances, with the exception of a few materials which are called the magnetic materials, as iron, cobalt, nickel, etc., in which it is very much higher, reaching sometimes and under certain conditions in iron values as high as $\mu = 6000$ and even as high as $\mu = 30{,}000$.

In these magnetic materials the permeability μ is not constant but varies with the magnetic flux density, or number of lines of magnetic force per unit section, $\mathcal{B}$, decreasing rapidly for high values of $\mathcal{B}$.

In such materials the use of the term μ is therefore inconvenient, and the inductance, L, is calculated by the relation between the magnetizing force as given in ampere-turns per unit length of magnetic circuit, or by "field intensity," and magnetic induction $\mathcal{B}$.

The magnetic induction $\mathcal{B}$ in magnetic materials is the sum of the "space induction" $\mathcal{H}$, corresponding to unit permeability, plus the "metallic induction" $\mathcal{B}'$, which latter reaches a finite limiting value. That is,

$$\mathcal{B} = \mathcal{H} + \mathcal{B}'. \tag{19}$$

The limiting values, or so-called "saturation values," of $\mathcal{B}'$ are approximately, in lines of magnetic force per square centimeter:

Iron	21,000
Cobalt	12,000
Nickel	6,000
Magnetite	5,000
Manganese alloys	up to 5,000

The inductance, L, therefore is a constant of the circuit if the space surrounding the conductor contains no magnetic material, and is more or less variable with the current, i, if magnetic material exists in the space surrounding the conductor. In the latter case, with increasing current, i, the inductance, L, first slightly increases, reaches a maximum, and then decreases, approaching as limiting value the value which it would have in the absence of the magnetic material.

7. The capacity, C, is proportional to the section and inversely proportional to the length of the electrostatic field of the conductor:

$$C = \frac{\kappa A}{l}, \tag{20}$$

where κ is a constant of the material filling the space surrounding the conductor, which is called the "dielectric constant," or the "specific capacity," or "permittivity."

Usually the section and the length of the different parts of the electrostatic circuit are different, and the capacity therefore has to be calculated piecemeal, or by integration.

The dielectric constant κ of different materials varies over a relative narrow range only. It is approximately:

$\kappa = 1$ in the vacuum, in air and in other gases,
$\kappa = 2$ to 3 in oils, paraffins, fiber, etc.,
$\kappa = 3$ to 4 in rubber and gutta-percha,
$\kappa = 3$ to 5 in glass, mica, etc.,

reaching values as high as 7 to 8 in organic compounds of heavy metals, as lead stearate, and about 12 in sulphur.

The dielectric constant, κ, is practically constant for all voltages e, up to that voltage at which the electrostatic field intensity, or the electrostatic gradient, that is, the "volts per centimeter," exceeds a certain value δ, which depends upon the material and which is called the "dielectric strength" or "disruptive strength" of the material. At this potential gradient the medium breaks down mechanically, by puncture, and ceases to insulate, but electricity passes and so equalizes the potential gradient.

The disruptive strength, δ, given in volts per centimeter is approximately:

Air: 30,000.
Oils: 250,000 to 1,000,000.
Mica: up to 4,000,000.

The capacity, C, of a circuit therefore is constant up to the voltage e, at which at some place of the electrostatic field the dielectric strength is exceeded, disruption takes place, and a part of the surrounding space therefore is made conducting, and by this increase of the effective size of the conductor the capacity C is increased.

8. Of the amount of energy consumed in creating the electric field of the circuit not all is returned at the disappearance of the electric field, but a part is consumed by conversion into heat in producing or in any other way changing the electric field. That is, the conversion of electric energy into and from the electromagnetic and electrostatic stress is not complete, but a loss of energy occurs, especially with the magnetic field in the so-called magnetic materials, and with the electrostatic field in unhomogeneous dielectrics.

The energy loss in the production and reconversion of the magnetic component of the field can be represented by an effective resistance r' which adds itself to the resistance r_0 of the conductor and more or less increases it.

The energy loss in the electrostatic field can be represented by an effective resistance r'', shunting across the circuit, and consuming an energy current i'', in addition to the current i in the conductor. Usually, instead of an effective resistance r'', its reciprocal is used, that is, the energy loss in the electrostatic field represented by a shunted conductance g.

In its most general form the electric circuit therefore contains the constants:

1. Inductance L, storing the energy, $\frac{i^2L}{2}$,
2. Capacity C, storing the energy, $\frac{e^2C}{2}$,
3. Resistance $r = r_0 + r'$, consuming the power, $i^2r = i^2r_0 + i^2r'$,
4. Conductance g, consuming the power, e^2g,

where r_0 is the resistance of the conductor, r' the effective resistance representing the power loss in the magnetic field L, and g represents the power loss in the electrostatic field C.

9. If of the three components of the electric field, the electromagnetic stress, electrostatic stress, and the power gradient, one equals zero, a second one must equal zero also. That is, either all of the three components exist or only one exists.

Electric systems in which the magnetic component of the field is absent, while the electrostatic component may be considerable, are represented for instance by an electric generator or a battery on open circuit, or by the electrostatic machine. In such systems the disruptive effects due to high voltage, therefore, are most pronounced, while the power is negligible, and phenomena of this character are usually called "static."

Electric systems in which the electrostatic component of the field is absent, while the electromagnetic component is considerable, are represented for instance by the short-circuited secondary coil of a transformer, in which no potential difference and, therefore, no electrostatic field exists, since the generated e.m.f. is consumed at the place of generation. Practically negligible also is the electrostatic component in all low-voltage circuits.

The effect of the resistance on the flow of electric energy in industrial applications is restricted to fairly narrow limits: as the resistance of the circuit consumes power and thus lowers the efficiency of the electric transmission, it is uneconomical to permit too high a resistance. As lower resistance requires a larger expenditure of conductor material, it is usually uneconomical to lower the resistance of the circuit below that which gives a reasonable efficiency.

As result hereof, practically always the relative resistance, that is, the ratio of the power lost in the resistance to the total power, lies between 2 per cent and 20 per cent.

It is different with the inductance L and the capacity C. Of the two forms of stored energy, the magnetic $\frac{i^2L}{2}$ and electrostatic $\frac{e^2C}{2}$, usually one is so small that it can be neglected compared with the other, and the electric circuit with sufficient approximation treated as containing resistance and inductance, or resistance and capacity only.

In the so-called electrostatic machine and its applications, frequently only capacity and resistance come into consideration.

In all lighting and power distribution circuits, direct current or alternating current, as the 110- and 220-volt lighting circuits, the 500-volt railway circuits, the 2000-volt primary distribution circuits, due to the relatively low voltage, the electrostatic energy $\frac{e^2C}{2}$ is still so very small compared with the electromagnetic energy, that the capacity C can for most purposes be neglected and the circuit treated as containing resistance and inductance only.

Of approximately equal magnitude is the electromagnetic energy $\frac{i^2L}{2}$ and the electrostatic energy $\frac{e^2C}{2}$ in the high-potential long-distance transmission circuit, in the telephone circuit, and in the condenser discharge, and so in most of the phenomena resulting from lightning or other disturbances. In these cases all three circuit constants, r, L, and C, are of essential importance.

10. In an electric circuit of negligible inductance L and negligible capacity C, no energy is stored, and a change in the circuit thus can be brought about instantly without any disturbance or intermediary transient condition.

In a circuit containing only resistance and capacity, as a static machine, or only resistance and inductance, as a low or medium voltage power circuit, electric energy is stored essentially in one form only, and a change of the circuit, as an opening of the circuit, thus cannot be brought about instantly, but occurs more or less gradually, as the energy first has to be stored or discharged.

In a circuit containing resistance, inductance, and capacity, and therefore capable of storing energy in two different forms, the mechanical change of circuit conditions, as the opening of a circuit, can be brought about instantly, the internal energy of the circuit adjusting itself to the changed circuit conditions by a transfer of energy between static and magnetic and inversely, that is, after the circuit conditions have been changed, a transient phenomenon, usually of oscillatory nature, occurs in the circuit by the readjustment of the stored energy.

These transient phenomena of the readjustment of stored electric energy with a change of circuit conditions require careful study wherever the amount of stored energy is sufficiently large to cause serious damage. This is analogous to the phenomena of the readjustment of the stored energy of mechanical motion: while it may be harmless to instantly stop a slowly moving light carriage, the instant stoppage, as by collision, of a fast railway train leads to the usual disastrous result. So also, in electric systems of small stored energy, a sudden change of circuit conditions may be safe, while in a high-potential power system of very great stored electric energy any change of circuit conditions requiring a sudden change of energy is liable to be destructive.

Where electric energy is stored in one form only, usually little danger exists, since the circuit protects itself against sudden change by the energy adjustment retarding the change, and only where energy is stored electrostatically and magnetically, the mechanical change of the circuit conditions, as the opening of the circuit, can be brought about instantly, and the stored energy then surges between electrostatic and magnetic energy.

In the following, first the phenomena will be considered which result from the stored energy and its readjustment in circuits storing energy in one form only, which usually is as electromagnetic energy, and then the general problem of a circuit storing energy electromagnetically and electrostatically will be considered.

CHAPTER II

INTRODUCTION

11. In the investigation of electrical phenomena, currents and potential differences, whether continuous or alternating, are usually treated as stationary phenomena. That is, the assumption is made that after establishing the circuit a sufficient time has elapsed for the currents and potential differences to reach their final or permanent values, that is, become constant, with continuous current, or constant periodic functions of time, with alternating current. In the first moment, however, after establishing the circuit, the currents and potential differences in the circuit have not yet reached their permanent values, that is, the electrical conditions of the circuit are not yet the normal or permanent ones, but a certain time elapses while the electrical conditions adjust themselves.

12. For instance, a continuous e.m.f., e_0, impressed upon a circuit of resistance r, produces and maintains in the circuit a current,

$$i_0 = \frac{e_0}{r}.$$

In the moment of closing the circuit of e.m.f. e_0 on resistance r, the current in the circuit is zero. Hence, after closing the circuit the current i has to rise from zero to its final value i_0. If the circuit contained only resistance but no inductance, this would take place instantly, that is, there would be no transition period. Every circuit, however, contains some inductance. The inductance L of the circuit means L interlinkages of the circuit with lines of magnetic force produced by unit current in the circuit, or iL interlinkages by current i. That is, in establishing current i_0 in the circuit, the magnetic flux i_0L must be produced. A change of the magnetic flux iL surrounding a circuit generates in the circuit an e.m.f.,

$$e = \frac{d}{dt}(iL).$$

This opposes the impressed e.m.f. e_0, and therefore lowers the e.m.f. available to produce the current, and thereby the current, which then cannot instantly assume its final value, but rises thereto gradually, and so between the starting of the circuit and the establishment of permanent condition a transition period appears. In the same manner and for the same reasons, if the impressed e.m.f. e_0 is withdrawn, but the circuit left closed, the current i does not instantly disappear but gradually dies out, as shown in Fig. 1, which gives the rise and the decay of a

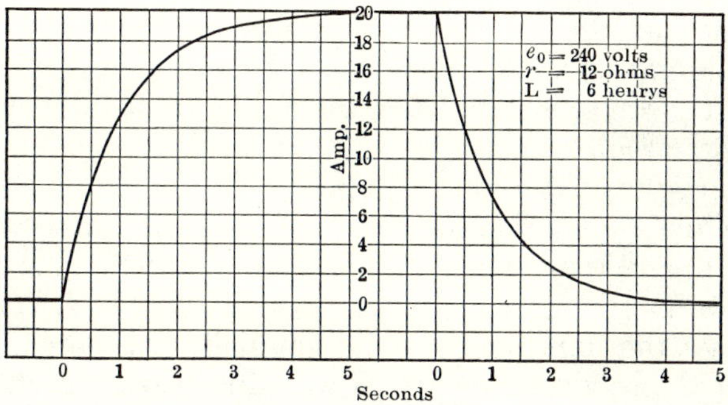

Fig. 1. Rise and decay of continuous current in an inductive circuit.

continuous current in an inductive circuit: the exciting current of an alternator field, or a circuit having the constants $r = 12$ ohms; $L = 6$ henrys, and $e_0 = 240$ volts; the abscissas being seconds of time.

13. If an electrostatic condenser of capacity C is connected to a continuous e.m.f. e_0, no current exists, in stationary condition, in this direct-current circuit (except that a very small current may leak through the insulation or the dielectric of the condenser), but the condenser is charged to the potential difference e_0, or contains the electrostatic charge

$$Q = Ce_0.$$

In the moment of closing the circuit of e.m.f. e_0 upon the capacity C, the condenser contains no charge, that is, zero potential difference exists at the condenser terminals. If there were no resistance and no inductance in the circuit in the moment of closing the circuit, an infinite current would exist charging the condenser instantly to the potential difference e_0. If r is the resistance of the direct-current circuit containing the condenser, and this circuit contains no inductance, the current starts at the value $i = \frac{e_0}{r}$, that is, in the first moment after closing the circuit all the impressed e.m.f. is consumed by the current in the resistance, since no charge and therefore no potential difference exists at the condenser. With increasing charge of the condenser, and therefore increasing potential difference at the condenser terminals, less and less e.m.f. is available for the resistance, and the current decreases, and ultimately becomes zero, when the condenser is fully charged.

If the circuit also contains inductance L, then the current cannot rise instantly but only gradually: in the moment after closing the circuit the potential difference at the condenser is still zero, and rises at such a rate that the increase of magnetic flux iL in the inductance produces an e.m.f. Ldi/dt, which consumes the impressed e.m.f. Gradually the potential difference at the condenser increases with its increasing charge, and the current and thereby the e.m.f. consumed by the resistance increases, and so less e.m.f. being available for consumption by the inductance, the current increases more slowly, until ultimately it ceases to rise, has reached a maximum, the inductance consumes no e.m.f., but all the impressed e.m.f. is consumed by the current in the resistance and by the potential difference at the condenser. The potential difference at the condenser continues to rise with its increasing charge; hence less e.m.f. is available for the resistance, that is, the current decreases again, and ultimately becomes zero, when the condenser is fully charged. During the decrease of current the decreasing magnetic flux iL in the inductance produces an e.m.f., which assists the impressed e.m.f., and so retards somewhat the decrease of current.

Fig. 2 shows the charging current of a condenser through an inductive circuit, as i, and the potential difference at the condenser terminals, as e, with a continuous impressed e.m.f. e_0, for the circuit constants $r = 250$ ohms; $L = 100$ mh.; $C = 10$ mf., and $e_0 = 1000$ volts.

If the resistance is very small, the current immediately after closing the circuit rises very rapidly, quickly charges the condenser, but at the moment where the condenser is fully charged to the impressed e.m.f. e_0, current still exists. This current cannot instantly stop, since the decrease of current and therewith the decrease of its magnetic flux iL generates an e.m.f.,

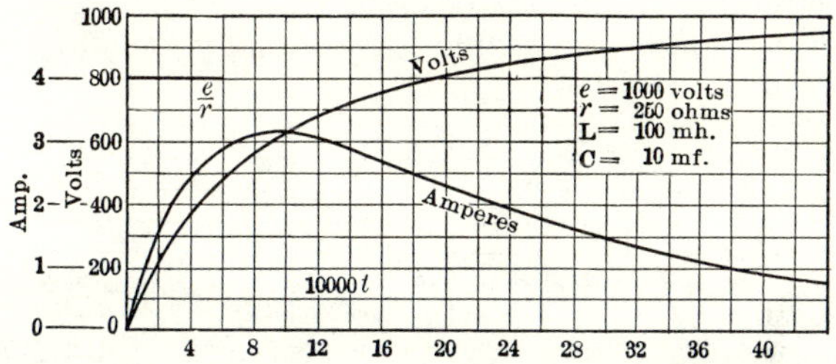

Fig. 2. Charging a condenser through a circuit having resistance and inductance. Constant potential. Logarithmic charge: high resistance.

which maintains the current, or retards its decrease. Hence electricity still continues to flow into the condenser for some time after it is fully charged, and when the current ultimately stops, the condenser is overcharged, that is, the potential difference at the condenser terminals is higher than the impressed e.m.f. e_0, and as result the condenser has partly to discharge again, that is, electricity begins to flow in the opposite direction, or out of the condenser. In the same manner this reverse current, due to the inductance of the circuit, overreaches and discharges the condenser farther than down to the impressed e.m.f. e_0, so that after the discharge current stops again a charging current — now less than the initial charging current — starts, and so by a series of oscillations, overcharges and undercharges, the condenser gradually charges itself, and ultimately the current dies out.

Fig. 3 shows the oscillating charge of a condenser through an inductive circuit, by a continuous impressed e.m.f. e_0. The current is represented by i, the potential difference at the condenser terminals by e, with the time as abscissas. The constants of the circuit are: $r = 40$ ohms; $L = 100$ mh.; $C = 10$ mf., and $e_0 = 1000$ volts.

In such a continuous-current circuit, containing resistance, inductance, and capacity in series to each other, the current at the moment of closing the circuit as well as the final current is zero, but a current exists immediately after closing the circuit, as a transient phenomenon; a temporary current, steadily increasing and then decreasing again to zero, or consisting of a number of alternations of successively decreasing amplitude: an oscillating current.

If the circuit contains no resistance and inductance, the current into the condenser would theoretically be infinite. That

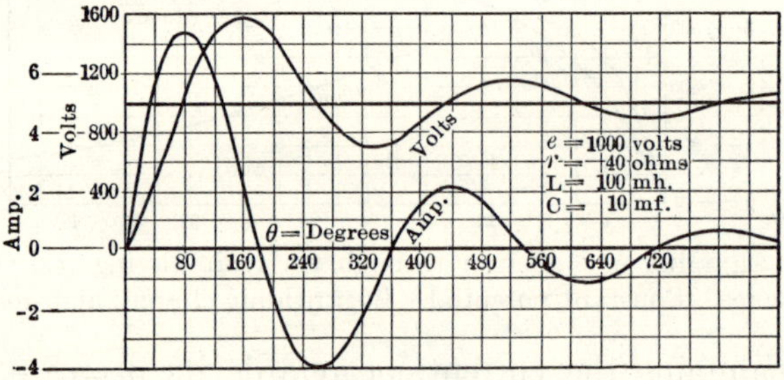

Fig. 3. Charging a condenser through a circuit having resistance and inductance. Constant potential. Oscillating charge: low resistance.

is, with low resistance and low inductance, the charging current of a condenser may be enormous, and therefore, although only transient, requires very serious consideration and investigation. If the resistance is very low and the inductance appreciable, the overcharge of the condenser may raise its voltage above the impressed e.m.f., e_0 sufficiently to cause disruptive effects.

14. If an alternating e.m.f.,

$$e = E \cos \theta,$$

is impressed upon a circuit of such constants that the current lags 45°, that is, the current is

$$i = I \cos (\theta - 45°),$$

and the circuit is closed at the moment $\theta = 45°$, at this moment the current should be at its maximum value. It is, however, zero, and since in a circuit containing inductance (that is, in practically any circuit) the current cannot change instantly, it follows that in this case the current gradually rises from zero as initial value to the permanent value of the sine wave i.

This approach of the current from the initial value, in the

present case zero, to the final value of the curve i, can either be gradual, as shown by the curve i_1 of Fig. 4, or by a series of oscillations of gradually decreasing amplitude, as shown by curve i_2 of Fig. 4.

15. The general solution of an electric current problem therefore includes besides the permanent term, constant or periodic,

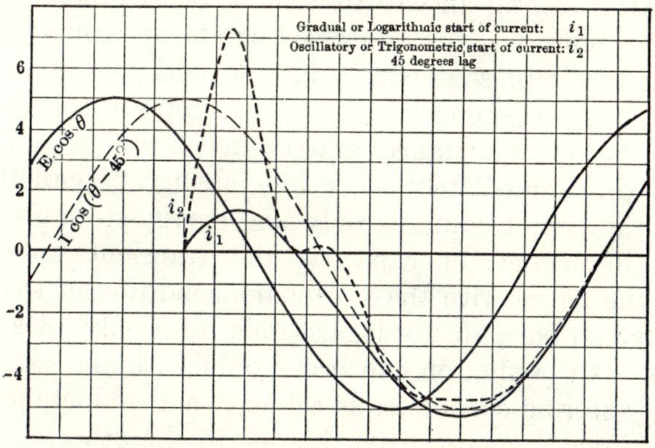

Fig. 4. Starting of an alternating-current circuit having inductance.

a transient term, which disappears after a time depending upon the circuit conditions, from an extremely small fraction of a second to a number of seconds.

These transient terms appear in closing the circuit, opening the circuit, or in any other way changing the circuit conditions, as by a change of load, a change of impedance, etc.

In general, in a circuit containing resistance and inductance only, but no capacity, the transient terms of current and voltage are not sufficiently large and of long duration to cause harmful nor even appreciable effects, and it is mainly in circuits containing capacity that excessive values of current and potential difference may be reached by the transient term, and therewith serious results occur. The investigation of transient terms therefore is largely an investigation of the effects of electrostatic capacity.

16. No transient terms result from the resistance, but only those circuit constants which represent storage of energy, magnetically by the inductance L, electrostatically by the capacity C, give rise to transient phenomena, and the more the resistance predominates, the less is therefore the severity and duration of the transient term.

When closing a circuit containing inductance or capacity or both, the energy stored in the inductance and the capacity has first to be supplied by the impressed e.m.f. before the circuit conditions can become stationary. That is, in the first moment after closing an electric circuit, or in general changing the circuit conditions, the impressed e.m.f., or rather the source producing the impressed e.m.f., has, in addition to the power consumed in maintaining the circuit, to supply the power which stores energy in inductance and capacity, and so a transient term appears immediately after any change of circuit condition. If the circuit contains only one energy-storing constant, as either inductance or capacity, the transient term, which connects the initial with the stationary condition of the circuit, necessarily can be a steady logarithmic term only, or a gradual approach. An oscillation can occur only with the existence of two energy-storing constants, as capacity and inductance, which permit a surge of energy from the one to the other, and therewith an overreaching.

17. Transient terms may occur periodically and in rapid succession, as when rectifying an alternating current by synchronously reversing the connections of the alternating impressed e.m.f. with the receiver circuit (as can be done mechanically or without moving apparatus by unidirectional conductors, as arcs). At every half wave the circuit reversal starts a transient term, and usually this transient term has not yet disappeared, frequently not even greatly decreased, when the next reversal again starts a transient term. These transient terms may predominate to such an extent that the current essentially consists of a series of successive transient terms.

18. If a condenser is charged through an inductance, and the condenser shunted by a spark gap set for a lower voltage than the impressed, then the spark gap discharges as soon as the condenser charge has reached a certain value, and so starts a transient term; the condenser charges again, and discharges, and so by the successive charges and discharges of the condenser a series of transient terms is produced, recurring at a frequency depending upon the circuit constants and upon the ratio of the disruptive voltage of the spark gap to the impressed e.m.f.

Such a phenomenon for instance occurs when on a high-potential alternating-current system a weak spot appears in the cable insulation and permits a spark discharge to pass to the ground, that is, in shunt to the condenser formed by the cable conductor and the cable armor or ground.

19. In most cases the transient phenomena occurring in electric circuits immediately after a change of circuit conditions are of no importance, due to their short duration. They require serious consideration, however,—

(*a*) In those cases where they reach excessive values. Thus in connecting a large transformer to an alternator the large initial value of current may do damage. In short-circuiting a large alternator, while the permanent or stationary short-circuit current is not excessive and represents little power, the very much larger momentary short-circuit current may be beyond the capacity of automatic circuit-opening devices and cause damage by its high power. In high-potential transmissions the potential differences produced by these transient terms may reach values so high above the normal voltage as to cause disruptive effects. Or the frequency or steepness of wave front of these transients may be so great as to cause destructive voltages across inductive parts of the circuits, as reactors, end turns of transformers and generators, etc.

(*b*) Lightning, high-potential surges, etc., are in their nature essentially transient phenomena, usually of oscillating character.

(*c*) The periodical production of transient terms of oscillating character is one of the foremost means of generating electric currents of very high frequency as used in wireless telegraphy, etc.

(*d*) In alternating-current rectifying apparatus, by which the direction of current in a part of the circuit is reversed every half wave, and the current so made unidirectional, the stationary condition of the current in the alternating part of the circuit is usually never reached, and the transient term is frequently of primary importance.

(*e*) In telegraphy the current in the receiving apparatus essentially depends on the transient terms, and in long-distance cable telegraphy the stationary condition of current is never approached, and the speed of telegraphy depends on the duration of the transient terms.

(*f*) Phenomena of the same character, but with space instead of time as independent variable, are the distribution of voltage and current in a long-distance transmission line; the phenomena occurring in multigap lightning arresters; the transmission of current impulses in telephony; the distribution of alternating current in a conductor, as the rail return of a single-phase railway; the distribution of alternating magnetic flux in solid magnetic material, etc.

Some of the simpler forms of transient terms are investigated and discussed in the following pages.

CHAPTER III.

INDUCTANCE AND RESISTANCE IN CONTINUOUS-CURRENT CIRCUITS.

20. In continuous-current circuits the inductance does not enter the equations of stationary condition, but, if e_0 = impressed e.m.f., r = resistance, L = inductance, the permanent value of current is $i_0 = \frac{e_0}{r}$.

Therefore less care is taken in direct-current circuits to reduce the inductance than in alternating-current circuits, where the inductance usually causes a drop of voltage, and direct-current circuits as a rule have higher inductance, especially if the circuit is used for producing magnetic flux, as in solenoids, electromagnets, machine-fields.

Any change of the condition of a continuous-current circuit, as a change of e.m.f., of resistance, etc., which leads to a change of current from one value i_0 to another value i_1, results in the appearance of a transient term connecting the current values i_0 and i_1, and into the equation of the transient term enters the inductance.

Count the time t from the moment when the change in the continuous-current circuit starts, and denote the impressed e.m.f. by e_0, the resistance by r, and the inductance by L.

$i_1 = \frac{e_0}{r}$ = current in permanent or stationary condition after the change of circuit condition.

Denoting by i_0 the current in circuit before the change, and therefore at the moment $t = 0$, by i the current during the change, the e.m.f. consumed by resistance r is

$$ir,$$

and the e.m.f. consumed by inductance L is

$$L \frac{di}{dt},$$

where i = current in the circuit.

Hence,
$$e_0 = ir + L\frac{di}{dt}, \tag{1}$$

or, substituting $e_0 = i_1 r$, and transposing,

$$-\frac{r}{L}dt = \frac{di}{i - i_1}. \tag{2}$$

This equation is integrated by

$$-\frac{r}{L}t = \log(i - i_1) - \log c,$$

where $-\log c$ is the integration constant, or,

$$i - i_1 = c\varepsilon^{-\frac{r}{L}t}.$$

However, for $t = 0$, $i = i_0$.

Substituting this, gives

$$i_0 - i_1 = c,$$

hence,
$$i = i_1 + (i_0 - i_1)\,\varepsilon^{-\frac{r}{L}t}, \tag{3}$$

the equation of current in the circuit.

The counter e.m.f. of self-inductance is

$$e_1 = -L\frac{di}{dt} = r(i_0 - i_1)\,\varepsilon^{-\frac{r}{L}t}, \tag{4}$$

hence a maximum for $t = 0$, thus:

$$e_1^0 = r(i_0 - i_1). \tag{5}$$

The e.m.f. of self-inductance e_1 is proportional to the change of current $(i_0 - i_1)$, and to the resistance r of the circuit after the change, hence would be ∞ for $r = \infty$, or when opening the circuit. That is, an inductive circuit cannot be opened instantly, but the arc following the break maintains the circuit for some time, and the voltage generated in opening an inductive circuit is the higher the quicker the break. Hence in a highly inductive circuit, as an electromagnet or a machine field, the insulation may be punctured by excessive generated e.m.f. when quickly opening the circuit.

As example, some typical circuits may be considered.

21. *Starting of a continuous-current lighting circuit, or non-inductive load.*

Let e_0 = 125 volts = impressed e.m.f. of the circuit, and i_1 = 1000 amperes = current in the circuit under stationary condition; then the effective resistance of the circuit is

$$r = \frac{e_0}{i_1} = 0.125 \text{ ohm.}$$

Assuming 10 per cent drop in feeders and mains, or 12.5 volts, gives a resistance, r_0 = 0.0125 ohm of the supply conductors. In such large conductor the inductance may be estimated as 10 mh. per ohm; hence, L = 0.125 mh. = 0.000125 henry.

The current at the moment of starting is $i_0 = 0$, and the general equation of the current in the circuit therefore is, by substitution in (3),

$$i = 1000\,(1 - \varepsilon^{-1000\,t}). \tag{6}$$

The time during which this current reaches half value, or i = 500 amperes, is given by substitution in (6)

$$500 = 1000\,(1 - \varepsilon^{-1000\,t}),$$

hence

$$\varepsilon^{-1000\,t} = 0.5,$$

$$t = 0.00069 \text{ seconds.}$$

The time during which the current reaches 90 per cent of its full value, or i = 900 amperes, is t = 0.0023 seconds, that is, the current is established in the circuit in a practically inappreciable time, a fraction of a hundredth of a second.

22. *Excitation of a motor field.*

Let, in a continuous-current shunt motor, e_0 = 250 volts = impressed e.m.f., and the number of poles = 8.

Assuming the magnetic flux per pole, Φ_0 = 12.5 megalines, and the ampere-turns per pole required to produce this magnetic flux as $\mathfrak{F}$ = 9000.

Assuming 1000 watts used for the excitation of the motor field gives an exciting current

$$i_1 = \frac{1000}{250} = 4 \text{ amperes,}$$

and herefrom the resistance of the total motor field circuit as

$$r = \frac{e_0}{i_1} = 62.5 \text{ ohms.}$$

To produce $\mathfrak{F} = 9000$ ampere-turns, with $i_1 = 4$ amperes, requires $\frac{\mathfrak{F}}{i_1} = 2250$ turns per field spool, or a total of $n = 18{,}000$ turns.

$n = 18{,}000$ turns interlinked with $\Phi_0 = 12.5$ megalines gives a total number of interlinkages for $i_1 = 4$ amperes of $n\Phi_0 = 225 \times 10^9$, or 562.5×10^9 interlinkages per unit current, or 10 amperes, that is, an inductance of the motor field circuit $L = 562.5$ henrys.

The constants of the circuit thus are $e_0 = 250$ volts; $r = 62.5$ ohms; $L = 562.5$ henrys, and $i_0 = 0 =$ current at time $t = 0$.

Hence, substituting in (3) gives the equation of the exciting current of the motor field as

$$i = 4\,(1 - \varepsilon^{-0.1111t}) \tag{7}$$

Half excitation of the field is reached after the time $t = 6.23$ seconds;

90 per cent of full excitation, or $i = 3.6$ amperes, after the time $t = 20.8$ seconds.

That is, such a motor field takes a very appreciable time after closing the circuit before it has reached approximately full value and the armature circuit may safely be closed.

Assume now the motor field redesigned, or reconnected so as to consume only a part, for instance half, of the impressed e.m.f., the rest being consumed in non-inductive resistance. This may be done by connecting the field spools by two in multiple.

In this case the resistance and the inductance of the motor field are reduced to one-quarter, but the same amount of external resistance has to be added to consume the impressed e.m.f., and the constants of the circuit then are: $e_0 = 250$ volts; $r = 31.25$ ohms; $L = 140.6$ henrys, and $i_0 = 0$.

The equation of the exciting current (3) then is

$$i = 8\,(1 - \varepsilon^{-0.2222t}), \tag{8}$$

that is, the current rises far more rapidly. It reaches 0.5 value after $t = 3.11$ seconds, 0.9 value after $t = 10.4$ seconds.

An inductive circuit, as a motor field circuit, may be made to respond to circuit changes more rapidly by inserting non-inductive resistance in series with it and increasing the impressed e.m.f., that is, the larger the part of the impressed e.m.f. consumed by non-inductive resistance, the quicker is the change.

Disconnecting the motor field winding from the impressed e.m.f. and short-circuiting it upon itself, as by leaving it connected in shunt with the armature (the armature winding resistance and inductance being negligible compared with that of the field winding), causes the field current and thereby the field magnetism to decrease at the same rate as it increased in (7) and (8), provided the armature instantly comes to a standstill, that is, its e.m.f. of rotation disappears. This, however, is usually not the case, but the motor armature slows down gradually, its momentum being consumed by friction and other losses, and while still revolving an e.m.f. of gradually decreasing intensity is generated in the armature winding; this e.m.f. is impressed upon the field.

The discharge of a motor field winding through the armature winding, after shutting off the power, therefore leads to the case of an inductive circuit with a varying impressed e.m.f.

23. *Discharge of a motor field winding.*

Assume that in the continuous-current shunt motor discussed under 22, the armature comes to rest $t_1 = 40$ seconds after the energy supply has been shut off by disconnecting the motor from the source of impressed e.m.f., while leaving the motor field winding still in shunt with the motor armature winding.

The resisting torque, which brings the motor to rest, may be assumed as approximately constant, and therefore the deceleration of the motor armature as constant, that is, the motor speed decreasing proportionally to the time.

If then S = full motor speed, $S\left(1 - \frac{t}{t_1}\right)$ is the speed of the motor at the time t after disconnecting the motor from the source of energy.

Assume the magnetic flux Φ of the motor as approximately proportional to the exciting current, at exciting current i the magnetic flux of the motor is $\Phi = \frac{i}{i_1}\Phi_0$, where $\Phi_0 = 12.5$ megalines is the flux corresponding to full excitation $i_1 = 4$ amperes.

The e.m.f. generated in the motor armature winding and thereby impressed upon the field winding is proportional to the magnetic flux of the field, Φ, and to the speed $S\left(1 - \frac{t}{t_1}\right)$, and since full speed S and full flux Φ_0 generate an e.m.f. $e_0 =$ 250 volts, the e.m.f. generated by the flux Φ and speed $S\left(1 - \frac{t}{t_1}\right)$, that is, at time t is

$$e = e_0 \frac{i}{i_1}\left(1 - \frac{t}{t_1}\right), \tag{9}$$

and since

$$\frac{e_0}{i_1} = r,$$

we have

$$e = ir\left(1 - \frac{t}{t_1}\right); \tag{10}$$

or for $r = 62.5$ ohms, and $t_1 = 40$ seconds, we have

$$e = 62.5\, i\, (1 - 0.025\, t). \tag{11}$$

Substituting this equation (10) of the impressed e.m.f. into the differential equation (1) gives the equation of current i during the field discharge,

$$ir\left(1 - \frac{t}{t_1}\right) = ir + L\frac{di}{dt}, \tag{12}$$

hence,

$$-\frac{rtdt}{t_1 L} = \frac{di}{i}, \tag{13}$$

integrated by

$$-\frac{rt^2}{2\, t_1 L} = \log ci,$$

where the integration constant c is found by

$$t = 0, \quad i = i_1, \quad \log ci_1 = 0, \quad c = \frac{1}{i_1},$$

hence,

$$-\frac{rt^2}{2\, t_1 L} = \log \frac{i}{i_1}, \tag{14}$$

or,

$$i = i_1 \varepsilon^{-\frac{rt^2}{2\, t_1 L}}, \tag{15}$$

This is the equation of the field current during the time in which the motor armature gradually comes to rest.

At the moment when the motor armature stops, or for

$$t = t_1,$$

it is

$$i_2 = i_1 \varepsilon^{-\frac{rt_1}{2L}}. \tag{16}$$

This is the same value which the current would have with the armature permanently at rest, that is, without the assistance of the e.m.f. generated by rotation, at the time $t = \frac{t_1}{2}$.

The rotation of the motor armature therefore reduces the decrease of field current so as to require twice the time to reach value i_2, that it would without rotation.

These equations cease to apply for $t > t_1$, that is, after the armature has come to rest, since they are based on the speed equation $S\left(1 - \frac{t}{t_1}\right)$, and this equation applies only up to $t = t_1$, but for $t > t_1$ the speed is zero, and not negative, as given by $S\left(1 - \frac{t}{t_1}\right)$.

That is, at the moment $t = t_1$ a break occurs in the field discharge curve, and after this time the current i decreases in accordance with equation (3), that is,

$$i = i_2 \varepsilon^{-\frac{r}{L}\left(t - t_1\right)}, \tag{17}$$

or, substituting (16),

$$i = i_1 \varepsilon^{-\frac{r}{L}\left(t - \frac{t_1}{2}\right)}. \tag{18}$$

Substituting numerical values in these equations gives:
for $t < t_1$,

$$i = 4\,\varepsilon^{-0.001388\,t^2}; \tag{19}$$

for $t = t_1 = 40$,

$$i = 0.436; \tag{20}$$

for $t > t_1$,

$$i = 4\,\varepsilon^{-0.1111\,(t - 20)}. \tag{21}$$

Hence, the field has decreased to half its initial value after the time $t = 22.15$ seconds, and to one tenth of its initial value after $t = 40.73$ seconds.

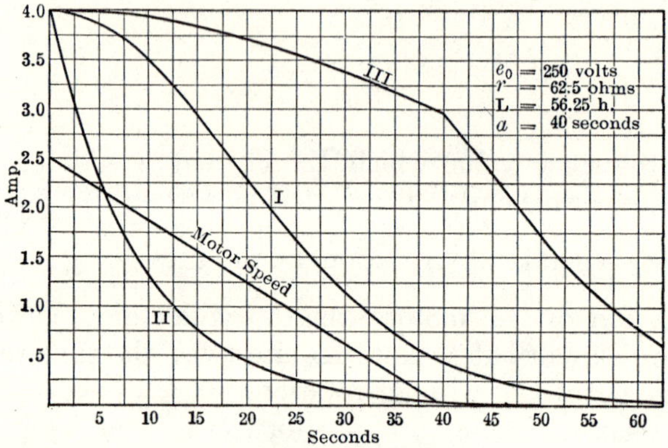

Fig. 5. Field discharge current.

Fig. 5 shows as curve I the field discharge current, by equations (19), (20), (21), and as curve II the current calculated by the equation

$$i = 4\,\varepsilon^{-0.1111t},$$

that is, the discharge of the field with the armature at rest, or when short-circuited upon itself and so not assisted by the e.m.f. of rotation of the armature.

The same Fig. 5 shows as curve III the beginning of the field discharge current for $L = 4200$, that is, the case that the field circuit has a much higher inductance, as given by the equation

$$i = 4\,\varepsilon^{-0.000185\,t^2}.$$

As seen in the last case, the decrease of field current is very slow, the field decreasing to half value in 47.5 seconds.

24. *Self-excitation of direct-current generator.*

In the preceding, the inductance L of the machine has been assumed as constant, that is, the magnetic flux Φ as proportional to the exciting current i. For higher values of Φ, this is not even approximately the case. The self-excitation of the direct-current generator, shunt or series wound, that is, the feature that the voltage of the machine after the start gradually builds up from the value given by the residual magnetism to its full value, depends upon the disproportionality of the magnetic flux with the magnetizing current. When considering this phenomenon, the inductance cannot therefore be assumed as constant.

When investigating circuits in which the inductance L is not constant but varies with the current, it is preferable not to use the term "inductance" at all, but to introduce the magnetic flux Φ.

The magnetic flux Φ varies with the magnetizing current i by an empirical curve, the magnetic characteristic or saturation curve of the machine. This can approximately, within the range considered here, be represented by a hyperbolic curve, as was first shown by Fröhlich in 1882:

$$\Phi = \frac{\phi i}{1 + bi}, \tag{22}$$

where ϕ = magnetic flux per ampere, in megalines, at low density.

$\frac{\phi}{b}$ = magnetic saturation value, or maximum magnetic flux, in megalines, and

$$\frac{i}{\Phi} = \frac{1 + bi}{\phi} \tag{23}$$

can be considered as the magnetic exciting reluctance of the machine field circuit, which here appears as linear function of the exciting current i.

Considering the same shunt-wound commutating machine as in (12) and (13), having the constants $r = 62.5$ ohms = field resistance; $\Phi_0 = 12.5$ megalines = magnetic flux per pole at normal m.m.f.; $\mathfrak{F} = 9000$ ampere-turns = normal m.m.f. per pole; $n = 18{,}000$ turns = total field turns (field turns per pole $= \frac{18{,}000}{8} = 2250$), and $i_1 = 4$ amperes = current for full excitation, or flux, $\Phi_0 = 12.5$ megalines.

Assuming that at full excitation, Φ_0, the magnetic reluctance has already increased by 50 per cent above its initial value, that is, that the ratio $\frac{\text{ampere-turns}}{\text{magnetic·flux}}$, or $\frac{i}{\Phi}$, at $\Phi = \Phi_0 = 12.5$ megalines and $i = i_1 = 4$ amperes, is 50 per cent higher than at low excitation, it follows that

$$\left.\begin{aligned} 1 + bi_1 &= 1.5, \\ \text{or} \qquad b &= 0.125. \end{aligned}\right\} \tag{24}$$

Since $i = i_1 = 4$ produces $\Phi = \Phi_0 = 12.5$, it follows, from (22) and (24)

$$\phi = 4.69.$$

That is, the magnetic characteristic (22) of the machine is approximated by

$$\Phi = \frac{4.69\, i}{1 + .125\, i}. \tag{25}$$

Let now e_c = e.m.f. generated by the rotation of the armature per megaline of field flux.

This e.m.f. e_c is proportional to the speed, and depends upon the constants of the machine. At the speed assumed in (12) and (13), $\Phi_0 = 12.5$ megalines, $e_0 = 250$ volts, that is,

$$e_c = \frac{e_0}{\Phi_0} = 20 \text{ volts.}$$

Then, in the field circuit of the machine, the impressed e.m.f., or e.m.f. generated in the armature by its rotation through the magnetic field is,

$$e = e_c\Phi = 20\Phi;$$

the e.m.f. consumed by the field resistance r is

$$ir = 62.5\, i;$$

the e.m.f. consumed by the field inductance, that is, generated in the field coils by the rise of magnetic flux Φ, is

$$n\frac{d\Phi}{dt}10^{-2} = 180\frac{d\Phi}{dt}.$$

(Φ being given in megalines, e_0 in volts.)

The differential equation of the field circuit therefore is (1)

$$e_c\Phi = ir + \frac{n}{100}\frac{d\Phi}{dt}. \tag{26}$$

Since this equation contains the differential quotient of Φ, it is more convenient to make Φ and not i the dependent variable; then substitute for i from equation (22),

$$i = \frac{\Phi}{\phi - b\Phi}, \tag{27}$$

which gives

$$e_c\Phi = \frac{\Phi r}{\phi - b\Phi} + \frac{n}{100}\frac{d\Phi}{dt}, \tag{28}$$

or, transposed,

$$\frac{100\,dt}{n} = \frac{(\phi - b\Phi)\,d\Phi}{\Phi\left\{(\phi e_c - r) - be_c\,\Phi\right\}}. \tag{29}$$

This equation is integrated by resolving into partial fraction by the identity

$$\frac{\phi - b\Phi}{\Phi\left\{(\phi e_c - r) - be_c\,\Phi\right\}} = \frac{A}{\Phi} + \frac{B}{\phi e_c - r - be_c\,\Phi}; \tag{30}$$

resolved, this gives

$$\phi - b\Phi = A\,(\phi e_c - r) - (Abe_c\,\Phi - B\,\Phi);$$

hence,

$$\left.\begin{aligned} A &= \frac{\phi}{\phi e_c - r}, \\ B &= \frac{br}{\phi e_c - r}, \end{aligned}\right\} \tag{31}$$

and

$$\frac{100\,dt}{n} = \frac{\phi d\,\Phi}{(\phi e_c - r)\,\Phi} + \frac{brd\Phi}{(\phi e_c - r)\,(\phi e_c - r - be_c\,\Phi)}. \tag{32}$$

This integrates by the logarithmic functions

$$\frac{100\,t}{n} = \frac{\phi}{\phi e_c - r}\log\Phi - \frac{r}{e_c(\phi e_c - r)}\log\,(\phi e_c - r - be_c\,\Phi) + C. \tag{33}$$

The integration constant C is calculated from the residual magnetic flux of the machine, that is, the remanent magnetism of the field poles at the moment of start.

Assume, at the time, $t = 0$, $\Phi = \Phi_r = 0.5$ megalines = residual magnetism and substituting in (33),

$$0 = \frac{\phi}{\phi e_c - r} \log \Phi_r - \frac{r}{e_c (\phi e_c - r)} \log (\phi e_c - r - b e_c \Phi_r) + C,$$

and herefrom calculate C.

C substituted in (33) gives

$$\frac{100\, t}{n} = \frac{\phi}{\phi e_c - r} \log \frac{\Phi}{\Phi_r} - \frac{r}{e_c (\phi e_c - r)} \log \frac{\phi e_c - r - b e_c \Phi}{\phi e_c - r - b e_c \Phi_r}, \qquad (34)$$

or,

$$t = \frac{n}{100\, e_c (\phi e_c - r)} \left\{ \phi e_c \log \frac{\Phi}{\Phi_r} - r \log \frac{\phi e_c - r - b e_c \Phi}{\phi e_c - r - b e_c \Phi_r} \right\}, \qquad (35)$$

substituting

$$e = e_c \Phi$$

and

$$e_m = e_c \Phi_r,$$

where e_m = e.m.f. generated in the armature by the rotation in the residual magnetic field,

$$t = \frac{n}{100\, e_c (\phi e_c - r)} \left\{ \phi e_c \log \frac{e}{e_m} - r \log \frac{\phi e_c - r - b e}{\phi e_c - r - b e_m} \right\}. \qquad (36)$$

This, then, is the relation between e and t, or the equation of the building up of a continuous-current generator from its residual magnetism, its speed being constant.

Substituting the numerical values $n = 18{,}000$ turns; $\phi = 4.69$ megalines; $b = 0.125$; $e_c = 20$ volts; $r = 62.5$ ohms; $\Phi_r = 0.5$ megaline, and $e_m = 10$ volts, we have

$$t = 26.8 \log \Phi - 17.9 \log (31.25 - 2.5\, \Phi) + 79.6 \qquad (37)$$

and

$$t = 26.8 \log e - 17.9 \log (31.25 - 0.125\, e) - 0.98. \qquad (38)$$

Fig. 6 shows the e.m.f. e as function of the time t. As seen, under the conditions assumed here, it takes several minutes before the e.m.f. of the machine builds up to approximately full value.

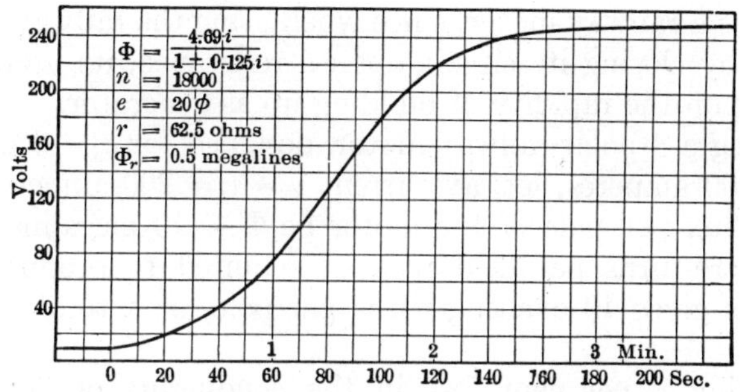

Fig. 6. Building-up curve of a shunt generator.

The phenomenon of self-excitation of shunt generators therefore is a transient phenomenon which may be of very long duration.

From equations (35) and (36) it follows that

$$e = \frac{\phi e_c - r}{b} = 250 \text{ volts} \tag{39}$$

is the e.m.f. to which the machine builds up at $t = \infty$, that is, in stationary condition.

To make the machine self-exciting, the condition

$$\phi e_c - r > 0 \tag{40}$$

must obtain, that is, the field winding resistance must be

$$\left.\begin{aligned} &r < \phi e_c, \\ \text{or,} \quad &r < 93.8 \text{ ohms}, \end{aligned}\right\} \tag{41}$$

or, inversely, e_c, which is proportional to the speed, must be

$$\left.\begin{aligned} &e_c > \frac{r}{\phi}, \\ \text{or,} \quad &e_c > 13.3 \text{ volts}. \end{aligned}\right\} \tag{42}$$

The time required by the machine to build up decreases with increasing e_c, that is, increasing speed; and increases with increasing r, that is, increasing field resistance.

25. *Self-excitation of direct-current series machine.*

Of interest is the phenomenon of self-excitation in a series machine, as a railway motor, since when using the railway motor as brake, by closing its circuit upon a resistance, its usefulness depends upon the rapidity of building up as generator.

Assuming a 4-polar railway motor, designed for $e_0 = 600$ volts and $i_1 = 200$ amperes, let, at current $i = i_1 = 200$ amperes, the magnetic flux per pole of the motor be $\Phi_0 = 10$ megalines, and 8000 ampere-turns per field pole be required to produce this flux. This gives 40 exciting turns per pole, or a total of $n = 160$ turns.

Estimating 8 per cent loss in the conductors of field and armature at 200 amperes, this gives a resistance of the motor circuit $r_0 = 0.24$ ohms.

To limit the current to the full load value of $i_1 = 200$ amperes, with the machine generating $e_0 = 600$ volts, requires a total resistance of the circuit, internal plus external, of

$$r = 3 \text{ ohms},$$

or an external resistance of 2.76 ohms.

600 volts generated by 10 megalines gives

$$e_c = 60 \text{ volts per megaline per field pole.}$$

Since in railway motors at heavy load the magnetic flux is carried up to high values of saturation, at $i_1 = 200$ amperes the magnetic reluctance of the motor field may be assumed as three times the value which it has at low density, that is, in equation (22),

$$1 + bi_1 = 3,$$

or,

$$b = 0.01,$$

and since for $i = 200$, $\Phi = 10$, we have in (22)

$$\phi = 0.15,$$

hence,

$$\Phi = \frac{0.15\, i}{1 + 0.01\, i} \tag{43}$$

represents the magnetic characteristic of the machine.

Assuming a residual magnetism of 10 per cent, or $\Phi_r =$ 1 megaline, hence $e_m = e_c \Phi_r = 60$ volts, and substituting in equation (36) gives $n = 160$ turns; $\phi = 0.15$ megaline; $b =$ 0.01; $e_c = 60$ volts; $r = 3$ ohms; $\Phi_r = 1$ megaline, and $e_m =$ 60 volts,

$$t = 0.04 \log e - 0.01333 \log (600 - e) - 0.08. \qquad (44)$$

This gives for $e = 300$, or 0.5 excitation, $t = 0.072$ seconds; and for $e = 540$, or 0.9 excitation, $t = 0.117$ seconds; that is, such a motor excites itself as series generator practically instantly, or in a small fraction of a second.

The lowest value of e_c at which self-excitation still takes place is given by equation (42) as

$$e_c = \frac{r}{\phi} = 20,$$

that is, at one-third of full speed.

If this series motor, with field and armature windings connected in generator position,—that is, reverse position,—short-circuits upon itself,

$$r = 0.24 \text{ ohms},$$

we have

$$t = 0.0274 \log e - 0.00073 \log (876 - e) - 0.1075, \qquad (45)$$

that is, self-excitation is practically instantaneous:

$$e = 300 \text{ volts is reached after } t = 0.044 \text{ seconds.}$$

Since for $e = 300$ volts, the current $i = \frac{e}{r} = 1250$ amperes, the power is $p = ei = 375$ kw., that is, a series motor short-circuited in generator position instantly stops.

Short-circuited upon itself, $r = 0.24$, this series motor still builds up at $e_c = \frac{r}{\phi} = 1.6$, and since at full load speed $e_c = 60$, $e_c = 1.6$ is 2.67 per cent of full load speed, that is, the motor acts as brake down to 2.67 per cent of full speed.

It must be considered, however, that the parabolic equation (22) is only an approximation of the magnetic characteristic, and the results based on this equation therefore are approximate only.

One of the most important transient phenomena of direct-current circuits is the reversal of current in the armature coil short-circuited by the commutator brush in the commutating machine. Regarding this, see *Theoretical Elements of Electrical Engineering*, Part II, Section B [McGraw-Hill edition, volume 3].

CHAPTER IV.

INDUCTANCE AND RESISTANCE IN ALTERNATING-CURRENT CIRCUITS.

26. In alternating-current circuits, the inductance L, or, as it is usually employed, the reactance $x = 2\,\pi f L$, where f = frequency, enters the expression of the transient as well as the permanent term.

At the moment $\theta = 0$, let the e.m.f. $e = E \cos(\theta - \theta_0)$ be impressed upon a circuit of resistance r and inductance L, thus inductive reactance $x = 2\,\pi f L$; let the time $\theta = 2\,\pi f t$ be counted from the moment of closing the circuit, and θ_0 be the phase of the impressed e.m.f. at this moment.

In this case the e.m.f. consumed by the resistance $= ir$, where i = instantaneous value of current.

The e.m.f. consumed by the inductance L is proportional to L and to the rate of change of the current, $\frac{di}{dt}$, thus, is $L\frac{di}{dt}$, or, by substituting $\theta = 2\,\pi f t$, $x = 2\,\pi f L$, the e.m.f. consumed by inductance is $x\frac{di}{d\theta}\cdot$

Since $e = E \cos(\theta - \theta_0)$ = impressed e.m.f.,

$$E \cos(\theta - \theta_0) = ir + x\frac{di}{d\theta} \qquad (1)$$

is the differential equation of the problem.

This equation is integrated by the function

$$i = I \cos(\theta - \delta) + A\varepsilon^{-a\theta}, \qquad (2)$$

where ε = basis of natural logarithms = 2.7183.

Substituting (2) in (1),

$$E \cos(\theta - \theta_0) = Ir \cos(\theta - \delta) + Ar\varepsilon^{-a\theta} - Ix \sin(\theta - \delta) - Aax\varepsilon^{-a\theta},$$

or, rearranged:

$$(E \cos\theta_0 - Ir\cos\delta - Ix\sin\delta)\cos\theta + (E\sin\theta_0 - Ir\sin\delta + Ix\cos\delta)\sin\theta - A\varepsilon^{-a\theta}(ax - r) = 0.$$

Since this equation must be fulfilled for any value of θ, if (2) is the integral of (1), the coefficients of $\cos \theta$, $\sin \theta$, $\varepsilon^{-a\theta}$ must vanish separately.

That is,

$$\left.\begin{aligned} E \cos \theta_0 - Ir \cos \delta - Ix \sin \delta &= 0, \\ E \sin \theta_0 - Ir \sin \delta + Ix \cos \delta &= 0, \\ \text{and} \qquad ax - r &= 0. \end{aligned}\right\} \tag{3}$$

Herefrom it follows that

$$a = \frac{r}{x} \cdot \tag{4}$$

Substituting in (3),

$$\left.\begin{aligned} \tan \theta_1 &= \frac{x}{r} \\ \text{and} \qquad z &= \sqrt{r^2 + x^2}, \end{aligned}\right\} \tag{5}$$

where θ_1 = lag angle and z = impedance of circuit, we have

$$\left.\begin{aligned} E \cos \theta_0 - Iz \cos (\delta - \theta_1) &= 0 \\ \text{and} \qquad E \sin \theta_0 - Iz \sin (\delta - \theta_1) &= 0, \end{aligned}\right\}$$

and herefrom

$$\left.\begin{aligned} I &= \frac{E}{z} \\ \text{and} \qquad \delta &= \theta_0 + \theta_1. \end{aligned}\right\} \tag{6}$$

Thus, by substituting (4) and (6) in (2), the integral equation becomes

$$i = \frac{E}{z} \cos (\theta - \theta_0 - \theta_1) + A\varepsilon^{-\frac{r}{x}\theta}, \tag{7}$$

where A is still indefinite, and is determined by the initial conditions of the circuit, as follows:

for $\qquad \theta = 0, \qquad i = 0;$

hence, substituting in (7).

$$0 = \frac{E}{z} \cos (\theta_0 + \theta_1) + A,$$

or,

$$A = -\frac{E}{z}\cos(\theta_0 + \theta_1), \tag{8}$$

and, substituted in (7),

$$i = \frac{E}{z}\left\{\cos(\theta - \theta_0 - \theta_1) - \varepsilon^{-\frac{r}{x}\theta}\cos(\theta_0 + \theta_1)\right\} \tag{9}$$

is the general expression of the current in the circuit.

If at the starting moment $\theta = 0$ the current is not zero but $= i_0$, we have, substituted in (7),

$$i_0 = \frac{E}{z}\cos(\theta_0 + \theta_1) + A,$$

$$A = i_0 - \frac{E}{z}\cos(\theta_0 + \theta_1),$$

$$i = \frac{E}{z}\left\{\cos(\theta - \theta_0 - \theta_1) - \left(\cos(\theta_0 + \theta_1) - \frac{i_0 z}{E}\right)\varepsilon^{-\frac{r}{x}\theta}\right\}. \tag{10}$$

27. The equation of current (9) contains a permanent term $\frac{E}{z}\cos(\theta - \theta_0 - \theta_1)$, which usually is the only term considered, and a transient term $\frac{E}{z}\varepsilon^{-\frac{r}{x}\theta}\cos(\theta_0 + \theta_1)$.

The greater the resistance r and smaller the reactance x, the more rapidly the term $\frac{E}{z}\varepsilon^{-\frac{r}{x}\theta}\cos(\theta_0 + \theta_1)$ disappears.

This transient term is a maximum if the circuit is closed at the moment $\theta_0 = -\theta_1$, that is, at the moment when the permanent value of current, $\frac{E}{z}\cos(\theta - \theta_0 - \theta_1)$, should be a maximum, and is then

$$\frac{E}{z}\varepsilon^{-\frac{r}{x}\theta}.$$

The transient term disappears if the circuit is closed at the moment $\theta_0 = 90° - \theta_1$, or when the stationary term of current passes the zero value.

As example is shown, in Fig. 7, the starting of the current under the conditions of maximum transient term, or $\theta_0 = -\theta_1$, in a circuit of the following constants: $\frac{x}{r} = 0.1$, corresponding approximately to a lighting circuit, where the permanent value

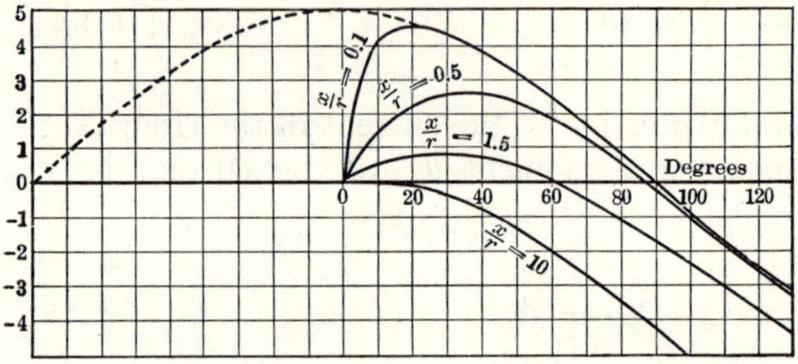

Fig. 7. Starting current of an inductive circuit.

of current is reached in a small fraction of a half wave; $\frac{x}{r} = 0.5$, corresponding to the starting of an induction motor with rheostat in the secondary circuit; $\frac{x}{r} = 1.5$, corresponding to an unloaded transformer, or to the starting of an induction motor with short-circuited secondary, and $\frac{x}{r} = 10$, corresponding to a reactive coil.

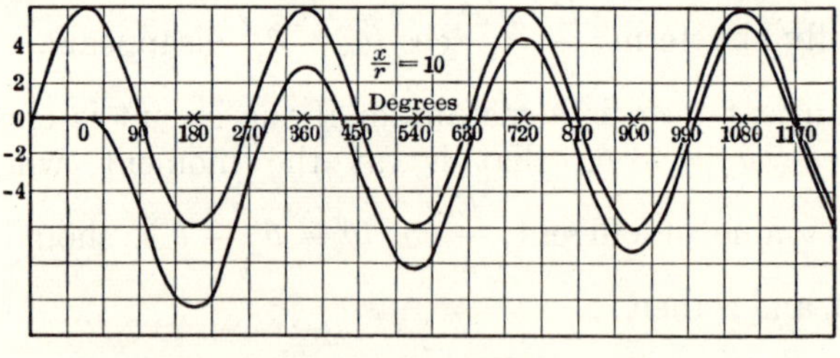

Fig. 8. Starting current of an inductive circuit.

Of the last case, $\frac{x}{r} = 10$, a series of successive waves are plotted in Fig. 8, showing the very gradual approach to permanent condition.

Fig. 9 shows, for the circuit $\frac{x}{r} = 1.5$, the current when closing the circuit 0°, 30°, 60°, 90°, 120°, 150° respectively behind the zero value of permanent current.

The permanent value of current is shown in Fig. 7 in dotted line.

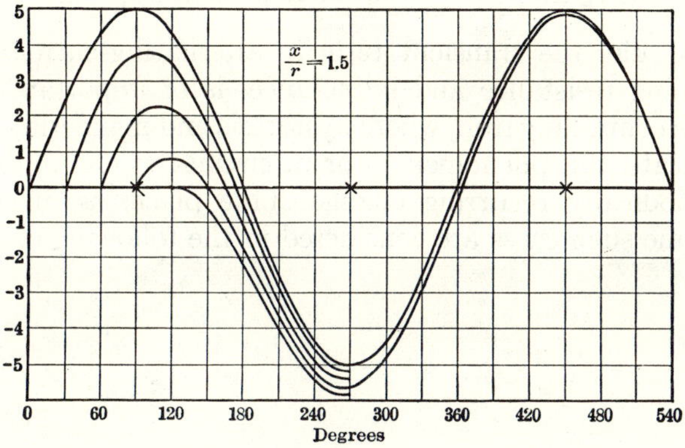

Fig. 9. Starting current of an inductive circuit.

28. Instead of considering, in Fig. 9, the current wave as consisting of the superposition of the permanent term $I \cos(\theta - \theta_0)$ and the transient term $-I\varepsilon^{-\frac{r}{x}\theta} \cos\theta_0$ the current wave can directly be represented by the permanent term

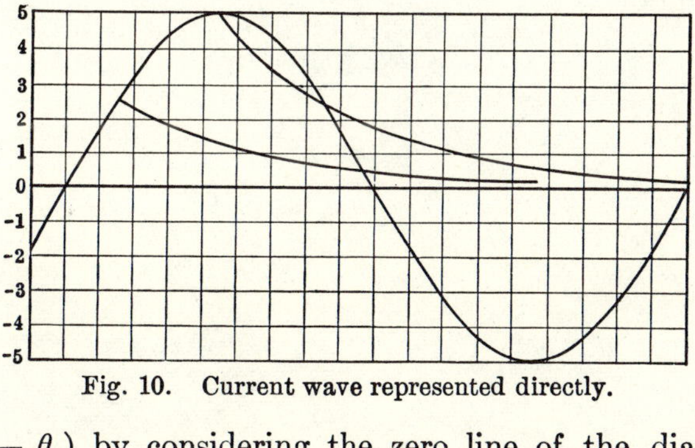

Fig. 10. Current wave represented directly.

$I \cos(\theta - \theta_0)$ by considering the zero line of the diagram as deflected exponentially to the curve $I\varepsilon^{-\frac{r}{x}\theta} \cos\theta_0$ in Fig. 10. That is, the instantaneous values of current are the vertical distances of the sine wave $I \cos (\theta - \theta_0)$ from the exponential curve $I\varepsilon^{-\frac{r}{x}\theta} \cos \theta_0$, starting at the initial value of permanent current.

In polar coordinates, in this case $I \cos (\theta - \theta_0)$ is the circle, $I\varepsilon^{-\frac{r}{x}\theta} \cos \theta_0$ the exponential or loxodromic spiral.

As a rule, the transient term in alternating-current circuits containing resistance and inductance is of importance only in circuits containing iron, where hysteresis and magnetic saturation complicate the phenomenon, or in circuits where unidirectional or periodically recurring changes take place, as in rectifiers, and some such cases are considered in the following chapters.

CHAPTER V

RESISTANCE, INDUCTANCE, AND CAPACITY IN SERIES. CONDENSER CHARGE AND DISCHARGE

29. If a continuous e.m.f. e is impressed upon a circuit containing resistance, inductance, and capacity in series, the stationary condition of the circuit is zero current, $i = o$, and the potential difference at the condenser equals the impressed e.m.f., $e_1 = e$, no permanent current exists, but only the transient current of charge or discharge of the condenser.

The capacity C of a condenser is defined by the equation

$$i = C\frac{de}{dt},$$

that is, the current into a condenser is proportional to the rate of increase of its e.m.f. and to the capacity.

It is therefore

$$de = \frac{1}{C}\,idt,$$

and

$$e = \frac{1}{C}\int idt \tag{1}$$

is the potential difference at the terminals of a condenser of capacity C with current i in the circuit to the condenser.

Let then, in a circuit containing resistance, inductance, and capacity in series, e = impressed e.m.f., whether continuous, alternating, pulsating, etc.; i = current in the circuit at time t; r = resistance; L = inductance, and C = capacity; then the e.m.f. consumed by resistance r is

$$ri;$$

the e.m.f. consumed by inductance L is

$$L\frac{di}{dt},$$

and the e.m.f. consumed by capacity C is

$$e_1 = \frac{1}{C}\int idt;$$

hence, the impressed e.m.f. is

$$e = ri + L\frac{di}{dt} + \frac{1}{C}\int idt, \tag{2}$$

and herefrom the potential difference at the condenser terminals is

$$e_1 = \frac{1}{C}\int idt = e - ri - L\,\frac{di}{dt}. \tag{3}$$

Equation (2) differentiated and rearranged gives

$$L\frac{d^2i}{dt^2} + r\frac{di}{dt} + \frac{1}{C}i = \frac{de}{dt} \tag{4}$$

as the general differential equation of a circuit containing resistance, inductance, and capacity in series.

30. If the impressed e.m.f. is constant,

$$e = \text{constant},$$

then

$$\frac{de}{dt} = 0,$$

and equation (4) assumes the form, for continuous-current circuits,

$$L\frac{d^2i}{dt^2} + r\frac{di}{dt} + \frac{1}{C}i = 0. \tag{5}$$

This equation is a linear relation between the dependent variable, i, and its differential quotients, and as such is integrated by an exponential function of the general form

$$i = A\varepsilon^{-at}. \tag{6}$$

(This exponential function also includes the trigonometric functions sine and cosine, which are exponential functions with imaginary exponent a.)

Substituting (6) in (5) gives

$$\left(a^2L - ar + \frac{1}{C}\right)A\varepsilon^{-at} = 0;$$

this must be an identity, irrespective of the value of t, to make (6) the integral of (5). That is,

$$a^2L - ar + \frac{1}{C} = 0. \tag{7}$$

A is still indefinite, and therefore determined by the terminal conditions of the problem.

From (7) follows

$$a = \frac{r \pm \sqrt{r^2 - \frac{4\,L}{C}}}{2\,L}, \tag{8}$$

hence the two roots,

$$\left.\begin{aligned} a_1 &= \frac{r - s}{2\,L} \\ \text{and}\qquad a_2 &= \frac{r + s}{2\,L}, \end{aligned}\right\} \tag{9}$$

where
$$s = \sqrt{r^2 - \frac{4\,L}{C}}\,. \tag{10}$$

Since there are two roots, a_1 and a_2, either of the two expresions (6), ε^{-a_1t} and ε^{-a_2t}, and therefore also any combination of these two expressions, satisfies the differential equation (5).

That is, the general integral equation, or solution of differential equation (5), is

$$i = A_1\varepsilon^{-\frac{r-s}{2L}t} + A_2\varepsilon^{-\frac{r+s}{2L}t}. \tag{11}$$

Substituting (11) and (9) in equation (3) gives the potential difference at the condenser terminals as

$$e_1 = e - \left\{\frac{r + s}{2}A_1\varepsilon^{-\frac{r-s}{2L}t} + \frac{r - s}{2}A_2\varepsilon^{-\frac{r+s}{2L}t}\right\} \tag{12}$$

31. Equations (11) and (12) contain two indeterminate constants, A_1 and A_2, which are the integration constants of the differential equation of second order, (5), and determined by the terminal conditions, the current and the potential difference at the condenser at the moment $t = 0$.

Inversely, since in a circuit containing inductance and capacity two electric quantities must be given at the moment of start of the phenomenon, the current and the condenser potential — representing the values of energy stored at the moment $t = 0$ as electromagnetic and as electrostatic energy, respectively — the equations must lead to two integration constants, that is, to a differential equation of second order.

Let $i = i_0$ = current and $e_1 = e_0$ = potential difference at condenser terminals at the moment $t = 0$; substituting in (11) and (12),

$$i_0 = A_1 + A_2$$

and

$$e_0 = e - \frac{r+s}{2} A_1 - \frac{r-s}{2} A_2;$$

hence,

$$\left.\begin{aligned} A_1 &= - \frac{e_0 - e + \dfrac{r-s}{2} i_0}{s} \\ \text{and} \qquad A_2 &= + \frac{e_0 - e + \dfrac{r+s}{2} i_0}{s}, \end{aligned}\right\} \tag{13}$$

and therefore, substituting in (11) and (12), the *current* is

$$i = \frac{e_0 - e + \dfrac{r+s}{2} i_0}{s} \varepsilon^{-\frac{r+s}{2L}t} - \frac{e_0 - e + \dfrac{r-s}{2} i_0}{s} \varepsilon^{-\frac{r-s}{2L}t}, \tag{14}$$

the *condenser potential* is

$$e_1 = e - \frac{1}{2}\left\{(r-s)\frac{e_0 - e + \dfrac{r+s}{2} i_0}{s} \varepsilon^{-\frac{r+s}{2L}t} - (r+s)\frac{e_0 - e + \dfrac{r-s}{2} i_0}{s} \varepsilon^{-\frac{r-s}{2L}t}\right\} \tag{15}$$

For *no condenser charge*, or $i_0 = 0,\ e_0 = 0,$ we have

$$A_1 = \frac{e}{s}$$

and

$$A_2 = -\frac{e}{s} = -A_1;$$

substituting in (11) and (12), we get the *charging current* as

$$i = \frac{e}{s}\left\{\varepsilon^{-\frac{r-s}{2L}t} - \varepsilon^{-\frac{r+s}{2L}t}\right\}. \tag{16}$$

The *condenser potential as*

$$e_1 = e\left\{1 - \frac{1}{2s}\left[(r+s)\,\varepsilon^{-\frac{r-s}{2L}t} - (r-s)\,\varepsilon^{-\frac{r+s}{2L}t}\right]\right\}. \tag{17}$$

For a *condenser discharge* or $i_0 = 0,\ e = e_0$, we have

$$A_1 = -\frac{e_0}{s}$$

and

$$A_2 = +\frac{e_0}{s} = -A_1;$$

hence, the *discharging current* is

$$i = -\frac{e_0}{s}\left\{\varepsilon^{-\frac{r-s}{2L}t} - \varepsilon^{-\frac{r+s}{2L}t}\right\}. \tag{18}$$

The *condenser potential* is

$$e_1 = \frac{e_0}{2s}\left\{(r+s)\,\varepsilon^{-\frac{r-s}{2L}t} - (r-s)\,\varepsilon^{-\frac{r+s}{2L}t}\right\}, \tag{19}$$

that is, in condenser discharge and in condenser charge the currents are the same, but opposite in direction, and the condenser potential rises in one case in the same way as it falls in the other.

32. As example is shown, in Fig. 11, the charge of a condenser of $C = 10$ mf. capacity by an impressed e.m.f. of

$e = 1000$ volts through a circuit of $r = 250$ ohms resistance and $L = 100$ mh. inductance; hence, $s = 150$ ohms, and the charging current is

$$i = 6.667 \left\{\varepsilon^{-500t} - \varepsilon^{-2000t}\right\} \text{ amperes.}$$

The condenser potential is

$$e_1 = 1000 \left\{1 - 1.333\, \varepsilon^{-500t} + 0.333\, \varepsilon^{-2000t}\right\} \text{ volts.}$$

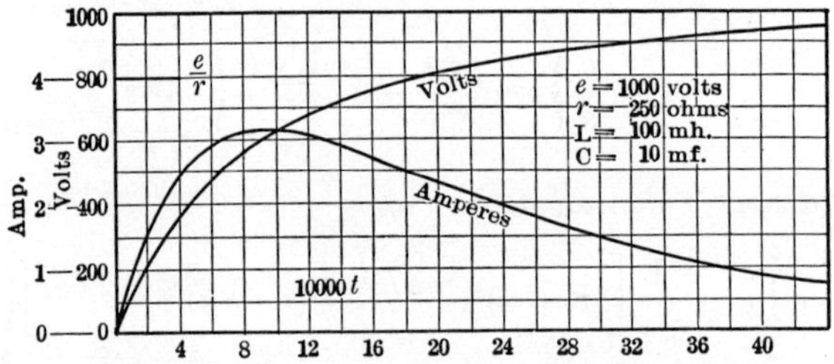

Fig. 11. Charging a condenser through a circuit having resistance and inductance. Constant potential. Logarithmic charge.

33. The equations (14) to (19) contain the square root,

$$s = \sqrt{r^2 - \frac{4L}{C}},$$

hence, they apply in their present form only when

$$r^2 > \frac{4L}{C}.$$

If $r^2 = \frac{4L}{C}$, these equations become indeterminate, or $= \frac{0}{0}$, and if $r^2 < \frac{4L}{C}$, s is imaginary, and the equations assume a complex imaginary form. In either case they have to be rearranged to assume a form suitable for application.

Three cases have thus to be distinguished:

(*a*) $r^2 > \frac{4L}{C}$, in which the equations of the circuit can be used in their present form. Since the functions are exponential or logarithmic, this is called the *logarithmic* case.

(*b*) $r^2 = \frac{4L}{C}$ is called the *critical* case, marking the transition between (*a*) and (*c*), but belonging to neither.

(*c*) $r^2 < \frac{4L}{C}$. In this case trigonometric functions appear; it is called the *trigonometric* case, or *oscillation.*

34. In the *logarithmic* case,

$$r^2 > \frac{4L}{C}$$

or,

$$4L < Cr^2,$$

that is, with high resistance, or high capacity, or low inductance, equations (14) to (19) apply.

The term $\varepsilon^{-\frac{r-s}{2L}t}$ is always greater than $\varepsilon^{-\frac{r+s}{2L}t}$, since the former has a lower coefficient in the exponent, and the difference of these terms, in the equations of condenser charge and discharge, is always positive. That is, the current rises from zero at $t = 0$, reaches a maximum and then falls again to zero at $t = \infty$, but it never reverses. The maximum of the current is less than $i = \frac{e}{s}\cdot$

The exponential term in equations (17) and (19) also never reverses. That is, the condenser potential gradually changes, without ever reversing or exceeding the impressed e.m.f. in the charge or the starting potential in the discharge.

Hence, in the case $r^2 > \frac{4L}{C}$, no abnormal voltage is produced in the circuit, and the transient term is of short duration, so that a condenser charge or discharge under these conditions is relatively harmless.

In charging or discharging a condenser, or in general a circuit containing capacity, the insertion of a resistance in series in the circuit of such value that $r^2 > \frac{4L}{C}$ therefore eliminates the danger from abnormal electrostatic or electromagnetic stresses.

In general, the higher the resistance of a circuit, compared with inductance and capacity, the more the transient term is suppressed.

35. In a circuit containing resistance and capacity but no inductance, $L = 0$, we have, substituting in (5),

$$r \frac{di}{dt} + \frac{1}{C} i = 0, \tag{20}$$

or, transposing,

$$\frac{di}{i} = - \frac{dt}{rC},$$

which is integrated by

$$i = c\varepsilon^{-\frac{t}{rC}}, \tag{21}$$

where c = integration constant.

Equation (21) gives for $t = 0$, $i = c$; that is, the current at the moment of closing the circuit must have a finite value, or must jump instantly from zero to c. This is not possible, but so also it is not possible to produce a circuit without any inductance whatever.

Therefore equation (21) does not apply for very small values of time, t, but for very small t the inductance, L, of the circuit, however small, determines the current.

The potential difference at the condenser terminals from (3) is

$$e_1 = e - ri$$

hence

$$e_1 = e - rc\varepsilon^{-\frac{t}{rC}} \tag{22}$$

The integration constant c cannot be determined from equation (21) at $t = 0$, since the current i makes a jump at this moment.

But from (22) it follows that if at the moment $t = 0$, $e_1 = e_0$, $e_0 = e - rc$,

hence,

$$c = \frac{e - e_0}{r},$$

and herefrom the equations of the non-inductive condenser circuit,

$$i = \frac{(e - e_0)\varepsilon^{-\frac{t}{rC}}}{r} \tag{23}$$

and

$$e_1 = e - (e - e_0)\varepsilon^{-\frac{t}{rC}}. \tag{24}$$

As seen, these equations do not depend upon the current i_0 in the circuit at the moment before $t = 0$.

36. These equations do not apply for very small values of t, but in this case the inductance, L, has to be considered, that is, equations (14) to (19) used.

For $L = 0$ the second term in (14) becomes indefinite, as it contains $\varepsilon^{\frac{0}{0}t}$, and therefore has to be evaluated as follows:

For $L = 0$, we have

$$s = r,$$

$$\frac{r+s}{2} = r,$$

and

$$\frac{r-s}{2} = 0$$

and, developed by the binomial theorem, dropping all but the first term,

$$r - s = r\left\{1 - \sqrt{1 - \frac{4L}{r^2C}}\right\}$$

$$= \frac{2L}{rC},$$

and

$$\frac{r-s}{2L} = \frac{1}{rC},$$

$$\frac{r+s}{2L} = \frac{r}{L}.$$

Substituting these values in equations (14) and (15) gives the current as

$$i = \frac{e - e_0}{r}\varepsilon^{-\frac{t}{rC}} - \frac{e - e_0 - ri_0}{r}\varepsilon^{-\frac{r}{L}t} \tag{25}$$

and the potential difference at the condenser as

$$e_1 = e - (e - e_0)\,\varepsilon^{-\frac{t}{rC}}; \tag{26}$$

that is, in the equation of the current, the term

$$-\frac{e - e_0 - ri_0}{r}\varepsilon^{-\frac{r}{L}t}$$

has to be added to equation (23). This term makes the transition from the circuit conditions before $t = 0$ to those after $t = 0$, and is of extremely short duration.

For instance, choosing the same constants as in § 32, namely: $e = 1000$ volts; $r = 250$ ohms; $C = 10$ mf., but choosing the inductance as low as possible, $L = 5$ mh., gives the equations of condenser charge, i.e., for $i_0 = 0$ and $e_0 = 0$,

$$i = 4 \left\{\varepsilon^{-400t} - \varepsilon^{-50,000t}\right\}$$

and

$$e_1 = 1000 \left\{1 - \varepsilon^{-400t}\right\}.$$

The second term in the equation of the current, $\varepsilon^{-50,000t}$, has decreased already to 1 per cent after $t = 17.3 \times 10^{-6}$ seconds, while the first term, ε^{-400t}, has during this time decreased only by 0.7 per cent, that is, it has not yet appreciably decreased.

37. In the *critical* case,

$$r^2 = \frac{4L}{C}$$

and

$$s = 0,$$

$$a_1 = a_2 = \frac{r}{2L},$$

$$A_1 = -A_2 = \frac{e - e_0 - \frac{r}{2} i_0}{s}.$$

Hence, substituting in equation (14) and rearranging,

$$i = \epsilon^{-\frac{r}{2L}t} \left\{ i_0 \left(\epsilon^{\frac{s}{2L}t} + \epsilon^{-\frac{s}{2L}t} \right) + \left(e - e_0 - \frac{r}{2} i_0 \right) \frac{\epsilon^{\frac{s}{2L}t} - \epsilon^{-\frac{s}{2L}t}}{s} \right\}. \tag{27}$$

The last term of this equation,

$$F = \frac{N}{D} = \frac{\varepsilon^{\frac{s}{2L}t} - \varepsilon^{-\frac{s}{2L}t}}{s} = \frac{0}{0},$$

that is, becomes indeterminate for $s = 0$, and therefore is evaluated by differentiation,

$$F = \frac{\frac{dN}{ds}}{\frac{dD}{ds}} = \frac{t}{L}. \tag{28}$$

Substituting (28) in (27) gives the equation of *current*,

$$i = \left[i_0 + \frac{t}{L}\left(e - e_0 - \frac{r}{2}\,i_0\right)\right]\epsilon^{-\frac{r}{2L}t}. \tag{29}$$

The condenser potential is found, by substituting in (15), to be

$$e_1 = e - \frac{1}{2}\epsilon^{-\frac{r}{2L}t}\left\{(e - e_0)\left(\epsilon^{\frac{s}{2L}t} + \epsilon^{-\frac{s}{2L}t}\right) - \frac{r}{s}\left(e - e_0 - \frac{ri_0}{2}\right)\left(\epsilon^{\frac{s}{2L}t} - \epsilon^{-\frac{s}{2L}t}\right)\right\} \tag{30}$$

The last term of this equation is, for $s = 0$:

$$\frac{r}{s}\left(e - e_0 - \frac{ri_0}{2}\right)\left(\epsilon^{\frac{s}{2L}t} - \epsilon^{-\frac{s}{2L}t}\right) = \frac{rt}{L}\left(e - e_0 - \frac{ri_0}{2}\right) \tag{31}$$

This gives the *condenser potential* as:

$$e_1 = e - \epsilon^{-\frac{r}{2L}t}\left\{(e - e_0) + \frac{rt}{2L}\left(e - e_0 - \frac{ri_0}{2}\right)\right\} \tag{32}$$

Herefrom it follows that for the condenser charge, $i_0 = 0$ and $e_0 = 0$,

$$i = \frac{t}{L}e\,\varepsilon^{-\frac{r}{2L}t}$$

and

$$e_1 = e\left\{1 - \left(1 + \frac{rt}{2L}\right)\varepsilon^{-\frac{r}{2L}t}\right\};$$

for the condenser discharge, $i_0 = 0$ and $e = 0$,

$$i = -\frac{t}{L} e_0 \varepsilon^{-\frac{r}{2L}t}$$

and

$$e_1 = \left(1 + \frac{rt}{2L}\right) e_0 \varepsilon^{-\frac{r}{2L}t}.$$

38. As an example are shown, in Fig. 12, the charging current and the potential difference at the terminals of the condenser,

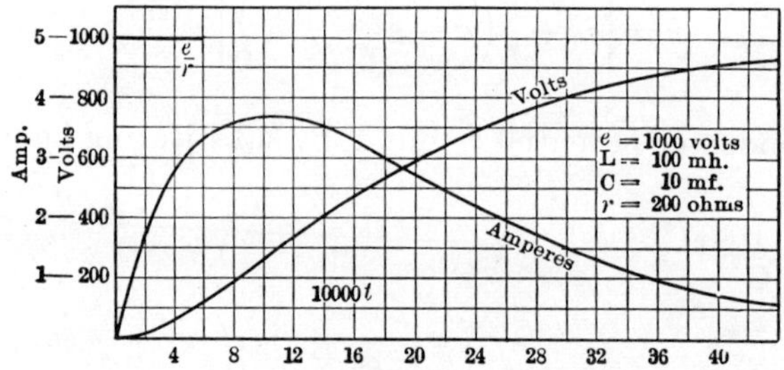

Fig. 12. Charging a condenser through a circuit having resistance and inductance. Constant potential. Critical charge.

in a circuit having the constants, $e = 1000$ volts; $C = 10$ mf.; $L = 100$ mh., and such resistance as to give the critical start, that is,

$$r = \sqrt{\frac{4L}{C}} = 200 \text{ ohms.}$$

In this case,

$$i = 10{,}000\, t\varepsilon^{-1000t}$$

and

$$e_1 = 1000 \left\{1 - (1 + 1000\,t)\, \varepsilon^{-1000t}\right\}.$$

39. In the *trigonometric* or *oscillating* case,

$$r^2 < \frac{4L}{C}.$$

The term under the square root (10) is negative, that is, the square root, s, is imaginary, and a_1 and a_2 are complex imaginary quantities, so that the equations (11) and (12) appear in imaginary form. They obviously can be reduced to real terms, since the phenomenon is real. Since an exponential function with imaginary exponents is a trigonometric function, and inversely, the solution of the equation thus leads to trigonometric functions, that is, the phenomenon is periodic or oscillating.

Substituting $s = jq$, we have

$$q = \sqrt{\frac{4L}{C} - r^2} \tag{33}$$

and

$$\left.\begin{aligned} a_1 &= \frac{r - jq}{2}, \\ a_2 &= \frac{r + jq}{2}. \end{aligned}\right\} \tag{34}$$

Substituting (34) in (11) and (12), and rearranging,

$$i = \varepsilon^{-\frac{r}{2L}t} \left\{ A_1 \varepsilon^{+\frac{jq}{2L}t} + A_2 \varepsilon^{-\frac{jq}{2L}t} \right\}, \tag{35}$$

$$e_1 = e - \varepsilon^{-\frac{r}{2L}t} \left\{ \frac{r + jq}{2} A_1 \varepsilon^{+\frac{jq}{2L}t} + \frac{r - jq}{2} A_2 \varepsilon^{-\frac{jq}{2L}t} \right\}. \tag{36}$$

Between the exponential function and the trigonometric functions exist the relations

$$\left.\begin{aligned} \varepsilon^{+jv} &= \cos v + j \sin v \\ \text{and} \qquad \varepsilon^{-jv} &= \cos v - j \sin v. \end{aligned}\right\} \tag{37}$$

Substituting (37) in (35), and rearranging, gives

$$i = \varepsilon^{-\frac{r}{2L}t} \left\{ (A_1 + A_2) \cos \frac{q}{2L} t + j (A_1 - A_2) \sin \frac{q}{2L} t \right\}.$$

Substituting the two new integration constants,

$$\left.\begin{aligned} B_1 &= A_1 + A_2 \\ \text{and} \qquad B_2 &= j (A_1 - A_2), \end{aligned}\right\} \tag{38}$$

gives

$$i = \varepsilon^{-\frac{r}{2L}t} \left\{ B_1 \cos \frac{q}{2L} t + B_2 \sin \frac{q}{2L} t \right\}. \tag{39}$$

In the same manner, substituting (37) in (36), rearranging, and substituting (38), gives

$$e_1 = e - \varepsilon^{-\frac{r}{2L}t} \left\{ \frac{rB_1 + qB_2}{2} \cos \frac{q}{2L} t + \frac{rB_2 - qB_1}{2} \sin \frac{q}{2L} t \right\}. \quad (40)$$

B_1 and B_2 are now the two integration constants, determined by the terminal conditions. That is, for $t = 0$, let $i = i_0$ = current and $e_1 = e_0$ = potential difference at condenser terminals, and substituting these values in (39) and (40) gives

$$i_0 = B_1$$

and

$$e_0 = e - \frac{rB_1 + qB_2}{2},$$

hence,

$$\left.\begin{aligned} B_1 &= i_0 \\ \text{and} \quad B_2 &= \frac{2(e - e_0) - ri_0}{q}. \end{aligned}\right\} \quad (41)$$

Substituting (41) in (39) and (40) gives the general *equations of condenser oscillation:*
the current is

$$i = \varepsilon^{-\frac{r}{2L}t} \left\{ i_0 \cos \frac{q}{2L} t + \frac{2(e - e_0) - ri_0}{q} \sin \frac{q}{2L} t \right\}, \quad (42)$$

and the potential difference at condenser terminals is

$$e_1 = e - \varepsilon^{-\frac{r}{2L}t} \left\{ (e - e_0) \cos \frac{q}{2L} t + \frac{r(e - e_0) - \frac{r^2 + q^2}{2} i_0}{q} \sin \frac{q}{2L} t \right\}. \quad (43)$$

Herefrom follow the equations of condenser charge and discharge, as special case:

For *condenser charge*, $i_0 = 0$; $e_0 = 0$, we have

$$i = \frac{2e}{q} \varepsilon^{-\frac{r}{2L}t} \sin \frac{q}{2L} t \quad (44)$$

and

$$e_1 = e\left\{1 - \varepsilon^{-\frac{r}{2L}t}\left(\cos\frac{q}{2L}t + \frac{r}{q}\sin\frac{q}{2L}t\right)\right\}, \tag{45}$$

and for *condenser discharge*, $i_0 = 0$, $e = 0$, we have

$$i = -\frac{2e_0}{q}\varepsilon^{-\frac{r}{2L}t}\sin\frac{q}{2L}t \tag{46}$$

and

$$e_1 = e_0\varepsilon^{-\frac{r}{2L}t}\left\{\cos\frac{q}{2L}t + \frac{r}{q}\sin\frac{q}{2L}t\right\}. \tag{47}$$

40. As an example is shown the oscillation of condenser charge in a circuit having the constants, $e = 1000$ volts; $L = 100$ mh., and $C = 10$ mf.

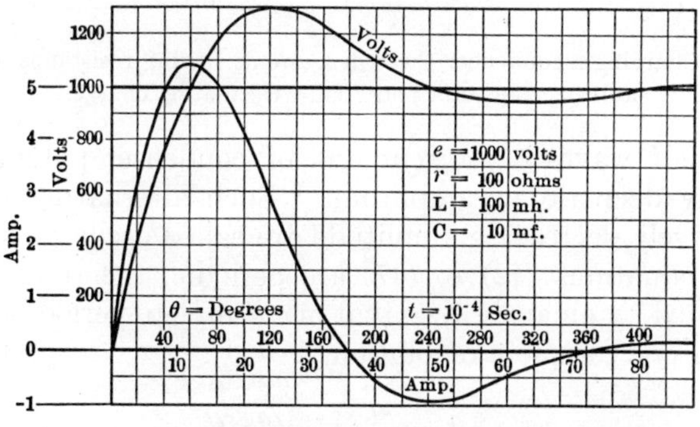

Fig. 13. **Charging a condenser through a circuit having resistance and inductance. Constant potential. Oscillating charge.**

(*a*) In Fig. 13, $r = 100$ ohms, hence, $q = 173$ and the current is

$$i = 11.55\,\varepsilon^{-500t}\sin 866\,t;$$

the condenser potential is

$$e_1 = 1000\{1 - \varepsilon^{-500t}(\cos 866\,t + 0.577\sin 866\,t)\}.$$

(*b*) In Fig. 14, $r = 40$ ohms, hence, $q = 196$ and the current is

$$i = 10.2\,\varepsilon^{-200t}\sin 980\,t;$$

the condenser potential is

$$e_1 = 1000\,\{1 - \varepsilon^{-200t}(\cos 980\,t + 0.21\sin 980\,t)\}.$$

41. Since the equations of current and potential difference (42) to (47) contain trigonometric functions, the phenomena are periodic or waves, similar to alternating currents. They differ from the latter by containing an exponential factor $\varepsilon^{-\frac{r}{2L}t}$, which steadily decreases with increase of t. That is, the suc-

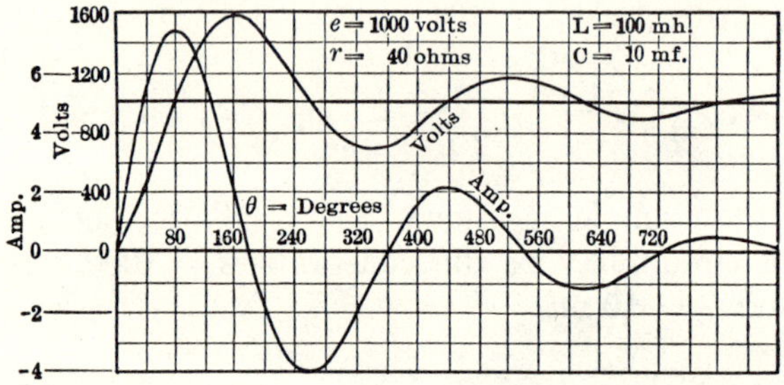

Fig. 14. Charging a condenser through a circuit having resistance and inductance. Constant potential. Oscillating charge.

cessive half waves of current and of condenser potential progressively decrease in amplitude. Such alternating waves of progressively decreasing amplitude are called oscillating waves.

Since equations (42) to (47) are periodic, the time t can be represented by an angle θ, so that one complete period is denoted by 2π or one complete revolution,

$$\theta = \frac{q}{2L}t = 2\pi ft. \tag{48}$$

$$2\pi f = \frac{q}{2L},$$

hence, the frequency of oscillation is

$$f = \frac{q}{4\pi L}, \tag{49}$$

or, substituting

$$q = \sqrt{\frac{4L}{C} - r^2}$$

gives the frequency of oscillation as

$$f = \frac{1}{2\pi}\sqrt{\frac{1}{LC} - \left(\frac{r}{2L}\right)^2}. \tag{50}$$

This frequency decreases with increasing resistance r, and becomes zero for $\left(\frac{r}{2L}\right)^2 = \frac{1}{LC}$, that is, $r^2 = \frac{4L}{C}$, or the critical case, where the phenomenon ceases to be oscillating.

If the resistance is small, so that the second term in equation (50) can be neglected, the frequency of oscillation is

$$f = \frac{1}{2\pi\sqrt{LC}}. \tag{51}$$

Substituting θ for t by equation (48)

$$t = \frac{2L}{q}\theta$$

in equations (42) and (43) gives the *general equations,*

$$i = \varepsilon^{-\frac{r}{q}\theta}\left\{i_0 \cos\theta + \frac{(e-e_0) - \frac{r}{2}i_0}{2q}\sin\theta\right\}, \tag{52}$$

$$e_1 = e - \varepsilon^{-\frac{r}{q}\theta}\left\{(e-e_0)\cos\theta + \frac{r(e-e_0) - \frac{r^2+q^2}{2}i_0}{q}\sin\theta\right\}, \tag{53}$$

$$\theta = 2\pi f t \tag{48}$$

and

$$f = \frac{q}{4\pi L} = \frac{1}{2\pi}\sqrt{\frac{1}{LC} - \left(\frac{r}{2L}\right)^2}. \tag{50}$$

42. If the resistance r can be neglected, that is, if r^2 is small compared with $\frac{4L}{C}$, the following equations are approximately exact:

$$q = 2\sqrt{\frac{L}{C}} \tag{54}$$

and

$$f = \frac{1}{2\pi\sqrt{LC}},$$

or,

$$2\pi f = \frac{1}{\sqrt{LC}}. \tag{55}$$

Introducing now $x = 2\,\pi f L$ = inductive reactance and $x' = \dfrac{1}{2\,\pi f C}$ = capacity reactance, and substituting (55), we have

$$x = \sqrt{\frac{L}{C}}$$

and

$$x' = \sqrt{\frac{L}{C}},$$

hence, $x' = x,$

that is, the frequency of oscillation of a circuit containing inductance and capacity, but negligible resistance, is that frequency f which makes the condensive reactance $x' = \dfrac{1}{2\,\pi f C}$ equal the inductive reactance $x = 2\,\pi f L$:

$$x' = x = \sqrt{\frac{L}{C}}\cdot \qquad (56)$$

Then (54),

$$q = 2\,x, \qquad (57)$$

and the general equations (52) and (53) are

$$i = \varepsilon^{-\frac{r}{2x}\theta}\left\{ i_0 \cos\theta + \frac{(e - e_0) - \frac{r}{2}\,i_0}{4\,x}\sin\theta \right\}; \qquad (58)$$

$$e_1 = e - \varepsilon^{-\frac{r}{2x}\theta}\left\{ (e - e_0)\cos\theta + \frac{r\,(e - e_0) - 2\,x^2 i_0}{2\,x}\sin\theta \right\}; \qquad (59)$$

$$x = \sqrt{\frac{L}{C}} \qquad (56)$$

and by (48) and (55):

$$\theta = \frac{t}{\sqrt{LC}}\cdot$$

43. Due to the factor $\varepsilon^{-\frac{r}{2L}t}$, successive half waves of oscillation decrease the more in amplitude, the greater the resistance r.

The ratio of the amplitude of successive half waves, or the decrement of the oscillation, is $\Delta = \varepsilon^{-\frac{r}{2L}t_1}$. where t_1 = duration of one half wave or one half cycle, $= \frac{1}{2f}$.

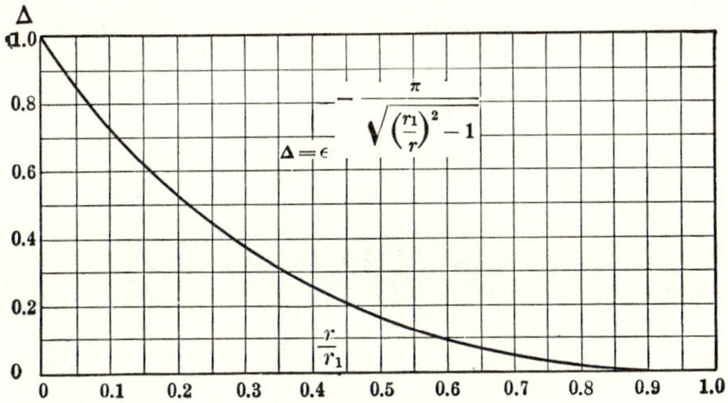

Fig. 15. Decrement of Oscillation.

Hence, from (50),

$$t_1 = \frac{\pi}{\sqrt{\frac{1}{LC} - \left(\frac{r}{2L}\right)^2}},$$

and

$$\Delta = \varepsilon^{-\delta} = \varepsilon^{-\frac{\pi}{\sqrt{\frac{4L}{r^2 C} - 1}}}. \tag{60}$$

Denoting the critical resistance as

$$r_1^2 = \frac{4L}{C}, \tag{61}$$

we have

$$\left.\begin{aligned} \Delta &= \varepsilon^{-\delta} = \varepsilon^{-\frac{\pi}{\sqrt{\left(\frac{r_1}{r}\right)^2 - 1}}}, \\ \text{or,}\qquad \delta &= \frac{\pi}{\sqrt{\left(\frac{r_1}{r}\right)^2 - 1}}; \end{aligned}\right\} \tag{62}$$

that is, the decrement of the oscillating wave, or the decay of the oscillation, is a function only of the ratio of the resistance of the circuit to its critical resistance, that is, the minimum resistance which makes the phenomenon non-oscillatory.

In Fig. 15 are shown the numerical values of the decrement Δ, for different ratios of actual to critical resistance $\frac{r}{r_1}$.

As seen, for $r > 0.21\, r_1$, or a resistance of the circuit of more than 21 per cent of its critical resistance, the decrement Δ is below 50 per cent, or the second half wave less than half the first one, etc.; that is, very little oscillation is left.

Where resistance is inserted into a circuit to eliminate the danger from oscillations, one-fifth of the critical resistance, or $r = 0.4 \sqrt{\frac{L}{C}}$, seems sufficient to practically dampen out the oscillation.

CHAPTER VI.

OSCILLATING CURRENTS.

44. The charge and discharge of a condenser through an inductive circuit produces periodic currents of a frequency depending upon the circuit constants.

The range of frequencies which can be produced by electrodynamic machinery is rather limited: synchronous machines or ordinary alternators can give economically and in units of larger size frequencies from 10 to 125 cycles. Frequencies below 10 cycles are available by commutating machines with low frequency excitation. Above 125 cycles the difficulties rapidly increase, due to the great number of poles, high peripheral speed, high power required for field excitation, poor regulation due to the massing of the conductors, which is required because of the small pitch per pole of the machine, etc., so that 1000 cycles probably is the limit of generation of constant potential alternating currents of appreciable power and at fair efficiency. For smaller powers, by using capacity for excitation, inductor alternators have been built and are in commercial service for wireless telegraphy and telephony, for frequencies up to 100,000 and even 200,000 cycles per second.

Still, even going to the limits of peripheral speed, and sacrificing everything for high frequency, a limit is reached in the frequency available by electrodynamic generation.

It becomes of importance, therefore, to investigate whether by the use of the condenser discharge the range of frequencies can be extended.

Since the oscillating current approaches the effect of an alternating current only if the damping is small, that is, the resistance low, the condenser discharge can be used as high frequency generator only by making the circuit of as low resistance as possible.

This, however, means limited power. When generating oscillating currents by condenser discharge, the load put on the circuit, that is, the power consumed in the oscillating-current circuit, represents an effective resistance, which increases the rapidity of the decay of the oscillation, and thus limits the power, and, when approaching the critical value, also lowers the frequency. This is obvious, since the oscillating current is the dissipation of the energy stored electrostatically in the condenser, and the higher the resistance of the circuit, the more rapidly is this energy dissipated, that is, the faster the oscillation dies out.

With a resistance of the circuit sufficiently low to give a fairly well sustained oscillation, the frequency is, with sufficient approximation,

$$f = \frac{1}{2\pi\sqrt{LC}}.$$

45. The constants, capacity, C, inductance, L, and resistance, r, have no relation to the size or bulk of the apparatus. For instance, a condenser of 1 mf., built to stand continuously a potential of 10,000 volts, is far larger than a 200-volt condenser of 100 mf. capacity. The energy which the former is able to store is $\frac{Ce^2}{2} = 50$ joules, while the latter stores only 2 joules, and therefore the former is 25 times as large.

A reactive coil of 0.1 henry inductance, designed to carry continuously 100 amperes, stores $\frac{Li^2}{2} = 500$ joules; a reactive coil of 1000 times the inductance, 100 henrys, but of a current-carrying capacity of 1 ampere, stores 5 joules only, therefore is only about one-hundredth the size of the former.

A resistor of 1 ohm, carrying continuously 1000 amperes, is a ponderous mass, dissipating 1000 kw.; a resistor having a resistance a million times as large, of one megohm, may be a lead pencil scratch on a piece of porcelain.

Therefore the size or bulk of condensers and reactors depends not only on C and L but also on the voltage and current which can be applied continuously, that is, it is approximately proportional to the energy stored, $\frac{Ce^2}{2}$ and $\frac{Li^2}{2}$, or since in electrical engineering energy is a quantity less frequently used than power, condensers and reactors are usually characterized by the power or rather apparent power which can be impressed upon them continuously by referring to a standard frequency, for which 60 cycles is generally used.

That means that reactors, condensers, and resistors are rated in kilowatts or kilovolt-amperes, just as other electrical apparatus, and this rating characterizes their size within the limits of design, while a statement like "a condenser of 10 mf." or "a reactor of 100 mh." no more characterizes the size than a statement like "an alternator of 100 amperes capacity" or "a transformer of 1000 volts."

A bulk of 1 cu. ft. in condenser can give about 5 to 10 kv-amp. at 60 cycles. Hence, 100 kv-amp. constitutes a very large size of condenser.

In the oscillating condenser discharge, the frequency of oscillation is such that the inductive reactance equals the condensive reactance. The same current is in both at the same terminal voltage. That means that the volt-amperes consumed by the inductance equal the volt-amperes consumed by the capacity.

The kilovolt-amperes of a condenser as well as of a reactor are proportional to the frequency. With increasing frequency, at constant voltage impressed upon the condenser, the current varies proportionally with the frequency; at constant alternating current through the reactor, the voltage varies proportionally with the frequency.

If then at the frequency of oscillation, reactor and condenser have the same kv-amp., they also have the same at 60 cycles.

A 100-kv-amp. condenser requires a 100-kv-amp. reactive coil for generating oscillating currents. A 100-kv-amp. reactive coil has approximately the same size as a 50-kw. transformer and can indeed be made from such a transformer, of ratio 1 : 1, by connecting the two coils in series and inserting into the magnetic circuit an air gap of such length as to give the rated magnetic density at the rated current.

A very large oscillating-current generator, therefore, would consist of 100-kv-amp. condenser and 100-kv-amp. reactor.

46. Assuming the condenser to be designed for 10,000 volts alternating impressed e.m.f. at 60 cycles, the 100 kv-amp. condenser consumes 10 amperes: its condensive reactance is $x_c = \frac{E}{I} = 1000$ ohms, and the capacity $C = \frac{1}{2\pi f_0 x_c} = 2.65$ mf.

Designing the reactor for different currents, and therewith different voltages, gives different values of inductance L, and therefore of frequency of oscillation f.

From the equations of the instantaneous values of the condenser discharge, (46) and (47), follow their effective values, or $\sqrt{\text{mean square}}$,

$$e_1 = \frac{e_0}{\sqrt{2}} \varepsilon^{-\frac{r}{2L}t}$$

and

$$i = \frac{e_0 \sqrt{2}}{q} \varepsilon^{-\frac{r}{2L}t} = \frac{e_0}{\sqrt{2}} \sqrt{\frac{C}{L}} \varepsilon^{-\frac{r}{2L}t}, \tag{63}$$

and thus the power,

$$p_1 = e_1 i = \frac{e_0^2}{2} \sqrt{\frac{C}{L}} \varepsilon^{-\frac{r}{L}t}, \tag{64}$$

since for small values of r

$$q = 2\sqrt{\frac{L}{C}}.$$

Herefrom would follow that the energy of each discharge is

$$W = \int_0^\infty p_1 dt = \frac{e_0^2}{2r} \sqrt{CL}. \tag{65}$$

Therefore, for 10,000 volts effective at 60 cycles at the condenser terminals, the e.m.f. is

$$e_0 = 10{,}000\sqrt{2},$$

and the condenser voltage is

$$e_1 = 10{,}000\, \varepsilon^{-\frac{r}{2L}t}.$$

Designing now the 100-kv-amp. reactive coil for different voltages and currents gives for an oscillation of 10,000 volts:

Reactive Coil.		React-ance.	Inductance.	Frequency of Oscillation.	Oscillating Current.	Oscillating Power.
Amp. i_0.	Volts, e_0.	$\frac{e_0}{i_0}=x_0$.	$\frac{x_0}{2\pi f_0}=L$.	$f=\frac{1}{2\pi\sqrt{LC}}$.	Amp., i.	Kv-amp., p_1.
1	100,000	10^5	265	6	1	10
10	10,000	10^3	2.65	60	10	100
100	1,000	10	2.65×10^{-2}	600	100	1,000
1,000	100	10^{-1}	2.65×10^{-4}	6,000	1,000	10,000
10,000	10	10^{-3}	2.65×10^{-6}	60,000	10,000	100,000
100,000	1	10^{-5}	2.65×10^{-8}	600,000	100,000	1,000,000
					$\times \varepsilon^{-\frac{r}{2L}t}$	$\times \varepsilon^{-\frac{r}{L}t}$

As seen, with the same kilovolt-ampere capacity of condenser and of reactive coil, practically any frequency of oscillation can be produced, from low commercial frequencies up to hundred thousands of cycles.

At frequencies between 500 and 2000 cycles, the use of iron in the reactive coil has to be restricted to an inner core, and at frequencies above this iron cannot be used, since hysteresis and eddy currents would cause excessive damping of the oscillation. The reactive coil then becomes larger in size.

47. Assuming 96 per cent efficiency of the reactive coil and 99 per cent of the condenser,

$$r = 0.05\, x,$$

gives

$$r = 0.05\sqrt{\frac{L}{C}},$$

since

$$x = 2\,\pi f L,$$

$$f = \frac{1}{2\,\pi\,\sqrt{LC}},$$

and the energy of the discharge, by (65), is

$$W = \frac{e_0^{\,2}}{2\,r}\sqrt{LC} = 10\, e_0^{\,2}\, C \text{ volt-ampere-seconds;}$$

thus the power factor is

$$\cos\theta_0 = 0.05.$$

Since the energy stored in the capacity is

$$W_0 = \frac{e_0^2 C}{2} \text{ joules,}$$

the critical resistance is

$$r_1 = 2\sqrt{\frac{L}{C}};$$

hence,

$$\frac{r}{r_1} = 0.025,$$

and the decrement of the oscillation is

$$\Delta = 0.92,$$

that is, the decay of the wave is very slow at no load.

Assuming, however, as load an external effective resistance equal to three times the internal resistance, that is, an electrical efficiency of 75 per cent, gives the total resistance as

$$r + r' = 0.2\, x;$$

hence,

$$\frac{r + r'}{r_1} = 0.1,$$

and the decrement is

$$\Delta = 0.73;$$

hence a fairly rapid decay of the wave.

At high frequencies, electrostatic, inductive, and radiation losses greatly increase the resistance, thus giving lower efficiency and more rapid decay of the wave.

48. The frequency of oscillation does not directly depend upon the size of apparatus, that is, the kilovolt-ampere capacity of condenser and reactor. Assuming, for instance, the size, in kilovolt-amperes, reduced to $\frac{1}{n}$, then, if designed for the same voltage, condenser and reactor, each takes $\frac{1}{n}$ the current, that is, the condensive reactance is n times as great, and therefore the capacity of the condenser, C, reduced to $\frac{1}{n}$, the inductance, L, is increased n-fold, so that the product CL, and thereby the frequency, remains the same; the power output, however, of the oscillating currents is reduced to $\frac{1}{n}$.

The limit of frequency is given by the mechanical dimensions. With a bulk of condenser of 10 to 20 cu. ft., the minimum length of the discharge circuit cannot well be less than 10 ft.; 10 ft. of conductor of large size have an inductance of at least 0.002 mh. $= 2 \times 10^{-6}$, and the frequency of oscillation would therefore be limited to about 60,000 cycles per second, even without any reactive coil, in a straight discharge path.

The highest frequency which can be reached may be estimated about as follows:

The minimum length of discharge circuit is the gap between the condenser plates.

The minimum condenser capacity is given by two spheres, since small plates give a larger capacity, due to the edges.

The minimum diameter of the spheres is 1.5 times their distance, since a smaller sphere diameter does not give a clean spark discharge, but a brush discharge precedes the spark.

With $e_0 = 10{,}000\sqrt{2}$, the spark gap length between spheres is $e = 0.3$ in., and the diameter of the spheres therefore 0.45 in. The oscillating circuit then consists of two spheres of 0.45 in., separated by a gap of 0.3 in.

This gives an approximate length of oscillating circuit of 0.5 in., or an inductance $L = \frac{0.3 \times 10^{-3}}{24{,}000} = 0.125 \times 10^{-7}$ henry.

The capacity of the spheres against each other may be estimated as $C = 50 \times 10^{-8}$ mf.; this gives the frequency of oscillation as

$$f = \frac{1}{2\pi\sqrt{LC}} = 2 \times 10^9,$$

or, 2 billion cycles.

At
$$e_0 = 10{,}000\sqrt{2} \text{ volts},$$

$$e_1 = 10{,}000\,\varepsilon^{-\frac{r}{2L}t} \text{ volts},$$

$$i = 2.83\,\varepsilon^{-\frac{r}{2L}t} \text{ amp.},$$

and
$$p_1 = 28.3\,\varepsilon^{-\frac{r}{L}t} \text{ kv-amp.}$$

Reducing the size and spacing of the spheres proportionally, and proportionally lowering the voltage, or increasing the dielectric strength of the gap by increasing the air pressure, gives still higher frequencies.

As seen, however, the power of the oscillation decreases with increasing frequency, due to the decrease of size and therewith of storage ability, of capacity, and of inductance.

With a frequency of billions of cycles per second, the effective resistance must be very large, and therefore the damping rapid.

Such an oscillating system of two spheres separated by a gap would have to be charged by induction, or the spheres charged separately and then brought near each other, or the spheres may be made a part of a series of spheres separated by gaps and connected across a high potential circuit, as in some forms of lightning arresters.

Herefrom it appears that the highest frequency of oscillation of appreciable power which can be produced by a condenser discharge reaches billions of cycles per second, thus is enormously higher than the highest frequencies which can be produced by electrodynamic machinery.

At five billion cycles per second, the wave length is about 6 cm., that is, the frequency only a few octaves lower than the lowest frequencies observed as heat radiation or ultra red light.

The average wave length of visible light, 55×10^{-6} cm., corresponding to a frequency of 5.5×10^{14} cycles, would require spheres 10^{-5} cm. in diameter, that is, approaching molecular dimensions.

OSCILLATING-CURRENT GENERATOR.

49. A system of constant impressed e.m.f., e, charging a condenser C through a circuit of inductance L and resistance r, with a discharge circuit of the condenser, C, comprising an air gap in series with a reactor of inductance L_0 and a resistor of resistance r_0, is a generator of oscillating current if the air gap is set for such a voltage e_0 that it discharges before the voltage of the condenser C has reached the maximum, and if the resistance r_0 is such as to make the condenser discharge oscillatory, that is,

$$r_0^2 < \frac{4 L_0}{C}.$$

In such a system, as shown diagrammatically in Fig. 16, as soon, during the charge of the condenser, as the terminal voltage at C and thereby at the spark gap has reached the value e_0, the condenser C discharges over this spark gap, its potential difference falls to zero, then it charges again up to potential difference e_0, discharges, etc. Thus a series of oscillating discharges

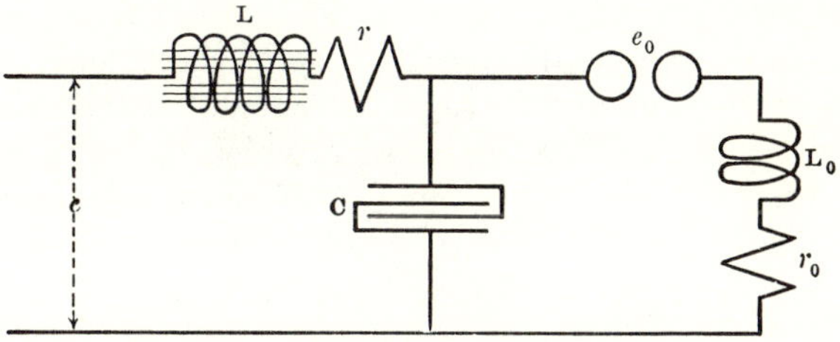

Fig. 16. Oscillating-current generator.

occur in the circuit, L_0, r_0, at intervals equal to the time required to charge condenser C over reactor L and resistor r, up to the potential difference e_0, with an impressed e.m.f. e.

The resistance, r, obviously should be as low as possible, to get good efficiency of transformation; the inductance, L, must be so large that the time required to charge condenser C to potential e_0 is sufficient for the discharge over L_0, r_0 to die out and also the spark gap e_0 to open, that is, the conducting products of the spark in the gap e_0 to dissipate. This latter takes a considerable time, and an air blast directed against the spark gap e_0, by carrying away the products of the discharge, permits a more rapid recurrence of the discharge. The velocity of the air blast (and therefore the pressure of the air) must be such as to carry the ionized air or the metal vapors which the discharge forms in the gap e_0 out of the discharge path faster than the condenser recharges.

Assuming, for instance, the spark gap, e_0, set for 20,000 volts, or about 0.75 in., the motion of the air blast during successive discharges then should be large compared with 0.75 in., hence at least 3 to 6 in. With 1000 discharges per second, this would require an air velocity of $v = 250$ to 500 feet per second, with 5000 discharges per second an air velocity of $v = 1250$ to 2500 feet per second, corresponding to an air pressure of approximately $p = 14.7 \{(1 + 2 v^2 10^{-7})^{3.5} - 1\}$ lb. per sq. in., or 0.66 to 2.75 lb. in the first, 23 to 230 lb. in the second case.

While the condenser charge may be oscillatory or logarithmic, efficiency requires a low value of r, that is, an oscillatory charge.

With a frequency of discharge in L_0, r_0 very high compared with the frequency of charge, the duration of the discharge is short compared with the duration of the charge, that is, the oscillating currents consist of a series of oscillations separated by relatively long periods of rest. Thus the current in L does not appreciably change during the time of the discharge, and at the end of the condenser charge the current in the reactor, L, is the same as the current in L, with which the next condenser charge starts. The charging current of the condenser, C, in L thus changes from i_0 at the beginning of the charge, or condenser e.m.f., $e_0 = 0$, to the same value i_0 at the end of the charge, or condenser e.m.f., $e_1 = e_0$.

50. Counting, therefore, the time, t, from the moment when the condenser charge begins, we have the terminal conditions:

$t = 0$, $i = i_0$, $e_1 = 0$ at the beginning of the condenser charge.
$t = t_0$, $i = i_0$, $e_1 = e_0$ at the end of the condenser charge.

In the condenser discharge, through circuit L_0, r_0, counting the time t' from the moment when the condenser discharge begins, that is, $t' = t - t_0$, we have

$$t' = 0, \quad i = 0, \quad e_1 = e_0 \text{ the terminal condition.}$$

e_0, thus, is that value of the voltage e_1 at which discharge takes place across the spark gap, and t_0 is the time elapsing between $e_1 = 0$ and $e_1 = e_0$, or the time required to build up the voltage e_1 sufficiently to break down the spark gap.

Under the assumption that the period of oscillation of the condenser charge through L, r, is large compared with the period of oscillation of the condenser discharge through L_0, r_0, the equations are:

(*A*) Condenser discharge:

$$i = \frac{2\,e_0}{q_0} \varepsilon^{-\frac{r_0}{2L_0}t'} \sin \frac{q_0}{2\,L_0} t', \tag{66}$$

$$e_1 = e_0 \varepsilon^{-\frac{r_0}{2L_0}t'} \left\{ \cos \frac{q_0}{2\,L_0} t' + \frac{r_0}{q_0} \sin \frac{q_0}{2\,L_0} t' \right\}, \tag{67}$$

where

$$q_0 = \sqrt{\frac{4\,L_0}{C} - r_0^2}. \tag{68}$$

(*B*) Condenser charge:

$$i = \varepsilon^{-\frac{r}{2L}t} \left\{ i_0 \cos \frac{q}{2L} t + \frac{2e - ri_0}{q} \sin \frac{q}{2L} t \right\}, \tag{69}$$

$$e_1 = e - \varepsilon^{-\frac{r}{2L}t} \left\{ e \cos \frac{q}{2L} t + \frac{re - \frac{r^2 + q^2}{2} i_0}{q} \sin \frac{q}{2L} t \right\}, \tag{70}$$

where

$$q = \sqrt{\frac{4L}{C} - r^2}. \tag{71}$$

Substituting in (69) and (70) the above discussed terminal conditions,

$$t = t_0, \quad i = i_0, \quad e_1 = e_0,$$

gives

$$i_0 = \varepsilon^{-\frac{r}{2L}t_0} \left\{ i_0 \cos \frac{q}{2L} t_0 + \frac{2e - ri_0}{q} \sin \frac{q}{2L} t_0 \right\} \tag{72}$$

and

$$e_0 = e - \varepsilon^{-\frac{r}{2L}t_0} \left\{ e \cos \frac{q}{2L} t_0 + \frac{re - \frac{r^2 + q^2}{2} i_0}{q} \sin \frac{q}{2L} t_0 \right\}. \tag{73}$$

Denoting, for convenience,

$$\left.\begin{aligned} \frac{r}{2L} t_0 &= s, \\ \frac{q}{2L} t_0 &= \phi, \\ \text{and} \qquad \frac{r}{q} &= a, \end{aligned}\right\} \tag{74}$$

and resolving (72) for i_0, gives

$$i_0 = \frac{2e}{q} \frac{\varepsilon^{-s} \sin \phi}{1 - \varepsilon^{-s} \cos \phi + a\varepsilon^{-s} \sin \phi}, \tag{75}$$

and substituting (75) in (73) and rearranging,

$$e_0 = e \frac{1 - 2\varepsilon^{-s} \cos \phi + \varepsilon^{-2s}}{1 - \varepsilon^{-s} \cos \phi + a\varepsilon^{-s} \sin \phi}. \tag{76}$$

The two equations (75), (76) permit the calculation of two of the three quantities i_0, e_0, t_0: the time, t_0, of condenser charge appears in the exponential function, in s, and in the trigonometric function, in ϕ.

Since in an oscillating-current generator of fair efficiency, that is, when r is as small as possible, s is a small quantity, ε^{-s} can be resolved into the series

$$\varepsilon^{-s} = 1 - s + \frac{s^2}{2} - + \cdots. \tag{77}$$

Substituting (77) in (75), and dropping all terms higher than s^2, gives

$$i_0 = \frac{2\,e}{q} \frac{\left(1 - s + \frac{s^2}{2}\right) \sin \phi}{1 - \cos \phi + s \cos \phi - \frac{s^2}{2} \cos \phi + a \sin \phi - as \sin \phi}.$$

Multiplying numerator and denominator by $\left(1 + \frac{s}{2}\right)$, and rearranging, gives

$$\left.\begin{aligned} i_0 &= \frac{2\,e}{q} \frac{\sin \phi}{\frac{2+s}{2-s} - \cos \phi + a \sin \phi} \\ &= \frac{2\,e}{q} \frac{\sin \phi}{\frac{2\,s}{2-s} + 2 \sin^2 \frac{\phi}{2} + a \sin \phi}. \end{aligned}\right\} \tag{78}$$

Substituting (77) in (76), dropping terms higher than s^2 and as, multiplying numerator and denominator by $\left(1 + \frac{s}{2}\right)$, and rearranging, gives

$$e_0 = 2\,e \frac{2 \sin^2 \frac{\phi}{2} + \frac{s^2}{2}}{\frac{2\,s}{2-s} + 2 \sin^2 \frac{\phi}{2} + a \sin \phi}. \tag{79}$$

Substituting t_0 in (78) and (79) gives

$$i_0 = \frac{2\,e}{q} \frac{\sin \frac{q}{2\,L}\, t_0}{\frac{2\,rt_0}{4\,L - rt_0} + 2 \sin^2 \frac{q}{4\,L}\, t_0 + \frac{r}{q} \sin \frac{q}{2\,L}\, t_0} \tag{80}$$

and

$$e_0 = 2\,e \frac{2 \sin^2 \frac{q}{4\,L}\, t_0 + \frac{r^2 t_0^{\,2}}{8\,L^2}}{\frac{2\,rt_0}{4\,L - rt_0} + 2 \sin^2 \frac{q}{4\,L}\, t_0 + \frac{r}{q} \sin \frac{q}{2\,L}\, t_0} \tag{81}$$

as approximate equations giving i_0 and e_0 as functions of t_0, or the time of condenser charge.

51. The time, t_0, during which the condenser charges, increases with increasing e_0, that is, increasing length of the spark gap in the discharge circuit, at first almost proportionally, then, as e_0 approaches $2\,e$, more slowly.

As long as e_0 is appreciably below $2\,e$, that is, about $e_0 < 1.75\,e$, t_0 is relatively short, and the charging current i, which increases from i_0 to a maximum, and then decreases again to i_0, does not vary much, but is approximately constant, with an average value very little above i_0, so that the power supplied by the impressed e.m.f., e, to the charging circuit can approximately be assumed as

$$p_0 = ei_0. \tag{82}$$

The condenser discharge is intermittent, consisting of a series of oscillations, with a period of rest between the oscillations, which is long compared with the duration of the oscillation, and during which the condenser charges again.

The discharge current of the condenser is, (66),

$$i = \frac{2\,e_0}{q_0} \varepsilon^{-\frac{r_0}{2\,L_0} t} \sin \frac{q_0}{2\,L_0}\, t, \text{ in amp.,}$$

and since such an oscillation recurs at intervals of t_0 seconds, the effective value, or square root of mean square of the discharge current, is

$$i_1 = \sqrt{\frac{1}{t_0} \int_0^{t_0} i^2\, dt}. \tag{83}$$

Long before $t = t_0$, i is practically zero, and as upper limit of the integral can therefore be chosen ∞ instead of t_0.

Substituting (66) in (83), and taking the constant terms out of the square root, gives the effective value of discharge current as

$$i_1 = \frac{2\,e_0}{q_0}\sqrt{\frac{1}{t_0}\int_0^{\infty}\varepsilon^{-\frac{r_0}{L_0}t}\sin^2\frac{q_0}{2\,L_0}t\,dt}$$

$$= \frac{2\,e_0}{q_0}\sqrt{\frac{1}{2\,t_0}\left\{\int_0^{\infty}\varepsilon^{-\frac{r_0}{L_0}t}\,dt - \int_0^{\infty}\varepsilon^{-\frac{r_0}{L_0}t}\cos\frac{q_0}{L_0}t\,dt\right\}}; \tag{84}$$

however,

$$\int_0^{\infty}\varepsilon^{-\frac{r_0}{L_0}t}\,dt = -\frac{L_0}{r_0}\left[\varepsilon^{-\frac{r_0}{L_0}t}\right]_0^{\infty} = \frac{L_0}{r_0},$$

and by fractional integration,

$$\int_0^{\infty}\varepsilon^{-\frac{r_0}{L_0}t}\cos\frac{q_0}{L_0}t\,dt$$

$$= \frac{\frac{L_0}{q_0}}{1+\left(\frac{r_0}{q_0}\right)^2}\left[\varepsilon^{-\frac{r_0}{L_0}t}\left\{\sin\frac{q_0}{L_0}t - \frac{r_0}{q_0}\cos\frac{q_0}{L_0}t\right\}\right]_0^{\infty}$$

$$= \frac{L_0 r_0}{r_0^{\,2} + q_0^{\,2}};$$

hence, substituting in (84),

$$i_1 = e_0\sqrt{\frac{2\,L_0}{t_0 r_0 (r_0^{\,2} + q_0^{\,2})}}. \tag{85}$$

Since

$$q^2 = \frac{4\,L}{C} - r^2,$$

we have, substituting in (85),

$$i_1 = e_0\sqrt{\frac{C}{2\,t_0 r_0}}, \tag{86}$$

and, denoting by

$$f_1 = \frac{1}{t_0},$$

the frequency of condenser charge, or the number of complete trains of discharge oscillations per second,

$$i_1 = e_0 \sqrt{\frac{Cf_1}{2\,r_0}}, \tag{87}$$

that is, the effective value of the discharge current is proportional to the condenser potential, e_0, proportional to the square root of the capacity, C, and the frequency of charge, f_1, and inversely proportional to the square root of the resistance, r_0, of the discharge circuit; but it does not depend upon the inductance L_0 of the discharge circuit, and therefore does not depend on the frequency of the discharge oscillation.

The power of the discharge is

$$p_1 = i_1^2 r_0 = f_1 \frac{e_0^2 C}{2}. \tag{88}$$

Since $\frac{e_0^2 C}{2}$ is the energy stored in the condenser of capacity C at potential e_0, and f_1 the frequency or number of discharges of this energy per second, equation (88) is obvious.

Inversely therefore, from equation (88), that is, the total energy stored in the condenser and discharging per second, the effective value of discharge current can be directly calculated as

$$i_1 = \sqrt{\frac{p_1}{r_0}} = e_0 \sqrt{\frac{Cf_1}{2\,r_0}}.$$

The ratio of effective discharge current, i_1, to mean charging current, i_0, is

$$\frac{i_1}{i_0} = \frac{e_0}{i_0} \sqrt{\frac{Cf_1}{2\,r_0}}, \tag{89}$$

and substituting (80) and (81) in (89),

$$\frac{i_1}{i_0} = q \sqrt{\frac{Cf_1}{2r_0}} \; \frac{2 \sin^2 \frac{q}{4\,L} t_0 + \frac{r^2 t_0^2}{8\,L}}{\sin \frac{q}{2\,L} t_0}. \tag{90}$$

The magnitude of this quantity can be approximated by neglecting r compared with $\frac{4L}{C}$, that is, substituting $q = \sqrt{\frac{4L}{C}}$ and replacing the sine-function by the arcs. This gives

$$\frac{i_1}{i_0} = \frac{1}{\sqrt{2 r_0 C f_1}}, \tag{91}$$

that is, the ratio of currents is inversely proportional to the square root of the resistance of the discharge circuit, of the capacity, and of the frequency of charge.

52. Example: Assume an oscillating-current generator, feeding a Tesla transformer for operating X-ray tubes, or directly supplying an iron arc (that is, a condenser discharge between iron electrodes) for the production of ultraviolet light.

The constants of the charging circuit are: the impressed e.m.f., $e = 15{,}000$ volts; the resistance, $r = 10{,}000$ ohms; the inductance, $L = 250$ henrys, and the capacity, $C = 2 \times 10^{-8}$ farads $= 0.02$ mf.

The constants of the discharge circuit are: (*a*) operating Tesla transformer, the estimated resistance, $r_0 = 20$ ohms (effective) and the estimated inductance, $L_0 = 60 \times 10^{-6}$ henry $= 0.06$ mh.; (*b*) operating ultraviolet arc, the estimated resistance, $r_0 = 5$ ohms (effective) and the estimated inductance, $L_0 = 4 \times 10^{-6}$ henry $= 0.004$ mh.

Therefore in the charging circuit,

$$q = 223{,}400 \text{ ohms}, \qquad \frac{r}{q} = 0.0448,$$

$$\frac{q}{2L} = 446.8, \qquad \frac{r}{2L} = 20,$$

$$\frac{L}{r} = 0.025;$$

then

$$\left.\begin{aligned} i_0 &= 0.1344 \frac{\sin 446.8\, t_0}{\dfrac{2 t_0}{0.1 - t_0} + 2 \sin^2 223.4\, t_0 + 0.0448 \sin 446.8\, t_0}, \\ &\text{and} \\ e_0 &= 30{,}000 \frac{2 \sin^2 223.4\, t_0 + 200\, t_0^2}{\dfrac{2 t_0}{0.1 - t_0} + 2 \sin^2 223.4\, t_0 + 0.0448 \sin 446.8\, t_0}. \end{aligned}\right\} \tag{92}$$

Fig. 17 shows i_0 and e_0 as ordinates, with the time of charge t_0 as abscissas.

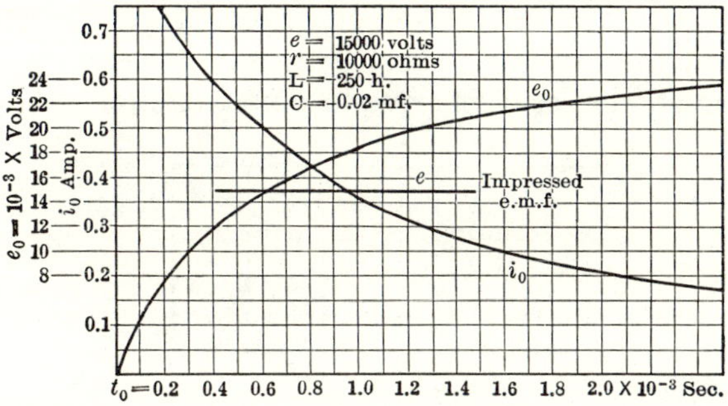

Fig. 17. Oscillating-current generator charge.

The frequency of the charging oscillation is

$$f = \frac{q}{4\,\pi L} = 71.2 \text{ cycles per sec.};$$

for

$$i_0 = 0.365 \text{ amp.},$$

substituting in equations (69) and (70) we have

$$\left.\begin{array}{l} i = \varepsilon^{-20t}\left\{0.365 \cos 446.8\,t + 0.118 \sin 446.8\,t\right\}, \text{ in amp.,} \\ \text{and} \\ e_1 = 15{,}000\left\{1 - \varepsilon^{-20t}\left[\cos 446.8\,t - 2.67 \sin 446.8\,t\right]\right\}, \text{ in volts,} \end{array}\right\} \quad (93)$$

the equations of condenser charge.

From these equations the values of i and e_1 are plotted in Fig. 18, with the time t as abscissas.

As seen, the value $i = i_0 = 0.365$ amp., is reached again at the time $t_0 = 0.0012$, that is, after 30.6 time-degrees or about $\frac{1}{12}$ of a period. At this moment the condenser e.m.f. is $e_1 = e_0 = 22{,}300$ volts; that is, by setting the spark gap for 22,300 volts the duration of the condenser charge is 0.0012 second, or in other words, every 0.0012 second, or 833 times per second, discharge oscillations are produced.

With this spark gap, the charging current at the beginning and at the end of the condenser charge is 0.365 amp., and the average charging current is 0.3735 amp. at 15,000 volts, consuming 5.6 kva.

Assume that the e.m.f. at the condenser terminals at the end of the charge is $e_0 = 22{,}300$ volts; then consider two cases, namely: (*a*) the condenser discharges into a Tesla transformer, and (*b*) the condenser discharges into an iron arc.

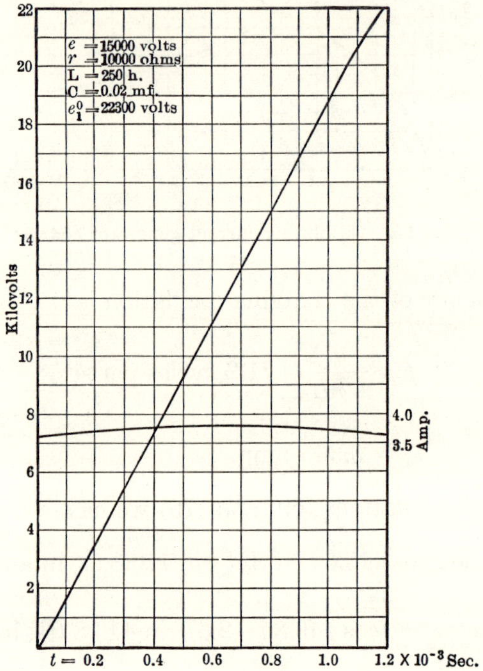

Fig. 18. Oscillating-current generator condenser charge.

(*a*) The Tesla transformer, that is, an oscillating-current transformer, has no iron, but a primary coil of very few turns (20) and a secondary coil of a larger number of turns (360), both immersed in oil.

While the actual ohmic resistance of the discharge circuit is only 0.1 ohm, the load on the secondary of the Tesla transformer, the dissipation of energy into space by brush discharge, etc., and the increase of resistance by unequal current distribution in the conductor, increase the effective resistance to many times the ohmic resistance. We can, therefore, assign the following estimated values: $r_0 = 20$ ohms; $L_0 = 60 \times 10^{-6}$ henry, and $C = 2 \times 10^{-8}$ farad.

Then

$$q_0 = 108 \text{ ohms}, \qquad \frac{r_0}{q_0} = 0.186,$$

$$\frac{q_0}{2L_0} = 0.898 \times 10^6, \qquad \frac{r_0}{2L_0} = 0.1667 \times 10^6,$$

which give

$$\left.\begin{aligned} i &= 415\, \varepsilon^{-0.1667 \times 10^6 t} \sin 0.898 \times 10^6 t, \text{ amp.} \\ &\text{and} \\ e_1 &= 22{,}300\, \varepsilon^{-0.1667 \times 10^6 t}\{\cos 0.898 \times 10^6 t + 0.186 \sin 0.898 \times 10^6 t\}, \\ &\quad \text{volts.} \end{aligned}\right\} \tag{94}$$

The frequency of oscillation is

$$f_0 = \frac{0.898 \times 10^6}{2\pi} = 143{,}000 \text{ cycles per sec.} \tag{95}$$

Fig. 19 shows the current i and the condenser potential e_1 during the discharge, with the time t as abscissas. As seen, the discharge frequency is very high compared with the fre-

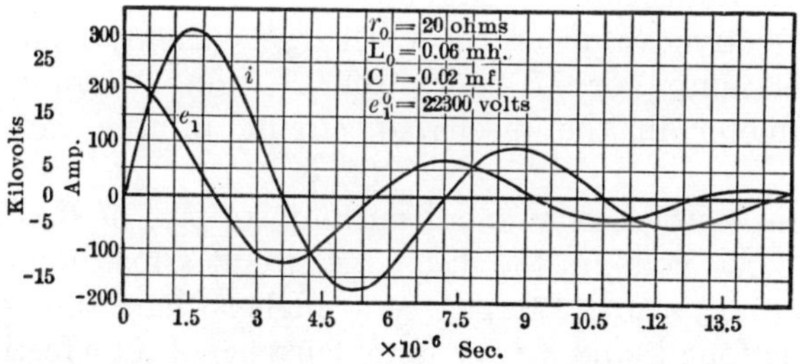

Fig. 19. Oscillating-current generator condenser discharge.

quency of charge, the duration of discharge very short, and the damping very great; a decrement of 0.55, so that the oscillation dies out very rapidly. The oscillating current, however, is enormous compared with the charging current; with a mean charging current of 0.3735 amp., and a maximum charging current of 0.378 amp. the maximum discharge current is 315 amp., or 813 times as large as the charging current.

The effective value of the discharge current, from equation (87), is $i_1 = 14.4$ amp., or nearly 40 times the charging current.

53. (*b*) When discharging the condenser directly, through an ultraviolet or iron arc, in a straight path, and estimating $r_0 = 5$ ohms and $L_0 = 4 \times 10^{-6}$ henry, we have

$$q_0 = 27.84 \text{ ohms}, \qquad \frac{r_0}{q_0} = 0.1795,$$

$$\frac{q_0}{2\,L_0} = 3.48 \times 10^6, \qquad \frac{r_0}{2\,L_0} = 0.625 \times 10^6;$$

then,

$$\left.\begin{aligned} &i = 1600\,\varepsilon^{-0.625 \times 10^6 t} \sin 3.48 \times 10^6\,t, \text{ in amp.,} \\ &\text{and} \\ &e_1 = 22{,}300\,\varepsilon^{-0.625 \times 10^6 t}\{\cos 3.48 \times 10^6\,t + 0.1795 \sin 3.48 \times 10^6\,t\}, \\ &\qquad \text{in volts,} \end{aligned}\right\} \tag{96}$$

and the frequency of oscillation is

$$f_0 = 562{,}000 \text{ cycles per sec.}; \tag{97}$$

that is, the frequency is still higher, over half a million cycles; the maximum discharge current over 1000 amperes; however, the duration of the discharge is still shorter, the oscillations dying out more rapidly.

The effective value of the discharge current, from (87), is $i_1 = 28.88$ amp., or 77 times the charging current. A hot wire ammeter in the discharge circuit in this case showed 29 amp.

As seen, with a very small current supply, of 0.3735 amp., at $e = 15{,}000$ volts, in the discharge circuit a maximum voltage of 22,300, or nearly 50 per cent higher than the impressed voltage, is found, and a very large current, of an effective value very many times larger than the supply current.

As a rule, instead of a constant impressed e.m.f., e, a low frequency alternating e.m.f. is used, since it is more conveniently generated by a step-up transformer. In this case the condenser discharges occur not at constant intervals of t_0 seconds, but only during that part of each half wave when the e.m.f. is sufficient to jump the gap e_0, and at intervals which are shorter at the maximum of the e.m.f. wave than at its beginning and end.

For instance, using a step-up transformer giving 17,400 volts effective (by the ratio of turns 1 : 150, with 118 volts impressed at 60 cycles), or a maximum of 24,700 volts, then during each half wave the first discharge occurs as soon as the voltage has reached 22,300, sufficient to jump the spark gap, and then a series of discharges occurs, at intervals decreasing with the increase of the impressed e.m.f., up to its maximum, and then with increasing intervals, until on the decreasing wave the e.m.f. has fallen below that which, during the charging oscillation, can jump the gap e_0, that is, about 13,000 volts. Then the oscillating discharges stop, and start again during the next half wave.

Hence the phenomenon is of the same character as investigated above for constant impressed e.m.f., except that it is intermittent, with gaps during the zero period of impressed voltage and unequal time intervals t_0 between the successive discharges.

54. An underground cable system can act as an oscillating-current generator, with the capacity of the cables as condenser, the internal inductance of the generators as reactor, and a short-circuiting arc as discharge circuit.

In a cable system where this phenomenon was observed the constants were approximately as follows: capacity of the cable system, $C = 102$ mf.; inductance of 30,000-kw. in generators, $L = 6.4$ mh.; resistance of generators and circuit up to the short-circuiting arc, $r = 0.1$ ohm and $r = 1.0$ ohm respectively; impressed e.m.f., 11,000 volts effective, and the frequency 25 cycles per second.

The frequency of charging oscillation in this case is

$$f = \frac{q}{4\pi L} = 197 \text{ cycles per sec.}$$

since

$$q = \sqrt{\frac{4L}{C} - r^2} = 15.8 \text{ ohms.}$$

Substituting these values in the preceding equations, and estimating the constants of the discharge circuit, gives enormous values of discharge current and e.m.f.

CHAPTER VII

RESISTANCE, INDUCTANCE, AND CAPACITY IN SERIES IN ALTERNATING-CURRENT CIRCUIT

55. Let, at time $t = 0$ or $\theta = 0$, the e.m.f.,

$$e = E \cos (\theta - \theta_0), \tag{1}$$

be impressed upon a circuit containing in series the resistance, r, the inductance, L, and the capacity, C.

$$\left.\begin{array}{ll} \text{The inductive reactance is} & x = 2\pi f L \\ \text{and the condensive reactance is} & x_c = \dfrac{1}{2\pi f C}, \end{array}\right\} \tag{2}$$

where f = frequency and $\theta = 2\pi f t$. (3)

Then the e.m.f. consumed by resistance is ri; the e.m.f. consumed by inductance is

$$L\frac{di}{dt} = x\frac{di}{d\theta},$$

and the e.m.f. consumed by capacity is

$$e_1 = \frac{1}{C}\int i\,dt = x_c \int i\,d\theta, \tag{4}$$

where i = instantaneous value of the current.

Hence,

$$e = ri + x\frac{di}{d\theta} + x_c \int i\,d\theta, \tag{5}$$

or,

$$E \cos (\theta - \theta_0) = ri + x\frac{di}{d\theta} + x_c \int i\,d\theta, \tag{6}$$

and hence, the difference of potential at the condenser terminals is

$$e_1 = x_c \int i\,d\theta = E \cos (\theta - \theta_0) - ri - x\frac{di}{d\theta}. \tag{7}$$

Equation (6) differentiated gives

$$E \sin (\theta - \theta_0) + x \frac{d^2 i}{d\theta^2} + r \frac{di}{d\theta} + x_c i = 0. \tag{8}$$

The integral of this equation (8) is of the general form

$$i = A\varepsilon^{-a\theta} + B \cos (\theta - \sigma). \tag{9}$$

Substituting (9) in (8), and rearranging, gives

$$A\varepsilon^{-a\theta} \{a^2 x - ar + x_c\} + \sin\theta \{E \cos \theta_0 - rB \cos\sigma - B (x - x_c) \sin\sigma\} \\ - \cos\theta \{E \sin\theta_0 - rB \sin\sigma + B (x - x_c) \cos\sigma\} = 0,$$

and, since this must be an identity,

$$\left.\begin{aligned} a^2 x - ar + x_c &= 0, \\ E \cos\theta_0 - rB \cos\sigma - B (x - x_c) \sin\sigma &= 0, \\ E \sin\theta_0 - rB \sin\sigma + B (x - x_c) \cos\sigma &= 0. \end{aligned}\right\} \tag{10}$$

Substituting

$$\left.\begin{aligned} s &= \sqrt{r^2 - 4\,x\,x_c}, \\ z_0 &= \sqrt{r^2 + (x - x_c)^2}, \\ \tan\gamma &= \frac{x - x_c}{r}, \end{aligned}\right\} \tag{11}$$

in equations (10) gives

$$\left.\begin{aligned} a &= \frac{r \pm s}{2\,x}, \\ B &= \frac{E}{z_0}, \\ \sigma &= \theta_0 + \gamma \\ \text{and} \qquad A &= \text{indefinite}, \end{aligned}\right\} \tag{12}$$

and the equation of current, (9), thus is

$$i = \frac{E}{z_0} \cos (\theta - \theta_0 - \gamma) + A_1 \varepsilon^{-\frac{r-s}{2x}\theta} + A_2 \varepsilon^{-\frac{r+s}{2x}\theta}, \tag{13}$$

and, substituting (12) in (7), and rearranging, the potential difference at the condenser terminals is

$$e_1 = \frac{Ex_c}{z_0} \sin (\theta - \theta_0 - \gamma) - \frac{r+s}{2} A_1 \epsilon^{-\frac{r-s}{2x}\theta} - \frac{r-s}{2} A_2 \epsilon^{-\frac{r+s}{2x}\theta}. \quad (14)$$

The two integration constants A_1 and A_2 are given by the terminal conditions of the problem.

Let, at the moment of start,

$$\theta = 0,$$

$$\left.\begin{aligned} i &= i_0 = \text{instantaneous value of current and} \\ e_1 &= e_0 = \text{instantaneous value of condenser potential difference.} \end{aligned}\right\} \quad (15)$$

Substituting in (13) and (14),

$$i_0 = \frac{E}{z_0} \cos (\theta_0 + \gamma) + A_1 + A_2$$

and

$$e_0 = -\frac{Ex_c}{z_0} \sin (\theta_0 + \gamma) - \frac{r+s}{2} A_1 - \frac{r-s}{2} A_2.$$

Therefore

$$\left.\begin{aligned} A_1 + A_2 &= i_0 - \frac{E}{z_0} \cos (\theta_0 + \gamma) \\ &\text{and} \\ A_1 - A_2 &= -\frac{ri_0 + 2e_0}{s} + \frac{E}{sz_0} \{r \cos (\theta_0 + \gamma) - 2x_c \sin (\theta_0 + \gamma)\}, \end{aligned}\right\} \quad (16)$$

or,

$$\left.\begin{aligned} A_1 &= -\frac{\frac{r-s}{2} i_0 + e_0}{s} + \frac{E}{sz_0} \left\{ \frac{r-s}{2} \cos (\theta_0 + \gamma) - x_c \sin (\theta_0 + \gamma) \right\} \\ &\text{and} \\ A_2 &= +\frac{\frac{r+s}{2} i_0 + e_0}{s} - \frac{E}{sz_0} \left\{ \frac{r+s}{2} \cos (\theta_0 + \gamma) - x_c \sin (\theta_0 + \gamma) \right\}. \end{aligned}\right\} \quad (17)$$

Substituting (17) in (13) and (14) gives the integral equations of the problem.

The *current* is

$$i = \frac{E}{z_0}\cos(\theta - \theta_0 - \gamma) + \frac{E}{sz_0}\left\{ \varepsilon^{-\frac{r-s}{2x}\theta}\left[\frac{r-s}{2}\cos(\theta_0+\gamma) - x_c\sin(\theta_0+\gamma)\right]\right.$$

$$\left. - \varepsilon^{-\frac{r+s}{2x}\theta}\left[\frac{r+s}{2}\cos(\theta_0+\gamma) - x_c\sin(\theta_0+\gamma)\right]\right\}$$

$$- \frac{1}{s}\left\{\varepsilon^{-\frac{r-s}{2x}\theta}\left[\frac{r-s}{2}i_0 + e_0\right] - \varepsilon^{-\frac{r+s}{2x}\theta}\left[\frac{r+s}{2}i_0 + e_0\right]\right\}, \tag{18}$$

and the *potential difference at the condenser terminals* is

$$e_1 = \frac{Ex_c}{z_0}\sin(\theta - \theta_0 - \gamma)$$

$$- \frac{E}{2sz_0}\left\{(r+s)\varepsilon^{-\frac{r-s}{2x}\theta}\left[\frac{r-s}{2}\cos(\theta_0+\gamma) - x_c\sin(\theta_0+\gamma)\right]\right.$$

$$\left. - (r-s)\varepsilon^{-\frac{r+s}{2x}\theta}\left[\frac{r+s}{2}\cos(\theta_0+\gamma) - x_c\sin(\theta_0+\gamma)\right]\right\}$$

$$+ \frac{1}{2s}\left\{(r+s)\varepsilon^{-\frac{r-s}{2x}\theta}\left[\frac{r-s}{2}i_0 + e_0\right] - (r-s)\varepsilon^{-\frac{r+s}{2x}\theta}\left[\frac{r+s}{2}i_0 + e_0\right]\right\}, \tag{19}$$

where

$$\left.\begin{aligned} z_0 &= \sqrt{r^2 + (x - x_c)^2}, \\ \tan\gamma &= \frac{x - x_c}{r}, \\ s &= \sqrt{r^2 - 4\,x\,x_c}. \end{aligned}\right\} \tag{11}$$

and

The expressions of i and e_1 consist of three terms each:

(1) The permanent term, which is the only one remaining after some time;

(2) A transient term depending upon the constants of the circuit, r, s, x_c, z_0, x, the impressed e.m.f., E, and its phase θ_0 at the moment of starting, but independent of the conditions existing in the circuit before the start; and

(3) A term depending, besides upon the constants of the circuit, upon the instantaneous values of current and potential difference, i_0 and e_0, at the moment of starting the circuit, and thereby upon the electrical conditions of the circuit before impressing the e.m.f., *e*. This term disappears if the circuit is dead before the start.

Equations (18) and (19) contain the term $s = \sqrt{r^2 - 4\,x\,x_c} = \sqrt{r^2 - 4\,\frac{L}{C}}$; hence apply only when $r^2 > 4\,x\,x_c$, but become indeterminate if $r^2 = 4\,x\,x_c$, and imaginary if $r^2 < 4\,x\,x_c$; in the latter cases they have to be rearranged so as to appear in real form, in manner similar to that in Chapter V.

56. In the *critical case*, $r^2 = 4\,x x_c$ and $s = 0$, equation (18), rearranged, assumes the form

$$i = \frac{E}{z_0}\cos(\theta - \theta_0 - \gamma) + \frac{E}{z_0}\varepsilon^{-\frac{r}{2x}\theta}$$

$$\left\{\left[\frac{r}{2}\cos(\theta_0 + \gamma) - x_c \sin(\theta_0 + \gamma)\right]\frac{\varepsilon^{+\frac{s}{2x}\theta} - \varepsilon^{-\frac{s}{2x}\theta}}{s} - \cos(\theta_0 + \gamma)\right\}$$

$$- \varepsilon^{-\frac{r}{2x}\theta}\left\{\left[\frac{r}{2}i_0 + e_0\right]\frac{\varepsilon^{+\frac{s}{2x}\theta} - \varepsilon^{-\frac{s}{2x}\theta}}{s} - i_0\right\}.$$

However, developing in a series, and canceling all but the first term as infinitely small, we have

$$\frac{\varepsilon^{+\frac{s}{2x}\theta} - \varepsilon^{-\frac{s}{2x}\theta}}{s} = \frac{\theta}{x};$$

hence the *current* is

$$i = \frac{E}{z_0}\cos(\theta - \theta_0 - \gamma) + \frac{E}{z_0}\varepsilon^{-\frac{r}{2x}\theta}$$

$$\left\{\left[\frac{r}{2}\cos(\theta_0 + \gamma) - x_c \sin(\theta_0 + \gamma)\right]\frac{\theta}{x} - \cos(\theta_0 + \gamma)\right\}$$

$$+ \varepsilon^{-\frac{r\theta}{2x}}\left\{i_0 - \left[\frac{r}{2}i_0 + e_0\right]\frac{\theta}{x}\right\}, \qquad (20)$$

and in the same manner the *potential difference at condenser terminals* is

$$e_1 = \frac{E x_c}{z_0} \sin(\theta - \theta_0 - \gamma) - \frac{E}{2 z_0} \varepsilon^{-\frac{r}{2x}\theta}$$

$$\left\{ \left[\frac{r^2}{2} \cos(\theta_0 + \gamma) - x_c r \sin(\theta_0 + \gamma) \right] \frac{\theta}{x} - 2 x_c \sin(\theta + \gamma) \right\}$$

$$+ \frac{1}{2} \varepsilon^{-\frac{r}{2x}\theta} \left\{ \left[\frac{r^2}{2} i_0 + r e_0 \right] \frac{\theta}{x} + 2 e_0 \right\}. \qquad (21)$$

Here again three terms exist, namely: a permanent term, a transient term depending only on E and θ_0, and a transient term depending on i_0 and e_0.

57. In the *trigonometric* or *oscillatory* case, $r^2 < 4 x x_c$, s becomes imaginary, and equations (18) and (19) therefore contain complex imaginary exponents, which have to be eliminated, since the complex imaginary form of the equation obviously is only apparent, the phenomenon being real.

Substituting

$$q = \sqrt{4 x x_c - r^2} = js \qquad (22)$$

in equations (13) and (14), and also substituting the trigonometric expressions

$$\left.\begin{aligned} \varepsilon^{+j\frac{q}{2x}\theta} &= \cos\frac{q}{2x}\theta + j \sin\frac{q}{2x}\theta \\ \text{and} \qquad \varepsilon^{-j\frac{q}{2x}\theta} &= \cos\frac{q}{2x}\theta - j \sin\frac{q}{2x}\theta, \end{aligned}\right\} \qquad (23)$$

and separating the imaginary and the real terms, gives

$$i = \frac{E}{z_0} \cos(\theta - \theta_0 - \gamma) + \varepsilon^{-\frac{r}{2x}\theta}$$

$$\left\{ (A_1 + A_2) \cos\frac{q}{2x}\theta - j (A_1 - A_2) \sin\frac{q}{2x}\theta \right\}$$

and

$$e_1 = \frac{E x_c}{z_0} \sin (\theta - \theta_0 - \gamma) - \varepsilon^{-\frac{r}{2x}\theta}$$

$$\left\{ \frac{A_1 + A_2}{2} \left[r \cos \frac{q}{2x} \theta - q \sin \frac{q}{2x} \theta \right] + j \frac{A_1 - A_2}{2} \right.$$

$$\left. \left[q \cos \frac{q}{2x} \theta + r \sin \frac{q}{2x} \theta \right] \right\};$$

then substituting herein the equations (16) and (22) the imaginary disappears, and we have the *current*,

$$i = \frac{E}{z_0} \cos (\theta - \theta_0 - \gamma) - \frac{E}{z_0} \varepsilon^{-\frac{r}{2x}\theta}$$

$$\left\{ \cos (\theta_0 + \gamma) \cos \frac{q}{2x} \theta + \left[\frac{2 x_c}{q} \sin (\theta_0 + \gamma) - \frac{r}{q} \cos (\theta_0 + \gamma) \right] \sin \frac{q}{2x} \theta \right\}$$

$$+ \varepsilon^{-\frac{r}{2x}\theta} \left\{ i_0 \cos \frac{q}{2x} \theta - \frac{2 e_0 + r i_0}{q} \sin \frac{q}{2x} \theta \right\}, \tag{24}$$

and the *potential difference at the condenser terminals*,

$$e_1 = \frac{E x_c}{z_0} \sin (\theta - \theta_0 - \gamma) + \frac{E x_c}{z_0} \varepsilon^{-\frac{r}{2x}\theta}$$

$$\left\{ \sin (\theta_0 + \gamma) \cos \frac{q}{2x} \theta + \left[\frac{r}{q} \sin (\theta_0 + \gamma) - \frac{2x}{q} \cos (\theta_0 + \gamma) \right] \sin \frac{q}{2x} \theta \right\}$$

$$+ \varepsilon^{-\frac{r}{2x}\theta} \left\{ e_0 \cos \frac{q}{2x} \theta + \frac{2 r e_0 + 4 x x_c i_0}{2q} \sin \frac{q}{2x} \theta \right\}. \tag{25}$$

Here the three component terms are seen also.

58. As examples are shown in Figs. 20 and 21, the starting of the current i, its permanent term i', and the two transient terms i_1 and i_2, and their difference, for the constants $E = 1000$ volts = maximum value of impressed e.m.f.; $r = 200$ ohms = resistance; $x = 75$ ohms = inductive reactance, and $x_c = 75$ ohms = condensive reactance. We have

$$4 x x_c = 22{,}500$$

and

$$r^2 = 40{,}000;$$

therefore

$$r^2 > 4 x x_c,$$

that is, the start is logarithmic, and $z_0 = 200$, $s = 132$, and $\gamma = 0$.

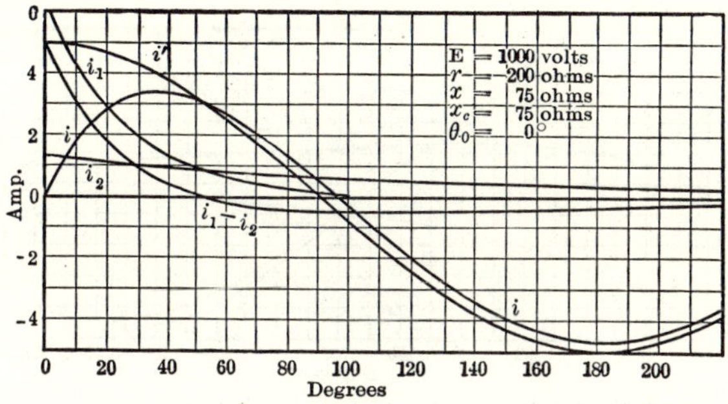

Fig. 20. Starting of an alternating-current circuit, having capacity, inductance and resistance in series. Logarithmic start.

In Fig. 20 the circuit is closed at the moment $\theta_0 = 0$, that is, at the maximum value of the impressed e.m.f., giving from the equations (18) and (19), since $i_0 = 0$, $e_0 = 0$,

$$i = 5 \left\{ \cos \theta - 1.26\, \varepsilon^{-2.22\,\theta} + 0.26\, \varepsilon^{-0.452\,\theta} \right\}$$

and

$$e_1 = 375 \left\{ \sin \theta + 0.57 \left(\varepsilon^{-2.22\,\theta} - \varepsilon^{-0.452\,\theta} \right) \right\}.$$

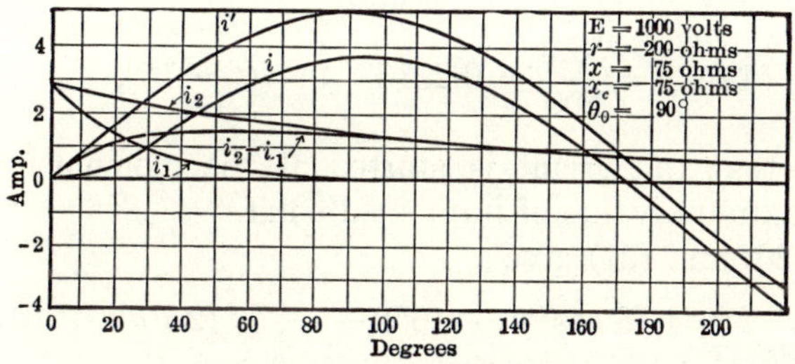

Fig. 21. Starting of an alternating-current circuit having capacity, inductance and resistance in series. Logarithmic start.

In Fig. 21 the circuit is closed at the moment $\theta_0 = 90°$, that is, at the zero value of the impressed e.m.f., giving the equations

$$i = 5 \left\{ \sin \theta + 0.57 \left(\varepsilon^{-2.22\,\theta} - \varepsilon^{-0.452\,\theta} \right) \right\}$$

and

$$e_1 = -\,375 \left\{ \cos \theta + 0.26\, \varepsilon^{-2.22\,\theta} - 1.26\, \varepsilon^{-0.452\,\theta} \right\}.$$

There exists no value of θ_0 which does not give rise to a transient term.

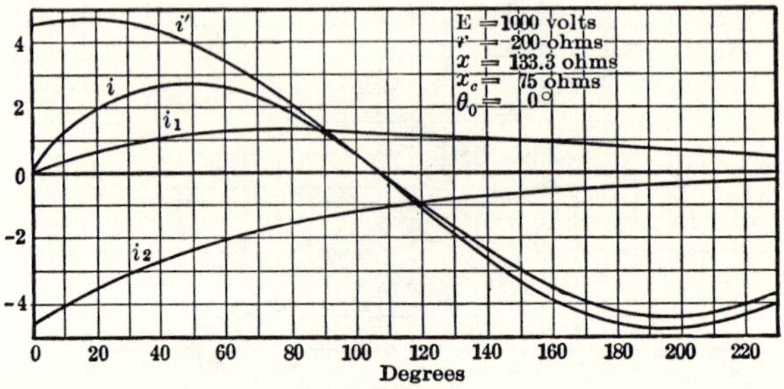

Fig. 22. Starting of an alternating-current circuit having capacity, inductance and resistance in series. Critical start.

In Fig. 22 the start of a circuit is shown, with the inductive reactance increased so as to give the critical condition,

$$r^2 = 4\,x\,x_c,$$

but otherwise the constants are the same as in Figs. 20 and 21, that is, $E = 1000$ volts; $r = 200$ ohms; $x = 133.3$ ohms, and $x_c = 75$ ohms;

therefore $\quad z_0 = 208.3,$

$$\tan\gamma = \frac{58.3}{200} = 0.2915, \quad \text{or} \quad \gamma = 16°,$$

assuming that the circuit is started at the moment $\theta_0 = 0$, or at the maximum value of impressed e.m.f.

Then (20) and (21) give

$$i = 4.78 \cos(\theta - 16°) + \varepsilon^{-0.75\,\theta}(2.7\,\theta - 4.6)$$

and

$$e_1 = 358 \sin(\theta - 16°) - \varepsilon^{-0.75\,\theta}(410\,\theta - 99).$$

Here also no value of θ_0 exists at which the transient term disappears.

59. The most important is the oscillating case, $r^2 < 4\,x\,x_c$, since it is the most common in electrical circuits, as underground cable systems and overhead high potential circuits, and also is practically the only one in which excessive currents or excessive voltages, and thereby dangerous phenomena, may occur.

If the condensive reactance x_c is high compared with the resistance r and the inductive reactance x, the equations of start for the circuit from dead condition, that is, $i_0 = 0$ and $e_0 = 0$, are found by substitution into the general equations (24) and (25), which give the *current* as

$$i = -\frac{E}{x_c}\left\{\sin(\theta - \theta_0) + \varepsilon^{-\frac{r}{2x}\theta}\left[\sin\theta_0 \cos\sqrt{\frac{x_c}{x}}\theta - \sqrt{\frac{x_c}{x}}\cos\theta_0 \sin\sqrt{\frac{x_c}{x}}\theta\right]\right\} \tag{26}$$

and the *potential difference at the condenser terminals* as

$$e_1 = E\left\{\cos(\theta - \theta_0) - \varepsilon^{-\frac{r}{2x}\theta}\left[\cos\theta_0 \cos\sqrt{\frac{x_c}{x}}\theta + \left(\frac{r}{2\sqrt{x\,x_c}}\cos\theta_0 + \sqrt{\frac{x}{x_c}}\sin\theta_0\right)\sin\sqrt{\frac{x_c}{x}}\theta\right]\right\}, \tag{27}$$

where

$$q = 2\sqrt{x\,x_c},\ z_0 = x_c,\ \text{and}\ \gamma = -90°. \tag{28}$$

In this case an oscillating term always exists whatever the value of θ_0, that is, the point of the wave, where the circuit is started.

The frequency of oscillation therefore is

$$f_0 = \frac{q}{2x}f = f\sqrt{\frac{x_c}{x} - \frac{r^2}{4x^2}}$$

or, approximately,

$$f_0 = \sqrt{\frac{x_c}{x}}f, \tag{29}$$

where f = fundamental frequency.

Substituting $x = 2\pi fL$ and $x_c = \frac{1}{2\pi fC}$, we have

$$f_0 = \frac{1}{2\pi}\sqrt{\frac{1}{CL} - \frac{r^2}{4L^2}},$$

or, approximately,

$$f_0 = \frac{1}{2\pi\sqrt{CL}}. \tag{30}$$

60. The oscillating start, or, in general, change of circuit conditions, is the most important, since in circuits containing capacity the transient effect is almost always oscillating.

The most common examples of capacity are distributed capacity in transmission lines, cables, etc., and capacity in the form of electrostatic condensers for neutralizing lagging currents, for constant potential-constant current transformation, etc.

(*a*) In transmission lines or cables the charging current is a fraction of full-load current i_0, and the e.m.f. of self-inductance consumed by the line reactance is a fraction of the impressed e.m.f. e_0. Since, however, the charging current is (approximately) $=\frac{e_0}{x_c}$ and the e.m.f. of self-inductance $= xi_0$, we have

$$\frac{e_0}{x_c} < i_0, \; xi_0 < e_0;$$

hence, multiplying,

$$\frac{x}{x_c} < 1 \text{ and } x < x_c.$$

The resistance r is of the same magnitude as x; thus

$$4\,x\,x_c > r^2.$$

For instance, with 10 per cent resistance drop, 30 per cent reactance voltage, and 20 per cent charging current in the line, assuming half the resistance and reactance as in series with the capacity (that is, representing the distributed capacity of the line by one condenser shunted across its center) and denoting

$$p = \frac{e_0}{i_0},$$

where e_0 = impressed voltage, i_0 = full-load current, we have

$$\left.\begin{aligned} x_c &= \frac{p}{0.2} = 5\,p, \\ x &= 0.5 \times 0.3\,p = 0.15\,p, \\ r &= 0.5 \times 0.1\,p = 0.05\,p, \end{aligned}\right\}$$

and

$$r \div x \div x_c = 1 \div 3 \div 100,$$

and

$$4\,x\,x_c \div r^2 = 1200 \div 1.$$

In this case, to make the start non-oscillating, we must have $x < \frac{1}{400} r$, or $x < 0.000125\, p$, which is not possible; or $r > p\sqrt{3}$, which can be done only by starting the circuit through a very large non-inductive resistance (of such size as to cut the starting current down to less than $\frac{1}{\sqrt{3}}$ of full-load current). Even in this case, however, oscillations would appear by a change of load, etc., after the start of the circuit.

(*b*) When using electrostatic condensers for producing wattless leading currents, the resistance in series with the condensers is made as low as possible, for reasons of efficiency.

Even with the extreme value of 10 per cent resistance, or $r \div x_c = 1 \div 10$, the non-oscillating condition is $x < \frac{1}{40} r$, or 0.25 per cent, which is not feasible.

In general, if

x consumes........	1	2	4	9	16	per cent of the condenser potential difference,
r must consume >	20	28.3	40	60	80	per cent of the condenser potential difference.

That is, a very high non-inductive resistance is required to avoid oscillations.

The frequency of oscillation is approximately $f_0 = \sqrt{\frac{x_c}{x}} f$ that is, is lower than the impressed frequency if $x_c < x$ (or the permanent current lags), and higher than the impressed frequency if $x_c > x$ (or the permanent current leads). In transmission lines and cables the latter is always the case.

Since in a transmission line $\frac{p}{x_c}$ is approximately the charging current, as fraction of full-load current, and $\frac{x}{p}$ half the line e.m.f. of self-inductance, or reactance voltage, as fraction of impressed voltage, the following is approximately true:

The frequency of oscillation of a transmission line is the impressed frequency divided by the square root of the product of charging current and of half the reactance voltage of the line, given respectively as fractions of full-load current and of impressed voltage. For instance, 10 per cent charging current, 20 per cent reactance voltage, gives an oscillation frequency

$$f_0 = \frac{f}{\sqrt{0.1 \times 0.1}} = 10 f.$$

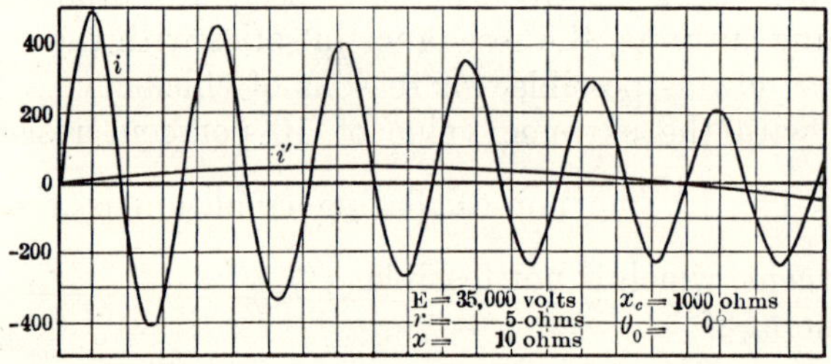

Fig. 23. Starting of an alternating-current circuit having capacity, inductance and resistance in series. Oscillating start of transmission line.

61. In Figs. 23 and 24 is given as example the start of current in a circuit having the constants, $E = 35{,}000 \cos(\theta - \theta_0)$; $r = 5$ ohms; $x = 10$ ohms, and $x_c = 1000$ ohms.

In Fig. 23 for $\theta_0 = 0°$, or approximately maximum oscillation,

$$i = -35 \left\{ \sin\theta - 10\, \varepsilon^{-0.25\,\theta} \sin 10\,\theta \right\}$$

and

$$e_1 = 35{,}000 \left\{ \cos\theta - \varepsilon^{-0.25\,\theta} \left[\cos 10\,\theta + 0.025 \sin 10\,\theta \right] \right\}.$$

In Fig. 24 for $\theta_0 = 90°$, or approximately minimum oscillation,

$$\left. \begin{aligned} i &= 35 \left\{ \cos\theta - \varepsilon^{-0.25\,\theta} \cos 10\,\theta \right\} \\ \text{and} \quad e_1 &= 35{,}000 \left\{ \sin\theta + 0.1\, \varepsilon^{-0.25\,\theta} \sin 10\,\theta \right\}. \end{aligned} \right\}$$

As seen, the frequency is 10 times the fundamental, and in starting the potential difference nearly doubles.

As further example, Fig. 25 shows the start of a circuit of a frequency of oscillation of the same magnitude as the fundamental, in resonance condition, $x = x_c$, and of high resistance.

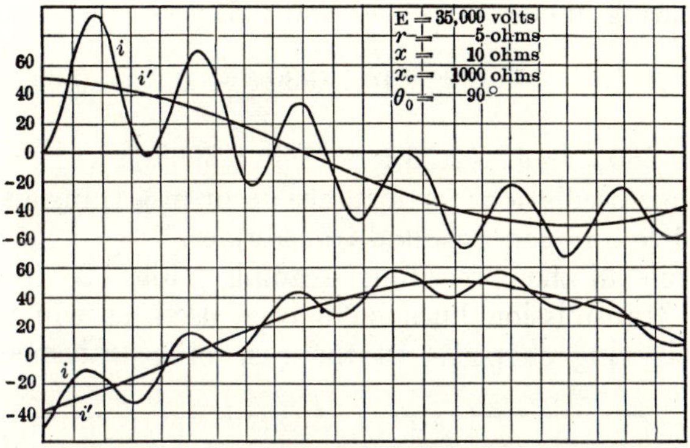

Fig. 24. **Starting of an alternating-current circuit having capacity, inductance and resistance in series. Oscillating start of transmission line.**

The circuit constants are $E = 1500$ volts; $r = 30$ ohms; $x = 20$ ohms; $x_c = 20$ ohms, and $\theta_0 = -\gamma$; which give $q = 26.46$; $z_0 = 30$; $\gamma = 0$, and $\theta_0 = 0$.

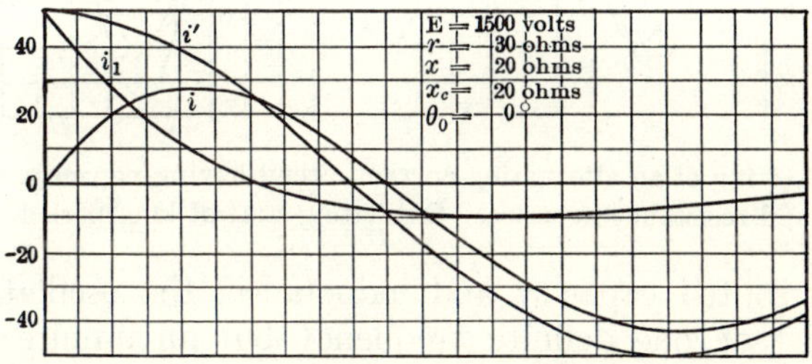

Fig. 25. **Starting of an alternating-current circuit having capacity, inductance and resistance in series. Oscillating start. High resistance.**

Substituting in equations (24) and (25) gives

$$\left.\begin{array}{l} i = 50 \left\{\cos\theta - \varepsilon^{-0.75\,\theta}[\cos 0.661\,\theta - 1.14 \sin 0.661\,\theta]\right\} \\ \text{and} \\ e_1 = 1000 \left\{\sin\theta - 1.51\,\varepsilon^{-0.75\,\theta} \sin 0.661\,\theta\right\}. \end{array}\right\}$$

As example of an oscillation of long wave, Fig. 26 represents the start of a circuit having the constants $E = 1500$ volts; $r = 10$ ohms; $x = 62.5$ ohms; $x_c = 10$ ohms, and $\theta_0 = -\gamma$; which give $q = 49$; $z_0 = 53.4$; $\gamma = 79°$, and $\theta_0 = -79°$.

Substituting in equations (24) and (25) gives

$$\left. \begin{aligned} i &= 28 \left\{\cos\theta - \varepsilon^{-0.08\,\theta}[\cos 0.39\,\theta - 0.2 \sin 0.39\,\theta]\right\} \\ \text{and} \quad & \\ e_1 &= 280 \left\{\sin\theta - 2.55\,\varepsilon^{-0.08\,\theta} \sin 0.396\,\theta\right\}. \end{aligned} \right\}$$

Such slow oscillations for instance occur in a transmission line connected to an open circuited transformer.

62. While in the preceding examples, Figs. 23 to 26, constants of transmission lines have been used, as will be shown in the following chapters, in the case of a transmission line

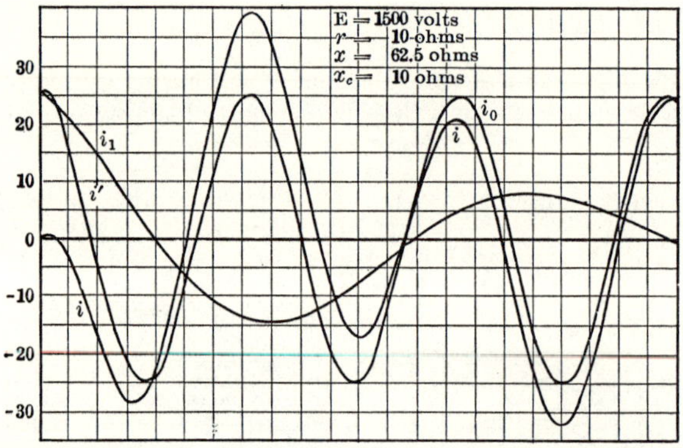

Fig. 26. Starting of an alternating-current circuit having capacity, inductance and resistance in series. Oscillating start of long period.

with distributed capacity and inductance, the oscillation does not consist of one definite frequency but an infinite series of frequencies, and the preceding discussion thus approximates only the fundamental frequency of the system. This, however, is the frequency which usually predominates in a high power low frequency surge of the system.

In an underground cable system the preceding discussion applies more closely, since in such a system capacity and inductance are more nearly localized: the capacity is in the underground cables, which are of low inductance, and the inductance is in the generating system, which has practically no capacity.

In an underground cable system the tendency therefore is

either towards a local, very high frequency oscillation, or traveling wave, of very limited power, in a part of the cables, or a low frequency high power surge, frequently of destructive magnitude, of the joint capacity of the cables, against the inductance of the generating system.

63. The physical meaning of the transient terms can best be understood by reviewing their origin.

In a circuit containing resistance and inductance only, but a single transient term appears of exponential nature. In such a circuit at any moment, and thus at the moment of start, the current should have a certain definite value, depending on the constants of the circuit. In the moment of start, however, the current may have a different value, depending on the preceding condition, as for instance the value zero if the circuit has been open before. The current thus adjusts itself from the initial value to the permanent value on an exponential curve, which disappears if the initial value happens to coincide with the final value, as for instance if the circuit is closed at the moment of the e.m.f. wave, when the permanent current should be zero. The approach of current to the permanent value is retarded by the inductance, accelerated by the resistance of the circuit.

In a circuit containing inductance and capacity, at any moment the current has a certain value and the condenser a certain charge, that is, potential difference. In the moment of start, current intensity and condenser charge have definite values, depending on the previous condition, as zero, if the circuit was open, and thus two transient terms must appear, depending upon the adjustment of current and of condenser e.m.f. to their permanent values.

Since at the moment when the current is zero the condenser e.m.f. is maximum, and inversely, in a circuit containing inductance and capacity, the starting of a circuit always results in the appearance of a transient term.

If the circuit is closed at the moment when the condenser e.m.f. should be zero, that is, about the maximum value of current, the transient term of current cannot exceed in amplitude its final value, since its maximum or initial value equals the value which the current should have at this moment. If, however, the circuit is closed at the moment where the current should be zero and the condenser e.m.f. maximum, the condenser being without charge acts in the first moment like a short circuit, that is, the current begins at a value corresponding to the impressed e.m.f. divided by the line impedance. Thus if we neglect the resistance and if the condenser reactance equals n^2 times line reactance, the current starts at n^2 times its final rate; thus it would, in a half wave, give n^2 times the full charge of the condenser, or in other words, charge the condenser in $\frac{1}{n}$ of the time of a half wave. That is, the period of the starting current is $\frac{1}{n}$ and the amplitude n times that of the final current. However, as soon as the condenser is charged, in $\frac{1}{n}$ of a period of the impressed e.m.f., the magnetic field of the charging current produces a return current, discharging the condenser again at the same rate.

Thus the normal condition of start is an oscillation of such a frequency as to give the full condenser charge at a rate which when continued up to full frequency would give an amplitude equal to the impressed e.m.f. divided by the line reactance. The effect of the line resistance is to consume e.m.f. and thus dampen the oscillation, until the resistance consumes during the condenser charge as much energy as the magnetic field would store up, and then the oscillation disappears and the start becomes exponential.

Analytically the double transient term appears as the result of the two roots of a quadratic equation, as seen above.

CHAPTER VIII.

LOW FREQUENCY SURGES IN HIGH POTENTIAL SYSTEMS.

64. In electric circuits of considerable capacity, that is, in extended high potential systems, as long distance transmission lines and underground cable systems, occasionally destructive high potential low frequency surges occur; that is, oscillations of the whole system, of the same character as in the case of localized capacity and inductance discussed in the preceding chapter.

While a system of distributed capacity has an infinite number of frequencies, which usually are the odd multiples of a fundamental frequency of oscillation, in those cases where the fundamental frequency predominates and the effect of the higher frequencies is negligible, the oscillation can be approximated by the equations of oscillation given in Chapters V and VII, which are far simpler than the equations of an oscillation of a system of distributed capacity.

Such low frequency surges take in the total system, not only the transmission lines but also the step-up transformers, generators, etc., and in an underground cable system in such an oscillation the capacity and inductance are indeed localized to a certain extent, the one in the cables, the other in the generating system. In an underground cable system, therefore, of the infinite series of frequencies of oscillations which theoretically exist, only the fundamental frequency and those very high harmonics which represent local oscillations of sections of cables can be pronounced, and the first higher harmonics of the fundamental frequency must be practically absent. That is, oscillations of an underground cable system are either

(*a*) Low frequency high power surges of the whole system, of a frequency of a few hundred cycles, frequently of destructive character, or,

(*b*) Very high frequency low power oscillations, local in character, so called "static," probably of frequencies of hundred thousands of cycles, rarely directly destructive, but indirectly harmful in their weakening action on the insulation and the possibility of their starting a low frequency surge.

The former ones only are considered in the present chapter. Their causes may be manifold, — changes of circuit conditions, as starting, opening a short circuit, existence of a flaring arc on the system, etc.

In the circuit from the generating system to the capacity of the transmission line or the underground cables, we have always $r^2 < \frac{4L}{C}$; that is, the phenomenon is always oscillatory, and equations (24) and (25), Chapter VII, apply, and for the *current* we have

$$i=\frac{E}{z_0}\cos(\theta-\theta_0-\gamma)+\varepsilon^{-\frac{r}{2x}\theta}\left\{\left[i_0-\frac{E}{z_0}\cos(\theta_0+\gamma)\right]\cos\frac{q}{2x}\theta\right.$$
$$\left.-\left[\frac{2e_0+ri_0}{q}+\frac{E}{qz_0}\Big(2x_c\sin(\theta_0+\gamma)-r\cos(\theta_0+\gamma)\Big)\right]\sin\frac{q}{2x}\theta\right\}, \quad (1)$$

and for the *condenser potential* we have

$$e_1=\frac{Ex_c}{z_0}\sin(\theta-\theta_0-\gamma)+\varepsilon^{-\frac{r}{2x}\theta}\left\{\left[e_0+\frac{Ex_c}{z_0}\sin(\theta_0+\gamma)\right]\cos\frac{q}{2x}\theta\right.$$
$$\left.+\left[\frac{2re_0+4xx_ci_0}{2q}+\frac{Ex_c}{qz_0}\Big(r\sin(\theta_0+\gamma)-2x\cos(\theta_0+\gamma)\Big)\right]\sin\frac{q}{2x}\theta\right\} \quad (2)$$

65. These equations (1) and (2) can be essentially simplified by neglecting terms of secondary magnitude.

x_c is in high potential transmission lines or cables always very large compared with r and x.

The full-load resistance and reactance voltage may vary from less than 5 per cent to about 20 per cent of the impressed e.m.f., the charging current of the line from 5 per cent to about 20 per cent of full-load current, at normal voltage and frequency.

In this case, x_c is from 25 to more than 400 times as large as r or x, and r and x thus negligible compared with x_c.

It is then, in close approximation:

$$\left.\begin{aligned} z_0 &= x_c, \\ q &= 2\sqrt{x\,x_c}, \\ \gamma &= -\frac{\pi}{2} = -90^\circ. \end{aligned}\right\} \qquad (3)$$

Substituting these values in equations (1) and (2) gives the *current* as

$$i = -\frac{E}{x_c}\sin(\theta-\theta_0) + \varepsilon^{-\frac{r}{2x}\theta}\left\{\left[i_0 - \frac{E}{x_c}\sin\theta_0\right]\cos\sqrt{\frac{x_c}{x}}\,\theta\right.$$
$$\left.-\left[\frac{2\,e_0+ri_0}{2\sqrt{x\,x_c}} - \frac{E}{2\sqrt{x\,x_c}}\left(2\cos\theta_0 + \frac{r}{x_c}\sin\theta_0\right)\right]\sin\sqrt{\frac{x_c}{x}}\,\theta\right\}, \qquad (4)$$

and the *potential difference at the condenser* as

$$e_1 = E\cos(\theta-\theta_0) + \varepsilon^{-\frac{r}{2x}\theta}\left\{[e_0 - E\cos\theta_0]\cos\sqrt{\frac{x_c}{x}}\,\theta\right.$$
$$+\left[\frac{2\,re_0 + 4\,xx_ci_0}{4\sqrt{xx_c}} - \frac{E}{4\sqrt{xx_c}}\right.$$
$$\left.\left.(2\,r\cos\theta_0 + 4\,x\sin\theta_0)\right]\sin\sqrt{\frac{x_c}{x}}\,\theta\right\}.$$

These equations consist of three terms:

$$\left.\begin{aligned} i &= i' + i'' + i''', \\ e_1 &= e_1' + e_1'' + e_1'''; \end{aligned}\right\} \qquad (5)$$

$$\left.\begin{aligned} i' &= -\frac{E}{x_c}\sin(\theta-\theta_0), \\ e_1' &= E\cos(\theta-\theta_0); \end{aligned}\right\} \qquad (6)$$

$$\left.\begin{aligned} i'' &= -\frac{E}{x_c}\varepsilon^{-\frac{r}{2x}\theta}\left\{\sin\theta_0\cos\sqrt{\frac{x_c}{x}}\theta - \left[\sqrt{\frac{x_c}{x}}\cos\theta_0 + \frac{r}{2\sqrt{xx_c}}\sin\theta_0\right]\sin\sqrt{\frac{x_c}{x}}\theta\right\}, \\ e_1'' &= -E\varepsilon^{-\frac{r}{2x}\theta}\left\{\cos\theta_0\cos\sqrt{\frac{x_c}{x}}\theta + \left(\frac{r}{2\sqrt{x x_c}}\cos\theta_0 + \sqrt{\frac{x}{x_c}}\sin\theta_0\right)\sin\sqrt{\frac{x_c}{x}}\theta\right\}; \end{aligned}\right\} \quad (7)$$

or, by dropping terms of secondary order,

$$\left.\begin{aligned} i'' &= \frac{E}{\sqrt{x x_c}}\varepsilon^{-\frac{r}{2x}\theta}\cos\theta_0\sin\sqrt{\frac{x_c}{x}}\theta, \\ e_1'' &= -E\varepsilon^{-\frac{r}{2x}\theta}\cos\theta_0\cos\sqrt{\frac{x_c}{x}}\theta; \end{aligned}\right\} \quad (8)$$

and:

$$\left.\begin{aligned} i''' &= \varepsilon^{-\frac{r}{2x}\theta}\left\{i_0\cos\sqrt{\frac{x_c}{x}}\theta - \frac{2e_0 + ri_0}{2\sqrt{x x_c}}\sin\sqrt{\frac{x_c}{x}}\theta\right\}, \\ e_1''' &= \varepsilon^{-\frac{r}{2x}\theta}\left\{e_0\cos\sqrt{\frac{x_c}{x}}\theta + \frac{2re_0 + 4xx_ci_0}{4\sqrt{xx_c}}\sin\sqrt{\frac{x_c}{x}}\theta\right\}; \end{aligned}\right\} \quad (9)$$

or, by dropping terms of secondary order,

$$\left.\begin{aligned} i''' &= \varepsilon^{\frac{r}{2x}\theta}\left\{i_0\cos\sqrt{\frac{x_c}{x}}\theta - \frac{e_0}{\sqrt{x x_c}}\sin\sqrt{\frac{x_c}{x}}\theta\right\}, \\ e_1''' &= \varepsilon^{-\frac{r}{2x}\theta}\left\{e_0\cos\sqrt{\frac{x_c}{x}}\theta + i_0\sqrt{x x_c}\sin\sqrt{\frac{x_c}{x}}\theta\right\}. \end{aligned}\right\} \quad (10)$$

Thus the *total current* is approximately

$$i = -\frac{E}{x_c}\sin(\theta - \theta_0) + \varepsilon^{-\frac{r}{2x}\theta}\left\{i_0\cos\sqrt{\frac{x_c}{x}}\theta - \frac{e_0 - E\cos\theta_0}{\sqrt{x x_c}}\sin\sqrt{\frac{x_c}{x}}\theta\right\},$$

and the *difference of potential at the condenser* is

$$e_1 = E\cos(\theta - \theta_0) + \varepsilon^{-\frac{r}{2x}\theta}\left\{(e_0 - E\cos\theta_0)\cos\sqrt{\frac{x_c}{x}}\theta + i_0\sqrt{x x_c}\sin\sqrt{\frac{x_c}{x}}\theta\right\}. \quad (11)$$

Of the three terms: i', e_1'; i'', e_1''; i''', e_1''', the first obviously represents the stationary condition of charging current and condenser potential, since the two other terms disappear for $t = \infty$.

The second term, i'', e_1'', represents that component of oscillation which depends upon the phase of impressed e.m.f., or the point of the impressed e.m.f. wave, at which the oscillation begins, while the third term, i''', e_1''', represents the component of oscillation which depends upon the instantaneous values of current and e.m.f. respectively, at the moment at which the oscillation begins. $\varepsilon^{-\frac{r}{2x}\theta}$ is the decrement of the oscillation.

66. The frequency of oscillation is

$$f_0 = \sqrt{\frac{x_c}{x}}\, f,$$

where f is the impressed frequency. That is, the frequency of oscillation equals the impressed frequency times the square root of the ratio of condensive reactance and inductive reactance of the circuit, or is the impressed frequency divided by the square root of inductance voltage and capacity current, as fraction of impressed voltage and full-load current.

Since

$$x_c = \frac{1}{2\,\pi f C} \quad \text{and} \quad x = 2\,\pi f L,$$

the frequency of oscillation is

$$f_0 = \frac{1}{2\,\pi\sqrt{CL}};$$

that is, is independent of the frequency of the impressed e.m.f.

Substituting

$$\theta = 2\,\pi f t, \quad x_c = \frac{1}{2\,\pi f C} \quad \text{and} \quad x = 2\,\pi f L$$

in equations (8), (10), and (11), we have

$$\left.\begin{aligned} i'' &= \sqrt{\frac{C}{L}}\, E\varepsilon^{-\frac{r}{2L}t} \cos\theta_0 \sin\frac{t}{\sqrt{CL}}, \\ e_1'' &= -\, E\varepsilon^{-\frac{r}{2L}t} \cos\theta_0 \cos\frac{t}{\sqrt{CL}}; \end{aligned}\right\} \quad (12)$$

$$\left.\begin{aligned} i''' &= \varepsilon^{-\frac{r}{2L}t}\left\{i_0 \cos\frac{t}{\sqrt{CL}} - e_0\sqrt{\frac{C}{L}}\sin\frac{t}{\sqrt{CL}}\right\}, \\ e_1''' &= \varepsilon^{-\frac{r}{2L}t}\left\{e_0 \cos\frac{t}{\sqrt{CL}} + i_0\sqrt{\frac{L}{C}}\sin\frac{t}{\sqrt{CL}}\right\}; \end{aligned}\right\} \quad (13)$$

$$\left.\begin{aligned} i &= -2\pi fCE\sin(\theta-\theta_0) + \varepsilon^{-\frac{r}{2L}t}\left\{i_0\cos\frac{t}{\sqrt{CL}} \right. \\ &\qquad \left. - (e_0 - E\cos\theta_0)\sqrt{\frac{C}{L}}\sin\frac{t}{\sqrt{CL}}\right\}, \\ e_1 &= E\cos(\theta-\theta_0) + \varepsilon^{-\frac{r}{2L}t}\left\{(e_0 - E\cos\theta_0)\cos\frac{t}{\sqrt{CL}} \right. \\ &\qquad \left. + i_0\sqrt{\frac{L}{C}}\sin\frac{t}{\sqrt{CL}}\right\}. \end{aligned}\right\} \quad (14)$$

The oscillating terms of these equations are independent of the impressed frequency. That is, the oscillating currents and potential differences, caused by a change of circuit conditions (as starting, change of load, or opening circuit), are independent of the impressed frequency, and thus also of the wave shape of the impressed e.m.f., or its higher harmonics (except as regards terms of secondary order).

The first component of oscillation, equation (12), depends not only upon the line constants and the impressed e.m.f., but principally upon the phase, or the point of the impressed e.m.f. wave, at which the oscillation starts; however, it does not depend upon the previous condition of the circuit. Therefore this component of oscillation is the same as the oscillation produced in starting the transmission line, that is, connecting it, unexcited, to the generator terminals.

There exists no point of the impressed e.m.f. wave where no oscillation occurs (while, when starting a circuit containing resistance and inductance only, at the point of the impressed e.m.f. wave where the final current passes zero the stationary condition is instantly reached).

With capacity in circuit, any change of circuit conditions involves an electric oscillation.

The maximum intensities of the starting oscillation occur near the value $\theta_0 = 0$, and are

$$\left.\begin{aligned} i'' &= \frac{E}{\sqrt{x\,x_c}}\,\varepsilon^{-\frac{r}{2x}\theta} \sin \sqrt{\frac{x_c}{x}}\,\theta \\ \text{and} \qquad & \\ e_1'' &= -E\,\varepsilon^{-\frac{r}{2x}\theta} \cos \sqrt{\frac{x_c}{x}}\,\theta. \end{aligned}\right\} \qquad (15)$$

Since

$$i' = -\frac{E}{x_c} \sin (\theta - \theta_0)$$

is the stationary value of charging current, it follows that the maximum intensity which the oscillating current, produced in starting a transmission line, may reach is $\sqrt{\frac{x_c}{x}}$ times the stationary charging current, or the initial current bears to the stationary value the same ratio as the frequency of oscillation to the impressed frequency.

The maximum oscillating e.m.f. generated in starting a transmission line is of the same value as the impressed e.m.f. Thus the maximum value of potential difference occurring in a transmission line at starting is less than twice the impressed e.m.f. and no excessive voltages can be generated in starting a circuit.

The minimum values of the starting oscillation occur near $\theta_0 = 90°$, and are, from equations (7),

$$\left.\begin{aligned} i'' &= -\frac{E}{x_c}\,\varepsilon^{-\frac{r}{2x}\theta} \cos \sqrt{\frac{x_c}{x}}\,\theta \\ \text{and} \qquad & \\ e_1'' &= -\sqrt{\frac{x}{x_c}}\,E\,\varepsilon^{-\frac{r}{2x}\theta} \sin \sqrt{\frac{x_c}{x}}\,\theta; \end{aligned}\right\} \qquad (16)$$

that is, the oscillating current is of the same intensity as the charging current, and the maximum rush of current thus is less than twice the stationary value. The potential difference in the circuit rises only little above the impressed e.m.f.

The second component of the oscillation, equation (13), does not depend upon the point of the impressed e.m.f. wave at which the oscillation starts, θ_0, nor upon the impressed e.m.f. as a whole, E, but, besides upon the constants of the circuit, it depends only upon the instantaneous values of current and of potential difference in the circuit at the moment when the oscillation starts, i_0 and e_0.

Thus, if $i_0 = 0$, $e_0 = 0$, or in starting a transmission line, unexcited, by connecting it to the impressed e.m.f., this term disappears. It is this component which may cause excessive potential differences. Two cases shall more fully be discussed, namely:

(*a*) Opening the circuit of a transmission line under load, and (*b*) rupturing a short-circuit on the transmission line.

67. (*a*) If i_0 is the instantaneous value of full-load current, e_0 the instantaneous value of difference of potential at the condenser, ri_0 is small compared with e_0, and $\sqrt{x\,x_c}\,i_0$ is of the same magnitude as e_0.

Writing

$$\tan\delta = \frac{e_0}{i_0\sqrt{x\,x_c}},$$

and substituting in equations (10), we have

$$\left.\begin{aligned} i''' &= \sqrt{i_0^{\,2} + \frac{e_0^{\,2}}{x\,x_c}}\,\varepsilon^{-\frac{r}{2x}\theta}\cos\left(\sqrt{\frac{x_c}{x}}\,\theta + \delta\right) \\ \text{and}\quad e_1''' &= \sqrt{e_0^{\,2} + i_0^{\,2}x\,x_c}\,\varepsilon^{-\frac{r}{2x}\theta}\sin\left(\sqrt{\frac{x_c}{x}}\,\theta + \delta\right); \end{aligned}\right\} \qquad (17)$$

that is, the amplitude of oscillation is $\sqrt{i_0^{\,2} + \frac{e_0^{\,2}}{x\,x_c}}$ for the current, and $\sqrt{e_0^{\,2} + i_0^{\,2}x\,x_c}$ for the e.m.f. Thus the generated e.m.f. can be larger than the impressed e.m.f., but is, as a rule, still of the same magnitude, except when x_c is very large.

In the expressions of the total current and potential difference at condenser, in equations (11), $(e_0 - E\cos\theta_0)$ is the difference between the potential difference at the condenser and the impressed e.m.f., at the instant of starting of the oscillation, or the voltage consumed by the line impedance, and this is small if the current is not excessive. Thus, neglecting the terms with $(e_0 - E \cos \theta_0)$, equations (11) assume the form

$$\left.\begin{aligned} i &= -\frac{E}{x_c} \sin(\theta - \theta_0) + i_0 \varepsilon^{-\frac{r}{2x}\theta} \cos \sqrt{\frac{x_c}{x}}\, \theta \\ \text{and} \quad & \\ e_1 &= E \cos(\theta - \theta_0) + i_0 \sqrt{x\, x_c}\, \varepsilon^{-\frac{r}{2x}\theta} \sin \sqrt{\frac{x_c}{x}}\, \theta; \end{aligned}\right\} \qquad (18)$$

that is, the oscillation of current is of the amplitude of full-load current, and the oscillation of condenser potential difference is of the amplitude $i_0 \sqrt{x\, x_c}$.

$x\, x_c$ is the ratio of inductance voltage to condenser current, in fractions of full-load voltage and current. We have, therefore,

$$i_0 \sqrt{x\, x_c} = i_0 \sqrt{\frac{L}{C}}.$$

Thus in circuits of very high inductance L and relatively low capacity C, $i_0 \sqrt{x\, x_c}$ may be much higher than the impressed e.m.f., and a serious rise of potential occur when opening the circuit under load, while in low inductance cables of high capacity $i_0 \sqrt{x\, x_c}$ is moderate; that is, the inductance, by tending to maintain the current, generates an e.m.f., producing a rise in potential, while capacity exerts a cushioning effect. Low inductance and high capacity thus are of advantage when breaking full-load current in a circuit.

68. (*b*) If a transmission line containing resistance, inductance, and capacity is short-circuited, and the short-circuit suddenly opened at time $t = 0$, we have, for $t < 0$,

$$\left.\begin{aligned} e_0 &= 0 \\ \text{and} \quad i &= \frac{E}{z} \cos(\theta - \theta_0 - \gamma), \\ \text{where} \quad z &= \sqrt{r^2 + x^2} \\ \text{and} \quad \tan \gamma &= \frac{x}{r}; \end{aligned}\right\} \qquad (19)$$

thus, at time $t = 0$,

$$i_0 = \frac{E}{z} \cos (\theta_0 + \gamma). \tag{20}$$

Substituting these values of e_0 and i_0 in equations (9) gives

$$i''' = \frac{E}{z} \cos (\theta_0 + \gamma) \varepsilon^{-\frac{r}{2x}\theta} \left\{ \cos \sqrt{\frac{x_c}{x}}\,\theta - \frac{r}{2\sqrt{x x_c}} \sin \sqrt{\frac{x_c}{x}}\,\theta \right\}$$

and

$$e_1''' = \frac{E}{z} \cos (\theta_0 + \gamma) \varepsilon^{-\frac{r}{2x}\theta} \sqrt{x x_c} \sin \sqrt{\frac{x_c}{x_c}}\,\theta,$$

or, neglecting terms of secondary magnitude,

$$\left.\begin{aligned} i''' &= \frac{E}{z} \varepsilon^{-\frac{r}{2x}\theta} \cos (\theta_0 + \gamma) \cos \sqrt{\frac{x_c}{x}}\,\theta \\ &\text{and} \\ e_1''' &= \frac{E\sqrt{x x_c}}{z} \varepsilon^{-\frac{r}{2x}\theta} \cos (\theta_0 + \gamma) \sin \sqrt{\frac{x_c}{x}}\,\theta; \end{aligned}\right\} \tag{21}$$

that is, i''' is of the magnitude of short-circuit current, and e_1''' of higher magnitude than the impressed e.m.f., since z is small compared with $\sqrt{x x_c}$.

The total values of current and condenser potential difference, from equation (11), are

$$\left.\begin{aligned} i &= -\frac{E}{x_c} \sin (\theta - \theta_0) + E\varepsilon^{-\frac{r}{2x}\theta} \left\{ \frac{\cos (\theta_0 + \gamma)}{z} \right. \\ &\qquad \left. \cos \sqrt{\frac{x_c}{x}}\,\theta + \frac{\cos \theta_0}{\sqrt{x x_c}} \sin \sqrt{\frac{x_c}{x}}\,\theta \right\} \\ &\text{and} \\ e_1 &= E \cos (\theta - \theta_0) - E\varepsilon^{-\frac{r}{2x}\theta} \left\{ \cos \theta_0 \cos \sqrt{\frac{x_c}{x}}\,\theta \right. \\ &\qquad \left. - \frac{\sqrt{x x_c} \cos (\theta_0 + \gamma)}{z} \sin \sqrt{\frac{x_c}{x}}\,\theta \right\}, \end{aligned}\right\} \tag{22}$$

or approximately, since all terms are negligible compared with i''' and e_1''',

$$\left.\begin{aligned} i &= \frac{E}{z}\varepsilon^{-\frac{r}{2x}\theta} \cos(\theta_0 + \gamma) \cos\sqrt{\frac{x_c}{x}}\,\theta \\ &\text{and} \\ e_1 &= \frac{E\sqrt{x x_c}}{z}\varepsilon^{-\frac{r}{2x}\theta} \cos(\theta_0 + \gamma) \sin\sqrt{\frac{x_c}{x}}\,\theta. \end{aligned}\right\} \quad (23)$$

These values are a maximum, if the circuit is opened at the moment $\theta_0 = -\gamma$, that is, at the maximum value of the short-circuit current, and are then

$$\left.\begin{aligned} i &= \frac{E}{z}\varepsilon^{-\frac{r}{2x}\theta} \cos\sqrt{\frac{x_c}{x}}\,\theta \\ &\text{and} \\ e_1 &= \frac{\sqrt{x x_c}}{z} E\varepsilon^{-\frac{r}{2x}\theta} \sin\sqrt{\frac{x_c}{x}}\,\theta. \end{aligned}\right\} \quad (24)$$

The amplitude of oscillation of the condenser potential difference is

$$\frac{\sqrt{x x_c}}{z} E,$$

or, neglecting the line resistance, as rough approximation, $x = z$,

$$\sqrt{\frac{x_c}{x}}\,E;$$

that is, the potential difference at the condenser is increased above the impressed e.m.f. in the proportion of the square root of the ratio of condensive reactance to inductive reactance, or inversely proportional to the square root of inductance voltage times capacity current, as fraction of the impressed voltage and the full-load current. Thus, in this case, the rise of voltage is excessive.

The minimum intensity of the oscillation due to rupturing short-circuit occurs if the circuit is broken at the moment

$\theta_0 = 90° - r$, that is, at the zero value of the short-circuit current. Then we have

$$\left.\begin{aligned} i &= -\frac{E}{x_c}\cos(\theta+\gamma) + \frac{E}{\sqrt{x x_c}}\varepsilon^{-\frac{r}{2x}\theta}\sin\gamma\sin\sqrt{\frac{x_c}{x}}\theta \\ \text{and} \\ e_1 &= E\sin(\theta+\gamma) - E\varepsilon^{-\frac{r}{2x}\theta}\sin\gamma\cos\sqrt{\frac{x_c}{x}}\theta; \end{aligned}\right\} \quad (25)$$

that is, the potential difference at the condenser is less than twice the impressed e.m.f.; therefore is moderate. Hence, a short-circuit can be opened safely only at or near the zero value of the short-circuit current.

The phenomenon ceases to be oscillating, and becomes an ordinary logarithmic discharge, if $\sqrt{r^2 - 4\,x x_c}$ is real, or

$$r > 2\sqrt{x x_c}.$$

Some examples may illustrate the phenomena discussed in the preceding paragraphs.

69. Let, in a transmission line carrying 100 amperes at full load, under an impressed e.m.f. of 20,000 volts, the resistance drop = 8 per cent, the inductance voltage = 15 per cent of the impressed voltage, and the charging current =8 per cent of full-load current. Assuming 1 per cent resistance drop in the step-up transformers, and a reactance voltage of 2½ per cent, the resistance drop between the constant potential generator terminals and the middle of the transmission line is then 5 per cent, or $r = 10$ ohms, and the inductance voltage is 10 per cent, or $x = 20$ ohms. The charging current of the line is 8 amperes, thus the condensive reactance $x_c = 2500$ ohms.

Then, assuming a sine wave of impressed e.m.f., we have

$$\begin{aligned} E &= 20{,}000\sqrt{2} = 28{,}280 \text{ volts}; \\ i' &= -\ 11.3\sin(\theta - \theta_0); \\ e_1' &= 28{,}280\cos(\theta - \theta_0); \\ i'' &= -\ 11.3\,\varepsilon^{-0.25\,\theta}[\sin\theta_0\cos 11.2\,\theta - 11.2\cos\theta_0\sin 11.2\,\theta], \end{aligned}$$

and $$\begin{aligned} e_1'' &= -28{,}280\,\varepsilon^{-0.25\theta}[\cos\theta_0\cos 11.2\,\theta + (0.0222\cos\theta_0 \\ &\quad + 0.0283\sin\theta_0)\sin 11.2\,\theta] \\ &\cong -28{,}280\,\varepsilon^{-0.25\,\theta}\cos\theta_0\cos 11.2\,\theta. \end{aligned}$$

Therefore the oscillations produced in starting the transmission line are

$$i = -11.3\,[\sin(\theta - \theta_0) + \varepsilon^{-0.25\,\theta}(\sin\theta_0 \cos 11.2\,\theta - 11.2 \cos\theta_0 \sin 11.2\,\theta)]$$

and
$$e_1 = 28{,}280\,\{\cos(\theta - \theta_0) - \varepsilon^{-0.25\,\theta}[\cos\theta_0 \cos 11.2\,\theta + (0.0222 \cos\theta_0 + 0.0283 \sin\theta_0) \sin 11.2\,\theta]\}$$
$$\cong 28{,}280\,[\cos(\theta - \theta_0) - \varepsilon^{-0.25\,\theta}\cos\theta_0 \cos 11.2\,\theta].$$

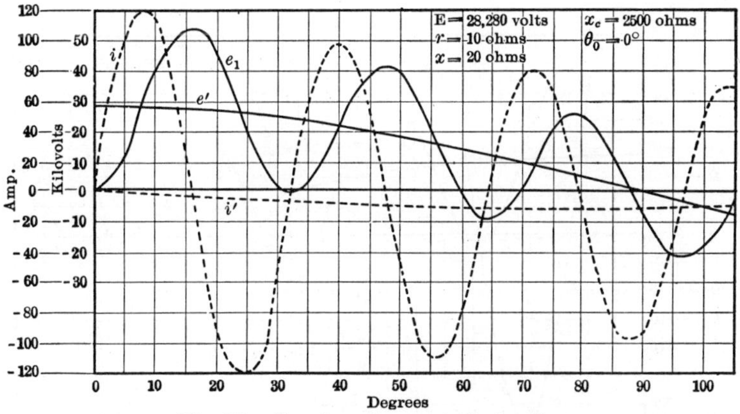

Fig. 27. Starting of a transmission line.

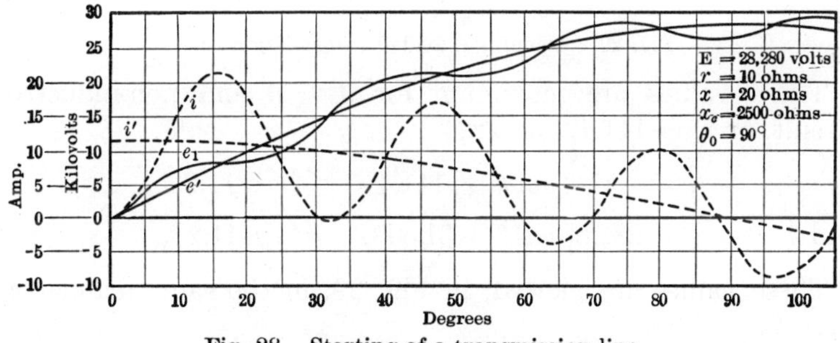

Fig. 28. Starting of a transmission line.

Hence the maximum values for $\theta_0 = 0$, are

$$i = -11.3\,(\sin\theta - 11.2\,\varepsilon^{-0.25\theta} \sin 11.2\,\theta)$$

and
$$e_1 = 28{,}280\,[\cos\theta - \varepsilon^{-0.25\,\theta}(\cos 11.2\,\theta + 0.0222 \sin 11.2\,\theta)]$$
$$\cong 28{,}280\,(\cos\theta - \varepsilon^{-0.25\,\theta} \cos 11.2\,\theta),$$

and the minimum values, for $\theta_0 = 90°$, are

$$i = 11.3\,(\cos\theta - \varepsilon^{-0.25\,\theta} \cos 11.2\,\theta)$$

and
$$e_1 = 28{,}280\,(\sin\theta - 0.0283\,\varepsilon^{-0.25\theta} \sin 11.2\,\theta)$$
$$\cong 28{,}280 \sin\theta$$

These values are plotted in Figs. 27 and 28, with the current, i, in dotted and the potential difference, e_1, in drawn line. The stationary values are plotted also, in thin lines, i and e', respectively.

(a) Opening the circuit under full load, we have

$$i = -11.3 \sin(\theta - \theta_0) + i_0 \varepsilon^{-0.25\theta} \cos 11.2\,\theta$$

and $e_1 = 28{,}280 \cos(\theta - \theta_0) + 224\, i_0 \varepsilon^{-0.25\theta} \sin 11.2\,\theta.$

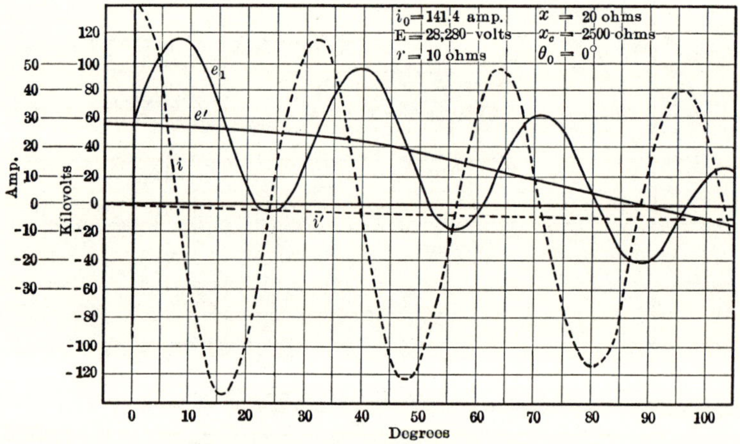

Fig. 29. Opening a loaded transmission line.

These values are maximum for $\theta_0 = 0$ and non-inductive circuit, or $i_0 = 141.4$, and are

$$i = -11.3 \sin\theta + 141.4\, \varepsilon^{-0.25\theta} \cos 11.2\,\theta$$

and $$e_1 = 28{,}280 \cos\theta + 31{,}600\, \varepsilon^{-0.25\theta} \sin 11.2\theta.$$

These values are plotted, in Fig. 29, in the same manner as Figs. 27 and 28.

(b) Rupturing the line under short-circuit, we have

$$z = 22.4$$

and $$i_0 = 1265 \cos(\theta_0 + \gamma);$$

and therefore

$$i = -11.3 \sin(\theta - \theta_0) + 1265\, \varepsilon^{-0.25\theta} [\cos(\theta_0 + \gamma) \cos 11.2\,\theta + 0.1 \cos\theta_0 \sin 11.2\,\theta]$$

and $e_1 = 28{,}280 \{\cos(\theta - \theta_0) - \varepsilon^{-0.25\,\theta}[\cos\theta_0 \cos 11.2\,\theta - 10\cos(\theta_0 + \gamma)\sin 11.2\,\theta]\}.$

These values are a maximum for $\theta_0 = -\gamma = -63°$, thus

$$i = -11.3 \sin(\theta + 63°) + 1265\,\varepsilon^{-0.25\,\theta}(\cos 11.2\,\theta + 0.044 \sin 11.2\,\theta)$$

and $e_1 = 28{,}280 \cos(\theta + 63°) - 282{,}800\,\varepsilon^{-0.25\,\theta}(0.044 \cos 11.2\,\theta - \sin 11.2\,\theta);$

that is, the potential difference rises about tenfold, to 282,800 volts. These values are plotted in Fig. 30.

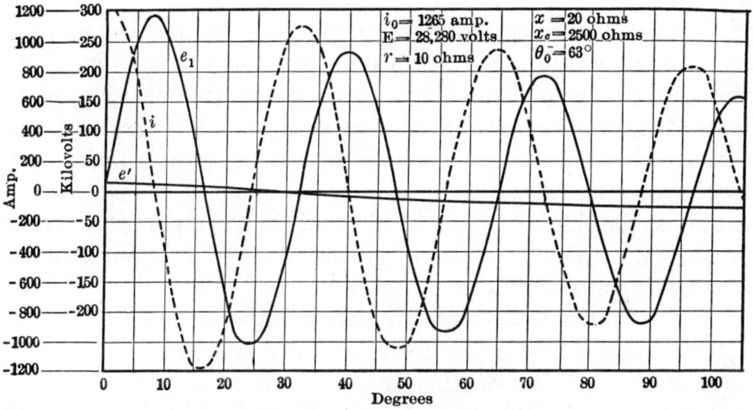

Fig. 30. Opening a short-circuited transmission line.

70. On an experimental 10,000-volt, 40-cycle line, when a destructive e.m.f. was produced by a short-circuiting arc, the author observed a drop in generator e.m.f. to about 5000 volts, due to the limited machine capacity. The resistance of the system was very low, about $r = 1$ ohm, while the inductive reactance may be estimated as $x = 10$ ohms, and the condensive reactance as $x_c = 20{,}000$ ohms. Therefore $\tan\gamma = 10$, or approximately, $\gamma = 90°$.

Herefrom it follows that

$$i = 707\,\varepsilon^{-0.05\,\theta} \cos 44.7\,\theta$$

and

$$e_1 = 316{,}000\,\varepsilon^{-0.05\,\theta} \sin 44.7\,\theta;$$

that is, the oscillation has a frequency of about 1800 cycles per second and a maximum e.m.f. of nearly one-third million volts, which fully accounts for its disruptive effects.

71. As conclusion, it follows herefrom:

1. A most important source of destructive high voltage phenomena in high potential circuits containing inductance and capacity are the electric oscillations produced by a change of circuit conditions, as starting, opening circuit, etc.

2. These phenomena are essentially independent of the frequency and the wave shape of the impressed e.m.f., but depend upon the conditions under which the circuit is changed, as the manner of change and the point of the impressed e.m.f. and current wave at which the change occurs.

3. The electric oscillations occurring in connecting a transmission line to the generator are not of dangerous potential, but the oscillations produced by opening the transmission circuit under load may reach destructive voltages, and the oscillations caused by interrupting a short-circuit are liable to reach voltages far beyond the strength of any insulation. Thus special precautions should be taken in opening a high potential circuit under load. But the most dangerous phenomenon is a low resistance short-circuit in open space.

4. The voltages produced by the oscillations in open-circuiting a transmission line under load or under short-circuit are moderate if the opening of the circuit occurs at a certain point of the e.m.f. wave. This point approximately coincides with the moment of zero current.

CHAPTER IX.

DIVIDED CIRCUIT.

72. A circuit consisting of two branches or multiple circuits 1 and 2 may be supplied, over a line or circuit 3, with an impressed e.m.f., e_0.

Let, in such a circuit, shown diagrammatically in Fig 31, r_1, L_1, C_1 and r_2, L_2, C_2 = resistance, inductance, and capacity, respectively, of the two branch circuits 1 and 2; r_0, L_0, C_0 = resistance, inductance, and capacity of the undivided part of the circuit, 3. Furthermore let e = potential difference at terminals of branch circuits 1 and 2, i_1 and i_2 respectively = currents in branch circuits 1 and 2, and i_3 = current in undivided part of circuit, 3.

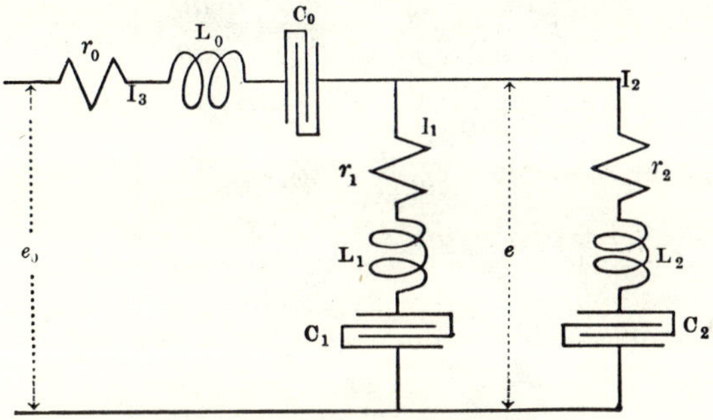

Fig. 31. Divided circuit.

Then
$$i_3 = i_1 + i_2 \tag{1}$$

and e.m.f. at the terminals of circuit 1 is

$$e = r_1 i_1 + L_1 \frac{di_1}{dt} + \frac{1}{C_1}\int i_1\, dt, \tag{2}$$

of circuit 2 is

$$e = r_2 i_2 + L_2 \frac{di_2}{dt} + \frac{1}{C_2}\int i_2\, dt, \tag{3}$$

and of circuit 3 is

$$e_0 = e + r_0 i_3 + L_0 \frac{di_3}{dt} + \frac{1}{C_0} \int i_3 \, dt. \tag{4}$$

Instead of the inductances, L, and capacities, C, it is usually preferable, even in direct-current circuits, to introduce the reactances, $x = 2\,\pi f L$ = inductive reactance, $x_c = \dfrac{1}{2\,\pi f C}$ = condensive reactance, referred to a standard frequency, such as $f = 60$ cycles per second. Instead of the time t, then, an angle

$$\theta = 2\,\pi f t \tag{5}$$

is introduced, and then we have

$$\left.\begin{aligned} L\frac{di}{dt} &= \frac{x}{2\,\pi f}\frac{di}{d\theta}\frac{d\theta}{dt} = x\frac{di}{d\theta} \\ \text{and} \qquad \frac{1}{C}\int i\,dt &= 2\,\pi f x_c \int \frac{i}{\frac{d\theta}{dt}} d\theta = x_c \int i\,d\theta, \end{aligned}\right\} \tag{6}$$

since

$$\frac{d\theta}{dt} = 2\,\pi f.$$

Hereby resistance, inductance, and capacity are expressed in the same units, ohms.

Time is expressed by an angle θ so that 360 degrees correspond to $\frac{1}{60}$ of a second, and the time effects thus are directly comparable with the phenomena on a 60-cycle circuit.

A better conception of the size or magnitude of inductance and capacity is secured. Since inductance and capacity are mostly observed and of importance in alternating-current circuits, a reactor having an inductive reactance of x ohms and i amperes conveys to the engineer a more definite meaning as regards size: it has a volt-ampere capacity of i^2x, that is, the approximate size of a transformer of half this capacity, or of a $\dfrac{i^2x}{2}$-watt transformer. A reactor having an inductance of L henrys and i amperes, however, conveys very little meaning to the engineer who is mainly familiar with the effect of inductance in alternating-current circuits.

Substituting therefore (5) and (6) in equations (2), (3), (4), gives the e.m.f. in circuit 1 as

$$e = r_1 i_1 + x_1 \frac{di_1}{d\theta} + x_{c_1} \int i_1 \, d\theta; \tag{7}$$

in circuit 2 as

$$e = r_2 i_2 + x_2 \frac{di_2}{d\theta} + x_{c_2} \int i_2 \, d\theta; \tag{8}$$

in circuit 3 as

$$e_0 = e + r_0 i_3 + x_0 \frac{di_3}{d\theta} + x_{c0} \int i_3 \, d\theta; \tag{9}$$

hence, the potential differences at the condenser terminals are

$$e_1 = x_{c_1} \int i_1 \, d\theta = e - r_1 i_1 - x_1 \frac{di_1}{d\theta}, \tag{10}$$

$$e_2 = x_{c_2} \int i_2 \, d\theta = e - r_2 i_2 - x_2 \frac{di_2}{d\theta}, \tag{11}$$

and

$$e_3 = x_{c0} \int i_3 \, d\theta = e_0 - e - r_0 i_3 - x_0 \frac{di_3}{d\theta}. \tag{12}$$

Differentiating equations (7), (8), and (9), to eliminate the integral, gives as differential equations of the divided circuit:

$$x_1 \frac{d^2 i_1}{d\theta^2} + r_1 \frac{di_1}{d\theta} + x_{c_1} i_1 = \frac{de}{d\theta}, \tag{13}$$

$$x_2 \frac{d^2 i_2}{d\theta^2} + r_2 \frac{di_2}{d\theta} + x_{c_2} i_2 = \frac{de}{d\theta}, \tag{14}$$

and

$$x_0 \frac{d^2 i_3}{d\theta^2} + r_0 \frac{di_3}{d\theta} + x_{c0} i_3 = \frac{de_0}{d\theta} - \frac{de}{d\theta}. \tag{15}$$

Subtracting (14) from (13) gives

$$\left(x_1 \frac{d^2 i_1}{d\theta^2} + r_1 \frac{di_1}{d\theta} + x_{c_1} i_1\right) - \left(x_2 \frac{d^2 i_2}{d\theta^2} + r_2 \frac{di_2}{d\theta} + x_{c_2} i_2\right) = 0. \tag{16}$$

Multiplying (15) by 2, and adding thereto (13) and (14), gives, by substituting (1), $i_3 = i_1 + i_2$,

$$(2\,x_0 + x_1)\,\frac{d^2 i_1}{d\theta^2} + (2\,r_0 + r_1)\,\frac{di_1}{d\theta} + (2\,x_{c_0} + x_{c_1})i_1 +$$

$$(2\,x_0 + x_2)\,\frac{d^2 i_2}{d\theta^2} + (2\,r_0 + r_2)\,\frac{di_2}{d\theta} + (2\,x_{c_0} + x_{c_2})i_2 = 2\,\frac{de_0}{d\theta}. \quad (17)$$

These two differential equations (16) and (17) are integrated by the functions

$$\left.\begin{aligned} i_1 &= i_1' + A_1\varepsilon^{-a\theta} \\ \text{and} \qquad i_2 &= i_2' + A_2\varepsilon^{-a\theta}, \end{aligned}\right\} \quad (18)$$

where i_1' and i_2' are the permanent values of current, and $i_1'' = A_1\varepsilon^{-a\theta}$ and $i_2'' = A_2\varepsilon^{-a\theta}$ are the transient current terms.

Substituting (18) in (16) and (17) gives

$$\left(x_1\,\frac{d^2 i_1'}{d\theta^2} + r_1\,\frac{di_1'}{d\theta} + x_{c_1} i_1'\right) - \left(x_2\,\frac{d^2 i_2'}{d\theta^2} + r_2\,\frac{di_2'}{d\theta} + x_{c_2} i_2'\right)$$

$$+ A_1\varepsilon^{-a\theta}(a^2 x_1 - a r_1 + x_{c_1}) - A_2\varepsilon^{-a\theta}(a^2 x_2 - a r_2 + x_{c_2}) = 0 \quad (19)$$

and

$$(2\,x_0 + x_1)\,\frac{d^2 i_1'}{d\theta^2} + (2\,r_0 + r_1)\,\frac{di_1'}{d\theta} + (2\,x_{c_0} + x_{c_1})i_1' + (2\,x_0 + x_2)$$

$$\frac{d^2 i_2'}{d\theta^2} + (2\,r_0 + r_2)\,\frac{di_2'}{d\theta} + (2\,x_{c_0} + x_{c_2})i_2 + A_1\varepsilon^{-a\theta}\{a^2(2\,x_0 + x_1)$$

$$- a\,(2\,r_0 + r_1) + (2\,x_{c_0} + x_{c_1})\} + A_2\varepsilon^{-a\theta}\,\{a^2\,(2\,x_0 + x_2)$$

$$- a\,(2\,r_0 + r_2) + (2\,x_{c_0} + x_{c_2})\} = 2\,\frac{de_0}{d\theta}. \quad (20)$$

73. For $\theta = \infty$, the exponential terms eliminate, and there remain the differential equations of the permanent terms i_1' and i_2', thus

$$\left(x_1\,\frac{d^2 i_1'}{d\theta^2} + r_1\,\frac{di_1'}{d\theta} + x_{c_1} i_1'\right) - \left(x_2\,\frac{d^2 i_2'}{d\theta^2} + r_2\,\frac{di_2'}{d\theta} + x_{c_2} i_2'\right) = 0 \quad (21)$$

and

$$(2\,x_0 + x_1)\,\frac{d^2 i_1'}{d\theta^2} + (2\,r_0 + r_1)\,\frac{di_1'}{d\theta} + (2\,x_{c_0} + x_{c_1})\,i_1' + (2\,x_0 + x_2)$$

$$\frac{d^2 i_2'}{d\theta^2} + (2\,r_0 + r_2)\,\frac{di_2'}{d\theta} + (2\,x_{c_0} + x_{c_2})\,i_2' = 2\,\frac{de_0}{d\theta}. \quad (22)$$

The solution of these equations (21) and (22) is the usual equation of electrical engineering, giving i_1' and i_2' as sine waves if the e.m.f., e_0, is a sine wave; giving i_1' and i_2' as constant quantities if e_0 is constant and x_{c_0} and either x_{c_1} or x_{c_2} or both vanish, and giving i_1' and $i_2' = 0$ if either x_{c_0} or both x_{c_1} and x_{c_2} differ from zero.

Subtracting (21) and (22) from (19) and (20) leaves as differential equations of the transient terms i_1'' and i_2'',

$$\varepsilon^{-a\theta} \{A_1 (a^2x_1 - ar_1 + x_{c_1}) - A_2 (a^2x_2 - ar_2 + x_{c_2})\} = 0 \qquad (23)$$

and

$$\varepsilon^{-a\theta} \{A_1 [a^2 (2 x_0 + x_1) - a (2 r_0 + r_1) + (2 x_{c_0} + x_{c_1})] + A_2 [a^2 (2 x_0 + x_2) - a (2r_0 + r_2) + (2 x_{c_0} + x_{c_2})]\} = 0. \qquad (24)$$

Introducing a new constant B, these equations give, from (23),

$$\left.\begin{aligned} A_1 &= B (a^2x_2 - ar_2 + x_{c_2}) \\ \text{and} \qquad A_2 &= B (a^2x_1 - ar_1 + x_{c_1}); \end{aligned}\right\} \qquad (25)$$

then substituting (25) in (24) gives

$$(a^2x_2 - ar_2 + x_{c_2}) [a^2(2 x_0 + x_1) - a (2 r_0 + r_1) + (2 x_{c_0} + x_{c_1})] + (a^2x_1 - ar_1 + x_{c_1})[a^2(2 x_0 + x_2) - a (2 r_0 + r_2) + (2 x_{c_0} + x_{c_2})] = 0, \qquad (26)$$

while B remains indeterminate as integration constant.

Quartic equation (26) gives four values of a, which may be all real, or two real and two conjugate imaginary, or two pairs of conjugate imaginary roots.

Rearranged, equation (26) gives

$$a^4 (x_0x_1 + x_0x_2 + x_1x_2) - a^3 \{r_0 (x_1 + x_2) + r_1 (x_0 + x_2) + r_2 (x_0 + x_1)\} + a^2 \{(r_0r_1 + r_0r_2 + r_1r_2) + x_{c_0} (x_1 + x_2) + x_{c_1} (x_0 + x_2) + x_{c_2} (x_0 + x_1)\} - a \{x_{c_0}(r_1 + r_2) + x_{c_1} (r_0 + r_2) + x_{c_2} (r_0 + r_1)\} + (x_{c_0}x_{c_1} + x_{c_0}x_{c_2} + x_{c_1}x_{c_2}) = 0. \qquad (27)$$

Let a_1, a_2, a_3, a_4 be the four roots of this quartic equation (27);

then

$$i_1 = i_1' + B_1 (a_1^2 x_2 - a_1 r_2 + x_{c_2}) \varepsilon^{-a_1\theta} + B_2 (a_2^2 x_2 - a_2 r_2 + x_{c_2}) \varepsilon^{-a_2\theta} + B_3 (a_3^2 x_2 - a_3 r_2 + x_{c_2}) \varepsilon^{-a_3\theta} + B_4 (a_4^2 x_2 - a_4 r_2 + x_{c_2}) \varepsilon^{-a_4\theta} \quad (28)$$

and

$$i_2 = i_2' + B_1 (a_1^2 x_1 - a_1 r_1 + x_{c_1}) \varepsilon^{-a_1\theta} + B_2 (a_2^2 x_1 - a_2 r_1 + x_{c_1}) \varepsilon^{-a_2\theta} + B_3 (a_3^2 x_1 - a_3 r_1 + x_{c_1}) \varepsilon^{-a_3\theta} + B_4 (a_4^2 x_1 - a_4 r_1 + x_{c_1}) \varepsilon^{-a_4\theta} \quad (29)$$

where the integration constants B_1, B_2, B_3 and B_4 are determined by the terminal conditions: the currents and condenser potentials at zero time, $\theta = 0$.

The quartic equation (27) usually has to be solved by approximation.

74. *Special Cases:* Continuous-current divided circuit, with resistance and inductance but no capacity, e_0 = constant.

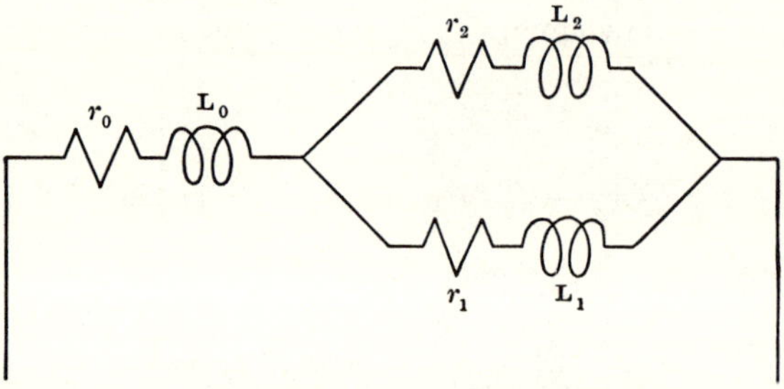

Fig. 32. Divided continuous-current circuit without capacity.

In such a circuit, shown diagrammatically in Fig. 32, equations (7), (8), and (9) are greatly simplified by the absence of the integral, and we have

$$e = r_1 i_1 + x_1 \frac{di_1}{d\theta}, \quad (30)$$

$$e = r_2 i_2 + x_2 \frac{di_2}{d\theta}, \quad (31)$$

and

$$e_0 = e + r_0 i_3 + x_0 \frac{di_3}{d\theta}. \quad (32)$$

(30) and (31) combined give

$$r_1 i_1 - r_2 i_2 + x_1 \frac{di_1}{d\theta} - x_2 \frac{di_2}{d\theta} = 0. \quad (33)$$

Substituting (1), $i_3 = i_1 + i_2$, in (32), multiplying it by 2 and adding thereto (30) and (31), gives

$$2\,e_0 = (2\,r_0 + r_1)\,i_1 + (2\,r_0 + r_2)\,i_2 + (2\,x_0 + x_1)\frac{di_1}{d\theta} + (2\,x_0 + x_2)\frac{di_2}{d\theta}. \tag{34}$$

Equations (33) and (34) are integrated by

$$\left.\begin{aligned} i_1 &= i_1' + A_1\varepsilon^{-a\theta} \\ \text{and}\quad i_2 &= i_2' + A_2\varepsilon^{-a\theta}. \end{aligned}\right\} \tag{35}$$

Substituting (35) in (33) and (34) gives

$$\left.\begin{aligned} &(r_1 i_1' - r_2 i_2') + \varepsilon^{-a\theta}\{A_1(r_1 - ax_1) - A_2(r_2 - ax_2)\} = 0 \\ &\text{and} \\ &2\,e_0 = (2\,r_0 + r_1)\,i_1' + (2\,r_0 + r_2)\,i_2' + \varepsilon^{-a\theta}\{A_1\,[(2\,r_0 + r_1) \\ &\qquad - a\,(2\,x_0 + x_1)] + A_2\,[(2\,r_0 + r_2) - a\,(2\,x_0 + x_2)]\}. \end{aligned}\right\}$$

These equations resolve into the equations of permanent state, thus

$$\left.\begin{aligned} r_1 i_1' - r_2 i_2' &= 0 \\ \text{and}\quad (2\,r_0 + r_1)\,i_1' + (2\,r_0 + r_2)\,i_2' &= 2\,e_0. \end{aligned}\right\}$$

Hence,

$$\left.\begin{aligned} i_1' &= e_0\frac{r_2}{r^2} \\ \text{and}\quad i_2' &= e_0\frac{r_1}{r^2}, \end{aligned}\right\} \tag{36}$$

where

$$r^2 = r_0 r_1 + r_0 r_2 + r_1 r_2, \tag{37}$$

and the transient equations having the coefficients

$$\left.\begin{aligned} &A_1\,(r_1 - ax_1) - A_2\,(r_2 - ax_2) = 0 \\ &\text{and} \\ &A_1\,[(2\,r_0 + r_1) - a\,(2\,x_0 + x_1)] + A_2\,[(2\,r_0 + r_2) \\ &\qquad - a\,(2\,x_0 + x_2)] = 0. \end{aligned}\right\}$$

Herefrom it follows that

$$\left.\begin{aligned} A_1 &= B\,(r_2 - a x_2) \\ \text{and} \qquad A_2 &= B\,(r_1 - a x_1), \end{aligned}\right\} \tag{38}$$

and

$$a^2\,(x_0 x_1 + x_0 x_2 + x_1 x_2) - a\,[r_0\,(x_1 + x_2) + r_1\,(x_0 + x_2) + r_2\,(x_0 + x_1)] + (r_0 r_1 + r_0 r_2 + r_1 r_2) = 0, \tag{39}$$

$$B = \text{indefinite.} \tag{40}$$

Substituting the abbreviations,

$$\left.\begin{aligned} & x_0 x_1 + x_0 x_2 + x_1 x_2 = x^2, \\ & r_0 r_1 + r_0 r_2 + r_1 r_2 = r^2, \\ \text{and} \quad & r_0\,(x_1 + x_2) + r_1\,(x_0 + x_2) + r_2\,(x_0 + x_1) = x_0\,(r_1 + r_2) \\ & \qquad + x_1\,(r_0 + r_2) + x_2\,(r_0 + r_1) = s^2, \end{aligned}\right\} \tag{41}$$

gives (39)

$$a^2 x^2 - a s^2 + r^2 = 0, \tag{42}$$

hence two roots,

$$\left.\begin{aligned} a_1 &= \frac{s^2 - q^2}{2\,x^2} \\ \text{and} \qquad a_2 &= \frac{s^2 + q^2}{2\,x^2}, \end{aligned}\right\} \tag{43}$$

where

$$q^2 = \sqrt{s^4 - 4\,r^2 x^2}. \tag{44}$$

The two roots of equation (42), a_1 and a_2, are always real, since in q^2

$$s^4 > 4\,r^2 x^2,$$

as seen by substituting (41) therein.

The final integral equations thus are

$$\left.\begin{aligned} i_1 &= e_0 \frac{r_2}{r^2} + (r_2 - a_1 x_2)\,B_1 \varepsilon^{-\frac{s^2 - q^2}{2\,x^2}\theta} + (r_2 - a_2 x_2)\,B_2 \varepsilon^{-\frac{s^2 + q^2}{2\,x^2}\theta} \\ \text{and} \qquad i_2 &= e_0 \frac{r_1}{r^2} + (r_1 - a_1 x_1)\,B_1 \varepsilon^{-\frac{s^2 - q^2}{2\,x^2}\theta} + (r_1 - a_2 x_1)\,B_2 \varepsilon^{-\frac{s^2 + q^2}{2\,x^2}\theta}. \end{aligned}\right\} \tag{45}$$

B_1 and B_2 are determined by the terminal conditions, as the currents i_1 and i_2 at the start, $\theta = 0$.

Let, at zero time, or $\theta = 0$,

$$\left.\begin{aligned} i_1 &= i_1{}^0 \\ \text{and} \qquad i_2 &= i_2{}^0; \end{aligned}\right\} \tag{46}$$

then, substituting in (45), we have

$$\left.\begin{aligned} i_1{}^0 &= e_0 \frac{r_2}{r^2} + (r_2 - a_1x_2)\,B_1 + (r_2 - a_2x_2)\,B_2 \\ \text{and} \qquad i_2{}^0 &= e_0 \frac{r_1}{r^2} + (r_1 - a_1x_1)\,B_1 + (r_1 - a_2x_1)\,B_2; \end{aligned}\right\} \tag{47}$$

and herefrom calculate B_1 and B_2.

75. For instance, in a continuous-current circuit, let the impressed e.m.f., $e_0 = 120$ volts; the resistance of the undivided part of the circuit, $r_0 = 20$ ohms; the reactance, $x_0 = 20$ ohms; the resistance of one of the branches, $r_1 = 20$ ohms; the reactance, $x_1 = 40$ ohms, and the resistance of the other branch, $r_2 = 5$ ohms, the reactance, $x_2 = 200$ ohms.

Thus one of the branches is of low resistance and high reactance, the other of high resistance and moderate reactance.

The permanent values of the currents, ($r^2 = 600$), are

$$\left.\begin{aligned} i_1' &= 1 \text{ amp.} \\ \text{and} \qquad i_2' &= 4 \text{ amp.} \end{aligned}\right\}$$

(a) Assuming now the resistance r_0 suddenly decreased from $r_0 = 20$ ohms to $r_0 = 15$ ohms, we have the permanent values of current as

$$\left.\begin{aligned} i_1' &= 1.265 \text{ amp.} \\ \text{and} \qquad i_2' &= 5.06 \text{ amp.} \end{aligned}\right\}$$

The previous values of currents, and thus the values of currents at the moment of start, $\theta = 0$, are

$$\left.\begin{aligned} i_1{}^0 &= 1 \text{ amp.} \\ \text{and} \qquad i_2{}^0 &= 4 \text{ amp.} \end{aligned}\right\}$$

therefrom follow the equations of currents, by substitution in the preceding,

$$\left.\begin{aligned} i_1 &= 1.265 + 0.455\,\varepsilon^{-0.0633\,\theta} - 0.720\,\varepsilon^{-0.586\,\theta} \\ \text{and} \qquad i_2 &= 5.06 - 1.038\,\varepsilon^{-0.0633\,\theta} - 0.022\,\varepsilon^{-0.586\,\theta}. \end{aligned}\right\}$$

(*b*) Assuming now the resistance r_0 suddenly raised again from $r_0 = 15$ ohms to $r_0 = 20$ ohms, leaving everything else the same, we have

$$\left.\begin{aligned} i_1^0 &= 1.265 \text{ amp.} \\ \text{and} \qquad i_2^0 &= 5.06 \text{ amp.}; \end{aligned}\right\}$$

and then

$$\left.\begin{aligned} i_1 &= 1 - 0.528\,\varepsilon^{-0.0697\,\theta} + 0.793\,\varepsilon^{-0.674\,\theta} \\ \text{and} \qquad i_2 &= 4 + 1.018\,\varepsilon^{-0.0697\,\theta} + 0.042\,\varepsilon^{-0.674\,\theta}. \end{aligned}\right\}$$

(*c*) Assuming now the resistance r_0 suddenly raised from $r_0 = 20$ ohms to $r_0 = 25$ ohms, gives

$$\left.\begin{aligned} i_1 &= 0.828 - 0.374\,\varepsilon^{-0.0743\,\theta} + 0.546\,\varepsilon^{-0.764\,\theta} \\ \text{and} \qquad i_2 &= 3.312 + 0.649\,\varepsilon^{-0.0743\,\theta} + 0.039\,\varepsilon^{-0.764\,\theta}. \end{aligned}\right\}$$

(*d*) Assuming now the resistance r_0 lowered again from $r_0 =$ 25 ohms to $r_0 = 20$ ohms, gives

$$\left.\begin{aligned} i_1 &= 1 + 0.342\,\varepsilon^{-0.0697\,\theta} - 0.514\,\varepsilon^{-0.674\,\theta} \\ \text{and} \qquad i_2 &= 4 - 0.660\,\varepsilon^{-0.0697\,\theta} - 0.028\,\varepsilon^{-0.674\,\theta}. \end{aligned}\right\}$$

76. In Fig. 33 are shown the variations of currents i_1 and i_2, resultant from a sudden variation of the resistance r_0 from 20 to 15, back to 20, to 25, and back again to 20 ohms. As seen, the readjustment of current i_2, that is, the current in the inductive branch of the circuit, to its permanent condition, is very slow and gradual. Current i_1, however, not only changes very rapidly with a change of r_0, but overreaches greatly; that is, a decrease of r_0 causes i_1 to increase rapidly to a temporary value far in excess of the permanent increase, and then gradually i_1

falls back to its normal, and inversely with an increase of r_0. Hence, any change of the main current is greatly exaggerated in the temporary component of current i_1; a permanent change of about 20 per cent in the total current results in a practically instantaneous change of the branch current i_1, by about 50 per cent in the present instance.

Thus, where any effect should be produced by a change of current, or of voltage, as a control of the circuit effected thereby, the action is made far more sensitive and quicker by shunting the operating circuit i_1, of as low inductance as possible, across

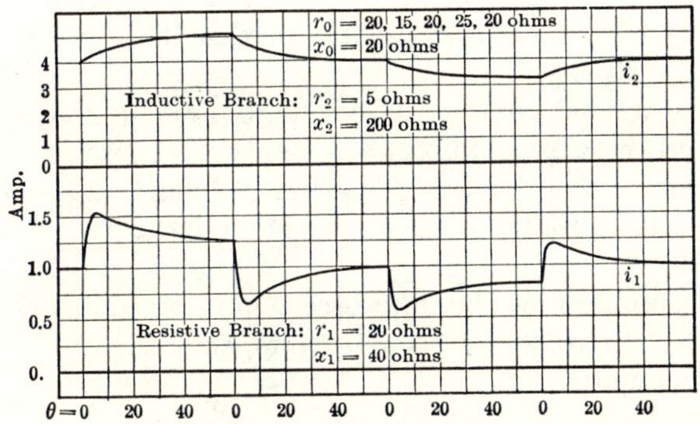

Fig. 33. Current in divided continuous-current circuit resulting from sudden variations in resistance.

a high inductance of as low resistance as possible. The sudden and temporary excess of the change of current i_1 takes care of the increased friction of rest in setting the operating mechanism in motion, and gives a quicker reaction than a mechanism operated directly by the main current.

This arrangement has been proposed for the operation of arc lamps of high arc voltage from constant potential circuits. The operating magnet, being in the circuit i_1, more or less anticipates the change of arc resistance by temporarily over-reaching.

77. The temporary increase of the voltage, e, across the branch circuit, i_1, corresponding to the temporary excess current of this circuit, may, however, result in harmful effects, as destruction of measuring instruments by the temporary excess voltage.

Let, for instance, in a circuit of impressed continuous e.m.f., $e_0 = 600$ volts, as an electric railway circuit, the resistance of the circuit equal 25 ohms, the inductive reactance 44 ohms. This gives a permanent current of $i' = 24$ amperes.

Let now a small part of the circuit, of resistance $r_2 = 1$ ohm, but including most of the reactance $x_2 = 40$ ohms — as a motor series field winding — be shunted by a voltmeter, and $r_1 = 1000$ ohms = resistance, $x_1 = 40$ ohms = reactance of the voltmeter circuit.

In permanent condition the voltmeter reads $\frac{1}{25} \times 600 = 24$ volts, but any change of circuit condition, as a sudden decrease or increase of supply voltage e_0, results in the appearance of a temporary term which may greatly increase the voltage impressed upon the voltmeter.

In this divided circuit, the constants are: undivided part of the circuit, $r_0 = 24$ ohms; $x_0 = 4$ ohms; first branch, voltmeter (practically non-inductive), $r_1 = 1000$ ohms, $x_1 = 40$ ohms; second branch, motor field, highly inductive, $r_2 = 1$ ohm, $x_2 = 40$ ohms.

(*a*) Assuming now the impressed e.m.f., e_0, suddenly dropped from $e_0 = 600$ volts to $e_0 = 540$ volts, that is, by 10 per cent, gives the equations

$$\left.\begin{aligned} i_1 &= 0.0216 - 0.0806\,\varepsilon^{-0.832\,\theta} + 0.0830\,\varepsilon^{-23.1\,\theta} \\ \text{and}\qquad i_2 &= 21.6 + 2.407\,\varepsilon^{-0.832\,\theta} - 0.007\,\varepsilon^{-23.1\,\theta}. \end{aligned}\right\}$$

(*b*) Assuming now the voltage, e_0, suddenly raised again from $e_0 = 540$ volts to $e_0 = 600$ volts, gives the equations

$$\left.\begin{aligned} i_1 &= 0.024 + 0.0806\,\varepsilon^{-0.832\,\theta} - 0.0830\,\varepsilon^{-23.1\,\theta} \\ \text{and}\qquad i_2 &= 24 - 2.407\,\varepsilon^{-0.832\,\theta} + 0.007\,\varepsilon^{-23.1\,\theta}. \end{aligned}\right\}$$

The voltage, e, across the voltmeter, or on circuit 1, is

$$e = r_1 i_1 + x_1 \frac{di_1}{d\theta} = 1000\, i_1' \mp 77.9\,\varepsilon^{-0.832\,\theta} \pm 6.2\,\varepsilon^{-23.1\,\theta},$$

where
$$i_1' = e\frac{r_1}{r^2}.$$

Hence, in case (*a*), drop of impressed voltage, e_0, by 10 per cent,

$$e = 21.6 - 77.9\,\varepsilon^{-0.832\,\theta} + 6.2\,\varepsilon^{-23.1\,\theta},$$

and in (*b*), rise of impressed voltage,

$$e = 24.0 + 77.9\,\varepsilon^{-0.832\,\theta} - 6.2\,\varepsilon^{-23.1\,\theta}.$$

This voltage, *e*, in the two cases, is plotted in Fig. 34. As seen, during the transition of the voltmeter reading from 21.6 to 24.0 volts, the voltage momentarily rises to 95.7 volts, or

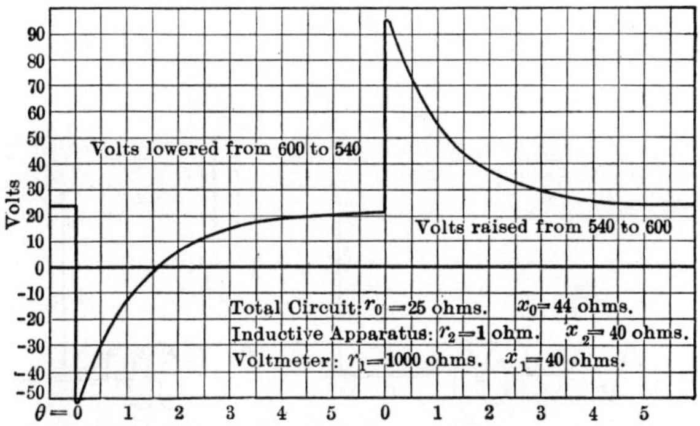

Fig. 34. Voltage across inductive apparatus in series with circuit of high resistance.

four times its permanent value, and during the decrease of permanent voltage from 24.0 to 21.6 volts the voltmeter momentarily reverses, going to 50.1 volts in reverse direction.

In a high voltage direct-current circuit, a voltmeter shunted across a low resistance, if this resistance is highly inductive, is in danger of destruction by any sudden change of voltage or current in the circuit, even if the permanent value of the voltage is well within the safe range of the voltmeter.

CAPACITY SHUNTING A PART OF A CONTINUOUS-CURRENT CIRCUIT.

78. A circuit of resistance r_1 and inductive reactance x_1 is shunted by the condensive reactance x_c, and supplied over the resistance r_0 and the inductive reactance x_0 by a continuous impressed e.m.f., e_0, as shown diagrammatically in Fig. 35.

In the undivided circuit,

$$e_0 = e + r_0 (i_1 + i_2) + x_0 \left(\frac{di_1}{d\theta} + \frac{di_2}{d\theta}\right). \tag{48}$$

In the inductive branch,

$$e = r_1 i_1 + x_1 \frac{di_1}{d\theta}. \tag{49}$$

In the condenser branch,

$$e = x_c \int i_2 \, d\theta. \tag{50}$$

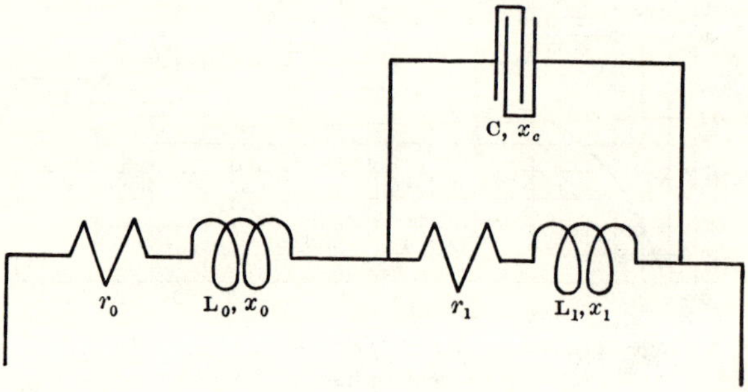

Fig. 35. Suppression of pulsations in direct-current circuits by series inductance and shunted capacity.

Eliminating e gives, from (48) and (49),

$$e_0 = (r_0 + r_1) i_1 + (x_0 + x_1)\frac{di_1}{d\theta} + r_0 i_2 + x_0 \frac{di_2}{d\theta}, \tag{51}$$

and from (49) and (50),

$$x_c \int i_2 \, d\theta = r_1 i_1 + x_1 \frac{di_1}{d\theta}. \tag{52}$$

Differentiating (52), to eliminate the integral,

$$x_c i_2 = r_1 \frac{di_1}{d\theta} + x_1 \frac{d^2 i_1}{d\theta^2}. \tag{53}$$

Substituting (53) in (51), and rearranging,

$$e_0 = (r_0 + r_1)\, i_1 + \frac{1}{x_c}\left\{(r_0 r_1 + x_c x_0 + x_c x_1)\frac{di_1}{d\theta} + (r_0 x_1 + r_1 x_0)\frac{d^2 i_1}{d\theta^2} + x_0 x_1 \frac{d^3 i_1}{d\theta^3}\right\}; \tag{54}$$

a differential equation of third order.

This resolves into the permanent term

$$e_0 = (r_0 + r_1)\, i_1',$$

hence,

$$i_1' = \frac{e_0}{r_0 + r_1}, \tag{55}$$

and a transient term

$$i_1'' = A\varepsilon^{-a\theta}; \tag{56}$$

that is,

$$i_1 = i_1' + A\varepsilon^{-a\theta} = \frac{e_0}{r_0 + r_1} + A\varepsilon^{-a\theta}. \tag{57}$$

Equation (57) substituted in (54) gives as equation of a,

$$x_c (r_0 + r_1) - a (r_0 r_1 + x_c x_0 + x_c x_1) + a^2 (r_0 x_1 + r_1 x_0) - a^3 x_0 x_1 = 0,$$

or

$$a^3 - a^2\left(\frac{r_0}{x_0} + \frac{r_1}{x_1}\right) + a\left(\frac{r_0 r_1}{x_0 x_1} + \frac{x_c}{x_0} + \frac{x_c}{x_1}\right) - \frac{x_c (r_0 + r_1)}{x_0 x_1} = 0, \tag{58}$$

while A remains indefinite as integration constant.

Equation (58) has three roots, a_1, a_2, and a_3, which either are all three real, when the phenomenon is logarithmic, or, one real and two imaginary, when the phenomenon is oscillating.

The integral equation for the current in branch 1 is

$$i_1 = \frac{e_0}{r_0 + r_1} + A_1\varepsilon^{-a_1\theta} + A_2\varepsilon^{-a_2\theta} + A_3\varepsilon^{-a_3\theta}; \tag{59}$$

the current in branch 2 is by (53)

$$i_2 = \frac{1}{x_c}\left(r_1 \frac{di_1}{d\theta} + x_1 \frac{d^2 i_1}{d\theta}\right)$$

$$= \frac{1}{x_c}\{- a_1 (r_1 - a_1 x_1) A_1\varepsilon^{-a_1\theta} - a_2 (r_1 - a_2 x_1) A_2\varepsilon^{-a_2\theta} - a_3 (r_1 - a_3 x_1) A_3\varepsilon^{-a_3\theta}\}, \tag{60}$$

and the potential difference at the condenser is

$$e = x_c \int i_2 \, d\theta = r_1 i_1 + x_1 \frac{di_1}{d\theta}$$

$$= \frac{r_1 e_0}{r_0 + r_1} + (r_1 - a_1 x_1) A_1 \varepsilon^{-a_1 \theta} + (r_1 - a_2 x_1) A_2 \varepsilon^{-a_2 \theta}$$

$$+ (r_1 - a_3 x_1) A_3 \varepsilon^{-a_3 \theta}. \qquad (61)$$

In the case of an oscillatory change, equations (59), (60), and (61) appear in complex imaginary form, and therefore have to be reduced to trigonometric functions.

The three integration constants, A_1, A_2, and A_3, are determined by the three terminal conditions, at $\theta = 0$, $i_1 = i_1^0$, $i_2 = i_2^0$, $e = e^0$.

79. As numerical example may be considered a circuit having the constants, $e_0 = 110$ volts; $r_0 = 1$ ohm; $x_0 = 10$ ohms; $r_1 = 10$ ohms; $x_1 = 100$ ohms, and $x_c = 10$ ohms.

In other words, a continuous e.m.f. of 110 volts supplies, over a line of $r_0 = 1$ ohm resistance, a circuit of $r_1 = 10$ ohms resistance. An inductive reactance $x_0 = 10$ ohms is inserted into the line, and an inductive reactance $x_1 = 100$ ohms in the load circuit, and the latter shunted by a condensive reactance of $x_c = 10$ ohms.

Then, substituting in equation (58),

$$a^3 - 0.2\, a^2 + 1.11\, a - 0.11 = 0.$$

This cubic equation gives by approximation one root, $a_1 = 0.1$, and, divided by $(a - 0.1)$, leaves the quadratic equation

$$a^2 - 0.1\, a + 1.1 = 0,$$

which gives the complex imaginary roots $a_2 = 0.05 - 1.047\, j$ and $a_3 = 0.05 + 1.047\, j$; then from the equation of current, by substituting trigonometric functions for the exponential functions with imaginary exponent, we get the equation for the *load current* as

$$i_1 = i_1' + A_1 \varepsilon^{-0.1\,\theta} + \varepsilon^{-0.05\,\theta} (B_1 \cos 1.047\,\theta + B_2 \sin 1.047\,\theta),$$

the *condenser potential* as

$$e = 10\, i_1' + \varepsilon^{-0.05\,\theta} \{(5\, B_1 + 104.7\, B_2) \cos 1.047\,\theta - (104.7\, B_1 - 5\, B_2) \sin 1.047\,\theta\},$$

and the *condenser current* as

$$i_2 = 10.9\,\varepsilon^{-0.05\,\theta}\{B_1 \cos 1.047\,\theta + B_2 \sin 1.047\,\theta\}.$$

At $e_0 = 110$ volts impressed, the permanent current is $i_1' = 10$ amp., the permanent condenser potential is $e' = 100$ volts, and the permanent condenser current is $i_2' = 0$.

Assuming now the voltage, e_0, suddenly dropped by 10 per cent, from $e_0 = 110$ volts to $e_0 = 99$ volts, gives the permanent current as $i_1' = 9$ amp. At the moment of drop of voltage, $\theta = 0$, we have, however, $i_1 = i_1^0 = 10$ amp.; $e = e' = 100$ volts, and $i_2 = 0$; hence, substituting these numerical values into the above equations of i_1, e, i_2, gives the three integration constants:

$$A_1 = 1;\ B_1 = 0, \text{ and } B_2 = 0.0955;$$

therefore the load current is

$$i_1 = 9 + \varepsilon^{-0.1\,\theta} + 0.0955\,\varepsilon^{-0.05\,\theta} \sin 1.047\,\theta,$$

the condenser current is

$$i_2 = 1.05\,\varepsilon^{-0.05\,\theta} \sin 1.047\,\theta,$$

and the condenser, or load, voltage is

$$e = 90 + \varepsilon^{-0.05\,\theta}\,(10 \cos 1.047\,\theta + 0.48 \sin 1.047\,\theta).$$

Without the condenser, the equation of current would be

$$i = 9 + \varepsilon^{-0.1\,\theta}.$$

In this combination of circuits with shunted condensive reactance x_c, at the moment of the voltage drop, or $\theta = 0$, the rate of change of the load current is, approximately,

$$\frac{di_1}{d\theta} = [-\,0.1\varepsilon^{-0.1\,\theta} + 0.0955 \times 1.047\varepsilon^{-0.05\,\theta} \cos 1.047\,\theta]_0 = 0,$$

while without the condenser it would be

$$\frac{di}{d\theta} = [-\,0.1\,\varepsilon^{-0.1\,\theta}]_0 = -\,0.1.$$

80. By shunting the circuit with capacity, the current in the circuit does not instantly begin to change with a change or fluctuation of impressed e.m.f.

In Fig. 36 is plotted, with θ as abscissas, the change of the current, i_1, in per cent, resulting from an instantaneous change of impressed e.m.f., e_0, of 10 per cent, with condenser in shunt to the load circuit, and without condenser.

As seen, at $\theta = 172° = 3.0$ radians, both currents, i_1 with the condenser and i without condenser, have dropped by the same

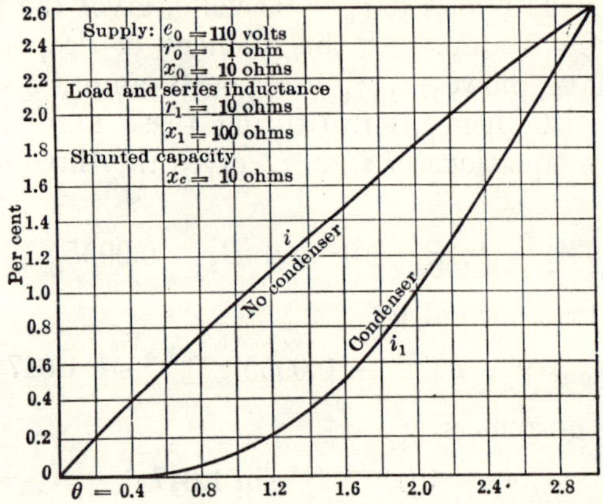

Fig. 36. Suppression of pulsations in direct-current circuits by series inductance and shunted capacity. Effect of 10 per cent drop of voltage.

amount, 2.6 per cent. But at $\theta = 57.3° = 1.0$ radian, i_1 has dropped only $\frac{1}{8}$ per cent., and i nearly 1 per cent, and at $\theta = 24°$, i_1 has not yet dropped at all, while i has dropped by 0.38 per cent.

That is, without condenser, all pulsations of the impressed e.m.f., e_0, appear in the load circuit as pulsations of the current, i, of a magnitude reduced the more the shorter the duration of the pulsation. After $\theta = 60°$, or $t = 0.00275$ seconds, the pulsation of the current has reached 10 per cent of the pulsation of impressed e.m.f.

With a condenser in shunt to the load circuit, the pulsation of current in the load circuit is still zero after $\theta = 24°$, or after 0.001 seconds, and reaches 1.25 per cent of the pulsation of impressed e.m.f., e_0, after $\theta = 60°$, or $t = 0.00275$ seconds.

A pulsation of the impressed e.m.f., e_0, of a frequency higher than 250 cycles, practically cannot penetrate to the load circuit, that is, does not appear at all in the load current i_1 regardless of how much a pulsation of the impressed e.m.f., e_0, it is, and a pulsation of impressed e.m.f., e_0, of a frequency of 120 cycles reappears in the load current i_1, reduced to 1 per cent of its value.

In cases where from a source of e.m.f., e_0, which contains a slight high frequency pulsation — as the pulsation corresponding to the commutator segments of a commutating machine — a current is desired showing no pulsation whatever, as for instance for the operation of a telephone exchange, a very high inductive reactance in series with the circuit, and a condensive reactance in shunt therewith, entirely eliminates all high frequency pulsations from the current, passing only harmless low frequency pulsations at a greatly reduced amplitude.

81. As a further example is shown in Fig. 37 the pulsation of a non-inductive circuit, $x_1 = 0$, of the resistance $r_1 = 4$ ohms, shunted by a condensive reactance $x_c = 10$ ohms, and supplied over a line of resistance $r_0 = 1$ ohm and inductive reactance $x_0 = 10$ ohms, by an impressed e.m.f., $e_0 = 110$ volts.

Due to $x_1 = 0$ equation (58) reduces to

$$a^2 - a\left(\frac{x_c}{r_1} + \frac{r_0}{x_0}\right) + \frac{x_c}{x_0}\left(1 + \frac{r_0}{r_1}\right) = 0;$$

or, substituting numerical values,

$$a^2 - 2.6\,a + 1.25 = 0$$

and $$a_1 = 0.637, \qquad a_2 = 1.963;$$

that is, both roots are real, or the phenomenon is logarithmic.

We now have

$$i_1 = i_1' + A_1\varepsilon^{-0.637\,\theta} + A_2\varepsilon^{-1.963\,\theta},$$
$$i_2 = -\,0.255\,A_1\varepsilon^{-0.637\,\theta} - 0.785\,A_2\varepsilon^{-1.963\,\theta},$$

and $$e = r_1 i_1 = 4\,(i_1' + A_1\varepsilon^{-0.637\,\theta} + A_2\varepsilon^{-1.963\,\theta}).$$

The load current is

$$i_1' = 22 \text{ amp.}$$

A reduction of the impressed e.m.f., e_0, by 10 per cent, or from 110 to 99 volts, gives the integration constants $A_1 = 3.26$ and $A_2 = -\,1.06$; hence,

$$i_1 = 19.8 + 3.26\,\varepsilon^{-0.637\,\theta} - 1.06\,\varepsilon^{-1.963\,\theta},$$
$$i_2 = -\,0.83\,(\varepsilon^{-0.637\,\theta} + \varepsilon^{-1.963\,\theta}),$$

and $$e = 4\,i_1.$$

Without a condenser, the equation of current would be

$$i = 19.8 + 2.2\,\varepsilon^{-0.5\,\theta}.$$

In Fig. 37 is shown, with θ as abscissas, the drop of current i_1 and i, in per cent.

Although here the change is logarithmic, while in the former paragraph it was trigonometric, the result is the same—a very great reduction, by the condenser, of the drop of current immediately after the change of e.m.f. However, in the present case

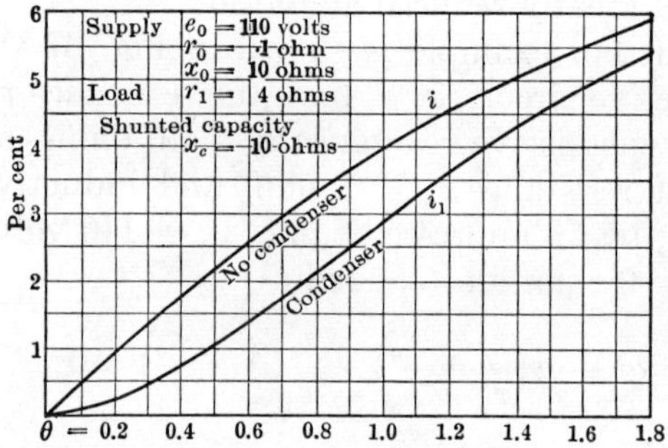

Fig. 37. Suppression of pulsations in non-inductive direct-current circuits by series inductance and shunted capacity. Effect of 10 per cent drop of voltage.

the change of the circuit is far more rapid than in the preceding case, due to the far lower inductive reactance of the present case. For instance, after $\theta = 0.1$, the drop of current, with condenser, is 0.045 per cent, without condenser, 0.5 per cent. At $\theta = 0.2$, the drop of current is 0.23 and 0.95 per cent respectively. For longer times or larger values of θ, the difference produced by the condenser becomes less and less.

This effect of a condenser across a direct-current circuit, of suppressing high frequency pulsations from reaching the circuit, requires a very large capacity.

CHAPTER X

MUTUAL INDUCTANCE

82. In the preceding chapters, circuits have been considered containing resistance, self-inductance, and capacity, but no mutual inductance; that is, the phenomena which take place in the circuit have been assumed as depending upon the impressed e.m.f. and the constants of the circuit, but not upon the phenomena taking place in any other circuit.

Of the magnetic flux produced by the current in a circuit and interlinked with this circuit, a part may be interlinked with a second circuit also, and so by its change generate an e.m.f. in the second circuit, and part of the magnetic flux produced by

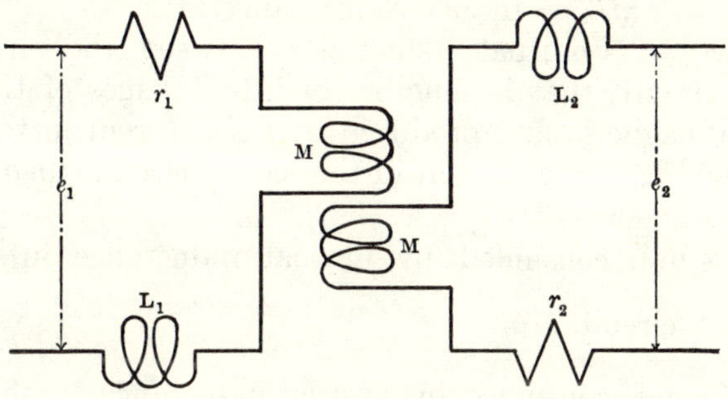

Fig. 38. Mutual inductance between circuits.

the current in a second circuit and interlinked with the second circuit may be interlinked also with the first circuit, and a change of current in the second circuit, that is, a change of magnetic flux produced by the current in the second circuit, then generates an e.m.f. in the first circuit.

Diagrammatically the mutual inductance between two circuits can be sketched as shown by M in Fig. 38, by two *coaxial* coils, while the self-inductance is shown by a single coil L, and the resistance by a *zigzag* line.

The presence of mutual inductance, with a second circuit, introduces into the equation of the circuit a term depending upon the current in the second circuit.

If i_1 = the current in the circuit and r_1 = the resistance of the circuit, then $r_1 i_1$ = the e.m.f. consumed by the resistance of the circuit. If L_1 = the inductance of the circuit, that is, total number of interlinkages between the circuit and the number of lines of magnetic force produced by unit current in the circuit, we have

$$L_1 \frac{di_1}{dt} = \text{e.m.f. consumed by the inductance,}$$

where, t = time.

If instead of time t an angle $\theta = 2\pi ft$ is introduced, where f is some standard frequency, as 60 cycles,

$$x_1 \frac{di_1}{d\theta} = \text{e.m.f. consumed by the inductance,}$$

where $x_1 = 2\pi fL_1$ = inductive reactance.

If now M = mutual inductance between the circuit and another circuit, that is, number of interlinkages of the circuit with the magnetic flux produced by unit current in the second circuit, and i_2 = the current in the second circuit, then

$$M \frac{di_2}{dt} = \text{e.m.f. consumed by mutual inductance in the first circuit,}$$

$$M \frac{di_1}{dt} = \text{e.m.f. consumed by mutual inductance in the second circuit.}$$

Introducing $x_m = 2\pi fM$ = mutual reactance between the two circuits, we have

$$x_m \frac{di_2}{d\theta} = \text{e.m.f. consumed by mutual inductance in the first circuit,}$$

$$x_m \frac{di_1}{d\theta} = \text{e.m.f. consumed by mutual inductance in the second circuit.}$$

If now e_1 = the e.m.f. impressed upon the first circuit and e_2 = the e.m.f. impressed upon the second circuit, the equations of the circuits are

$$e_1 = r_1 i_1 + x_1 \frac{di_1}{d\theta} + x_m \frac{di_2}{d\theta} + x_{c_1} \int i_1 \, d\theta \tag{1}$$

and

$$e_2 = r_2 i_2 + x_2 \frac{di_2}{d\theta} + x_m \frac{di_1}{d\theta} + x_{c_2} \int i_2 d\theta, \tag{2}$$

where r_1 = the resistance, $x_1 = 2\pi f L_1$ = the inductive reactance, and $x_{c_1} = \dfrac{1}{2\pi f C_1}$ = the condensive reactance of the first circuit; r_2 = the resistance, $x_2 = 2\pi f L_2$ = the inductive reactance, $x_{c_2} = \dfrac{1}{2\pi f C_2}$ = the condensive reactance of the second circuit, and $x_m = 2\pi f M$ = mutual inductive reactance between the two circuits.

83. In these equations, x_1 and x_2 are the total inductive reactance, L_1 and L_2 the total inductance of the circuit, that is, the number of magnetic interlinkages of the circuit with the total flux produced by unit current in the circuit, the self-inductive flux as well as the mutual inductive flux, and not merely the self-inductive reactance and inductance respectively.

In induction apparatus, such as transformers and induction machines, it is usually preferable to separate the total reactance x, into the self-inductive reactance, x_s, referring to the magnetic flux interlinked with the inducing circuit only, but with no other circuit, and the mutual inductive reactance, x_m, usually represented as a susceptance, which refers to the mutual inductive component of the total inductance; in which case $x = x_s + x_m$. This is not done in the present case.

Furthermore it is assumed that the circuits are inductively related to each other symmetrically, or reduced thereto; that is, where the mutual inductance is due to coils enclosed in the first circuit, interlinked magnetically with coils enclosed in the second circuit, as the primary and the secondary coils of a transformer, or a shunt and a series field winding of a generator, the two coils are assumed as of the same number of turns, or reduced thereto.

If $a = \frac{n_2}{n_1} = \frac{\text{No. turns second circuit}}{\text{No. turns first circuit}}$, the currents in the second circuit are multiplied, the e.m.fs. divided by a, the resistances and reactances divided by a^2, to reduce the second circuit to the first circuit, in the manner customary in dealing with transformers and especially induction machines.

If the ratio of the number of turns is introduced in the equations, that is, in the first equation $\frac{n_2}{n_1} x_m$ substituted for x_m, in the second equation $\frac{n_1}{n_2} x_m$ for x_m, and the equations then are

$$e_1 = r_1 i_1 + x_1 \frac{di_1}{d\theta} + \frac{n_2}{n_1} x_m \frac{di_2}{d\theta} + x_{c_1} \int i_1 \, d\theta \tag{3}$$

and

$$e_2 = r_2 i_2 + x_2 \frac{di_2}{d\theta} + \frac{n_1}{n_2} x_m \frac{di_1}{d\theta} + x_{c_2} \int i_2 \, d\theta. \tag{4}$$

Since the solution and further investigation of these equations (3), (4) are the same as in the case of equations (1) and (2), except that n_1 and n_2 appear as factors, it is preferable to eliminate n_1 and n_2 by reducing one circuit to the other by the ratio of turns $a = \frac{n_2}{n_1}$, and then use the simpler equations (1), (2).

(A) CIRCUITS CONTAINING RESISTANCE, INDUCTANCE, AND MUTUAL INDUCTANCE BUT NO CAPACITY.

84. In such a circuit, shown diagrammatically in Fig. 38, we have

$$e_1 = r_1 i_1 + x_1 \frac{di_1}{d\theta} + x_m \frac{di_2}{d\theta} \tag{5}$$

and

$$e_2 = r_2 i_2 + x_2 \frac{di_2}{d\theta} + x_m \frac{di_1}{d\theta}. \tag{6}$$

Differentiating (6) gives

$$\frac{de_2}{d\theta} = r_2 \frac{di_2}{d\theta} + x_2 \frac{d^2 i_2}{d\theta^2} + x_m \frac{d^2 i_1}{d\theta^2}; \tag{7}$$

from (5) follows

$$\frac{di_2}{d\theta} = \frac{e_1 - r_1 i_1 - x_1 \dfrac{di_1}{d\theta}}{x_m}, \tag{8}$$

and, differentiated,

$$\frac{d^2 i_2}{d\theta^2} = \frac{1}{x_m}\left\{\frac{de_1}{d\theta} - r_1 \frac{di_1}{d\theta} - x_1 \frac{d^2 i_1}{d\theta^2}\right\}. \tag{9}$$

Substituting (8) and (9) in (7) gives

$$r_2 e_1 + x_2 \frac{de_1}{d\theta} - x_m \frac{de_2}{d\theta} = r_1 r_2 i_1 + (r_1 x_2 + r_2 x_1) \frac{di_1}{d\theta} + (x_1 x_2 - x_m{}^2) \frac{d^2 i_1}{d\theta^2}, \tag{10}$$

and analogously,

$$r_1 e_2 + x_1 \frac{de_2}{d\theta} - x_m \frac{de_1}{d\theta} = r_1 r_2 i_2 + (r_1 x_2 + r_2 x_1) \frac{di_2}{d\theta} + (x_1 x_2 - x_m{}^2) \frac{d^2 i_2}{d\theta^2}. \tag{11}$$

Equations (10) and (11) are the two differential equations of second order, of currents i_1 and i_2.

If e_1', i_1' and e_2', i_2' are the permanent values of impressed e.m.fs. and of currents in the two circuits, and e_1'', i_1'' and e_2'', i_2'' are their transient terms, we have,

$$e_1 = e_1' + e_1'', \qquad i_1 = i_1' + i_1'',$$
$$e_2 = e_2' + e_2'', \qquad i_2 = i_2' + i_2''.$$

Since the permanent terms must fulfill the differential equations (10) and (11),

$$r_2 e_1' + x_2 \frac{de_1'}{d\theta} - x_m \frac{de_2'}{d\theta} = r_1 r_2 i_1' + (r_1 x_2 + r_2 x_1) \frac{di_1'}{d\theta} + (x_1 x_2 - x_m{}^2) \frac{d^2 i_1'}{d\theta^2} \tag{12}$$

and

$$r_1 e_2' + x_1 \frac{de_2'}{d\theta} - x_m \frac{de_1'}{d\theta} = r_1 r_2 i_2' + (r_1 x_2 + r_2 x_1) \frac{di_2'}{d\theta} + (x_1 x_2 - x_m{}^2) \frac{d^2 i_2'}{d\theta^2}, \tag{13}$$

subtracting equations (12) and (13) from (10) and (11) gives the differential equations of the transient terms,

$$r_2 e_1'' + x_2 \frac{de_1''}{d\theta} - x_m \frac{de_2''}{d\theta} = r_1 r_2 i_1'' + (r_1 x_2 + r_2 x_1) \frac{di_1''}{d\theta} + (x_1 x_2 - x_m^2) \frac{d^2 i_1''}{d\theta^2} \tag{14}$$

and

$$r_1 e_2'' + x_1 \frac{de_2''}{d\theta} - x_m \frac{de_1''}{d\theta} = r_1 r_2 i_2'' + (r_1 x_2 + r_2 x_1) \frac{di_2''}{d\theta} + (x_1 x_2 - x_m^2) \frac{d^2 i_2''}{d\theta^2}. \tag{15}$$

These differential equations of the transient terms are the same as the general differential equations (10) and (11) and the differential equations of the permanent terms (12) and (13).

85. If, as is usually the case, the impressed e.m.fs. contain no transient term, that is, the transient terms of current do not react upon the sources of supply of the impressed e.m.fs. and affect them, we have

$$e_1'' = 0 \quad \text{and} \quad e_2'' = 0;$$

hence, the differential equations of the transient terms are

$$0 = r_1 r_2 i + (r_1 x_2 + r_2 x_1) \frac{di}{d\theta} + (x_1 x_2 - x_m^2) \frac{d^2 i}{d\theta^2} \tag{16}$$

and are the same for both currents i_1'' and i_2'', that is, the transient terms of currents differ only by their integration constants, or the terminal conditions.

Equation (16) is integrated by the function

$$i = A \varepsilon^{-a\theta}. \tag{17}$$

Substituting (17) in (16) gives

$$A \varepsilon^{-a\theta} \{r_1 r_2 - a (r_1 x_2 + r_2 x_1) + a^2 (x_1 x_2 - x_m^2)\} = 0;$$

hence,

A = indefinite, as integration constant, and

$$a^2 - \frac{r_1 x_2 + r_2 x_1}{x_1 x_2 - x_m^2} a + \frac{r_1 r_2}{x_1 x_2 - x_m^2} = 0. \tag{18}$$

The exponent a is given by a quadratic equation (18). This quadratic equation (18) always has two real roots, and in this respect differs from the quadratic equation appearing in a circuit containing capacity, which latter may have two imaginary roots and so give rise to an oscillation.

Mutual induction in the absence of capacity thus always gives a logarithmic transient term; thus,

$$a = \frac{(r_1 x_2 + r_2 x_1) \pm \sqrt{(r_1 x_2 - r_2 x_1)^2 + 4\, r_1 r_2 x_m^2}}{2\,(x_1 x_2 - x_m^2)} \,. \quad (19)$$

As seen, the term under the radical in (19) is always positive, that is, the two roots a_1 and a_2 always real and always positive, since the square root is smaller than the term outside of it.

Herefrom then follows the integral equation of one of the currents, for instance i_1, as

$$i_1 = i_1' + A_1 \varepsilon^{-a_1 \theta} + A_2 \varepsilon^{-a_2 \theta}, \quad (20)$$

and eliminating from the two equations (5) and (6) the term $\frac{di_2}{d\theta}$ gives

$$i_2 = \frac{1}{r_2 x_m} \left\{ r_1 x_2 i_1 + (x_1 x_2 - x_m^2) \frac{di_1}{d\theta} + x_m e_2 - x_2 e_1 \right\}, \quad (21)$$

leaving the two integration constants A_1 and A_2 to be determined by the terminal conditions, as $\theta = 0$,

$$i_1 = i_1^0 \quad \text{and} \quad i_2 = i_2^0.$$

86. If the impressed e.m.fs. e_1 and e_2 are constant, we have

$$\frac{de_1}{d\theta} = 0 \quad \text{and} \quad \frac{de_2}{d\theta} = 0;$$

hence, the equations of the permanent terms (12) and (13) give

$$i_1' = \frac{e_1}{r_1} \quad \text{and} \quad i_2' = \frac{e_2}{r_2}; \quad (22)$$

thus:

$$i_1 = \frac{e_1}{r_1} + A_1 \varepsilon^{-a_1 \theta} + A_2 \varepsilon^{-a_2 \theta}$$

and

$$i_2 = \frac{e_2}{r_2} + A_1' \varepsilon^{-a_1 \theta} + A_2' \varepsilon^{-a_2 \theta}, \quad (23)$$

where, A_1' and A_2' follow from A_1 and A_2 by equation (21).

If the mutual inductance between the two circuits is perfect, that is,

$$x_m{}^2 = x_1 x_2, \tag{24}$$

equation (18) becomes, by multiplication with $\dfrac{x_1x_2 - x_m{}^2}{r_1x_2 + r_2x_1}$,

$$a = \frac{r_1 r_2}{r_1x_2 + r_2x_1}; \tag{25}$$

that is, only one transient term exists.

As example may be considered a circuit having the following constants: $e_1 = 100$ volts; $e_2 = 0$; $r_1 = 5$ ohms; $r_2 = 5$ ohms; $x_1 = 100$ ohms; $x_2 = 100$ ohms, and $x_m = 80$ ohms. This gives

$$i_1' = 20 \text{ amp. and } i_2' = 0,$$

and

$$a^2 - 0.278\,a + 0.00695 = 0;$$

the roots are $\quad a_1 = 0.0278 \quad \text{and} \quad a_2 = 0.251$

and

$$i_1 = 20 + A_1\varepsilon^{-0.0278\,\theta} + A_2\varepsilon^{-0.251\,\theta}.$$

By equation (21),

$$i_2 = -\,25 + 1.25\,i_1 + 9\frac{di_1}{d\theta};$$

hence,

$$i_2 = A_1\varepsilon^{-0.0278\,\theta} - A_2\varepsilon^{-0.251\,\theta}.$$

For $\theta = 0$ let $i_1{}^0 = 18$ amp., or the current 10 per cent below the normal, and $i_2{}^0 = 0$; then substituted, gives:

$$18 = 20 + A_1 + A_2 \quad \text{and} \quad 0 = A_1 - A_2,$$

hence, $\quad A_1 = A_2 = -\,1;$

and we have

$$i_1 = 20 - (\varepsilon^{-0.0278\,\theta} + \varepsilon^{-0.251\,\theta})$$

and $\quad i_2 = -\,(\varepsilon^{-0.0278\,\theta} - \varepsilon^{-0.251\,\theta}).$

87. An interesting application of the preceding is the investigation of the building up of an overcompounded direct-current generator, with sudden changes of load, or the building up, or down, of a compound wound direct-current booster.

While it would be desirable that a generator or booster, under sudden changes of load, should instantly adjust its voltage to the change so as to avoid a temporary fluctuation of voltage, actually an appreciable time must elapse.

A 600-kw. 8-pole direct-current generator overcompounds from 500 volts at no load to 600 volts at terminals at full load of 1000 amperes. The circuit constants are: resistance of armature winding, $r_0 = 0.01$ ohm; resistance of series field winding, $r_2' = 0.003$ ohm; number of turns per pole in shunt field winding, $n_1 = 1000$, and magnetic flux per pole at 500 volts, $\Phi = 10$ megalines. At 600 volts full load *terminal voltage* (or voltage from brush to brush) the generated e.m.f. is $e + ir_0 = 610$ volts.

From the saturation curve or magnetic characteristics of the machine, we have:

At no load and 500 volts:

5000 ampere-turns, 10 megalines and 5 amp. in shunt field circuit.

At no load and 600 volts:

7000 ampere-turns and 12 megalines.

At no load and 610 volts:

7200 ampere-turns and 12.2 megalines.

At full load and 600 volts:

8500 ampere-turns, 12.2 megalines and 6 amp. in shunt field.

Hence the demagnetizing force of the armature, due to the shift of brushes, is 1300 ampere-turns per pole.

At 600 volts and full load the shunt field winding takes 6 amperes, and gives 6000 ampere-turns, so that the series field winding has to supply 2500 ampere-turns per pole, of which 1300 are consumed by the armature reaction and 1200 magnetize.

At 1000 amp. full load the series field winding thus has 2.5 turns per pole, of which 1.3 neutralize the armature reaction and $n_2 = 1.2$ turns are effective magnetizing turns.

The ratio of effective turns in series field winding and in shunt field winding is $a = \frac{n_2}{n_1} = 1.2 \times 10^{-3}$. This then is the reduction factor of the shunt circuit to the series circuit.

It is convenient to reduce the phenomena taking place in the shunt field winding to the same number of turns as the series field winding, by the factors a and a^2 respectively.

If then e = terminal voltage of the armature, or voltage impressed upon the main circuit consisting of series field winding and external circuit, the same voltage is impressed upon the shunt field winding and reduced to the main circuit by factor a, gives

$$e_1 = ae = 1.2 \times 10^{-3}\, e.$$

Since at 500 volts impressed the shunt field current is 5 amperes, the field rheostat must be set so as to give to the shunt field circuit the total resistance of $r_1' = \frac{500}{5} = 100$ ohms.

Reduced to the main circuit by the square of the ratio of turns, this gives the resistance,

$$r_1 = a^2 r_1' = 144 \times 10^{-6} \text{ ohms}.$$

An increase of ampere-turns from 5000 to 7000, corresponding to an increase of current in the shunt field winding by 2 amperes, increases the generated e.m.f. from 500 to 600 volts, and the magnetic flux from 10 to 12, or by 2 megalines per pole. In the induction range covered by the overcompounding from 500 to 600 volts, 1 ampere increase in the shunt field increases the flux by 1 megaline per pole, and so, with $n_1 = 1000$ turns, gives 10^9 magnetic interlinkages per pole, or 8×10^9 interlinkages with 8 poles, per ampere, hence 80×10^9 interlinkages per unit current or 10 amperes, that is, an inductance of 80 henrys. Reduced to the main circuit this gives an inductance of $1.2^2 \times 10^{-6} \times 80 = 115.2 \times 10^{-6}$ henrys. This is the inductance due to the magnetic flux in the field poles, which interlinks with shunt and series coil, or the mutual inductance, $M = 115.2 \times 10^{-6}$ henrys.

Assuming the total inductance L_1 of the shunt field winding as 10 per cent higher than the mutual inductance M, that is, assuming 10 per cent stray flux, we have

$$L_1 = 1.1\, M = 126.7 \times 10^{-6} \text{ henrys}.$$

In the main circuit, full load is 1000 amp. at 600 volts. This gives the effective resistance of the main circuit as $r = 0.6$ ohm.

The quantities referring to the main circuit may be denoted without index.

The total inductance of the main circuit depends upon the character of the load. Assuming an average railway motor load, the inductance may be estimated as about $L = 2000 \times 10^{-6}$ henrys.

In the present problem the impressed e.m.fs. are not constant but depend upon the currents, that is, the sum $i + i_1$, where i_1 = shunt field current reduced to the main circuit by the ratio of turns.

The impressed e.m.f., e, is approximately proportional to the magnetic flux Φ, hence less than proportional to the current, in consequence of magnetic saturation. Thus we have

$$e = 500 \text{ volts for } 5000 \text{ ampere-turns,}$$

$$\text{or } i + i_1 = \frac{5000}{1.2} = 4170 \text{ amp. and}$$

$$e = 600 \text{ volts for } 7200 \text{ ampere-turns,}$$

$$\text{or } i + i_1 = \frac{7200}{1.2} = 6000 \text{ amp.;}$$

hence, 1830 amp. produce a rise of voltage of 100, or 1 amp. raises the voltage by $\frac{100}{1830} = \frac{1}{18.3}$.

At 6000 amp. the voltage is $\frac{6000}{18.3} = 328$ volts higher than at 0 amp., that is, the voltage in the range of saturation between 500 and 600 volts, when assuming the saturation curve in this range as straight line, is given by the equation

$$e = 272 + \frac{i + i_1}{18.3}.$$

The impressed e.m.f. of the shunt field is the same, hence, reduced to the main circuit by the ratio of turns, $a = 1.2 \times 10^{-3}$, is

$$e_1 = \left(272 + \frac{i + i_1}{18.3}\right) 1.2 \times 10^{-3}.$$

Assuming now as standard frequency, $f = 60$ cycles per sec., the constants of the two mutually inductive circuits shown diagrammatically in Fig. 38 are:

	Main Circuit.	Shunt Field Circuit.
Current..........	i amp.	i_1 amp.
Impressed e.m.f...	$e = 272 + \frac{i+i_1}{18.3}$ volts	$e_1 = \left(272+\frac{i+i_1}{18.3}\right) 1.2\times 10^{-3}$ volts
Resistance........	$r = .6$ ohms	$r_1 = 0.144\times 10^{-3}$ ohms
Inductance.......	$L = 2000\times 10^{-6}$ henrys	$L_1 = 126.7\times 10^{-6}$ henrys
Reactance, $2\pi fL$..	$x = 755\times 10^{-3}$ ohms	$x_1 = 47.8\times 10^{-3}$ ohms
Mutual inductance	$M = 115.2 \times 10^{-6}$ henrys	
Mutual reactance.	$x_m = 43.5 \times 10^{-3}$ ohms	

This gives the differential equations of the problem as

$$272 + \frac{i+i_1}{18.3} = 0.6\, i + 0.755 \frac{di}{d\theta} + 0.0435 \frac{di_1}{d\theta} \tag{26}$$

and

$$1.2\left(272 + \frac{i+i_1}{18.3}\right) = 0.144\, i_1 + 47.8 \frac{di_1}{d\theta} + 43.5 \frac{di}{d\theta}. \tag{27}$$

88. Eliminating $\frac{di}{d\theta}$ from equations (26) and (27) gives

$$\frac{di_1}{d\theta} = 0.695\, i - 0.0712\, i_1 - 338. \tag{28}$$

Equation (28) substituted in (26) gives

$$i_1 = 13.07 \frac{di}{d\theta} + 9.95\, i - 4950. \tag{29}$$

Equation (29) substituted in (28) gives

$$\frac{di_1}{d\theta} = -0.93 \frac{di}{d\theta} - 0.015\, i + 15. \tag{30}$$

Equation (29) differentiated, and equated with (30), gives

$$\frac{d^2i}{d\theta^2} + 0.828 \frac{di}{d\theta} + 0.00115\, i - 1.15 = 0. \tag{31}$$

Equation (31) is integrated by

$$i = i_0 + A\varepsilon^{-a\theta}.$$

Substituting this in (31) gives

$$A\varepsilon^{-a\theta}\{a^2 - 0.828\,a + 0.00115\} + \{0.00115\,i_0 - 1.15\} = 0,$$

hence, $i_0 = 1000$, A is indefinite, as integration constant, and

$$a^2 - 0.828\,a + 0.00115 = 0;$$

thus

$$a = 0.414 \pm 0.4126,$$

and the roots are

$$a_1 = 0.0014 \quad \text{and} \quad a_2 = 0.827.$$

Therefore

$$i = 1000 + A_1\varepsilon^{-0.0014\,\theta} + A_2\varepsilon^{-0.827\,\theta}. \tag{32}$$

Substituting (32) in (29) gives

$$i_1 = 5000 + 9.932\,A_1\varepsilon^{-0.0014\,\theta} - 0.85\,A_2\varepsilon^{-0.827\,\theta}. \tag{33}$$

Substituting in (32) and (33) the terminal conditions $\theta = 0$, $i = 0$, and $i_1 = 4170$, gives

$$A_1 + A_2 = -1000 \quad \text{and} \quad 9.932\,A_1 - 0.85\,A_2 = -830,$$

that is,

$$A_1 = -156 \quad \text{and} \quad A_2 = -844.$$

Therefore

$$i = 1000 - 156\,\varepsilon^{-0.0014\,\theta} - 844\,\varepsilon^{-0.827\,\theta} \tag{34}$$

and

$$i_1 = 5000 - 1550\,\varepsilon^{-0.0014\,\theta} + 720\,\varepsilon^{-0.827\,\theta}; \tag{35}$$

or the shunt field current i_1 reduced back to the number of turns of the shunt field by the factor $a = 1.2 \times 10^{-3}$ is

$$i_1' = 6 - 1.86\,\varepsilon^{-0.0014\,\theta} + 0.86\,\varepsilon^{-0.827\,\theta}, \tag{36}$$

and the terminal voltage of the machine is

$$e = 272 + \frac{i + i_1}{18.3},$$

or,

$$e = 600 - 93.2\, \varepsilon^{-0.0014\,\theta} - 6.8\, \varepsilon^{-0.827\,\theta}. \tag{37}$$

As seen, of the two exponential terms one disappears very quickly, the other very slowly.

Introducing now instead of the angle $\theta = 2\,\pi f t$ the time, t, gives the *main current* as

$$i = 1000 - 156\, \varepsilon^{-0.53\,t} - 844\, \varepsilon^{-311\,t},$$

the *shunt field current* as

$$i_1' = 6 - 1.86\, \varepsilon^{-0.53\,t} + 0.86\, \varepsilon^{-311\,t}, \tag{38}$$

and the *terminal voltage* as

$$e = 600 - 93.2\, \varepsilon^{-0.53\,t} - 6.8\, \varepsilon^{-311\,t}.$$

89. Fig. 39 shows these three quantities, with the time, t, as abscissas.

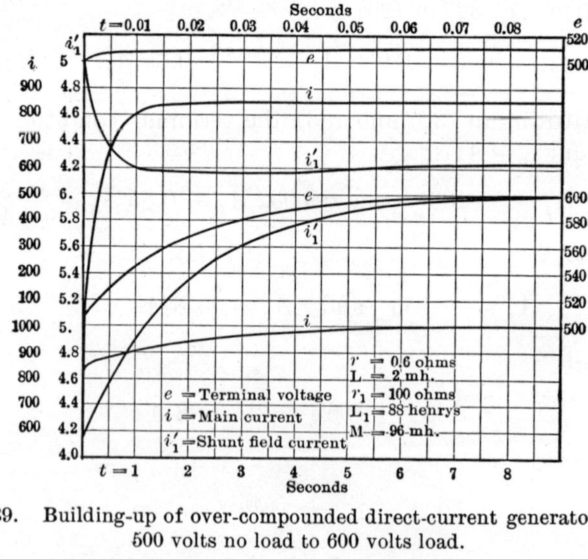

Fig. 39. Building-up of over-compounded direct-current generator from 500 volts no load to 600 volts load.

The upper part of Fig. 39 shows the first part of the curve with 100 times the scale of abscissas as the lower part. As seen, the transient phenomenon consists of two distinctly different periods: first a very rapid change covering a part of the range of current or e.m.f., and then a very gradual adjustment to the final condition.

So the main current rises from zero to 800 amp. in 0.01 sec., but requires for the next 100 amp., or to rise to a total of 900 amp., about a second, reaching 95 per cent of full value in 2.25 sec. During this time the shunt field current first falls very rapidly, from 5 amp. at start to 4.2 amp. in 0.01 sec., and then, after a minimum of 4.16 amp., at $t = 0.015$, gradually and very slowly rises, reaching 5 amp., or its starting point, again after somewhat more than a second. After 2.5 sec. the shunt field current has completed half of its change, and after 5.5 sec. 90 per cent of its change.

The terminal voltage first rises quickly by a few volts, and then rises slowly, completing 50 per cent of its change in 1.2 sec., 90 per cent in 4.5 sec., and 95 per cent in 5.5 sec.

Physically, this means that the terminal voltage of the machine rises very slowly, requiring several seconds to approach stationary conditions. First, the main current rises very rapidly, at a rate depending upon the inductance of the external circuit, to the value corresponding to the resistance of the external circuit and the initial or no load terminal voltage, and during this period of about 0.01 sec. the magnetizing action of the main current is neutralized by a rapid drop of the shunt field current. Then gradually the terminal voltage of the machine builds up, and the shunt field current recovers to its initial value in 1.15 sec., and then rises, together with the main current, in correspondence with the rising terminal voltage of the machine.

It is interesting to note, however, that a very appreciable time elapses before approximately constant conditions are reached.

90. In the preceding example, as well as in the discussion of the building up of shunt or series generators in Chapter III, the e.m.fs. and thus currents produced in the iron of the magnetic field by the change of the field magnetization have not been considered. The results therefore directly apply to a machine with laminated field, but only approximately to one with solid iron poles.

In machines with solid iron in the magnetic circuit, currents produced in the iron act as a second electric circuit in inductive relation to the field exciting circuit, and the transition period thus is slower.

As example may be considered the excitation of a series booster with solid and with laminated poles; that is, a machine with series field winding, inserted in the main circuit of a feeder, for the purpose of introducing into the circuit a voltage proportional to the load, and thus to compensate for the increasing drop of voltage with increase of load.

Due to the production of eddy currents in the solid iron of the field magnetic circuit, the magnetic flux density is not uniform throughout the whole field section during a change of the magnetic field, since the outer shell of the field iron is magnetized by the field coil only, while the central part of the iron is acted upon by the impressed m.m.f. of the field coil and the m.m.f. of the eddy currents in the outer part of the iron, and the change of magnetic flux density in the interior thus lags behind that of the outside of the iron. As result hereof the eddy currents in the different layers of the structure differ in intensity and in phase.

A complete investigation of the distribution of magnetism in this case leads to a transient phenomenon in space, and is discussed in "Transients in Space" [Dover Volume III, pp. 59–221]. For the present purpose, where the total m.m.f. of the eddy currents is small compared with that of the main field, we can approximate the effect of eddy currents in the iron by a closed circuit secondary conductor, that is, can assume uniform intensity and phase of secondary currents in an outer layer of the iron, that is, consider the outer layer of the iron, up to a certain depth, as a closed circuit secondary.

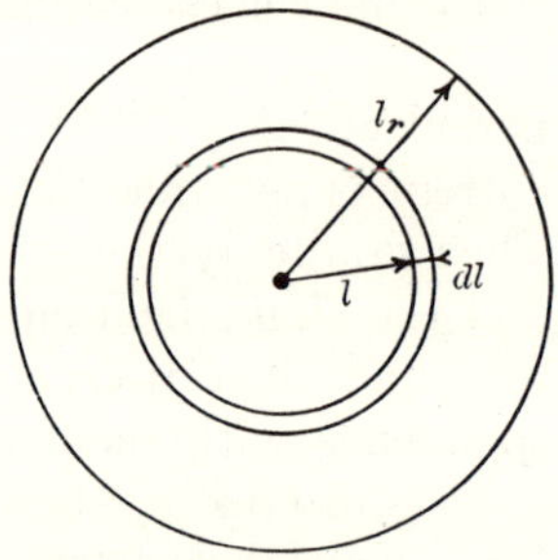

Fig. 40. Section of a magnetic circuit.

Let Fig. 40 represent a section of the magnetic circuit of the machine, and assume uniform flux density. If Φ = the total magnetic flux, l_r = the radius of the field section, then at a distance l from the center, the magnetic flux enclosed by a circle with radius l is $\left(\frac{l}{l_r}\right)^2 \Phi$, and the e.m.f. generated in the zone at distance l from the center is proportional to $\left(\frac{l}{l_r}\right)^2 \Phi$, that is, $e = a\left(\frac{l}{l_r}\right)^2 \Phi$. The current density of the eddy currents in this zone, which has the length $2\pi l$, is therefore proportional to $\frac{e}{2\pi l}$, or is $i = \frac{bl}{l_r^2} \Phi$. This current density acts as a m.m.f. upon the space enclosed by it, that is, upon $\left(\frac{l}{l_r}\right)^2$ of the total field section, and the magnetic reaction of the secondary current at distance l from the center therefore is proportional to $i\left(\frac{l}{l_r}\right)^2$, or is $\mathfrak{F} = \frac{cl^3}{l_r^4} \Phi$, and therefore the total magnetic reaction of the eddy currents is

$$i_0 = \int_0^{l_r} \mathfrak{F}\, dl = \frac{c}{4} \Phi.$$

At the outer periphery of the field iron, the generated e.m.f. is $e_1 = a\Phi$, the current density therefore $i_1 = \frac{b}{l_r} \Phi$, and the magnetic reaction $\mathfrak{F}_1 = \frac{c}{l_r} \Phi$, and therefore

$$i_0 = \frac{l_r}{4} \mathfrak{F}_1;$$

that is, the magnetic reaction of the eddy currents, assuming uniform flux density in the field poles, is the same as that of the currents produced in a closed circuit of a thickness $\frac{l_r}{4}$, or one-fourth the depth of the pole iron, of the material of the field pole and surrounding the field pole, that is, fully induced and fully magnetizing.

The eddy currents in the solid material of the field poles thus can be represented by a closed secondary circuit of depth $\frac{l_r}{4}$ surrounding the field poles.

The magnitude of the depth of the field copper on the spools is probably about one-fourth the depth of the field poles. Assuming then the width of the band of iron which represents the eddy current circuit as about twice the width of the field coils — since eddy currents are produced also in the yoke of the machine, etc. — and the conductivity of the iron as about 0.1 that of the field copper, the effective resistance of the eddy current circuit, reduced to the field circuit, approximates five times that of the field circuit.

Hence, if r_2 = resistance of main field winding, $r_1 = 5\,r_2$ = resistance of the secondary short circuit which represents the eddy currents.

Since the eddy currents extend beyond the space covered by the field coils, and considerably down into the iron, the self-inductance of the eddy current circuit is considerably greater than its mutual inductance with the main field circuit, and thus may be assumed as twice the latter.

91. As example, consider a 200-kw. series booster covering the range of voltage from 0 to 200, that is, giving a full load value of 1000 amperes at 200 volts. Making the assumptions set forth in the preceding paragraph, the following constants are taken: the armature resistance = 0.008 ohms and the series field winding resistance = 0.004 ohm; hence, the short circuit — or eddy current resistance — $r_1 = 0.02$ ohm. Furthermore let $M = 900 \times 10^{-6}$ henry = mutual inductance between main field and short-circuited secondary; hence, $x_m = 0.34$ ohm = mutual reactance, and therefore, assuming a leakage flux of the secondary equal to the main flux, $L_1 = 1800 \times 10^{-6}$ henry and $x_1 = 0.68$ ohm.

The booster is inserted into a constant potential circuit of 550 volts, so as to raise the voltage from 550 volts no load to 750 volts at 1000 amperes.

The total resistance of the circuit at full load, including main circuit and booster, therefore is $r = 0.75$ ohm.

The inductance of the external circuit may be assumed as $L = 4500 \times 10^{-6}$ henrys; hence, the reactance at $f = 60$ cycles per sec. is $x = 1.7$ ohms. The impressed e.m.f. of the circuit is $e = 550 + e'$, e' being the e.m.f. generated in the booster. Since at no load, for $i = 0$, $e' = 0$, and at full load, for $i = 1000$, $e' = 200$, assuming a straight line magnetic characteristic or saturation curve, that is, assuming the effect of magnetic saturation as negligible within the working range of the booster, we have

$$e = 550 + 0.2\,(i + i_1).$$

This gives the following constants:

	Main Circuit.	Eddy Current Circuit.
Current..............	i amp.	i_1 amp.
Impressed e.m.f......	$e = 550 + 0.2\,(i + i_1)$ volts.	0 volts.
Resistance...........	$r = 0.75$ ohm.	$r_1 = 0.02$ ohm.
Inductance..........	$L = 4500 \times 10^{-6}$ henrys.	$L_1 = 1800 \times 10^{-6}$ henrys.
Reactance...........	$x = 1.7$ ohms.	$x_1 = 0.68$ ohm.
Mutual inductance...	$M = 900 \times 10^{-6}$ henrys.	
Mutual reactance....	$x_m = 0.34$ ohm.	

This gives the differential equations of the problem as

$$550 - 0.55\,i + 0.2\,i_1 - 1.7\,\frac{di}{d\theta} - 0.34\,\frac{di_1}{d\theta} = 0 \tag{39}$$

and

$$0.02\,i_1 + 0.34\,\frac{di}{d\theta} + 0.68\,\frac{di_1}{d\theta} = 0. \tag{40}$$

Adding 2 times (39) to (40) gives

$$1100 - 1.1\,i + 0.42\,i_1 - 3.06\,\frac{di}{d\theta} = 0, \tag{41}$$

or

$$i_1 = 7.28\,\frac{di}{d\theta} + 2.62\,i - 2620, \tag{42}$$

herefrom:

$$0.02\,i_1 = 0.1456\,\frac{di}{d\theta} + 0.0524\,i - 52.4, \tag{43}$$

and

$$0.68\,\frac{di_1}{d\theta} = 4.95\,\frac{d^2i}{d\theta^2} + 1.78\,\frac{di}{d\theta}, \tag{44}$$

substituting the last two equations into (40),

$$\frac{d^2i}{d\theta^2} + 0.458\,\frac{di}{d\theta} + 0.0106\,i - 10.6 = 0. \tag{45}$$

If

$$i = i_0 + A\varepsilon^{-a\theta}, \tag{46}$$

then

$$A\varepsilon^{-a\theta}(a^2 - 0.458\,a + 0.0106) + 0.0106\,i_0 - 10.6 = 0.$$

As transient and permanent terms must each equal zero,

$$i_0 = 1000 \quad \text{and} \quad a^2 - 0.458\, a + 0.0106 = 0,$$

wherefrom $\qquad a = 0.229 \pm 0.205;$

the roots are $\qquad a_1 = 0.024 \quad \text{and} \quad a_2 = 0.434;$

then we have

$$i = 1000 + A_1 \varepsilon^{-0.024\,\theta} + A_2 \varepsilon^{-0.434\,\theta} \tag{47}$$

and

$$i_1 = 2.45\, A_1 \varepsilon^{-0.024\,\theta} - 0.55\, A_2 \varepsilon^{-0.434\,\theta}. \tag{48}$$

With terminal conditions $\theta = 0$, $i = 0$, and $i_1 = 0$,

$$A_1 = -183 \quad \text{and} \quad A_2 = -817.$$

If $\theta = 2\,\pi f t = 377.5$, we have

$$\left.\begin{aligned} i &= 1000 - 183\, \varepsilon^{-9.07t} - 817\, \varepsilon^{-164\,t}, \\ i_1 &= -450 \left\{ \varepsilon^{-9.07t} - \varepsilon^{-164\,t} \right\}, \\ \text{and} \qquad e &= 750 - 127\, \varepsilon^{-9.07t} - 73^{-164\,t}. \end{aligned}\right\} \tag{49}$$

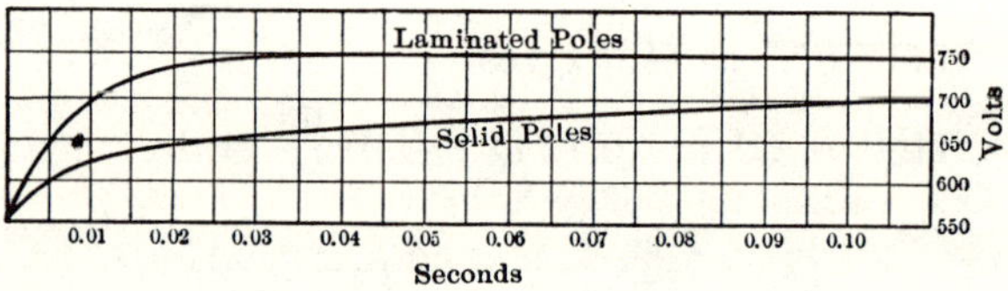

Fig. 41. Building up of feeder voltage by series booster.

In the absence of a secondary circuit, or with laminated field poles, equation (39) would assume the form $i_1 = 0$, or

$$550 + 0.2\, i = 0.75\, i + 1.7 \frac{di}{d\theta}; \tag{50}$$

hence, $\qquad \dfrac{di}{d\theta} = 0.323\,(1000 - i)$

and $\qquad i = 1000\,(1 - \varepsilon^{-0.323\,\theta});$

$$\left.\begin{aligned} \text{or} \qquad i &= 1000\,(1 - \varepsilon^{-122\,t}) \\ \text{and} \qquad e &= 750 - 200\, \varepsilon^{-122\,t}; \end{aligned}\right\} \tag{51}$$

that is, the e.m.f., e, approaches final conditions at a more rapid rate.

Fig. 41 shows the curves of the e.m.f., e, for the two conditions, namely, solid field poles, (49), and laminated field poles, (51).

(B) MUTUAL INDUCTANCE IN CIRCUITS CONTAINING SELF-INDUCTANCE AND CAPACITY.

92. The general eqations of such a pair of circuits, (1) and (2), differentiated to eliminate the integral give

$$\frac{de_1}{d\theta} = x_{c_1} i_1 + r_1 \frac{di_1}{d\theta} + x_1 \frac{d^2 i_1}{d\theta^2} + x_m \frac{d^2 i_2}{d\theta^2} \tag{52}$$

and

$$\frac{de_2}{d\theta} = x_{c_2} i_2 + r_2 \frac{di_2}{d\theta} + x_2 \frac{d^2 i_2}{d\theta^2} + x_m \frac{d^2 i_1}{d\theta^2}, \tag{53}$$

and the potential differences at the condensers, from (1) and (2), are

$$e_1' = x_{c_1} \int i_1 \, d\theta = e_1 - r_1 i_1 - x_1 \frac{di_1}{d\theta} - x_m \frac{di_2}{d\theta} \tag{54}$$

and

$$e_2' = x_{c_2} \int i_2 \, d\theta = e_2 - r_2 i_2 - x_2 \frac{di_2}{d\theta} - x_m \frac{di_1}{d\theta}. \tag{55}$$

If now the impressed e.m.fs., e_1 and e_2, contain no transient term, that is, if the transient values of currents i_1 and i_2 exert no appreciable reaction on the source of e.m.f., and if i_1' and i_2' are the permanent terms of current, then, substituting i_1' and i_2' in equations (52) and (53), and subtracting the result of this substitution from (52) and (53), gives the equations of the transient terms of the currents i_1 and i_2, thus:

$$0 = x_{c_1} i_1 + r_1 \frac{di_1}{d\theta} + x_1 \frac{d^2 i_1}{d\theta^2} + x_m \frac{d^2 i_2}{d\theta^2} \tag{56}$$

and

$$0 = x_{c_2} i_2 + r_2 \frac{di_2}{d\theta} + x_2 \frac{d^2 i_2}{d\theta^2} + x_m \frac{d^2 i_1}{d\theta^2}. \tag{57}$$

If the impressed e.m.fs., e_1 and e_2, are constant, $\frac{de_1}{d\theta}$ and $\frac{de_2}{d\theta}$

equal zero, and equations (52) and (53) assume the form (56) and (57); that is, equations (56) and (57) are the differential equations of the transient terms, for the general case of any e.m.fs., e_1 and e_2, which have no transient terms, and are the general differential equations of the case of constant impressed e.m.fs., e_1 and e_2.

From (56) it follows that

$$x_m \frac{d^2 i_2}{d\theta^2} = - x_{c_1} i_1 - r_1 \frac{di_1}{d\theta} - x_1 \frac{d^2 i_1}{d\theta^2} \cdot \tag{58}$$

Differentiating equation (57) twice, and substituting therein (58), gives

$$(x_1 x_2 - x_m{}^2) \frac{d^4 i}{d\theta^4} + (r_1 x_2 + r_2 x_1) \frac{d^3 i}{d\theta^3} + (x_{c_1} x_2 + x_{c_2} x_1 + r_1 r_2) \frac{d^2 i}{d\theta^2}$$
$$+ (x_{c_1} r_2 + x_{c_2} r_1) \frac{di}{d\theta} + x_{c_1} x_{c_2} i = 0. \tag{59}$$

This is a differential equation of fourth order, symmetrical in $r_1 x_1 x_{c_1}$ and $r_2 x_2 x_{c_2}$, which therefore applies to both currents, i_1 and i_2.

The expressions of the two currents i_1 and i_2 therefore differ only by their integration constants, as determined by the terminal conditions.

Equation (59) is integrated by

$$i = A \varepsilon^{-a\theta} \tag{60}$$

and substituting (60) in (59) gives for the determination of the exponent a the quartic equation

$$(x_1 x_2 - x_m{}^2)\, a^4 - (r_1 x_2 + r_2 x_1)\, a^3 + (x_{c_1} x_2 + x_{c_2} x_1 + r_1 r_2) a^2$$
$$- (x_{c_1} r_2 + x_{c_2} r_1)\, a + x_{c_1} x_{c_2} = 0,$$

or

$$a^4 - \frac{r_1 x_2 + r_2 x_1}{x_1 x_2 - x_m{}^2} a^3 + \frac{x_{c_1} x_2 + x_{c_2} x_1 + r_1 r_2}{x_1 x_2 - x_m{}^2} a^2$$
$$- \frac{x_{c_1} r_2 + x_{c_2} r_1}{x_1 x_2 - x_m{}^2} a + \frac{x_{c_1} x_{c_2}}{x_1 x_2 - x_m{}^2} = 0. \tag{61}$$

The solution of this quartic equation gives four values of a, and thus gives

$$i = A_1 \varepsilon^{-a_1 \theta} + A_2 \varepsilon^{-a_2 \theta} + A_3 \varepsilon^{-a_3 \theta} + A_4 \varepsilon^{-a_4 \theta}. \tag{62}$$

The roots, a, may be real, or two real and two imaginary, or all imaginary, and the solution of the equation by approximation therefore is difficult.

In the most important case, where the resistance, r, is small compared with the reactances x and x_c — and which is the only case where the transient terms are prominent in intensity and duration, and therefore of interest — as in the transformer and the induction coil or Ruhmkorff coil, the equation (61) can be solved by a simple approximation.

In this case, the roots, a, are two pairs of conjugate imaginary numbers, and the phenomenon oscillatory.

The real components of the roots, a, must be positive, since the exponential $\epsilon^{-a\theta}$ must decrease with increasing θ.

The four roots thus can be written:

$$\left.\begin{aligned} a_1 &= \alpha_1 - j\beta_1, \\ a_2 &= \alpha_1 + j\beta_1, \\ a_3 &= \alpha_2 - j\beta_2, \\ a_4 &= \alpha_2 + j\beta_2, \end{aligned}\right\} \qquad (63)$$

where α and β are positive numbers.

In the equation (61), the coefficients of a^3 and a are small, since they contain the resistances as factor, and this equation thus can be approximated by

$$a^4 + \frac{x_{c_1}x_2 + x_{c_2}x_1}{x_1x_2 - x_m^2}a^2 + \frac{x_{c_1}x_{c_2}}{x_1x_2 - x_m^2} = 0; \qquad (64)$$

hence,

$$a^2 = -\frac{1}{2}\left\{\frac{x_{c_1}x_2 + x_{c_2}x_1}{x_1x_2 - x_m^2} \pm \sqrt{\left(\frac{x_{c_1}x_2 + x_{c_2}x_1}{x_1x_2 - x_m^2}\right)^2 - \frac{4\,x_{c_1}x_{c_2}}{x_1x_2 - x_m^2}}\right\}; \qquad (65)$$

that is, a^2 is negative, having two roots,

$$b_1 = -\beta_1^2 \quad \text{and} \quad b_2 = -\beta_2^2.$$

This gives the four imaginary roots of a as first approximation:

$$\left.\begin{aligned} a &= \pm j\beta_1 \\ &\quad \pm j\beta_2. \end{aligned}\right\} \qquad (66)$$

If a_1, a_2, a_3, a_4 are the four roots of equation (61), this equation can be written

$$f(a) = (a - a_1)(a - a_2)(a - a_3)(a - a_4) = 0;$$

or, substituting (63),

$$f(a) = \{(a - \alpha_1)^2 + \beta_1^2\}\{(a - \alpha_2)^2 + \beta_2^2\} = 0, \quad (67)$$

and comparing (67) with (61) gives as coefficients of a^3 and of a,

$$\left.\begin{aligned} 2(\alpha_1 + \alpha_2) &= \frac{r_1 x_2 + r_2 x_1}{x_1 x_2 - x_m^2} \\ \text{and} \qquad & \\ 2(\alpha_1 \beta_2^2 + \alpha_2 \beta_1^2) &= \frac{x_{c_1} r_2 + x_{c_2} r_1}{x_1 x_2 - x_m^2}, \end{aligned}\right\} \quad (68)$$

and since β_1^2 and β_2^2 are given by (65) and (66) as roots of equation (64), $\alpha_1, \alpha_2, \beta_1, \beta_2$, and hereby the four roots a_1, a_2, a_3, a_4 of equation (61) are approximated by (64), (65), (66), (68).

The integration constants A_1, A_2, A_3, A_4 now follow from the terminal conditions.

93. As an example may be considered the operation of an inductorium, or Ruhmkorff coil, by make and break of a direct-current battery circuit, with a condenser shunting the break, in the usual manner.

Let $e_1 = 10$ volts = impressed e.m.f.; $r_1 = 0.4$ ohm = resistance of primary circuit, giving a current, at closed circuit and in stationary condition, of $i_0 = 25$ amp.; $r_2 = 0.2$ ohm = resistance of secondary circuit, reduced to the primary by the square of the ratio of primary ÷ secondary turns; $x_1 = 10$ ohms = primary inductive reactance; $x_2 = 10$ ohms = secondary inductive reactance, reduced to primary; $x_m = 8$ ohms = mutual inductive reactance; $x_{c_1} = 4000$ ohms = primary condensive reactance of the condenser shunting the break of the interrupter in the battery circuit, and $x_{c_2} = 6000$ ohms = secondary condensive reactance, due to the capacity of the terminals and the high tension winding.

Substituting these values, we have

$$\left.\begin{aligned} & e_1 = 10 \text{ volts} \qquad i_0 = 25 \text{ amp.} \\ r_1 = 0.4 \text{ ohm} \quad & x_1 = 10 \text{ ohms} \quad x_{c_1} = 4000 \text{ ohms} \\ r_2 = 0.2 \text{ ohm} \quad & x_2 = 10 \text{ ohms} \quad x_{c_2} = 6000 \text{ ohms} \\ & x_m = 8 \text{ ohms.} \end{aligned}\right\} \quad (69)$$

These values in equation (61) give

$$f(a) = a^4 - 0.167\, a^3 + 2780\, a^2 - 89\, a + 667{,}000 = 0, \tag{70}$$

and in equation (64) they give

$$f_1(a) = a^4 + 2780\, a^2 + 667{,}000 = 0$$

and $$a^2 = -(1390 \pm 1125)$$

$$= -2515,$$

or $$= -265;$$

hence, $$\beta_1 = 50.15$$

and $$\beta_2 = 16.28.$$

From (68) it follows that

$$\alpha_1 + \alpha_2 = 0.0833$$

and $$265\, \alpha_1 + 2515\, \alpha_2 = 44.5;$$

hence, $$\alpha_1 = 0.073,$$

$$\alpha_2 = 0.010.$$

Introducing for the exponentials with imaginary exponents the trigonometric functions give

$$\left.\begin{aligned} i_1 &= \varepsilon^{-0.073\,\theta}\{A_1 \cos 50.15\,\theta + A_2 \sin 50.15\,\theta\} \\ &\quad + \varepsilon^{-0.010\,\theta}\{B_1 \cos 16.28\,\theta + B_2 \sin 16.28\,\theta\} \\ i_2 &= \varepsilon^{-0.073\,\theta}\{C_1 \cos 50.15\,\theta + C_2 \sin 50.15\,\theta\} \\ &\quad + \varepsilon^{-0.010\,\theta}\{D_1 \cos 16.28\,\theta + D_2 \sin 16.28\,\theta\}, \end{aligned}\right\} \tag{71}$$

where the constants C and D depend upon A and B by equations (56), (57), or (58), thus:

Substituting (71) into (58),

$$8\frac{d^2 i_2}{d\theta^2} + 4000\, i_1 + 0.4\frac{di_1}{d\theta} + 10\frac{d^2 i_1}{d\theta^2} = 0 \tag{58}$$

gives an identity, from which, by equating the coefficients of $\varepsilon^{-a\theta} \cos b\theta$ and $\varepsilon^{-a\theta} \sin b\theta$ to zero, result four equations, in the coefficients

$$A_1\ B_1\ C_1\ D_1$$

$$A_2\ B_2\ C_2\ D_2,$$

from which follows, with sufficient approximation,

$$\left.\begin{aligned} A_1 &= -0.95\,C_1 \\ A_2 &= -0.98\,C_2 \\ B_1 &= +1.57\,D_1 \\ B_2 &= +1.57\,D_2; \end{aligned}\right\} \qquad (72)$$

hence,

$$\begin{aligned} i_1 = &-0.96\,\varepsilon^{-0.073\,\theta}\{C_1 \cos 50.15\,\theta + C_2 \sin 50.15\,\theta\} \\ &+ 1.57\,\varepsilon^{-0.010\,\theta}\{D_1 \cos 16.28\,\theta + D_2 \sin 16.28\,\theta\} \end{aligned} \qquad (73)$$

and substituting (71) and (73) in the equations of the condenser potential, (54) and (55), gives

$$\left.\begin{aligned} e_1' = {} & 10 + 79\,\varepsilon^{-0.073\,\theta}\{C_2 \cos 50.15\,\theta - C_1 \sin 50.15\,\theta\} \\ & - 385\,\varepsilon^{-0.010\,\theta}\{D_2 \cos 16.28\,\theta - D_1 \sin 16.28\,\theta\} \\ e_2' = {} & 118\,\varepsilon^{-0.073\,\theta}\{C_2 \cos 50.15\,\theta - C_1 \sin 50.15\,\theta\} \\ & + 367\,\varepsilon^{-0.010\,\theta}\{D_2 \cos 16.28\,\theta - D_1 \sin 16.28\,\theta\} \end{aligned}\right\} \qquad (74)$$

94. Substituting now the terminal conditions of the circuit:

At the moment where the interrupter opens the primary circuit the current in this circuit is $i_1 = \frac{e_0}{r_1} = 25$ amp. The condenser in the primary circuit, which is shunted across the break, was short-circuited before the break, hence of zero potential difference. The secondary circuit was dead. This then gives the conditions $\theta = 0$; $i_1 = 25$, $i_2 = 0$, $e_1' = 0$, and $e_2' = 0$.

Substituting these values in equations (71), (73), (74) gives

$$\left.\begin{aligned} 25 &= -0.95\,C_1 + 1.58\,D_1 \\ 0 &= C_1 + D_1 \\ 0 &= 10 + 79\,C_2 - 385\,D_2 \\ 0 &= 118\,C_2 + 367\,D_2 \end{aligned}\right\}$$

hence

$$\begin{aligned} C_1 &= -10 \\ C_2 &= -0.05 \cong 0 \\ D_1 &= +10 \\ D_2 &= +0.016 \cong 0, \end{aligned}$$

and

$$\left.\begin{aligned} i_1 &= 9.6\ \varepsilon^{-0.073\,\theta} \cos 50.15\,\theta + 15.7\ \varepsilon^{-0.010\,\theta} \cos 16.28\,\theta \\ i_2 &= -10\ \varepsilon^{-0.073\,\theta} \cos 50.15\,\theta + 10\ \varepsilon^{-0.010\,\theta} \cos 16.28\,\theta \\ e_1' &= 10 + 790\ \varepsilon^{-0.073\,\theta} \sin 50.15\,\theta + 3850\ \varepsilon^{-0.010\,\theta} \sin 16.28\,\theta \\ e_2' &= 1180\ \varepsilon^{-0.073\,\theta} \sin 50.15\,\theta - 3670\ \varepsilon^{-0.010\,\theta} \sin 16.28\,\theta \end{aligned}\right\} \quad (75)$$

Approximately therefore we have

$$\begin{aligned} i_1 &= 9.6\ \varepsilon^{-0.073\,\theta} \cos 50.15\,\theta + 15.7\ \varepsilon^{-0.010\,\theta} \cos 16.28\,\theta \\ i_2 &= -10\left\{\varepsilon^{-0.073\,\theta} \cos 50.15\,\theta - \varepsilon^{-0.010\,\theta} \cos 16.28\,\theta\right\} \\ e_1' &= 3850\ \varepsilon^{-0.010\,\theta} \sin 16.28\,\theta \\ e_2' &= -3670\ \varepsilon^{-0.010\,\theta} \sin 16.28\,\theta. \end{aligned}$$

The two frequencies of oscillation are 3009 and 977 cycles per sec., hence rather low.

The secondary terminal voltage has a maximum of nearly 4000, reduced to the primary, or 400 times as large as corresponds to the ratio of turns.

In this particular instance, the frequency 3009 is nearly suppressed, and the main oscillation is of the frequency 977.

CHAPTER XI.

GENERAL SYSTEM OF CIRCUITS

(A) CIRCUITS CONTAINING RESISTANCE AND INDUCTANCE ONLY.

95. Let, upon a general system or network of circuits connected with each other directly or inductively, and containing resistance and inductance, but no capacity, a system of e.m.fs., e_γ be impressed. These e.m.fs. may be of any frequency or wave shape, or may be continuous or anything else, but are supposed to be given by their equations. They may be free of transient terms, or may contain transient terms depending upon the currents in the system. In the latter case, the dependency of the e.m.f. upon the currents must obviously be given.

Then, in each branch circuit,

$$e - ri - L\frac{di}{dt} - \sum^{\delta} M_\delta \frac{di_\delta}{dt} = 0, \tag{1}$$

where e = total impressed e.m.f.; r = resistance; L = inductance, of the circuit or branch of circuit traversed by current i, and M_δ = mutual inductance of this circuit with any circuit in inductive relation thereto and traversed by current i_δ.

The currents in the different branch circuits of the system depend upon each other by Kirchhoff's law,

$$\sum i = 0 \tag{2}$$

at every branching point of the system.

By equation (2) many of the currents can be eliminated by expressing them in terms of the other currents, but a certain number of independent currents are left.

Let n = the number of independent currents, denoting these currents by i_κ, where $\kappa = 1, 2, \ldots n$. (3)

Usually, from physical considerations, the number of independent currents of the system, n, can immediately be given.

For these n currents i_κ, n independent differential equations of form (1) can be written down, between the impressed e.m.fs. e_γ or their combinations, and currents which are expressed by the n independent currents i_κ. They are given by applying equation (1) to a closed circuit or ring in the system.

These equations are of the form

$$\left.\begin{aligned} e_q - \sum_1^n {}^\kappa\, b_\kappa^{\,q}\, i_\kappa - \sum_1^n {}^\kappa\, c_\kappa^{\,q} \frac{di_\kappa}{dt} = 0, \\ \text{where} \qquad q = 1, 2, \ldots n, \end{aligned}\right\} \tag{4}$$

where the n^2 coefficients $b_\kappa^{\,q}$ are of the dimension of *resistance* and the n^2 coefficients $c_\kappa^{\,q}$ of the dimension of *inductance*. (5)

These n simultaneous differential equations of n variables i_κ are integrated by the equations

$$i_\kappa = i_\kappa' + \sum_1^m {}^i\, A_i^\kappa \varepsilon^{-a_i t}, \tag{6}$$

where i_κ' is the stationary value of current i_κ, reached for $t = \infty$.

Substituting (6) in (4) gives

$$e_q - \sum_1^n {}^\kappa\, b_\kappa^{\,q}\, i_\kappa' - \sum_1^n {}^\kappa\, c_\kappa^{\,q} \frac{di_\kappa'}{dt} - \sum_1^n {}^\kappa\, b_\kappa^{\,q} \sum_1^m {}^i\, A_i^\kappa \varepsilon^{-a_i t} + \sum_1^n {}^\kappa\, c_\kappa^{\,q} \sum_1^m {}^i\, a_i A_i^\kappa \varepsilon^{-a_i t} = 0. \tag{7}$$

For $t = \infty$, this equation becomes

$$e_q - \sum_1^n {}^\kappa\, b_\kappa^{\,q}\, i_\kappa' - \sum_1^n {}^\kappa\, c_\kappa^{\,q} \frac{di_\kappa'}{dt} = 0. \tag{8}$$

These n equations (8) determine the stationary components of the n currents, i_κ'.

Subtracting (8) from (7) gives, for the transient components of currents i_κ,

$$i_\kappa'' = \sum_1^m {}^i\, A_i^\kappa \varepsilon^{-a_i t}, \tag{9}$$

the n equations

$$\sum_1^n {}^\kappa\, b_\kappa^{\,q} \sum_1^m {}^i\, A_i^\kappa \varepsilon^{-a_i t} - \sum_1^n {}^\kappa\, c_\kappa^{\,q} \sum_1^m {}^i\, a_i A_i^\kappa \varepsilon^{-a_i t} = 0. \tag{10}$$

Reversing the order of summation in (10) gives

$$\sum_1^m{}^i \varepsilon^{-a_i t} \sum_1^n{}^\kappa A_i^\kappa (b_\kappa^q - a_i c_\kappa^q) = 0. \tag{11}$$

The n equations (11) must be identities, that is, the coefficients of $\varepsilon^{-a_i t}$ must individually disappear. Each equation (11) thus gives m equations between the constants a, A, b, c, for $i = 1, 2, \ldots m$, and since n equations (11) exist, we get altogether mn equations of the form

$$\left.\begin{aligned} &\sum_1^n{}^\kappa A_i^\kappa (b_\kappa^q - a_i c_\kappa^q) = 0, \\ &\text{where} \\ &q = 1, 2, 3, \ldots n \quad \text{and} \quad i = 1, 2, 3, \ldots m. \end{aligned}\right\} \tag{12}$$

In addition hereto, the n terminal conditions, or values of current i_κ'' for $t = 0$: i_κ^0, give by substitution in (9) n further equations,

$$i_\kappa^0 = \sum_1^m{}^i A_i^\kappa. \tag{13}$$

There thus exist $(mn + n)$ equations for the determination of the mn constants A_i^κ and the m constants a_i, or altogether $(mn + m)$ constants. That is,

$$m = n \tag{14}$$

and

$$i_\kappa = i_\kappa' + \sum_1^n{}^i A_i^\kappa \varepsilon^{-a_i t}, \tag{15}$$

where

$$\sum_1^n{}^\kappa A_i^\kappa (b_\kappa^q - a_i c_\kappa^q) = 0; \tag{16}$$

$$\sum_1^n{}^i A_i^\kappa = i_\kappa^0; \tag{17}$$

and

$$\left.\begin{aligned} q &= 1, 2, \ldots n, \\ \kappa &= 1, 2, \ldots n, \\ i &= 1, 2, \ldots n. \end{aligned}\right\} \tag{18}$$

Each of the n sets of n linear homogeneous equations in A_i^κ (16) which contains the same index i gives by elimination of A_i^κ the same determinant:

$$\|b_\kappa^q - a_i c_\kappa^q\| = \begin{vmatrix} b_1^1 - a_i c_1^1, & b_1^2 - a_i c_1^2, & b_1^3 - a_i c_1^3 & \ldots & b_1^n - a_i c_1^n \\ b_2^1 - a_i c_2^1, & b_2^2 - a_i c_2^2, & b_2^3 - a_i c_2^3 & \ldots & b_2^n - a_i c_2^n \\ b_3^1 - a_i c_3^1, & b_3^2 - a_i c_3^2, & b_3^3 - a_i c_3^3 & \ldots & b_3^n - a_i c_3^n \\ \cdot & \cdot & \cdot & \cdots & \cdot \\ b_n^1 - a_i c_n^1, & b_n^2 - a_i c_n^2, & b_n^3 - a_i c_n^3 & \ldots & b_n^n - a_i c_n^n \end{vmatrix} = 0. \quad (19)$$

Thus the n values of a_i are the n roots of the equation of nth degree (19), and determined by solving this equation.

Substituting these n values of a_i in the equations (16) gives n^2 linear homogeneous equations in A_i^κ, of which $n\ (n - 1)$ are independent equations, and these $n\ (n - 1)$ independent equations together with the n equations (17) give the n^2 linear equations required for the determination of the n^2 constants A_i^κ.

The problem of determining the equations of the phenomena in starting, or in any other way changing the circuit conditions, in a general system containing only resistance and inductance, with n independent currents and such impressed e.m.fs., e_γ, that the equations of stationary condition,

$$i_\kappa' = f_\kappa\ (t),$$

can be solved, still depends upon the solution of an equation of nth degree, in the exponents a_i of the exponential functions which represent the transient term.

96. As an example of the application of this method may be considered the following case, sketched diagrammatically in Fig. 42:

An alternator of e.m.f. $E \cos\ (\theta - \theta_0)$ feeds over resistance r_1 the primary of a transformer of mutual reactance x_m. The secondary of this transformer feeds over resistances r_2 and r_3 the primary of a second transformer of mutual reactance x_{m_0}, and the secondary of this second transformer is closed by resistance r_4. Across the circuit between the two transformers and the two resistances r_2 and r_3, is connected a continuous-current e.m.f., e_0, as a battery, in series with an inductive reactance x. The transformers obviously must be such as not to be saturated magnetically by the component of continuous current which traverses them, must for instance be open core transformers.

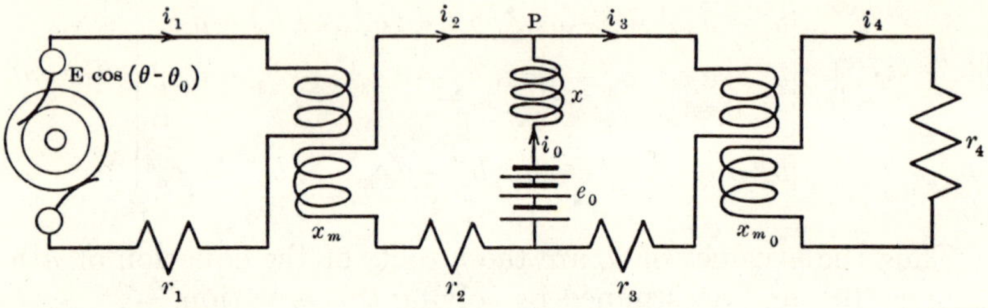

Fig. 42. Alternating-current circuit containing mutual and self-inductive reactance, resistance and continuous e.m.f.

Let i_1, i_2, i_0, i_3, i_4 = currents in the different circuits; then, at the dividing point P, by equation (2) we have

$$i_0 + i_2 - i_3 = 0;$$

hence,

$$i_0 = i_3 - i_2, \tag{20}$$

leaving four independent currents i_1, i_2, i_3, i_4.

This gives four equations (4):

$$\left.\begin{aligned} E\cos(\theta - \theta_0) - r_1 i_1 - x_m \frac{di_2}{d\theta} &= 0, \\ -e_0 - r_2 i_2 - x_m \frac{di_1}{d\theta} + x\left(\frac{di_3}{d\theta} - \frac{di_2}{d\theta}\right) &= 0, \\ e_0 - r_3 i_3 - x_{m0}\frac{di_4}{d\theta} - x\left(\frac{di_3}{d\theta} - \frac{di_2}{d\theta}\right) &= 0, \\ \text{and} \qquad - r_4 i_4 - x_{m0}\frac{di_3}{d\theta} &= 0. \end{aligned}\right\} \tag{21}$$

If now i_1', i_2', i'_3, i_4' are the permanent terms of current, by substituting these into (21) and subtraction, the equations of the transient terms rearranged are:

$$\left.\begin{array}{llllll}
q: & \kappa = 1 & 2 & 3 & 4 & \\
1 & r_1 i_1 & + x_m \dfrac{di_2}{d\theta} & & & = 0, \\
2 & x_m \dfrac{di_1}{d\theta} & + r_2 i_2 + x \dfrac{di_2}{d\theta} & - x \dfrac{di_3}{d\theta} & & = 0, \\
3 & & - x \dfrac{di_2}{d\theta} & + r_3 i_3 + x \dfrac{di_3}{d\theta} & + x_m \dfrac{di_4}{d\theta} & = 0, \\
4 & & & x_m \dfrac{di_3}{d\theta} & + r_4 i_4 & = 0.
\end{array}\right\} \quad (22)$$

These equations integrated by

$$i_\kappa'' = \sum_1^4 {}^i A_i^\kappa \varepsilon^{-\alpha_i \theta} \quad (23)$$

give for the determination of the exponents a_i the determinant (19):

$$\begin{vmatrix} r_1 & -ax_m & 0 & 0 \\ -ax_m & r_2 - ax & ax & 0 \\ 0 & ax & r_3 - ax & -ax_{m_0} \\ 0 & 0 & -ax_{m_0} & r_4 \end{vmatrix} = 0; \quad (24)$$

or, resolved,

$$f = a^4 x_m^2 x_{m_0}^2 + a^3 x (x_m^2 r_4 + x_{m_0}^2 r_1) - a^2 (x_m^2 r_3 r_4 + x_{m_0}^2 r_1 r_2) - axr_1 r_4 (r_2 + r_3) + r_1 r_2 r_3 r_4 = 0. \quad (25)$$

Assuming now the numerical values,

$$\left.\begin{array}{ll}
r_1 = 1 & x_m = 10 \\
r_2 = 1 & x_{m_0} = 10 \\
r_3 = 1 & x = 100 \\
r_4 = 10 &
\end{array}\right\} \quad (26)$$

equation (25) gives

$$f = a^4 + 11\, a^3 - 0.11\, a^2 - 0.2\, a + 0.001 = 0. \quad (27)$$

The sixteen coefficients,

$$A_i^\kappa, \quad i = 1, 2, 3, 4, \quad k = 1, 2, 3, 4,$$

are now determined by the 16 independent linear equations (12) and (13).

(B) CIRCUITS CONTAINING RESISTANCE, SELF-INDUCTANCE, MUTUAL INDUCTANCE AND CAPACITY.

97. The general method of dealing with such a system is the same as in (A).

Kirchhoff's equation (1) is of the form

$$e - ri - L\frac{di}{dt} - \sum^{\delta} M_{\delta}\frac{di_{\delta}}{dt} - \frac{1}{C}\int i\,dt = 0. \tag{28}$$

Eliminating now all the currents which can be expressed in terms of other currents, by means of equation (2), leaves n independent currents:

$$i_{\kappa}, \quad \kappa = 1, 2, \ldots n.$$

Substituting these currents i_{κ} in equations (28) gives n independent equations of the form

$$e_q - \sum_1^n{}^{\kappa}\, b_{\kappa}^{q} i_{\kappa} - \sum_1^n{}^{\kappa}\, c_{\kappa}^{q}\frac{di_{\kappa}}{dt} - \sum_1^n{}^{\kappa}\, g_{\kappa}^{q}\int i_{\kappa}\,dt = 0. \tag{29}$$

Resolving these equations for $\int i_{\kappa}\,dt$ gives

$$e_{\kappa}' = \frac{1}{C_{\kappa}}\int i_{\kappa}\,dt = \sum e + \sum bi + \sum c\,\frac{di}{dt} \tag{30}$$

as the equations of the potential differences at the condensers.

Differentiating (29) gives

$$\frac{de_q}{dt} - \sum_1^n{}^{\kappa}\, g_{\kappa}^{q}\, i_{\kappa} - \sum_1^n{}^{\kappa}\, b_{\kappa}^{q}\frac{di_{\kappa}}{dt} - \sum_1^n{}^{\kappa}\, c_{\kappa}^{q}\frac{d^2 i_{\kappa}}{dt^2} = 0, \tag{31}$$

where $$q = 1, 2, \ldots n.$$

By the same reasoning as before, the solution of these equations (31) can be split into two components, a permanent term,

$$i_{\kappa}' = f(t), \tag{32}$$

and a transient term, which disappears for $t = \infty$, and is given by the n simultaneous differential equations of second order, thus:

$$\sum_1^n{}^{\kappa}\left\{ g_{\kappa}^{q} i_{\kappa} + b_{\kappa}^{q}\frac{di_{\kappa}}{dt} + c_{\kappa}^{q}\frac{d^2 i_{\kappa}}{dt^2}\right\} = 0. \tag{33}$$

These equations are integrated by

$$i_\kappa = \sum_1^m {}^i A_i^\kappa \varepsilon^{-a_i t}. \tag{34}$$

Substituting (34) in (33) gives

$$\sum_1^n {}^\kappa \sum_1^m {}^i A_i^\kappa \varepsilon^{-a_i t} \left\{ g_\kappa^q - a_i b_\kappa^q + a_i^2 c_\kappa^q \right\}, \tag{35}$$

$$\left.\begin{aligned} \text{where} \qquad q &= 1, 2, \ldots n, \\ \kappa &= 1, 2, \ldots n, \\ \text{and} \qquad i &= 1, 2, \ldots m. \end{aligned}\right\} \tag{36}$$

Reversing in these n equations the order of summation,

$$\sum_1^m {}^i \varepsilon^{-a_i t} \sum_1^n {}^\kappa A_i^\kappa \left\{ g_\kappa^q - a_i b_\kappa^q + a_i^2 c_\kappa^q \right\} = 0, \tag{37}$$

and this gives, as identity, the mn equations for the determination of the constants:

$$\left.\begin{aligned} &\sum_1^n {}^\kappa A_i^\kappa \left\{ g_\kappa^q - a_i b_\kappa^q + a_i^2 c_\kappa^q \right\} = 0, \\ \text{where} \\ &q = 1, 2, \ldots n \quad \text{and} \quad i = 1, 2, \ldots m. \end{aligned}\right\} \tag{38}$$

In addition to these mn equations (38), two sets of terminal conditions exist, depending respectively on the instantaneous current and the instantaneous condenser potential at the moment of start.

The current is

$$i_\kappa = i_\kappa' + \sum_1^m {}^i A_i^\kappa \varepsilon^{-a_i t} \tag{39}$$

and the condenser potential of the circuit q is

$$e_q' = \sum_1^n {}^\kappa g_\kappa^q \int i_\kappa \, dt = e_q - \sum_1^n {}^\kappa b_\kappa^q i_\kappa - \sum_1^n {}^\kappa c_\kappa^q \frac{di_\kappa}{dt}; \tag{40}$$

hence, for $t = 0$,

$$i_\kappa^0 = i_\kappa' + \sum_1^m {}^i A_i^\kappa, \tag{41}$$

where $\kappa = 1, 2, \ldots n,$

and
$$e_q^{\,0} = e_q - \sum_1^n{}^{\kappa}\, b_\kappa^{\,q} i_\kappa^{\,0} - \sum_1^n{}^{\kappa}\, c_\kappa^{\,q} \frac{di_\kappa^{\,0}}{dt}, \tag{42}$$

where, $q = 1, 2 \ldots n;$

or, substituting (39) in (40), and then putting $t = 0$,

$$e_q^{\,0} = \left[e_q - \sum_1^n{}^{\kappa} \left\{ b_\kappa^{\,q} i_\kappa' - c_\kappa^{\,q} \frac{di_\kappa'}{dt} \right\} \right]_{t=0} - \sum_1^n{}^{\kappa} \sum_1^m{}^{i}\, A_i^{\kappa} (b_\kappa^{\,q} - a_i c_\kappa^{\,q}). \tag{43}$$

As seen, in (41) and (43), the first term is the instantaneous value of the permanent current i'_κ and condenser potential e_q'.

These two sets of n equations each, given by the terminal conditions of the current, $i'_\kappa = i_\kappa^{\,0}$ (42), and condenser potential, $e_q' = e_q^{\,0}$ (43), together with the mn equations (38), give a total of $(mn + 2\,n)$ equations for the determination of the mn constants A_i^{κ} and the m constants a_i, that is, a total of $(mn + m)$ constants.

From
$$mn + 2\,n = mn + m$$

it follows that
$$m = 2\,n. \tag{44}$$

We have, then, $2\,n$ constants, a_i, giving the coefficients in the exponents of the $2\,n$ exponential transient terms, and $2\,n^2$ coefficients, A_i^{κ}, and for their determination $2\,n^2$ equations,

$$\sum_1^n{}^{\kappa}\, A_i^{\kappa}\, (g_\kappa^{\,q} - a_i b_\kappa^{\,q} + a_i^2 c_\kappa^{\,q}) = 0, \tag{45}$$

n equations,

$$\sum_1^{2n}{}^{i}\, A_i^{\kappa} = i_\kappa^{\,0}, \tag{46}$$

and n equations,

$$\sum_1^n{}^{\kappa} \sum_1^{2n}{}^{i}\, A_i^{\kappa}\, (b_\kappa^{\,q} - a_i c_\kappa^{\,q}) = k_q^{\,0}, \tag{47}$$

where $$k_q^0 = \left[e_q - \sum_1^n{}^\kappa g_\kappa^q \int i_\kappa \, dt \right]_{t=o}; \tag{48}$$

or the difference between the condenser potential required by the permanent term and the actual condenser potential at time $t = 0$, where

$$\left.\begin{aligned} q &= 1, 2, 3, \ldots n, \\ \kappa &= 1, 2, 3, \ldots n, \\ \text{and} \qquad i &= 1, 2, 3, \ldots 2\,n. \end{aligned}\right\} \tag{49}$$

Eliminating A_i^κ from the equations (45) gives for each of the $2\,n$ sets of n equations which have the same a_i the determinant:

$$\|g_\kappa^q - a_i b_\kappa^q + a_i^2 c_\kappa^q\| =$$

$$\begin{vmatrix} g_1^1 - a_i b_1^1 + a_i^2 c_1^1, & g_1^2 - a_i b_1^2 + a_i^2 c_1^2, & \ldots & g_1^n - a_i b_1^n + a_i^2 c_1^n \\ g_2^1 - a_i b_2^1 + a_i^2 c_2^1, & g_2^2 - a_i b_2^2 + a_i^2 c_2^2, & \ldots & g_2^n - a_i b_2^n + a_i^2 c_2^n \\ g_3^1 - a_i b_3^1 + a_i^2 c_3^1, & g_3^2 - a_i b_3^2 + a_i^2 c_3^2, & \ldots & g_3^n - a_i b_3^n + a_i^2 c_3^n \\ \cdot & \cdot & \cdot & \cdot \\ \cdot & \cdot & \cdot & \cdot \\ g_n^1 - a_i b_n^1 + a_i^2 c_n^1, & g_n^2 - a_i b_n^2 + a_i^2 c_n^2, & \ldots & g_n^n - a_i b_n^n + a_i^2 c_n^n \end{vmatrix} = 0. \tag{50}$$

The $2\,n$ values of a_i thus are the roots of an equation of $2\,n$th order.

Substituting these values of a_i in equations (45), (46), (47), leaves $2\,n\,(n - 1)$ independent equations (45) and $2\,n$ independent equations (46) and (47), or a total of $2\,n^2$ linear equations, for the determination of the $2\,n^2$ constants A_i^κ, which now can easily be solved.

The roots of equation (50) may either be real or may be complex imaginary, and in the latter case each pair of conjugate roots gives by elimination of the imaginary form an electric oscillation.

That is, the solution of the problem of n independent circuits leads to n transient terms, each of which may be either an oscillation or a pair of exponential functions.

98. The preceding discussion gives the general method of the determination of the transient phenomena occurring in any system or net work of circuits containing resistances, self-inductances and mutual inductances and capacities, and impressed and counter e.m.fs. of any frequency or wave shape, alternating or continuous.

It presupposes, however,

(1) That the solution of the system for the permanent terms of currents and e.m.fs. is given.

(2) That, if the impressed e.m.fs. contain transient terms depending upon the currents in the system, these transient terms of impressed or counter e.m.fs. are given as linear functions of the currents or of their differential coefficients, that is, the rate of change of the currents.

(3) That resistance, inductance, and capacity are constant quantities, and for instance magnetic saturation does not appear.

The determination of the transient terms requires the solution of an equation of 2 nth degree, which is lowered by one degree for every independent circuit which contains no capacity.

Thus, for instance, a divided circuit having capacity in either branch leads to a quartic equation. A transmission line loaded with inductive or non-inductive load, when representing the capacity of the line by a condenser shunted across its middle, leads to a cubic equation.

CHAPTER XII.

MAGNETIC SATURATION AND HYSTERESIS IN ALTERNATING-CURRENT CIRCUITS

99. If an alternating e.m.f. is impressed upon a circuit containing resistance and inductance, the current and thereby the magnetic flux produced by the current assume their final or permanent values immediately only in case the circuit is closed at that point of the e.m.f. wave at which the permanent current is zero. Closing the circuit at any other point of the e.m.f. wave produces a transient term of current and of magnetic flux. So for instance, if the circuit is closed when the current i should have its negative maximum value $-I_0$, and therefore the magnetic flux and the magnetic flux density also be at their negative maximum value $-\Phi_0$ and $-\mathfrak{B}_0$ — that is, in an inductive circuit, near the zero value of the decreasing e.m.f. wave — during the first half wave of e.m.f. the magnetic flux, which generates the counter e.m.f., should vary from $-\Phi_0$ to $+\Phi_0$, or by $2\,\Phi_0$; hence, starting with 0, to generate the same counter e.m.f., it must rise to $+2\,\Phi_0$, that is, twice its permanent value, and so the current i also rises, at constant inductance L, from zero to twice its maximum permanent value, $2\,I_0$. Since the e.m.f. consumed by the resistance during the variation from 0 to $2\,I_0$ is greater than during the normal variation from $-I_0$ to $+I_0$, less e.m.f. is to be generated by the change of magnetic flux, that is, the magnetic flux does not quite rise to $2\,\Phi_0$, but remains below this value the more, the higher the resistance of the circuit. During the next half wave the e.m.f. has reversed, but the current is still mostly in the previous direction, and the generated e.m.f. thus must give the resistance drop, that is, the total variation of magnetic flux must be greater than $2\,\Phi_0$, the more, the higher the resistance. That is, starting at a value somewhat below $2\,\Phi_0$, it decreases below zero, and reaches a negative value. During the third half wave the magnetic flux, starting not at zero as in the first half wave, but at a negative value, thus reaches a lower positive maximum, and thus gradually, at a rate depending upon the resistance of the circuit, the waves of magnetic flux Φ, and thereby current i, approach their final permanent or symmetrical cycles.

100. In the preceding, the assumption has been made that the magnetic flux, Φ, or the flux density, $\mathfrak{B}$, is proportional to the current, or in other words, that the inductance, L, is constant. If the magnetic circuit interlinked with the electric circuit contains iron, and especially if it is an iron-clad or closed magnetic circuit, as that of a transformer, the current is not proportional to the magnetic flux or magnetic flux density, but increases for high values of flux density more than proportional, that is, the flux density in the iron reaches a finite limiting value. In the case illustrated above, the current corresponding to double the normal maximum magnetic flux, Φ_0, or flux density, $\mathfrak{B}_0$, may be many times greater than twice the normal maximum current, I_0. For instance, if the maximum permanent current is $I_0 = 4.5$ amperes, the maximum permanent flux density, $\mathfrak{B}_0 = 10{,}000$, and the circuit closed, as above, at that point of the e.m.f. wave where the flux density should have its negative maximum, $-\mathfrak{B}_0 = -10{,}000$, but the actual flux density is 0, during the first half wave of e.m.f., the flux density, when neglecting the resistance of the electric circuit, should rise from 0 to $2\mathfrak{B}_0 = 20{,}000$, and at this high value of saturation the corresponding current maximum would be, by the magnetic cycle, Fig. 43, 200 amperes, that is, not twice but 44.5 times the normal value. With such excessive values of current, the e.m.f. consumed by resistance would be in general considerable, and the e.m.f. consumed by inductance, and therefore the variation of magnetic flux density, considerably decreased, that is, the maximum magnetic flux density would not rise to 20,000, but remain considerably below this value. The maximum current, however, would be still very much greater than twice the normal maximum. That is, in an iron-clad circuit, in starting, the transient term of current may rise to values relatively very much higher than in air magnetic circuits. While in the latter it is limited to twice the normal value, in the iron-clad circuit, if the magnetic flux density reaches into the range of magnetic saturation, very much higher values of transient current are found. Due to the far greater effect of the resistance with such excessive values of current, the transient term of current during the first half waves decreases at a more rapid rate; due to the lack of proportionality between current and magnetic flux density, the transient term does not follow the exponential law any more.

101. In an iron-clad magnetic circuit, the current is not only not proportional to the magnetic flux density, but the same magnetic flux density can be produced by different currents, or with the same current the flux density can have very different values, depending on the point of the hysteresis cycle. Therefore the magnetic flux density for zero current may equal zero, or, on the decreasing branch of the hysteresis cycle, Fig. 43, may be $+ 7600$, or, on the increasing branch, $- 7600$. Thus, when closing the electric circuit energizing an iron-clad magnetic circuit, as a transformer, at the moment of zero current, the magnetic flux density may not be zero, but may still have a high value, as remanent magnetism. For instance, closing the circuit at the point of the e.m.f. wave where the permanent wave of magnetic flux density would have its negative maximum value, $- \mathfrak{B}_0 = - 10{,}000$, the actual density at this moment may be $\mathfrak{B}_r = + 7600$, the remanent magnetism of the cycle. During the first half wave of impressed e.m.f. the variation of flux density by $2\,\mathfrak{B}_0$, as required to generate the counter e.m.f., when neglecting the resistance, would bring the positive maximum of flux density up to $\mathfrak{B}_r + 2\,\mathfrak{B}_0 = 27{,}600$, requiring 1880 amperes maximum current, or 420 times the normal current. Obviously, no such rise could occur, since the resistance of the circuit would consume a considerable part of the e.m.f., and so lower the flux density by reducing the e.m.f. consumed by inductance.

It is obvious, however, that excessive values of transient current may occur in transformers and other iron-clad magnetic circuits.

102. When disconnecting a transformer, its current becomes zero, that is, the magnetic flux density is left at the value of the remanent magnetism $\pm \mathfrak{B}_r$, and during the period of rest more or less decreases spontaneously towards zero. Hence, in connecting a transformer into circuit its flux density may be anywhere between $+ \mathfrak{B}_r$ and $- \mathfrak{B}_r$. The maximum magnetic flux density during the first half cycle of impressed e.m.f. therefore is produced if the circuit is closed at the moment where the permanent value of the flux density should be a maximum, $\pm \mathcal{B}_0$, and the actual density in this moment is the remanent magnetism in opposite direction, $\mp \mathcal{B}_r$, and the maximum value of density which could occur then is $\pm (\mathcal{B}_r + 2 \mathcal{B}_0)$. If therefore the maximum magnetic flux density $\mathcal{B}_0$ in the transformer is such that $\mathcal{B}_r + 2 \mathcal{B}_0$ is still below saturation, the transient term of current cannot reach abnormal values. At $\mathcal{B} = 16{,}000$, the flux density is about at the bend of the saturation curve, and the current still moderate. Estimating $\mathcal{B}_r = 0.75 \mathcal{B}_0$ as approximate value, $\mathcal{B}_r + 2 \mathcal{B}_0 = 16{,}000$ thus gives $\mathcal{B} = 5800$, or 37,500 lines of magnetic flux per square inch.

Such low maximum density is uneconomical. However, for $B_r = 0$, which probably more nearly represents the starting conditions of a transformer, which has been disconnected for some time, the limit is $B_o = 8{,}000$ or 51,600 lines per square inch, and at least, at 60 cycles, in well designed transformers, the maximum densities do not very much exceed this value. With a large starting current, not only the resistance of the current consumes voltage, but the self-inductive or leakage flux of the transformer, which is essentially an air flux, and as such not limited by saturation, also consumes voltage. Furthermore, the terminal voltage usually, more or less, drops by the impedance between transformer and generating system, and as a result, at least in 60 cycle circuits, this phenomenon is not serious.

103. Since the relation between the current, i, and the magnetic flux density, $\mathcal{B}$, is empirically given by the magnetic cycle of the material, and cannot be expressed with sufficient accuracy by a mathematical equation, the problem of determining the transient starting current of a transformer is investigated by constructing the curves of current and magnetic flux density.

Let the normal magnetic cycle of a transformer be represented by the dotted curve in Figs. 43 and 44; the characteristic points are: the maximum values, $\pm \mathcal{B}_0 = \pm 10{,}000$; the remanent values, $\pm \mathcal{B}_r = \pm 7600$, and the maximum exciting current, $i_m = \pm 4.5$ amp.

At very high values of flux density an appreciable part of the total magnetic flux Φ may be carried through space, outside of the iron, depending on the construction of the transformer. The most convenient way of dealing with such a case is to resolve the magnetic flux density, $\mathcal{B}$, in the iron into the "metallic

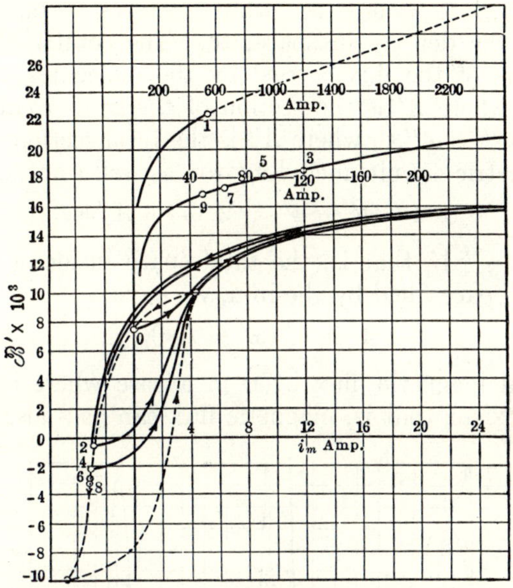

Fig. 43. Magnetic cycle of a transformer starting with low stray field.

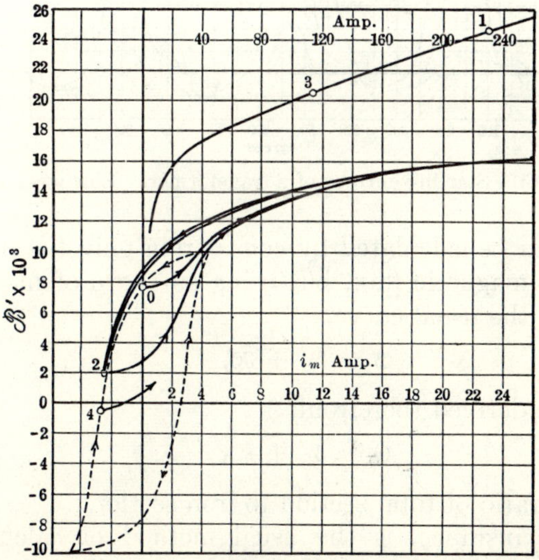

Fig. 44. Magnetic cycle of a transformer starting with high stray field.

flux density," $\mathcal{B}' = \mathcal{B} - \mathcal{H}$, which reaches a finite limiting value, and the density in space, $\mathcal{H}$. The total magnetic flux then consists of the flux carried by the molecules of the iron, $\Phi' = A'\mathcal{B}'$, where A' is the section of the iron circuit, and the space flux, $\Phi'' = A''\mathcal{H}$, where A'' is the total section interlinked with the electric circuit, including iron as well as other space.

$$\Phi = \Phi' + \Phi'' = A'\mathcal{B}' + A''\mathcal{H}.$$

If then $A'' = kA'$, that is, the total space inside of the coil is k times the space filled by the iron, we have

$$\Phi = A' (\mathcal{B}' + k\mathcal{H}),$$

or the total magnetic flux even in a case where considerable stray field exists, that is, magnetic flux can pass also outside of

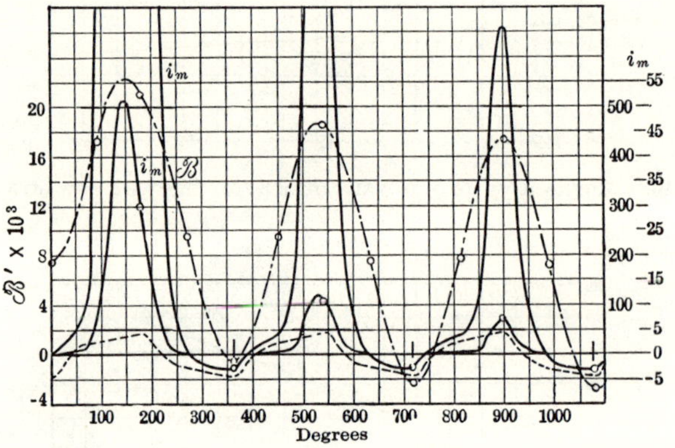

Fig. 45. Starting current of a transformer. Low stray field.

the iron, can be calculated by considering only the iron section as carrying magnetic flux, but using as curve of magnetic flux density not the usual curve,

$$\mathcal{B} = \mathcal{B}' + \mathcal{H},$$

but a curve derived therefrom,

$$\mathcal{B} = \mathcal{B}' + k\mathcal{H},$$

where k = ratio of total section to iron section.

This, for instance, is the usual method of calculating the m.m.f. consumed in the armature teeth of commutating machines at very high saturations.

In investigating the transient transformer starting current, the magnetic density curve thus is corrected for the stray field.

Figs. 43 and 45 correspond to $k = 3$, or a total effective air section equal to three times the iron section, that is, $\mathcal{B} = \mathcal{B}' + 3\,\mathcal{H}$.

Figs. 44 and 46 correspond to $k = 25$, or a section of stray field equal to 25 times the iron section, that is, $\mathcal{B} = \mathcal{B}' + 25\,\mathcal{H}$.

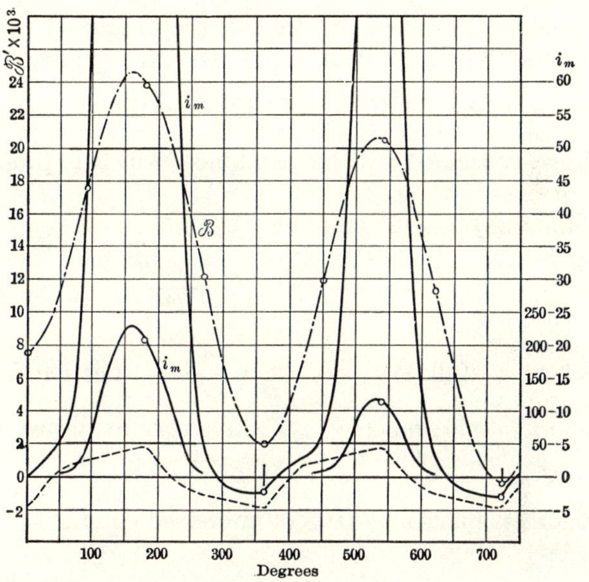

Fig. 46. Starting current of a transformer. High stray field.

104. At very high values of current the resistance consumes a considerable voltage, and thus reduces the e.m.f. generated by the magnetic flux, and thereby the maximum magnetic flux and transient current. The resistance, which comes into consideration here, is the total resistance of the transformer primary circuit plus leads and supply lines, back to the point where the voltage is kept constant, as generator, busbars, or supply main.

Assuming then at full load of $i_m = 50$ amperes effective in the transformer, a resistance drop of 8 per cent, or the voltage consumed by the resistance, as $e_r = 0.08$ of the impressed e.m.f.

Let now the remanent magnetic flux density be $\mathcal{B}_r = +\,7600$, and the circuit be closed at the moment $\theta = 0$, where the flux density should be $\mathcal{B} = -\mathcal{B}_0 = -10{,}000$; then the impressed e.m.f. is given by

$$e = -E \sin \theta = E \frac{d}{d\theta} (\cos \theta). \tag{1}$$

It is, however,

$$e = A \frac{d\mathcal{B}}{d\theta} + Ci, \tag{2}$$

where A and C are constants; that is, the impressed e.m.f., e, is consumed by the self-inductance, or the e.m.f. generated by the changing magnetic density, which is proportional to $\frac{d\mathcal{B}}{d\theta}$, and by the voltage consumed by the resistance, which is proportional to the current i.

Combining (1) and (2) gives

$$A \frac{d\mathcal{B}}{d\theta} + Ci = E \frac{d \cos \theta}{d\theta}. \tag{3}$$

However, at full load, we have $\frac{E}{\sqrt{2}}$ = effective impressed e.m.f. and i_m = 50 amperes = effective current; hence

$$Ci_m = 50\,C = \text{e.m.f. consumed by resistance},$$

and since this equals $e_r = 0.08$ of impressed e.m.f.,

$$Ci_m = \frac{e_r}{\sqrt{2}},$$

or

$$50\,C = \frac{0.08\,E}{\sqrt{2}},$$

or

$$\frac{C}{E} = \frac{e_r}{Ei_m\sqrt{2}} = \frac{0.08}{50\sqrt{2}} = 0.00113. \tag{4}$$

From (3) follows

$$d\mathcal{B} = \frac{E}{A} d \cos \theta - \frac{Ci}{A} d\theta \tag{5}$$

and

$$\int_0^{\frac{\pi}{2}} d\mathcal{B} = \frac{E}{A} \int_0^{\frac{\pi}{2}} d \cos \theta - \frac{C}{A} \int_0^{\frac{\pi}{2}} i\, d\theta;$$

hence, for $i = 0$, or negligible resistance drop, that is, permanent condition,

$$\mathfrak{B}_0 = \frac{E}{A} = 10{,}000. \tag{6}$$

Multiplying (4) and (6) gives

$$\frac{C}{A} = \frac{e_r \mathfrak{B}_0}{E i_m \sqrt{2}} = 11.3, \tag{7}$$

and substituting (6) and (7) in (5) gives

$$\begin{aligned} d\mathfrak{B} &= \mathfrak{B}_0 d \cos \theta - i \frac{e_r \mathfrak{B}_0}{E i_m \sqrt{2}} d\theta \\ &= 10{,}000\, d \cos \theta - 11.3\, i\, d\theta. \end{aligned} \tag{8}$$

Changing now from differential to difference, that is, replacing, as approximation, d by Δ, gives

$$\begin{aligned} \Delta\mathfrak{B} &= \mathfrak{B}_0 \Delta \cos \theta - i \frac{e_r \mathfrak{B}_0}{E i_m \sqrt{2}} \Delta\theta \\ &= 10{,}000\, \Delta \cos \theta - 11.3\, i \Delta\theta. \end{aligned} \tag{9}$$

Assuming now

$$\Delta\theta = 10° = 0.175 \tag{10}$$

gives for the increment of magnetic flux density during 10° change of angle the value

$$\Delta\mathfrak{B} = 10{,}000\, \Delta \cos \theta - 2\, i \tag{11}$$

and

$$\begin{aligned} \mathfrak{B} &= \mathfrak{B}' + \Delta\mathfrak{B} \\ &= \mathfrak{B}' + 10{,}000\, \Delta \cos \theta - 2\, i. \end{aligned} \tag{12}$$

From equation (12) the instantaneous values of magnetic flux density $\mathfrak{B}$, and therefrom, by the magnetic cycles, Figs. 43 and 44, respectively, the values of current i are calculated, by starting, for $\theta = 0$, with the remanent density $\mathfrak{B}' = \mathfrak{B}_r = 7600$, adding thereto the change of cosine, $10{,}000\, \Delta \cos \theta$, which gives a value $\mathfrak{B}_1 = \mathfrak{B}' + 10{,}000\, \Delta \cos \theta$, taking the corresponding value of i from the hysteresis cycle, Figs. 43 and 44, subtracting $2\, i$ from $\mathfrak{B}_1$, and then correcting i for the value corresponding to $\mathfrak{B} = \mathfrak{B}_1 - 2\, i$.

The quantity $2\, i$ is appreciable only during the range of the curve where i is very large.

105. The following table is given to illustrate the beginning of the calculation of the curve for low stray field.

STARTING CURRENT OF A TRANSFORMER.

$\theta°$	$\cos\theta$	$\Delta\mathfrak{B}_1 = 10\,\Delta\cos\theta$	$\mathfrak{B}_1 = \mathfrak{B} + \Delta\mathfrak{B}_1$	i	$D_1 = 2i \times 10^{-3}$	$\mathfrak{B}$
0	+1.00	....	7.6	0	....	....
10	0.98	+0.2	7.8	1.0	....	....
20	0.94	0.4	8.2	1.9	....	....
30	0.87	0.7	8.9	2.9	....	....
40	0.77	1.0	9.9	3.8	....	....
50	0.64	1.3	11.2	5.2	....	....
60	0.50	1.4	12.6	7.3	....	....
70	0.34	1.6	14.2	12.0	....	....
80	+0.17	1.7	15.9	27	....	....
90	0	1.7	17.6	70	0.1	17.5
100	−0.17	1.7	19.2	138	0.3	18.9
110	0.34	1.7	20.6	220	0.45	20.15
120	0.50	1.6	21.75	270	0.55	21.2
130	0.64	1.4	22.6	450	0.9	21.7
140	0.77	1.3	23.0	510	1.0	22.0
150	0.87	1.0	23.0	510	1.0	22.0
160	0.94	0.7	22.7	440	0.9	21.8
170	0.98	0.4	22.2	350	0.7	21.5
180	1.00	+0.2	21.7	250	0.5	21.2
190	0.98	−0.2	21.0	200	0.4	20.6
200	0.94	0.4	20.2	180	0.35	19.85
210	0.87	0.7	19.15	130	0.25	18.9
220	0.77	1.0	17.9	70	0.15	17.75
230	0.64	1.3	16.45	40	0.2	16.25
240	0.50	1.4	14.85	14	0.05	14.8
250	0.34	1.6	13.2	7	....	....
260	−0.17	1.7	11.5	3.2	....	....
270	0	1.7	9.8	−1.0	....	....
280	+0.17	1.7	8.1	− .4	....	....
290	0.34	1.7	6.4	−1.3	....	....
300	0.50	1.6	4.8	−1.9	....	....
310	0.64	1.4	3.4	−2.2	....	....
320	0.77	1.3	2.1	−2.5	....	....
330	0.87	1.0	1.1	−2.6	....	....
340	0.94	0.7	.4	−2.75	....	....
350	0.98	0.4	0	−2.8	....	....
360	+1.00	−0.2	− .2	−2.8	....	....
370	0.98	+0.2	0	−1.3	....	....
380	0.94	0.4	+0.4	−0.5	....	....
390	0.87	0.7	1.1	+0.3	....	....
400	0.77	1.0	2.1	1.0	....	....
410	0.64	1.3	3.5	1.7	....	....
420	0.50	1.4	4.9	2.2	....	....
430	0.34	1.6	6.5	2.7	....	....
440	+0.17	1.7	8.2	3.3	....	....
450	0	1.7	9.9	4.3	....	....
460	−0.17	1.7	11.6	6.2	....	....
470	−0.34	1.7	13.3	9.5	....	....
480	−0.50	1.6	14.9	16.5	....	....

The first column gives angle θ,
The second column gives $\cos \theta$,
The third column gives $\Delta \mathfrak{B}_1 = 10 \Delta \cos \theta$, in kilolines per sq. cm.,
The fourth column gives $\mathfrak{B}_1 = \mathfrak{B} + \Delta \mathfrak{B}_1$,
The fifth column gives i,
The sixth column gives $D_1 = 2 i \times 10^{-3}$, and
The seventh column gives $\mathfrak{B} = \mathfrak{B}_1 - D_1$,
i in the fifth column being chosen, by trial, so as to correspond, on the hysteresis cycles, not to $\mathfrak{B}_1$, but to $\mathfrak{B} = \mathfrak{B}_1 - D_1$.

These values are recorded as magnetic cycles on Figs. 43 and 44, and as waves of flux density, current, etc., in Figs. 45 and 46.

The maximum values of successive half waves are:

A. Low Stray Field. $k = 3$			B. High Stray Field. $k = 25$		
θ	$\mathfrak{B}$	i_m	θ	$\mathfrak{B}$	i_m
0	7.6	0	0	7.6	0
145°	22.0	510	160°	24.6	230
360°	− .2	−2.8	360°	+2.0	−2.4
530°	18.6	120	530°	20.7	117
720°	−2.2	−2.9	720°	− .4	−2.7
900°	18.0	92	....	....	....
1080°	−2.8	−3.0	....	....	....
1260°	17.4	66	....	....	....
1440°	−3.1	−3.0	....	....	....
1620°	16.9	50	....	....	....
Permanent	±10.0	±4.5	....	±10.0	±4.5

As seen, the maximum value of current during the first cycle, 510, is more than one hundred times the final value 4.5, and more than 7 times the maximum value of the full-load current, $50\sqrt{2} = 70.7$ amperes, and the transient current falls below full-load current only in the fourth cycle. That is, the excessive value of transient current in an ironclad circuit lasts for a considerable number of cycles.

In the presence of iron in the magnetic field of electric circuits, transient terms of current may thus occur which are very large compared with the transient terms in ironless reactors, which do not follow the exponential curve, can usually not be calculated by general equations, but require numerical investigation by the use of the magnetic cycles of the iron.

These transient terms lead to excessive current values only if the normal magnetic flux density exceeds half the saturation value of the iron, and so are most noticeable in 25-cycle circuits.

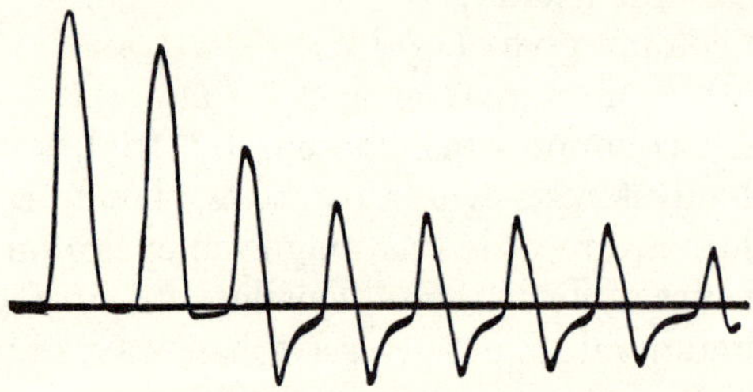

Fig. 47. Starting current of a 25-cycle transformer.

As illustration is shown, in Fig. 47, an oscillogram of the starting current of a 25-cycle transformer having a resistance in the supply circuit somewhat smaller than in the above instance, thus causing a still longer duration of the transient term of excessive current.

These starting transients of the ironclad inductance at high density are unsymmetrical waves, that is, successive half waves have different shapes, and when resolved into a trigonometric series, would give even harmonics as well as the odd harmonics.

Thus the first wave of Fig. 45 can, when neglecting the transient factor, be represented by the series:

$$\begin{aligned} i = &+ 108.3 - 183.8 \cos (\theta + 28.0°) \\ &+ 112.4 \cos 2 (\theta + 29.8°) - 53.1 \cos 3 (\theta + 33.3°) \\ &+ 27.2 \cos 4 (\theta + 39.1°) - 18.4 \cos 5 (\theta + 38.1°) \\ &+ 13.6 \cos 6 (\theta + 33.4°) - 8.1 \cos 7 (\theta + 32.7°) \end{aligned}$$

or, substituting: $\theta = \beta + 150°$, gives:

$$e = E \sin (\beta + 150°)$$

$$\begin{aligned} i = 108.3 &+ 183.8 \cos (\beta - 2.0°) + 112.4 \cos 2 (\beta - 0.2°) \\ &+ 53.1 \cos 3 (\beta + 3.3°) + 27.2 \cos 4 (\beta + 9.1°) \\ &+ 18.4 \cos 5 (\beta + 8.1°) + 13.6 \cos 6 (\beta + 3.4°) \\ &+ 8.1 \cos 7 (\beta + 2.7°). \end{aligned}$$

106. An approximate estimate of the initial value of the starting current of the transformer, at least of its magnitude under conditions where it is very large, can be made by separately considering the iron flux, that is the flux B_0 carried by the iron proper, which reaches a finite saturation value of about $S = 21{,}000$ lines per cm.², and the air flux or space flux, which is proportional to the current. Also neglecting the remanent magnetism—which usually is small—that is, assuming the initial magnetic cycle performed between zero and a maximum of flux density.

Let

i = maximum value of current

Φ = total maximum magnetic flux

n = number of turns of transformer circuit

$Z_1 = r_1 + jx_1$ = impedance of transformer circuit, where

x_1 = reactance of leakage flux

$Z_2 = r_2 + jx_2$ = impedance of circuit between transformer and source of constant voltage e_0.

e_0 = effective value of this constant voltage.

The voltage consumed by a variation of current between 0 and i, corresponding to an effective value of

$$i' = \frac{i}{2\sqrt{2}}$$

then is:

in the supply lines:

$$E_2 = i'(r_2 + jx_2) = \frac{i}{2\sqrt{2}}(r_2 + jx_2)$$

in the transformer impedance:

$$\dot{E}_1 = i'(r_1 + jx_1) = \frac{i}{2\sqrt{2}}(r_1 + jx_1)$$

by the magnetic flux Φ, at frequency f:

$$\dot{E} = j\frac{\pi f n \Phi}{\sqrt{2}} 10^{-8}$$

thus the total supply voltage:

$$\dot{E}_0 = \dot{E} + \dot{E}_1 + \dot{E}_2 = (r_1 + r_2)\frac{i}{2\sqrt{2}} + j\left[(x_1 + x_2)\frac{i}{2\sqrt{2}} + \frac{\pi f n \Phi}{\sqrt{2}} 10^{-8}\right]$$

and, absolute:

$$e_0^2 = \frac{1}{8}\{(r_1 + r_2)^2 i^2 + [(x_1 + x_2) i + 2 \pi f n \Phi 10^{-8}]^2\} \quad (1)$$

If now:

$$s_1 = \text{magnetic iron section, in cm.}^2,$$

the magnetic flux in the iron proper is:

$$\Phi_1 = s_1 S = 21{,}000 s$$

if

$$s_2 = \text{section, in cm.}^2,$$

$$l_2 = \text{length, in cm.},$$

of the total magnetic circuit, including iron and air space, the magnetic space flux is

$$\Phi_2 = 0.4 \pi n i \frac{s_2}{l_2}$$

hence, the total magnetic flux:

$$\Phi = \Phi_1 + \Phi_2$$
$$= s_1 S + 0.4 \pi n i \frac{s_2}{l_2} \quad (2)$$

Substituting (2) into (1) gives:

$$e_0^2 = \frac{1}{8}\left\{ (r_1 + r_2)^2 i^2 + \left[\left(x_1 + x_2 + 0.8 \pi^2 f n^2 \frac{s_2}{l_2} 10^{-8}\right) i + 2 \pi f n s_1 S\right]^2 \right\} \quad (3)$$

From equation (3) follows the value of i, the maximum initial or starting current of the transformer or reactor.

107. An approximate calculation, giving an idea of the shape of the transient of the ironclad magnetic circuit, can be made by neglecting the difference between the rising and decreasing magnetic characteristic, and using the approximation of the magnetic characteristic given by Fröhlich's formula:

$$\mathcal{B} = \frac{\mathcal{H}}{\alpha + \sigma \mathcal{H}}, \quad (1)$$

which is usually represented in the form given by Kennelly:

$$\rho = \frac{\mathcal{H}}{\mathcal{B}} = \alpha + \sigma \mathcal{H}; \quad (2)$$

that is, the reluctivity is a linear function of the field intensity. It gives a fair approximation for higher magnetic densities.

This formula is based on the fairly rational assumption that the permeability of the iron is proportional to its remaining magnetizability. That is, the magnetic-flux density $\mathcal{B}$ consists of a component $\mathcal{H}$, the field intensity, which is the flux density in space, and a component $\mathcal{B}' = \mathcal{B} - \mathcal{H}$, which is the additional flux density carried by the iron. $\mathcal{B}'$ is frequently called the "metallic-flux density." With increasing $\mathcal{H}$, $\mathcal{B}'$ reaches a finite limiting value, which in iron is about

$$\mathcal{B}_\infty' = 20{,}000 \text{ lines per cm}^2.$$

At any density $\mathcal{B}'$, the remaining magnetizability then is $\mathcal{B}_\infty' - \mathcal{B}'$, and, assuming the (metallic) permeability as proportional hereto, gives

$$\mu = c(\mathcal{B}_\infty' - \mathcal{B}'),$$

and, substituting

$$\mu = \frac{\mathcal{B}'}{\mathcal{H}'},$$

gives

$$\mathcal{B}' = \frac{c\mathcal{B}_\infty'\mathcal{H}'}{1 + c\mathcal{H}'},$$

or, substituting

$$\frac{1}{c\mathcal{B}_\infty'} = \alpha, \frac{1}{\mathcal{B}_\infty'} = \sigma,$$

gives equation (1).

For $\mathcal{H} = 0$ in equation (1), $\dfrac{\mathcal{B}}{\mathcal{H}} = \dfrac{1}{\alpha}$; for $\mathcal{H} = \infty$, $\mathcal{B} = \dfrac{1}{\sigma}$; that is, in equation (1), $\dfrac{1}{\alpha}$ = initial permeability, $\dfrac{1}{\sigma}$ = saturation value of magnetic density.

If the magnetic circuit contains an air gap, the reluctance of the iron part is given by equation (2), that of the air part is constant, and the total reluctance thus is

$$\rho = \beta + \sigma\mathcal{H},$$

where $\beta = \alpha$ plus the reluctance of the air gap. Equation (1), therefore, remains applicable, except that the value of α is increased.

In addition to the metallic flux given by equation (1), a greater or smaller part of the flux always passes through the air or through space in general, and then has constant permeance, that is, is given by

$$\mathcal{B} = c\mathcal{H}.$$

In general, the flux in an ironclad magnetic circuit can, therefore, be represented as function of the current by an expression of the form

$$\Phi = \frac{ai}{1 + bi} + ci, \tag{3}$$

where $\frac{ai}{1 + bi} = \Phi'$ is that part of the flux which passes through the iron and whatever air space may be in series with the iron, and ci is the part of the flux passing through nonmagnetic material.

Denoting now

$$\left.\begin{aligned} L_1 &= na\, 10^{-8}, \\ L_2 &= nc\, 10^{-8}, \end{aligned}\right\} \tag{4}$$

where n = number of turns of the electric circuit, which is interlinked with the magnetic circuit, L_2 is the inductance of the air part of the magnetic circuit, L_1 the (virtual) initial inductance, that is, inductance at very small currents, of the iron part of the magnetic circuit, and $\frac{a}{b}$ the saturation value of the flux in the iron. That is, for $i = 0, \frac{n\Phi'}{i} = L_1$; and for $i = \infty$, $\Phi' = \frac{a}{b}$.

If r = resistance, the duration of the component of the transient resulting from the air flux would be

$$T_2 = \frac{L_2}{r} = \frac{nc\, 10^{-8}}{r}, \tag{5}$$

and the duration of the transient which would result from the initial inductance of the iron flux would be

$$T_1 = \frac{L_1}{r} = \frac{na\, 10^{-8}}{r}. \tag{6}$$

108. The differential equation of the transient is: induced voltage plus resistance drop equal zero; that is,

$$n \frac{d\Phi}{dt} 10^{-8} + ri = 0.$$

Substituting (3) and differentiating gives

$$\frac{na\, 10^{-8}}{(1 + bi)^2} \frac{di}{dt} + nc\, 10^{-8} \frac{di}{dt} + ri = 0,$$

and, substituting (5) and (6),

$$\left\{ \frac{T_1}{(1 + bi)^2} + T_2 \right\} \frac{di}{dt} + i = 0;$$

hence, separating the variables,

$$\frac{T_1 di}{i(1 + bi)^2} + \frac{T_2 di}{i} + dt = 0. \tag{7}$$

The first term is integrated by resolving into partial fractions:

$$\frac{1}{i(1 + bi)^2} = \frac{1}{i} - \frac{b}{1 + bi} - \frac{b}{(1 + bi)^2},$$

and the integration of differential equation (7) then gives

$$T_1 \log \frac{i}{1 + bi} + T_2 \log i + \frac{T_1}{1 + bi} + t + C = 0. \tag{8}$$

If, then, for the time $t = t_0$, the current is $i = i_0$, these values substituted in (8) give the integration constant C:

$$T_1 \log \frac{i_0}{1 + bi_0} + T_2 \log i_0 + \frac{T_1}{1 + bi_0} + t_0 + C = 0, \tag{9}$$

and, subtracting (8) from (9), gives

$$t - t_0 = T_1 \log \frac{i_0(1 + bi)}{i(1 + bi_0)} + T_2 \log \frac{i_0}{i} + T_1 \left\{ \frac{1}{1 + bi_0} - \frac{1}{1 + bi} \right\} \tag{10}$$

This equation is so complex in i that it is not possible to calculate from the different values of t the corresponding values of i; but inversely, for different values of i the corresponding values of t can be calculated, and the corresponding values of i and t, derived in this manner, can be plotted as a curve, which gives the single-energy transient of the ironclad magnetic circuit.

Such is done in Fig. 48, for the values of the constants:

$$r = .3,$$
$$a = 4 \times 10^5,$$
$$c = 4 \times 10^4,$$
$$b = .6,$$
$$n = 300.$$

This gives

$$T_1 = 4,$$
$$T_2 = .4.$$

Assuming $i_0 = 10$ amperes for $t_0 = 0$, gives from (10) the equation:

$$T = 2.92 - \left\{ 9.21 \log^{10} \frac{i}{1 + 0.6\,i} + 0.921 \log^{10} i + \frac{4}{1 + 0.6\,i} \right\}.$$

Herein, the logarithms have been reduced to the base 10 by division with $log^{10}\epsilon = 0.4343$.

For comparison is shown, in dotted line, in Fig. 48, the transient of a circuit containing no iron, and of such constants as to give about the same duration:

$$t = 1.085 \log^{10} i - 0.507.$$

As seen, in the ironclad transient the current curve is very much steeper in the range of high currents, where magnetic sat-

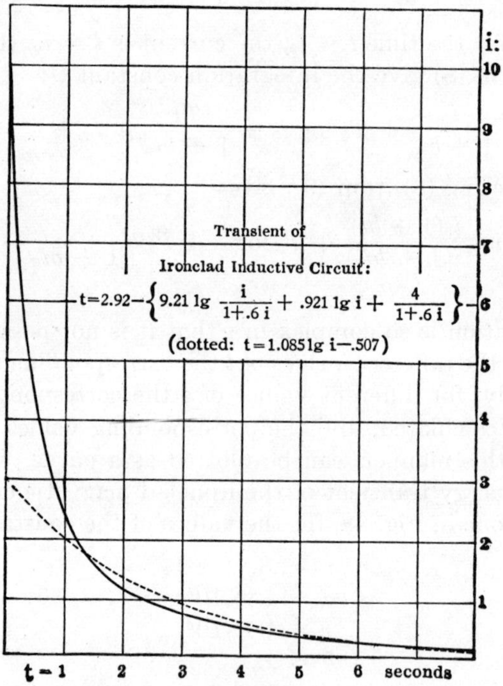

Fig. 48.

uration is reached, but the current is lower in the range of medium magnetic densities.

Thus, in ironclad transients very high-current values of short duration may occur, and such transients, as those of the starting current of alternating-current transformers, may therefore be of serious importance by their excessive current values.

CHAPTER XIII.

TRANSIENT TERM OF THE ROTATING FIELD.

109. The resultant of n_p equal m.m.fs. equally displaced from each other in space angle and in time-phase is constant in intensity, and revolves at constant synchronous velocity. When acting upon a magnetic circuit of constant reluctance in all directions, such a polyphase system of m.m.fs. produces a revolving magnetic flux, or a rotating field. ("Alternating Current Phenomena" Dover edition Volume II, Part I.) That is, if n_p equal magnetizing coils are arranged under equal space angles of $\frac{360}{n_p}$ electrical degrees, and connected to a symmetrical n_p phase electrical degrees, and connected to a symmetrical n_p phase system, that is, to n_p equal e.m.fs. displaced in time-phase by $\frac{360}{n_p}$ degrees, the resultant m.m.f. of these n_p coils is a constant and uniformly revolving m.m.f., of intensity $\mathfrak{F}_0 = \frac{n_p}{2}\mathfrak{F}$, where $\mathfrak{F}$ is the maximum value $\left(\text{hence } \frac{\mathfrak{F}}{\sqrt{2}} \text{ the effective value}\right)$ of the m.m.f. of each coil.

In starting, that is, when connecting such a system of magnetizing coils to a polyphase system of e.m.fs., a transient term appears, as the resultant magnetic flux first has to rise to its constant value. This transient term of the rotating field is the resultant of the transient terms of the currents and therefore the m.m.fs. of the individual coils.

If, then, $\mathfrak{F} = nI$ = maximum value of m.m.f. of each coil, where n = number of turns, and I = maximum value of current, and τ = space-phase angle of the coil, the instantaneous value of the m.m.f. of the coil, under permanent conditions, is

$$f' = \mathfrak{F} \cos(\theta - \tau), \tag{1}$$

and if the time θ is counted from the moment of closing the circuit, the transient term is, by Chapter IV,

$$f'' = -\mathfrak{F}\varepsilon^{-\frac{r}{x}\theta}\cos\tau, \tag{2}$$

where

$$Z = r - jx.$$

The complete value of m.m.f. of one coil is

$$f_1 = f' + f'' = \mathfrak{F}\left\{\cos(\theta - \tau) - \varepsilon^{-\frac{r}{x}\theta}\cos\tau\right\}. \tag{3}$$

In an n_p-phase system, successive e.m.fs. and therefore currents are displaced from each other by $\frac{1}{n_p}$ of a period, or an angle $\frac{2\pi}{n_p}$, and the m.m.f. of coil, i, thus is

$$f_i = \mathfrak{F}\left\{\cos\left(\theta - \tau - \frac{2\pi}{n_p}i\right) - \varepsilon^{-\frac{r}{x}\theta}\cos\left(\tau + \frac{2\pi}{n_p}i\right)\right\}. \tag{4}$$

The resultant of n_p such m.m.fs. acting together in the same direction would be

$$\sum_1^{n_p}{}_i f_i = \mathfrak{F}\sum_1^{n_p}{}_i \cos\left(\theta - \tau - \frac{2\pi}{n_p}i\right)$$
$$- \mathfrak{F}\varepsilon^{-\frac{r}{x}\theta}\sum_1^{n_p}{}_i \cos\left(\tau + \frac{2\pi}{n_p}i\right) = 0; \tag{5}$$

that is, the sum of the instantaneous values of the permanent terms as well as the transient terms of all the phases of a symmetrical polyphase system equals zero.

In the polyphase field, however, these m.m.fs. (4) do not act in the same direction, but in directions displaced from each other by a space angle $\frac{2\pi}{n_p}$ equal to the time angle of their phase displacement.

The component of the m.m.f., f_1, acting in the direction $(\theta_0 - \tau)$, thus is

$$f_1' = f_1\cos\left(\theta_0 - \tau - \frac{2\pi}{n_p}i\right), \tag{6}$$

and the sum of the components of all the n_p m.m.fs., in the direction $(\theta_0 - \tau)$, that is, the component of the resultant m.m.f. of the polyphase field, in the direction $(\theta_0 - \tau)$, is

$$\begin{aligned} f &= \sum_1^{np} {}^i f_i' \\ &= \mathfrak{F} \sum_1^{n_p} {}^i \left\{ \cos\left(\theta - \tau - \frac{2\pi}{n_p} i\right) - \varepsilon^{-\frac{r}{x}\theta} \cos\left(\tau + \frac{2\pi}{n_p} i\right)\right\}, \\ &\cos\left(\theta_0 - \tau - \frac{2\pi}{n_p} i\right). \end{aligned} \tag{7}$$

Transformed, this gives

$$\begin{aligned} f = \frac{\mathfrak{F}}{2} \Bigg\{ &\sum_1^{np} {}^i \cos\left(\theta + \theta_0 - 2\tau - \frac{4\pi}{n_p} i\right) + \sum_1^{n_p} {}^i \cos(\theta - \theta_0) \\ &- \varepsilon^{-\frac{r}{x}\theta} \sum_1^{n_p} {}^i \cos\theta_0 - \varepsilon^{-\frac{r}{x}\theta} \sum_1^{n_p} {}^i \cos\left(\theta_0 - 2\tau - \frac{4\pi}{n_p} i\right)\Bigg\}, \end{aligned}$$

and as the sums containing $\frac{4\pi}{n_p} i$ equal zero, we have

$$f = \frac{n_p}{2} \mathfrak{F} \left\{ \cos(\theta - \theta_0) - \varepsilon^{-\frac{r}{x}\theta} \cos\theta_0 \right\}, \tag{8}$$

and for $\theta = \infty$, that is as permanent term, this gives

$$f_0 = \frac{n_p}{2} \mathfrak{F} \cos(\theta - \theta_0); \tag{9}$$

hence, a maximum, and equal to $\frac{n_p}{2} \mathfrak{F}$, that is, constant, for $\theta_0 = \theta$, that is, uniform synchronous rotation. That is, the resultant of a polyphase system of m.m.fs., in permanent condition, rotates at constant intensity and constant synchronous velocity.

Before permanent condition is reached, however, the resultant m.m.f. in the direction $\theta_0 = \theta$, that is, in the direction of the synchronously rotating vector, in which in permanent condition the m.m.f. is maximum and constant, is given during the transient period, from equation (8), by

$$f_0 = \frac{n_p}{2} \mathfrak{F} \left\{ 1 - \varepsilon^{-\frac{r}{x}\theta} \cos \theta \right\}, \tag{10}$$

that is, it is not constant but periodically varying.

As example is shown, in Fig. 49, the resultant m.m.f. f_0 in the direction of the synchronously revolving vector, $\theta_0 = \theta$, for the

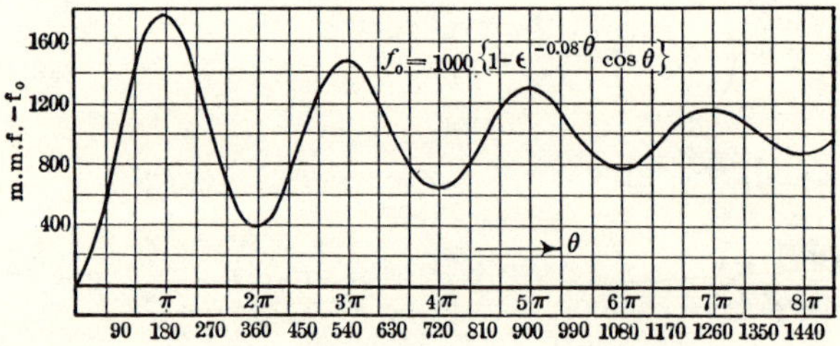

Fig. 49. Transient term of polyphase magnetomotive force.

constants $n_p = 3$, or a three-phase system; $\mathfrak{F} = 667$, and $Z = r - jx = 0.32 - 4j$; hence,

$$f_0 = 1000 \left(1 - \varepsilon^{-1.08\theta} \cos \theta\right),$$

with θ as abscissas, showing the gradual oscillatory approach to constancy.

110. The direction, $\theta_0 = \theta$, is, however, not the direction in which the resultant m.m.f. in equation (8) is a maximum, but the maximum is given by

$$\frac{df}{d\theta_0} = 0, \tag{11}$$

this gives

$$\sin (\theta - \theta_0) + \varepsilon^{-\frac{r}{x}\theta} \sin \theta_0 = 0, \tag{12}$$

hence,

$$\cot \theta_0 = \frac{\cos \theta - \varepsilon^{-\frac{r}{x}\theta}}{\sin \theta}; \tag{13}$$

that is, the resultant maximum m.m.f. of the polyphase system does not revolve synchronously, in the starting condition, but revolves with a varying velocity, alternately running ahead and dropping behind the position of uniform synchronous rotation, by equation (13), and only for $\theta = \infty$, equation (12) becomes $\cot \theta_0 = \cot \theta$, or $\theta_0 = \theta$, that is, uniform synchronous rotation.

The speed of rotation of the maximum m.m.f. is given from equation (12) by differentiation as

$$S = \frac{d\theta_0}{d\theta} = - \frac{\frac{dQ}{d\theta}}{\frac{dQ}{d\theta_0}},$$

where
$$Q = \sin (\theta - \theta_0) + \varepsilon^{-\frac{r}{x}\theta} \sin \theta_0;$$

hence,
$$S = \frac{\cos (\theta - \theta_0) - \frac{r}{x} \varepsilon^{-\frac{r}{x}\theta} \sin \theta_0}{\cos (\theta - \theta_0) - \varepsilon^{-\frac{r}{x}\theta} \cos \theta_0}, \tag{14}$$

or approximately,

$$S = \frac{1 - \frac{r}{x} \varepsilon^{-\frac{r}{x}\theta} \sin \theta_0}{1 - \varepsilon^{-\frac{r}{x}\theta} \cos \theta_0}. \tag{15}$$

For $\theta = \infty$, equation (14) becomes $S = 1$, or uniform synchronous rotation, but during the starting period the speed alternates between below and above synchronism.

From (13) follows

$$\left.\begin{aligned} \cos \theta_0 &= \frac{\cos \theta - \varepsilon^{-\frac{r}{x}\theta}}{\mathfrak{R}} \\ \text{and} \qquad \sin \theta_0 &= \frac{\sin \theta}{\mathfrak{R}}, \end{aligned}\right\} \tag{16}$$

where

$$\mathfrak{R} = \sqrt{\left(\cos \theta - \varepsilon^{-\frac{r}{x}\theta}\right)^2 + \sin^2 \theta} = \sqrt{1 - 2\varepsilon^{-\frac{r}{x}\theta} \cos \theta + \varepsilon^{-2\frac{r}{x}\theta}}. \tag{17}$$

The maximum value of the resultant m.m.f., at time-phase θ, and thus of direction θ_0 as given by equation (13) or (16), (17), is derived by substituting (16), (17) into (8), as:

$$f_m = \frac{n_p}{2} \mathfrak{F} \mathfrak{R}$$

$$= \frac{n_p}{2} \mathfrak{F} \sqrt{1 - 2 \varepsilon^{-\frac{r}{x}\theta} \cos\theta + \varepsilon^{-2\frac{r}{x}\theta}}, \tag{18}$$

hence is not constant, but pulsates periodically, with gradually decreasing amplitude of pulsation, around the mean value $\frac{n_p}{2}\mathfrak{F}$.

For $\theta = 0$, or at the moment of start, it is, by (13),

$$\cot \theta_0{}' = \frac{\cos\theta - \varepsilon^{-\frac{r}{x}\theta}}{\sin\theta} = \frac{0}{0},$$

hence, differentiating numerator and denominator,

$$\cot \theta_0{}' = \frac{-\sin\theta + \frac{r}{x}\varepsilon^{-\frac{r}{x}\theta}}{\cos\theta} = \frac{r}{x}$$

and

$$\tan \theta_0{}' = \frac{x}{r};$$

that is, the position of maximum resultant m.m.f. starts from angle $\theta_0{}'$ ahead of the permanent position, where $\theta_0{}'$ is the time-phase angle of the electric magnetizing circuit. The initial value of the resultant m.m.f., for $\theta = 0$, is $f_m = 0$, that is, the revolving m.m.f. starts from zero.

Substituting (16) in (15) gives the speed as function of time

$$S = \frac{1 - \varepsilon^{-\frac{r}{x}\theta}\left(\cos\theta - \frac{r}{x}\sin\theta\right)}{1 + \varepsilon^{-2\frac{r}{x}\theta} - 2\varepsilon^{-\frac{r}{x}\theta}\cos\theta} \tag{19}$$

for $\theta = 0$ this gives the starting speed of the rotating field

$$S_0 = \frac{0}{0}, \text{ or, indefinite;}$$

hence, after differentiating numerator and denominator twice, this value becomes definite.

$$S_0 = \frac{1}{2}; \qquad (20)$$

that is, the rotating field starts at half speed.

As illustration are shown, in Fig. 50, the maximum value of the resultant polyphase m.m.f., f_m, and its displacement in

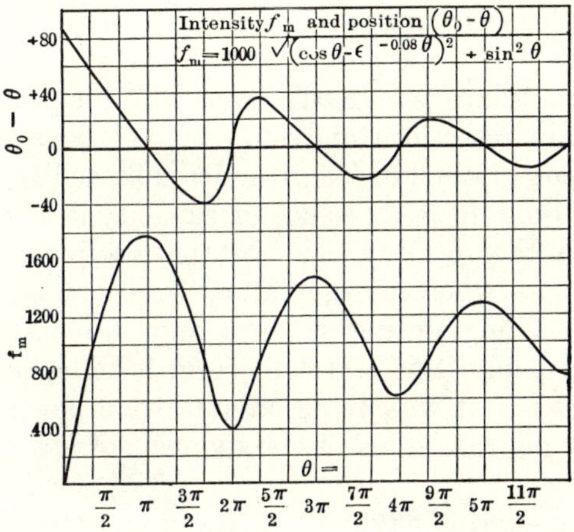

Fig. 50. Start of rotating field.

position from that of uniform synchronous rotation, $\theta_0 - \theta$, for the same constants as before, namely: $n_p = 3$; $\mathfrak{F} = 667$, and $Z = r - jx = 0.32 - 4j$; hence,

$$f_m = 1000\sqrt{1 - 2\,\varepsilon^{-0.08\theta}\cos\theta + \varepsilon^{-0.16\theta}},$$

with the time-phase angle θ as abscissas, for the first three cycles.

111. As seen, the resultant maximum m.m.f. of the polyphase system, under the assumed condition, starting at zero in the moment of closing the three-phase circuit, rises rapidly — within 60 time-degrees — to its normal value, overreaches and exceeds it by 78 per cent, then drops down again below normal, by 60 per cent, rises 47 per cent above normal, drops 37 per cent below normal, rises 28 per cent above normal, and thus by a series of oscillations approaches the normal value. The maximum value of the resultant m.m.f. starts in position

85 time-degrees ahead, in the direction of rotation, but has in half a period dropped back to the normal position, that is, the position of uniform synchronous rotation, then drops still further back to the maximum of 40 deg., runs ahead to 34 deg., drops 23 deg. behind, etc.

It is interesting to note that the transient term of the rotating field, as given by equations (10), (13), (18), does not contain the phase angle, that is, does not depend upon the point of the wave, $\theta = \tau$, at which the circuit is closed, while in all preceding investigations the transient term depended upon the point of the wave at which the circuit was closed, and that this transient term is oscillatory. In the preceding chapter, in circuits containing only resistance and inductance, the transient term has always been gradual or logarithmic, and oscillatory phenomena occurred only in the presence of capacity in addition to inductance. In the rotating field, or the polyphase m.m.f., we thus have a case where an oscillatory transient term occurs in a circuit containing only resistance and inductance but not capacity, and where this transient term is independent of the point of the wave at which the circuits were closed, that is, is always the same, regardless of the moment of start of the phenomenon.

The transient term of the polyphase m.m.f. thus is independent of the moment of start, and oscillatory in character, with an amplitude of oscillation depending only on the reactance factor, $\frac{x}{r}$, of the circuit.

CHAPTER XIV.

SHORT-CIRCUIT CURRENTS OF ALTERNATORS.

112. The short-circuit current of an alternator is limited by armature reaction and armature self-inductance; that is, the current in the armature represents a m.m.f. which with lagging current, as at short circuit, is demagnetizing or opposing the impressed m.m.f. of field excitation, and by combining therewith to a resultant m.m.f. reduces the magnetic flux from that corresponding to the field excitation to that corresponding to the resultant of field excitation and armature reaction, and thus reduces the generated e.m.f. from the nominal generated e.m.f., e_0, to the virtual generated e.m.f., e_1. The armature current also produces a local magnetic flux in the armature iron and pole-faces which does not interlink with the field coils, but is a true self-inductive flux, and therefore is represented by a reactance x_1. Combined with the effective resistance, r_1, of the armature winding, this gives the self-inductive impedance $Z_1 = r_1 - jx_1$, or $z_1 = \sqrt{r_1^2 + x_1^2}$. Vectorially subtracted from the virtual generated e.m.f., e_1, the voltage consumed by the armature current in the self-inductive impedance Z_1 then gives the terminal voltage, e.

At short circuit, the virtual generated e.m.f., e_1, is consumed by the armature self-inductive impedance, z_1. As the effective armature resistance, r_1, is very small compared with its self-inductive reactance, x_1, it can be neglected compared thereto, and the short-circuit current of the alternator, in permanent condition, thus is

$$i = \frac{e_1}{x_1}.$$

The armature reaction can be represented by an equivalent, or effective reactance, x_2, and the self-inductive reactance, x_1, and the effective reactance of armature reaction, x_2, combine to form the synchronous reactance, $x_0 = x_1 + x_2$, and the short-circuit current of the alternator, in permanent condition, therefore can be expressed by

$$i = \frac{e_0}{x_0},$$

where e_0 = nominal generated e.m.f.

113. The effective reactance of armature reaction, x_2, differs, however, essentially from the true self-inductive reactance, x_1, in that x_1 is instantaneous in its action, while the effective reactance of armature reaction, x_2, requires an appreciable time to develop: x_2 represents the change of the magnetic field flux produced by the armature m.m.f. The field flux, however, cannot change instantaneously, as it interlinks with the field exciting coil, and any change of the field flux generates an e.m.f. in the field coils, changing the field current so as to retard the change of the field flux. Hence, at the first moment after a change of armature current, the current change meets only the reactance, x_1, but not the reactance x_2. Thus, when suddenly short-circuiting an alternator from open circuit, in the moment before the short circuit, the field flux is that corresponding to the impressed m.m.f. of field excitation and the voltage in the armature, i.e., the nominal generated e.m.f., e_0 (corrected for magnetic saturation). At the moment of short circuit, a counter m.m.f., that of the armature reaction of the short-circuit current, is opposed to the impressed m.m.f. of the field excitation, and the magnetic flux, therefore, begins to decrease at such a rate that the e.m.f. generated in the field coils by the decrease of field flux increases the field current and therewith the m.m.f. so that when combined with the armature reaction it gives a resultant m.m.f. producing the instantaneous value of field flux. Immediately after short circuit, while the field flux still has full value, that is, before it has appreciably decreased, the field m.m.f. thus must have increased by a value equal to the counter m.m.f. of armature reaction. As the field is still practically unchanged, the generated e.m.f. is the nominal generated voltage, e_0, and the short-circuit current is

$$i'^0 = \frac{e_0}{x_1},$$

and from this value gradually dies down, with a decrease of the field flux and of the generated e.m.f., to

$$i = \frac{e_1}{x_1} = \frac{e_0}{x_0}.$$

Hence, approximately, when short-circuiting an alternator, in the first moment the short-circuit current is

$$i'^0 = \frac{e_0}{x_1},$$

while the field current has increased from its normal value i_0 to the value

$$i_0 \times \frac{\text{Field excitation} + \text{Armature reaction}}{\text{Field excitation}};$$

gradually the armature current decreases to

$$i = \frac{e_1}{x_1} = \frac{e_0}{x_0},$$

and the field current again to the normal value i_0.

Therefore, the momentary short-circuit current of an alternator bears to the permanent short-circuit current the ratio

$$\frac{i'^0}{i} = \frac{x_0}{x_1} = \frac{x_1 + x_2}{x_1},$$

that is,

$$\frac{\text{Armature self-inductance} + \text{Armature reaction}}{\text{Armature self-inductance}}.$$

In machines of high self-inductance and low armature reaction, as uni-tooth high frequency alternators, this increase of the momentary short-circuit current over the permanent short-circuit current is moderate, but may reach enormous values in machines of low self-inductance and high armature reaction, as large low frequency turbo alternators.

114. Superimposed upon this transient term, resulting from the gradual adjustment of the field flux to a change of m.m.f., is the transient term of armature reaction. In a polyphase alternator, the resultant m.m.f. of the armature in permanent conditions is constant in intensity and revolves with regard to the armature at uniform synchronous speed, hence is stationary with regard to the field. In the first moment, however, the resultant armature m.m.f. is changing in intensity and in velocity, approaching its constant value by a series of oscillations, as discussed in Chapter XIII. Hence, with regard to the field, the transient term of armature reaction is pulsating in intensity and oscillating in position, and therefore generates in the field coils

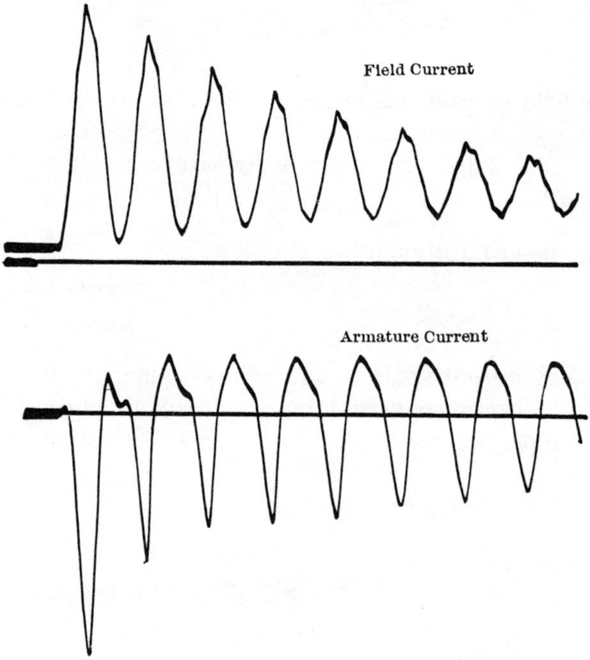

Fig. 51. Three-phase short-circuit current of a turbo-alternator.

an e.m.f. and causes a corresponding pulsation in the field current and field terminal voltage, of the same frequency as the armature current, as shown by the oscillogram of such a three-phase short-circuit, in Fig. 51. This pulsation of field current is independent of the point in the wave, at which the short-circuit occurs, and dies out gradually, with the dying out of the transient term of the rotating m.m.f.

In a single-phase alternator, the armature reaction is alternating with regard to the armature, hence pulsating, with double frequency, with regard to the field, varying between zero and its maximum value, and therefore generates in the field coils a double frequency e.m.f., producing a pulsation of field current of double frequency. This double-frequency pulsation of the field current and voltage at single-phase short-circuit is proportional to the armature current, and does not disappear with the disappearance of the transient term, but persists also after the permanent condition of short-circuit has been reached,

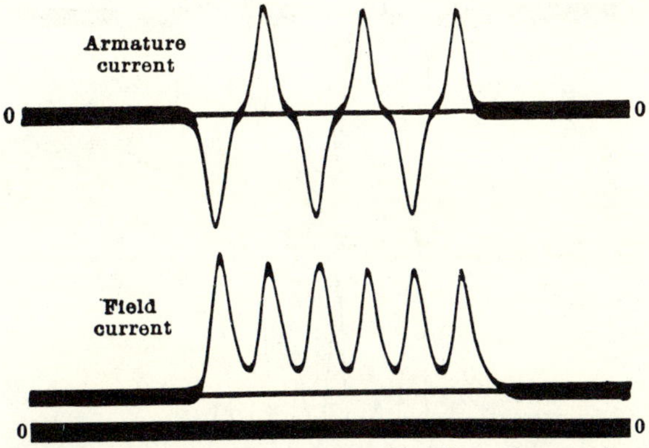

Fig. 52. Single-phase short-circuit current of a three-phase turbo-alternator.

merely decreasing with the decrease of the armature current. It is shown in the oscillogram of a single-phase short-circuit on a three-phase alternator, Fig. 52.

Superimposed on this double frequency pulsation is a single-frequency pulsation due to the transient term of the armature current, that is, the same as on polyphase short-circuit. With single-phase short-circuit, however, this normal frequency pulsation of the field depends on the point of the wave at which the short-circuit occurs, and is zero, if the circuit is closed at the moment when the short-circuit current is zero, as in Fig. 51, and a maximum when the short-circuit starts at the maximum point of the current wave. As this normal frequency pulsation gradually disappears, it causes the successive waves of the double frequency pulsation to be unequal in size at the beginning of the transient term, and gradually become equal, as shown in the oscillogram, Fig. 53.

The calculation of the transient term of the short-circuit current of alternators thus involves the transient term of the armature and the field current, as determined by the self-inductance of armature and of field circuit, and the mutual inductance between the armature circuits and the field circuit, and the impressed or generated voltage; therefore is rather complicated; but a simpler approximate calculation can be

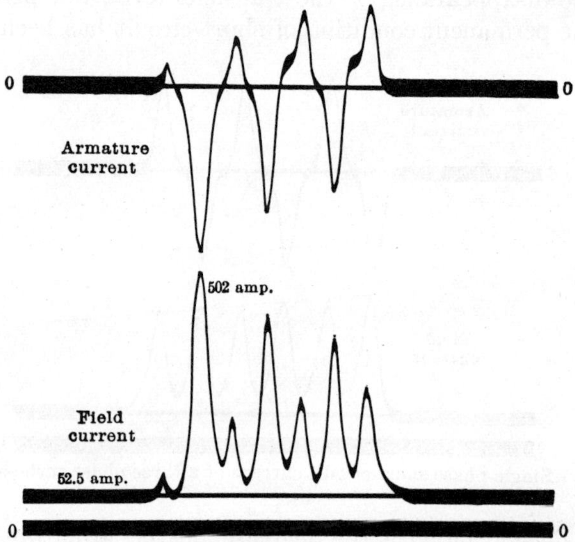

Fig. 53. Single-phase short-circuit current of a three-phase turbo-alternator.

given by considering that the duration of the transient term is short compared with that of the armature reaction on the field.

(*A*) *Polyphase alternator*.

115. Let n_p = number of phases; $\theta = 2\pi ft$ = time-phase angle; n_0 = number of field turns in series per pole; n_1 = number of armature turns in series per pole; $Z_0 = r_0 - jx_0$ = self-inductive impedance of field circuit; $Z_1 = r_1 - jx_1$ = self-inductive impedance of armature circuit; p = permeance of field magnetic circuit; $a = 2\pi fn_1 10^{-8}$ = induction coefficient of armature; E_0 = exciter voltage; $I_0 = \frac{E_0}{r_0}$ = field exciting current, in permanent condition; i_0 = field exciting current at time θ; i_0^0 = field exciting current immediately after short-circuit; i = armature current at time θ, and $k_t = \frac{2}{n_p}\frac{n_0}{n_1}$ = transformation ratio of field to resultant armature. Counting the time angle θ from the moment of short circuit, $\theta = 0$, and letting θ' = time-phase angle of one of the generator circuits at the moment of short circuit, we have,

$\mathfrak{F}_0 = n_0 I_0$ = field excitation, in permanent or stationary condition, (1)

$\Phi_0 = p\mathfrak{F}_0 = pn_0 I_0$ = magnetic flux corresponding thereto,

and

$$e_0^0 = ap\mathfrak{F}_0 = apn_0 I_0. \tag{2}$$

= nominal generated voltage, maximum value, at $\theta = 0$.

Hence,
$$I^0 = \frac{e_0^0}{x_1} = \frac{apn_0}{x_1} I_0 \tag{3}$$

= momentary short-circuit current at time $\theta = 0$, and

$$\mathfrak{F}_1^0 = \frac{n_p}{2} n_1 I^0 = \frac{n_p apn_1 n_0 I_0}{2\, x_1} \tag{4}$$

= resultant armature reaction thereof.

Assume this armature reaction as opposite to the field excitation,

$$\mathfrak{F}_0^0 = n_0 i_0^0, \tag{5}$$

as is the case at short circuit.

The resultant m.m.f. of the magnetic circuit at the moment of short-circuit is

$$\mathfrak{F}^0 = \mathfrak{F}_0^0 - \mathfrak{F}_1^0. \tag{6}$$

At this moment, however, the field flux is still Φ_0, and the resultant m.m.f. is given by (1) as

$$\mathfrak{F}^0 = \mathfrak{F}_0 = n_0 I_0. \tag{7}$$

Substituting (4), (5), (7) in (6) gives

$$n_0 I_0 = n_0 i_0^0 - \frac{n_p apn_1 n_0 I_0}{2\, x_1},$$

hence,
$$i_0^0 = \frac{x_1 + \dfrac{n_p apn_1}{2}}{x_1} I_0. \tag{8}$$

Writing $$x_2 = \frac{n_p a p n_1}{2}, \tag{9}$$

we have $$i_0^0 = \frac{x_1 + x_2}{x_1} I_0; \tag{10}$$

that is, at the moment of short circuit the field exciting current rises from I_0 to i_0^0, and then gradually dies down again to I_0 at a rate depending on the field impedance Z_0, that is, by $\varepsilon^{-\frac{r_0}{x_0}\theta}$, as discussed in preceding chapters. Hence, it can be represented by

$$i_0 = \frac{x_1 + x_2 \varepsilon^{-\frac{r_0}{x_0}\theta}}{x_1} I_0. \tag{11}$$

The resultant armature m.m.f., or armature reaction, is

$$\frac{n_p n_1 I^0}{2};$$

thus the magnetic flux which would be produced by it is

$$\frac{p n_p n_1 I^0}{2},$$

and therefore the voltage generated by this flux is

$$\frac{a p n_p n_1 I^0}{2};$$

hence,

$$x_2 = \frac{a p n_p n_1}{2}$$

$$= \frac{\text{Voltage corresponding to the m.m.f. of armature current}}{\text{Armature current}};$$

that is, x_2 is the equivalent or effective reactance of armature reaction.

In equations (10) and (11) the external self-inductance of the field circuit, that is, the reactance of the field circuit outside of the machine field winding, has been neglected. This would introduce a negative transient term in (11), thus giving equation (11) the approximate form

$$i_0 = \frac{x_1 + x_2\left(\varepsilon^{-\frac{r_0}{x_0}\theta} - \varepsilon^{-\frac{r_0}{x_3}\theta}\right)}{x_1} I_0, \tag{12}$$

where x_3 = self-inductive reactance of the field circuit outside of alternator field coils.

The more complete expression requires consideration when x_3 is very large, as when an external reactive coil is inserted in the field circuit.

In reality, x_2 is a mutual inductive reactance, and x_3 can be represented approximately by a corresponding increase of x_1.

116. If I = maximum value of armature current, we have

$$\mathfrak{F}_1 = \frac{n_p n_1 I}{2} = \text{armature m.m.f.},$$

hence,

$$\mathfrak{F} = n_0 i_0 - \frac{n_p n_1 I}{2} \tag{13}$$

= resultant m.m.f.,

and

$$E = ap\mathfrak{F}$$

= e.m.f. maximum generated thereby,

and

$$I = \frac{E}{x_1} = \frac{ap}{x_1}\mathfrak{F} \tag{14}$$

= armature current, maximum.

Substituting (13) in (14) gives

$$x_1 I = apn_0 i_0 - \frac{n_p apn_1}{2} I$$

and

$$I = \frac{apn_0 i_0}{x_1 + \frac{n_p apn_1}{2}} = \frac{apn_0 i_0}{x_1 + x_2}; \tag{15}$$

or, by (9),

$$I = \frac{2\, n_0}{n_p\, n_1} i_0 \frac{x_2}{x_1 + x_2}, \tag{16}$$

where $\frac{2}{n_p}\frac{n_0}{n_1} = k_t$ = transformation ratio of field turns to resultant armature turns; hence,

$$I = k_t i_0 \frac{x_2}{x_1 + x_2}. \tag{17}$$

Substituting (11) in (17) thus gives the *maximum value of the armature current* as

$$I = k_t I_0 \frac{x_2}{x_1 + x_2} \frac{x_1 + x_2 \varepsilon^{-\frac{r_0}{x_0}\theta}}{x_1}, \tag{18}$$

the *instantaneous value of the armature current* as

$$i = k_t I_0 \frac{x_2 \left(x_1 + x_2 \varepsilon^{-\frac{r_0}{x_0}\theta}\right)}{x_1 (x_1 + x_2)} \left\{ \cos(\theta - \theta') - \varepsilon^{-\frac{r_1}{x_1}\theta} \cos \theta' \right\}, \tag{19}$$

and by equation (10) of Chapter XIII, the *armature reaction* as

$$f = \frac{n_p n_1}{2} I_0 \frac{x_2 \left(x_1 + x_2 \varepsilon^{-\frac{r_0}{x_0}\theta}\right)}{(x_1 + x_2)} \left\{ 1 - \varepsilon^{-\frac{r_1}{x_1}\theta} \cos \theta \right\}, \tag{20}$$

where $x_1 + x_2 = x_0$ is the synchronous reactance of the alternator.

For $\theta = \infty$, or in permanent condition, equations (18), (19), (20) assume the usual form:

$$\left.\begin{aligned} I &= k_t I_0 \frac{x_2}{x_0}, \\ i &= k_t I_0 \frac{x_2}{x_0} \cos(\theta - \theta') \\ \text{and} \qquad f &= \frac{n_p n_1}{2} k_t I_0 \frac{x_2}{x_0}. \end{aligned}\right\} \tag{21}$$

117. As an example is calculated, the instantaneous value of the transient short-circuit current of a three-phase alternator, with the time angle θ as abscissas, and for the constants: the field turns, $n_0 = 100$; the normal field current, $I_0 = 200$ amp.; the field impedance, $Z_0 = r_0 - jx_0 = 1.28 - 160\,j$ ohms; the armature turns, $n_1 = 25$, and the armature impedance, $Z_1 =$

$r_1 - jx_1 = 0.4 - 5j$ ohms. For the phase angle, $\theta' = 0$, the transformation ratio then is

$$k_t = \frac{2}{n_p}\frac{n_0}{n_1} = \frac{8}{3} = 2.67,$$

and the equivalent impedance of armature reaction is

$$x_2 = \frac{n_p}{2}\left(\frac{n_1}{n_0}\right)^2 x_0$$
$$= 15,$$

and we have

$$I = 400\,(1 + 3\,\varepsilon^{-0.008\,\theta}), \tag{18}$$
$$i = 400\,(1 + 3\,\varepsilon^{-0.008\,\theta})\,(\cos\theta - \varepsilon^{-0.08\,\theta}), \tag{19}$$

and

$$f = 15{,}000\,(1 + 3\,\varepsilon^{-0.008\,\theta})\,(1 - \varepsilon^{-0.08\,\theta}\cos\theta). \tag{20}$$

(*B*) *Single-phase alternator.*

118. In a single-phase alternator, or in a polyphase alternator with one phase only short-circuited, the armature reaction is pulsating.

The m.m.f. of the armature current,

$$i = I\cos(\theta - \theta'), \tag{22}$$

of a single-phase alternator, is, with regard to the field,

$$f_1 = n_1 I \cos(\theta - \theta')\cos(\theta_0 - \theta');$$

hence, for position angle θ_0 = time angle θ, or synchronous rotation,

$$f_1 = \frac{n_1}{2} I\,\{1 + \cos 2\,(\theta - \theta')\}; \tag{23}$$

that is, of double frequency, with the average value,

$$\mathfrak{F}_1 = \frac{n_1}{2} I, \tag{24}$$

pulsating between 0 and twice the average value.

The average value (24) is the same as the value of the polyphase machine, for $n_p = 1$.

Using the same denotations as in (*A*), we have:

$$(1)\quad \mathfrak{F}_0 = n_0 I_0, \tag{25}$$

$$(4)\quad \mathfrak{F}_1^{\,0} = \frac{apn_1n_0}{2\,x_1} I_0\{1 + \cos 2\,(\theta - \theta')\} = \frac{apn_1n_0}{2\,x_1} I_0\{1 + \cos 2\,\theta'\}. \tag{26}$$

Denoting the effective reactance of armature reaction thus:

$$x_2 = \frac{apn_1}{2}, \tag{27}$$

and substituting (27) in (26) we obtain

$$\mathfrak{F}_1{}^0 = \frac{x_2}{x_1} n_0 I_0 \{1 + \cos 2(\theta - \theta')\} = \frac{x_2}{x_1} n_0 I_0 \{1 + \cos 2\theta'\}; \tag{28}$$

hence, by (6),

$$n_0 I_0 = n_0 i_0{}^0 - \frac{x_2}{x_1} n_0 I_0 \{1 + \cos 2\theta'\}$$

and

$$i_0{}^0 = \frac{x_1 + x_2}{x_1} I_0 \left\{1 + \frac{x_2}{x_1 + x_2} \cos 2\theta' \right\}, \tag{29}$$

and the *field current*,

$$i_0 = \frac{x_1 + x_2 \varepsilon^{-\frac{r_0}{x_0}\theta}}{x_1} I_0 \left\{1 + \frac{x_2}{x_1 + x_2} \cos 2(\theta - \theta_0) \right\} \tag{30}$$

119. If I = maximum value of armature current,

$$\mathfrak{F}_1 = \frac{n_1}{2} I \{1 + \cos 2(\theta - \theta')\} \tag{31}$$

$$= \text{armature m.m.f.};$$

hence,

$$\mathfrak{F} = n_0 i_0 - \mathfrak{F}_1 \tag{32}$$

$$= \text{resultant m.m.f.}$$

Since, however,

$$\Phi = p\mathfrak{F},$$

$$e = a\Phi = ap\mathfrak{F},$$

$$I = \frac{e}{x_1} = \frac{ap\mathfrak{F}}{x_1}, \tag{33}$$

and, by (27),

$$ap = \frac{2}{n_1} x_2,$$

we have, by (33)

$$\mathfrak{F} = \frac{x_1}{ap} I = \frac{n_1}{2} \frac{x_1}{x_2} I. \tag{34}$$

Substituting (30), (31), and (34) into (32) gives

$$\frac{n_1}{2} \frac{x_1}{x_2} I = n_0 I_0 \frac{x_1 + x_2 \varepsilon^{-\frac{r_0}{x_0}\theta}}{x_1} \left\{ 1 + \frac{x_2}{x_1 + x_2} \cos 2 (\theta - \theta') \right\}$$

$$- \frac{n_1}{2} I \{1 + \cos 2 (\theta - \theta')\};$$

or, substituting,

$$k_t = 2 \frac{n_0}{n_1} = \text{transformation ratio}, \tag{35}$$

and rearranging, gives

$$I = k_t I_0 \frac{x_2}{x_1 + x_2} \frac{x_1 + x_2 \varepsilon^{-\frac{r_0}{x_0}\theta}}{x_1} \tag{36}$$

as the *maximum value of the armature current.*

This is the same expression as found in (18) for the polyphase machine. except that now the reactances have different values.

Herefrom it follows that the *instantaneous value of the armature current* is

$$i = k_t I_0 \frac{x_2 \left(x_1 + x_2 \varepsilon^{-\frac{r_0}{x_0}\theta}\right)}{x_1 (x_1 + x_2)} \left\{ \cos (\theta - \theta') - \varepsilon^{-\frac{r_1}{x_1}\theta} \cos \theta' \right\}, \tag{37}$$

and, by (31), the *armature reaction* is

$$\mathfrak{F}_1 = \frac{n_1}{2} k_t I_0 \frac{x_2 \left(x_1 + x_2 \varepsilon^{-\frac{r_0}{x_0}\theta}\right)}{x_1 (x_1 + x_2)} \left\{ 1 + \cos 2 (\theta - \theta') \right\}. \tag{38}$$

For $\theta = \infty$, or permanent condition, equations (30), (36), (37), and (38) give

$$\left.\begin{aligned} i_0 &= I_0 \left\{ 1 + \frac{x_2}{x_1 + x_2} \cos 2\,(\theta - \theta') \right\}, \\ I &= k_t I_0 \frac{x_2}{x_1 + x_2}, \\ i &= k_t I_0 \frac{x_2}{x_1 + x_2} \cos\,(\theta - \theta'), \\ \text{and} \qquad \mathfrak{F}_1 &= \frac{n_1 k_t}{2} I_0 \frac{x_2}{x_1 + x_2} \{1 + \cos 2\,(\theta - \theta')\}. \end{aligned}\right\} \quad (39)$$

As seen, the field current i_0 is pulsating even in permanent condition, the more so the higher the armature reaction x_2 compared with the armature self-inductive reactance x_1.

120. Choosing the same example as in Fig. 53, paragraph 117, but assuming only one phase short-circuited, that is, a single-phase short circuit between two terminals, we have the effective armature series turns, $n_1 = 25\sqrt{3} = 43.3$; the armature impedance, $Z_1 = r_1 - jx_1 = 0.8 - 10\,j$; $\theta' = 0$; the transformation ratio, $k_t = 4.62$, and the effective reactance of armature reaction

$$x_2 = \frac{3}{32}\,x_0 = 15; \text{ herefrom,}$$

$$I = 555\,(1 + 1.5\,\varepsilon^{-0.008\,\theta}), \qquad (36)$$

$$i = 555\,(1 + 1.5\,\varepsilon^{-0.008\,\theta})\,(\cos\theta - \varepsilon^{-0.08\,\theta}), \qquad (37)$$

and

$$f = 12{,}000\,(1 + 1.5\,\varepsilon^{-0.008\,\theta})\,(1 + \cos 2\,\theta); \qquad (38)$$

and the field current is

$$i_0 = 200\,(1 + 1.5\,\varepsilon^{-0.008\,\theta})\,(1 + 0.6\cos 2\,\theta). \qquad (30)$$

In this case, in the open-circuited phase of the machine, a high third harmonic voltage is generated by the double frequency pulsation of the field, and to some extent also appears in the short-circuit current.

121. It must be considered, however, that the effective self-inductive reactance of the armature under momentary short-circuit conditions is not the same as the self-inductive reactance under permanent short circuit conditions, and is not constant, but varying, is a *transient reactance.*

The armature magnetic circuit is in inductive relation with the field magnetic circuit. Under permanent conditions, the resultant m.m.f. of the armature currents with regard to the field is constant, and the mutual inductance between field and armature circuits thus exerts no inductive effect. In the moment of short circuit, however—and to a lesser extent in the moment of any change of armature condition—the resultant armature m.m.f. is pulsating in intensity and direction with regard to the field, as seen in the preceding chapter, and appears as a transient term of the armature self-inductance.

The relations between armature magnetic circuit, field magnetic circuit and mutual magnetic circuit in alternators are similar as the relations in the alternating current transformer, between primary leakage flux, secondary leakage flux and mutual magnetic flux, except that in the transformer the inductive action of the mutual magnetic flux is permanent, while in the alternator it exists only in the moment of change of armature condition, and gradually disappears, thus must be represented by a transient effective reactance. See the chapter on "Reactance of Transformers" in Dover edition, Volume I, Part III.

For a more complete study, such as required for the predetermination of short circuit currents by the numerical calculation of the constants from the design of the machine, the reader must be referred to the literature.

APPENDIX

ALGEBRA OF COMPLEX IMAGINARY QUANTITIES

INTRODUCTION

122. The system of numbers, of which the science of algebra treats, finds its ultimate origin in experience. Directly derived from experience, however, are only the absolute integral numbers; fractions, for instance, are not directly derived from experience, but are abstractions expressing relations between different classes of quantities. Thus, for instance, if a quantity is divided in two parts, from one quantity two quantities are derived, and denoting these latter as halves expresses a relation, namely, that two of the new kinds of quantities are derived from, or can be combined to one of the old quantities.

123. Directly derived from experience is the *operation of counting* or of *numeration*,

$$a, a+1, a+2, a+3 \ . \ . \ . \ .$$

Counting by a given number of integers,

$$a + \underbrace{1 + 1 + 1 \ . \ . \ . + 1}_{b \text{ integers}} = c,$$

introduces the operation of *addition*, as multiple counting,

$$a + b = c.$$

It is

$$a + b = b + a;$$

that is, the terms of addition, or addenda, are interchangeable.

Multiple addition of the same terms,

$$\underbrace{a + a + a + \ . \ . \ . + a}_{b \text{ equal numbers}} = c,$$

introduces the operation of *multiplication*,

$$a \times b = c.$$

It is

$$a \times b = b \times a,$$

that is, the terms of multiplication, or factors, are interchangeable.

Multiple multiplication of the same factors,

$$\underbrace{a \times a \times a \times \ . \ . \ . \times a}_{b \text{ equal numbers}} = c$$

introduces the operation of *involution*,

$$a^b = c.$$

Since

$$a^b \text{ is not equal to } b^a,$$

the terms of involution are not interchangeable.

124. The reverse operation of addition introduces the operation of *subtraction*.

If

$$a + b = c,$$

it is

$$c - b = a.$$

This operation cannot be carried out in the system of absolute numbers, if

$$b > c.$$

Thus, to make it possible to carry out the operation of subtraction under any circumstances, the system of absolute numbers has to be expanded by the introduction of the *negative number*,

$$- a = (- 1) \times a,$$

where

$$(- 1) \text{ is the } \textit{negative unit}.$$

Thereby the system of numbers is subdivided in the positive and negative numbers, and the operation of subtraction possible for all values of subtrahend and minuend. From the definition of addition as multiple numeration, and subtraction as its inverse operation, it follows:

$$c - (- b) = c + b,$$

thus:

$$(- 1) \times (- 1) = 1;$$

that is, the negative unit is defined by $(- 1)^2 = 1$.

125. The reverse operation of multiplication introduces the operation of *division*.

If

$$a \times b = c,$$

it is

$$\frac{c}{b} = a.$$

In the system of integral numbers this operation can only be carried out if b is a factor of c.

To make it possible to carry out the operation of division under any circumstances, the system of integral numbers has to be expanded by the introduction of the *fraction*,

$$\frac{c}{b} = c \times \left(\frac{1}{b}\right),$$

where $\frac{1}{b}$ is the integer fraction, and is defined by

$$\left(\frac{1}{b}\right) \times b = 1.$$

126. The reverse operation of involution introduces two new operations, since in the involution,

$$a^b = c,$$

the quantities a and b are not reversible.

Thus

$$\sqrt[b]{c} = a, \text{ the } \textit{evolution},$$
$$\log_a c = b, \text{ the } \textit{logarithmation}.$$

The operation of evolution of terms, c, which are not complete powers, makes a further expansion of the system of numbers necessary, by the introduction of the *irrational number* (endless decimal fraction), as for instance,

$$\sqrt{2} = 1.414213\ldots$$

127. The operation of evolution of negative quantities, c, with even exponents, b, as for instance,

$$\sqrt[2]{-a},$$

makes a further expansion of the system of numbers necessary, by the introduction of the *imaginary unit*

$$\sqrt{-1}$$

Thus

$$\sqrt[2]{-a} = \sqrt[2]{-1} \times \sqrt[2]{a},$$

where:

$\sqrt{-1}$ is denoted by j.

Thus, the imaginary unit, j, is defined by

$$j^2 = -1.$$

By addition and subtraction of real and imaginary units, compound numbers are derived of the form,

$$a + jb,$$

which are denoted as *complex imaginary numbers*, or *general numbers*.

No further system of numbers is introduced by the operation of evolution.

The operation of logarithmation introduces the irrational and imaginary and complex imaginary numbers also, but no further system of numbers.

128. Thus, starting from the absolute integral numbers of experience, by the two conditions:

1st. Possibility of carrying out the algebraic operations and their reverse operations under all conditions,

2d. Permanence of the laws of calculation,

the expansion of the system of numbers has become necessary, into

positive and negative numbers,
integral numbers and fractions,
rational and irrational numbers,
real and imaginary numbers and complex imaginary numbers.

Therewith closes the field of algebra, and all the algebraic operations and their reverse operations can be carried out irrespective of the values of terms entering the operation.

Thus within the range of algebra no further extension of the system of numbers is necessary or possible, and the most general number is

$$a + jb,$$

where a and b can be integers or fractions, positive or negative, rational or irrational.

Any attempt to extend the system of numbers beyond the complex quantity, leads to numbers, in which the factors of a product are not interchangeable, in which one factor of a product may be zero without the product being zero, etc., and which thus cannot be treated by the usual methods of algebra, that is, are extra-algebraic numbers. Such for instance are the double frequency vector products of Chapter XIV [Dover Volume I, Part II].

Algebraic Operations with General Numbers

129. *Definition of imaginary unit:*

$$j_2 = -1.$$

Complex imaginary number:

$$\underset{\cdot}{A} = a + jb.$$

Substituting:

$$a = r \cos \beta,$$
$$b = r \sin \beta,$$

it is

$$\underset{\cdot}{A} = r(\cos \beta + j \sin \beta),$$

where

$$r = \sqrt{a^2 + b^2},$$
$$\tan \beta = \frac{b}{a},$$

r = vector,
β = amplitude of general number, $\underset{\cdot}{A}$.

Substituting:

$$\cos \beta = \frac{\epsilon^{j\beta} + \epsilon^{-j\beta}}{2},$$
$$\sin \beta = \frac{\epsilon^{j\beta} - \epsilon^{-j\beta}}{2j},$$

it is

$$\underset{\cdot}{A} = r\epsilon^{j\beta},$$

where $\epsilon = \lim_{n = \infty} \left(1 + \frac{1}{n}\right)^n = \sum_0^\infty \frac{1}{1 \times 2 \times 3 \times \ldots \times k}$, is the basis of the natural logarithms.

Conjugate numbers are called:

$$a + jb = r(\cos \beta + j \sin \beta) = r\epsilon^{j\beta},$$

and $a - jb = r(\cos [-\beta] + j \sin [-\beta]) = r(\cos \beta - j \sin \beta) = r\epsilon^{-j\beta}$, it is

$$(a + jb)(a - jb) = a^2 + b^2 = r^2.$$

Associate numbers are called:

$$a + jb = r(\cos\beta + j\sin\beta) = r\epsilon^{j\beta},$$

and

$$b + ja = r\left(\cos\left[\frac{\pi}{2} - \beta\right] + j\sin\left[\frac{\pi}{2} - \beta\right]\right) = r\epsilon^{j\left(\frac{\pi}{2} - \beta\right)},$$

it is

$$(a + jb)(b + ja) = j(a^2 + b^2) = jr^2.$$

If

$$a + jb = a' + jb',$$

it is

$$a = a',$$
$$b = b'.$$

If

$$a + jb = 0;$$

it is

$$a = 0,$$
$$b = 0.$$

130. *Addition and Subtraction:*

$$(a + jb) \pm (a' + jb') = (a + a') + j(b + b').$$

Multiplication:

$$(a + jb)(a' + jb') = (aa' - bb') + j(ab' + ba');$$

or $\quad r(\cos\beta + j\sin\beta) \times r'(\cos\beta' + j\sin\beta') = rr'(\cos[\beta + \beta'] + j\sin[\beta + \beta']);$

or $\quad re^{j\beta} \times r'e^{j\beta'} = rr'e^{j(\beta + \beta')}$

Division:

Expansion of complex imaginary fraction, for rationalization of denominator or numerator, by multiplication with the conjugate quantity:

$$\frac{a + jb}{a' + jb'} = \frac{(a + jb')(a' - jb')}{(a' + jb')(a' - jb')} = \frac{(aa' + bb') + j(ba' - ab')}{a'^2 + b'^2}$$
$$= \frac{(a + jb)(a - jb)}{(a' + jb')(a - jb)} = \frac{a^2 + b^2}{(aa' + bb') + j(ab' - ba')};$$

or,

$$\frac{r(\cos\beta + j\sin\beta)}{r'(\cos\beta' + j\sin\beta')} = \frac{r}{r'}(\cos[\beta - \beta'] + j\sin[\beta - \beta']);$$

or,

$$\frac{re^{j\beta}}{r'e^{j\beta'}} = \frac{r}{r'}e^{j(\beta - \beta')}$$

involution:

$$(a + jb)^n = \{r(\cos\beta + j\sin\beta)\}^n = \{re^{j\beta}\}^n$$
$$= r^n(\cos n\beta + j\sin n\beta) = r^n e^{jn\beta};$$

evolution:

$$\sqrt[n]{a + jb} = \sqrt[n]{r(\cos\beta + j\sin\beta)} = \sqrt[n]{re^{j\beta}}$$
$$= \sqrt[n]{r}\left(\cos\frac{\beta}{n} + j\sin\frac{\beta}{n}\right) = \sqrt[n]{r}e^{j\frac{\beta}{n}}.$$

131. *Roots of the Unit:*

$$\sqrt[2]{1} = +1,\ -1;$$
$$\sqrt[3]{1} = +1,\ \frac{-1 + j\sqrt{3}}{2},\ \frac{-1 - j\sqrt{3}}{2};$$
$$\sqrt[4]{1} = +1,\ -1,\ +j,\ -j;$$
$$\sqrt[6]{1} = +1,\ \frac{+1+j\sqrt{3}}{2},\ \frac{-1+j\sqrt{3}}{2},\ -1,\ \frac{-1-j\sqrt{3}}{2},\ \frac{+1-j\sqrt{3}}{2};$$
$$\sqrt[8]{1} = +1,\ -1,\ +j,\ -j,\ \frac{+1+j}{\sqrt{2}},\ \frac{+1-j}{\sqrt{2}},\ \frac{-1+j}{\sqrt{2}},\ \frac{-1-j}{\sqrt{2}};$$
$$\sqrt[n]{1} = \cos\frac{2\pi k}{n} + j\sin\frac{2\pi k}{n} = e^{\frac{2\pi jk}{n}},\ k = 0, 1, 2 \ldots\ldots n-1.$$

132. *Rotation:*

In the complex imaginary plane, multiplication with

$$\sqrt[n]{1} = \cos\frac{2\pi}{n} + j\sin\frac{2\pi}{n} = e^{\frac{2\pi j}{n}}$$

means rotation, in positive direction, by $\frac{1}{n}$ of a revolution,

multiplication with (-1) means reversal, or rotation by 180°,
multiplication with $(+j)$ means positive rotation by 90°,
multiplication with $(-j)$ means negative rotation by 90°.

133. *Complex Imaginary Plane:*

While the positive and negative numbers can be represented by the points of a line, the complex imaginary numbers or general numbers are represented by the points of a plane, with the horizontal axis, $A'OA$, as real axis, the vertical axis, $B'OB$, as imaginary axis. Thus all

the positive real numbers are represented by the points of half-axis $\overline{OA}$ toward the right;

the negative real numbers are represented by the points of half-axis $\overline{OA'}$ toward the left;

the positive imaginary numbers are represented by the points of half-axis $\overline{OB}$ upward;

the negative imaginary numbers are represented by the points of half-axis $\overline{OB'}$ downward;

the complex imaginary or general numbers are represented by the points outside of the coordinate axes.

INDEX TO VOLUME II

SOME DOVER SCIENCE BOOKS

SOME DOVER SCIENCE BOOKS

WHAT IS SCIENCE?,
Norman Campbell

This excellent introduction explains scientific method, role of mathematics, types of scientific laws. Contents: 2 aspects of science, science & nature, laws of science, discovery of laws, explanation of laws, measurement & numerical laws, applications of science. 192pp. 5⅜ x 8. 60043-2 Paperbound $1.25

FADS AND FALLACIES IN THE NAME OF SCIENCE,
Martin Gardner

Examines various cults, quack systems, frauds, delusions which at various times have masqueraded as science. Accounts of hollow-earth fanatics like Symmes; Velikovsky and wandering planets; Hoerbiger; Bellamy and the theory of multiple moons; Charles Fort; dowsing, pseudoscientific methods for finding water, ores, oil. Sections on naturopathy, iridiagnosis, zone therapy, food fads, etc. Analytical accounts of Wilhelm Reich and orgone sex energy; L. Ron Hubbard and Dianetics; A. Korzybski and General Semantics; many others. Brought up to date to include Bridey Murphy, others. Not just a collection of anecdotes, but a fair, reasoned appraisal of eccentric theory. Formerly titled *In the Name of Science*. Preface. Index. x + 384pp. 5⅜ x 8.
20394-8 Paperbound $2.00

PHYSICS, THE PIONEER SCIENCE,
L. W. Taylor

First thorough text to place all important physical phenomena in cultural-historical framework; remains best work of its kind. Exposition of physical laws, theories developed chronologically, with great historical, illustrative experiments diagrammed, described, worked out mathematically. Excellent physics text for self-study as well as class work. Vol. 1: Heat, Sound: motion, acceleration, gravitation, conservation of energy, heat engines, rotation, heat, mechanical energy, etc. 211 illus. 407pp. 5⅜ x 8. Vol. 2: Light, Electricity: images, lenses, prisms, magnetism, Ohm's law, dynamos, telegraph, quantum theory, decline of mechanical view of nature, etc. Bibliography. 13 table appendix. Index. 551 illus. 2 color plates. 508pp. 5⅜ x 8.
60565-5, 60566-3 Two volume set, paperbound $5.50

THE EVOLUTION OF SCIENTIFIC THOUGHT FROM NEWTON TO EINSTEIN,
A. d'Abro

Einstein's special and general theories of relativity, with their historical implications, are analyzed in non-technical terms. Excellent accounts of the contributions of Newton, Riemann, Weyl, Planck, Eddington, Maxwell, Lorentz and others are treated in terms of space and time, equations of electromagnetics, finiteness of the universe, methodology of science. 21 diagrams. 482pp. 5⅜ x 8.
20002-7 Paperbound $2.50

COLLEGE ALGEBRA, *H. B. Fine*
Standard college text that gives a systematic and deductive structure to algebra; comprehensive, connected, with emphasis on theory. Discusses the commutative, associative, and distributive laws of number in unusual detail, and goes on with undetermined coefficients, quadratic equations, progressions, logarithms, permutations, probability, power series, and much more. Still most valuable elementary-intermediate text on the science and structure of algebra. Index. 1560 problems, all with answers. x + 631pp. 5⅜ x 8. 60211-7 Paperbound $2.75

HIGHER MATHEMATICS FOR STUDENTS OF CHEMISTRY AND PHYSICS, *J. W. Mellor*
Not abstract, but practical, building its problems out of familiar laboratory material, this covers differential calculus, coordinate, analytical geometry, functions, integral calculus, infinite series, numerical equations, differential equations, Fourier's theorem, probability, theory of errors, calculus of variations, determinants. "If the reader is not familiar with this book, it will repay him to examine it," *Chem. & Engineering News*. 800 problems. 189 figures. Bibliography. xxi + 641pp. 5⅜ x 8. 60193-5 Paperbound $3.50

TRIGONOMETRY REFRESHER FOR TECHNICAL MEN, *A. A. Klaf*
A modern question and answer text on plane and spherical trigonometry. Part I covers plane trigonometry: angles, quadrants, trigonometrical functions, graphical representation, interpolation, equations, logarithms, solution of triangles, slide rules, etc. Part II discusses applications to navigation, surveying, elasticity, architecture, and engineering. Small angles, periodic functions, vectors, polar coordinates, De Moivre's theorem, fully covered. Part III is devoted to spherical trigonometry and the solution of spherical triangles, with applications to terrestrial and astronomical problems. Special time-savers for numerical calculation. 913 questions answered for you! 1738 problems; answers to odd numbers. 494 figures. 14 pages of functions, formulae. Index. x + 629pp. 5⅜ x 8. 20371-9 Paperbound $3.00

CALCULUS REFRESHER FOR TECHNICAL MEN, *A. A. Klaf*
Not an ordinary textbook but a unique refresher for engineers, technicians, and students. An examination of the most important aspects of differential and integral calculus by means of 756 key questions. Part I covers simple differential calculus: constants, variables, functions, increments, derivatives, logarithms, curvature, etc. Part II treats fundamental concepts of integration: inspection, substitution, transformation, reduction, areas and volumes, mean value, successive and partial integration, double and triple integration. Stresses practical aspects! A 50 page section gives applications to civil and nautical engineering, electricity, stress and strain, elasticity, industrial engineering, and similar fields. 756 questions answered. 556 problems; solutions to odd numbers. 36 pages of constants, formulae. Index. v + 431pp. 5⅜ x 8. 20370-0 Paperbound $2.25

INTRODUCTION TO THE THEORY OF GROUPS OF FINITE ORDER, *R. Carmichael*
Examines fundamental theorems and their application. Beginning with sets, systems, permutations, etc., it progresses in easy stages through important types of groups: Abelian, prime power, permutation, etc. Except 1 chapter where matrices are desirable, no higher math needed. 783 exercises, problems. Index. xvi + 447pp. 5⅜ x 8. 60300-8 Paperbound $3.00

NUMERICAL SOLUTIONS OF DIFFERENTIAL EQUATIONS,
H. Levy & E. A. Baggott

Comprehensive collection of methods for solving ordinary differential equations of first and higher order. All must pass 2 requirements: easy to grasp and practical, more rapid than school methods. Partial contents: graphical integration of differential equations, graphical methods for detailed solution. Numerical solution. Simultaneous equations and equations of 2nd and higher orders. "Should be in the hands of all in research in applied mathematics, teaching," *Nature.* 21 figures. viii + 238pp. 5⅜ x 8. 60168-4 Paperbound $1.85

ELEMENTARY STATISTICS, WITH APPLICATIONS IN MEDICINE AND THE BIOLOGICAL SCIENCES, *F. E. Croxton*

A sound introduction to statistics for anyone in the physical sciences, assuming no prior acquaintance and requiring only a modest knowledge of math. All basic formulas carefully explained and illustrated; all necessary reference tables included. From basic terms and concepts, the study proceeds to frequency distribution, linear, non-linear, and multiple correlation, skewness, kurtosis, etc. A large section deals with reliability and significance of statistical methods. Containing concrete examples from medicine and biology, this book will prove unusually helpful to workers in those fields who increasingly must evaluate, check, and interpret statistics. Formerly titled "Elementary Statistics with Applications in Medicine." 101 charts. 57 tables. 14 appendices. Index. vi + 376pp. 5⅜ x 8. 60506-X Paperbound $2.25

INTRODUCTION TO SYMBOLIC LOGIC,
S. Langer

No special knowledge of math required—probably the clearest book ever written on symbolic logic, suitable for the layman, general scientist, and philosopher. You start with simple symbols and advance to a knowledge of the Boole-Schroeder and Russell-Whitehead systems. Forms, logical structure, classes, the calculus of propositions, logic of the syllogism, etc. are all covered. "One of the clearest and simplest introductions," *Mathematics Gazette.* Second enlarged, revised edition. 368pp. 5⅜ x 8. 60164-1 Paperbound $2.25

A SHORT ACCOUNT OF THE HISTORY OF MATHEMATICS,
W. W. R. Ball

Most readable non-technical history of mathematics treats lives, discoveries of every important figure from Egyptian, Phoenician, mathematicians to late 19th century. Discusses schools of Ionia, Pythagoras, Athens, Cyzicus, Alexandria, Byzantium, systems of numeration; primitive arithmetic; Middle Ages, Renaissance, including Arabs, Bacon, Regiomontanus, Tartaglia, Cardan, Stevinus, Galileo, Kepler; modern mathematics of Descartes, Pascal, Wallis, Huygens, Newton, Leibnitz, d'Alembert, Euler, Lambert, Laplace, Legendre, Gauss, Hermite, Weierstrass, scores more. Index. 25 figures. 546pp. 5⅜ x 8.
20630-0 Paperbound $2.75

INTRODUCTION TO NONLINEAR DIFFERENTIAL AND INTEGRAL EQUATIONS,
Harold T. Davis

Aspects of the problem of nonlinear equations, transformations that lead to equations solvable by classical means, results in special cases, and useful generalizations. Thorough, but easily followed by mathematically sophisticated reader who knows little about non-linear equations. 137 problems for student to solve. xv + 566pp. 5⅜ x 8½. 60971-5 Paperbound $2.75

AN INTRODUCTION TO THE GEOMETRY OF N DIMENSIONS, *D. H. Y. Sommerville*

An introduction presupposing no prior knowledge of the field, the only book in English devoted exclusively to higher dimensional geometry. Discusses fundamental ideas of incidence, parallelism, perpendicularity, angles between linear space; enumerative geometry; analytical geometry from projective and metric points of view; polytopes; elementary ideas in analysis situs; content of hyper-spacial figures. Bibliography. Index. 60 diagrams. 196pp. 5⅜ x 8.
60494-2 Paperbound $1.50

ELEMENTARY CONCEPTS OF TOPOLOGY, *P. Alexandroff*

First English translation of the famous brief introduction to topology for the beginner or for the mathematician not undertaking extensive study. This unusually useful intuitive approach deals primarily with the concepts of complex, cycle, and homology, and is wholly consistent with current investigations. Ranges from basic concepts of set-theoretic topology to the concept of Betti groups. "Glowing example of harmony between intuition and thought," David Hilbert. Translated by A. E. Farley. Introduction by D. Hilbert. Index. 25 figures. 73pp. 5⅜ x 8. 60747-X Paperbound $1.25

ELEMENTS OF NON-EUCLIDEAN GEOMETRY, *D. M. Y. Sommerville*

Unique in proceeding step-by-step, in the manner of traditional geometry. Enables the student with only a good knowledge of high school algebra and geometry to grasp elementary hyperbolic, elliptic, analytic non-Euclidean geometries; space curvature and its philosophical implications; theory of radical axes; homothetic centres and systems of circles; parataxy and parallelism; absolute measure; Gauss' proof of the defect area theorem; geodesic representation; much more, all with exceptional clarity. 126 problems at chapter endings provide progressive practice and familiarity. 133 figures. Index. xvi + 274pp. 5⅜ x 8. 60460-8 Paperbound $2.00

INTRODUCTION TO THE THEORY OF NUMBERS, *L. E. Dickson*

Thorough, comprehensive approach with adequate coverage of classical literature, an introductory volume beginners can follow. Chapters on divisibility, congruences, quadratic residues & reciprocity. Diophantine equations, etc. Full treatment of binary quadratic forms without usual restriction to integral coefficients. Covers infinitude of primes, least residues. Fermat's theorem. Euler's phi function, Legendre's symbol, Gauss's lemma, automorphs, reduced forms, recent theorems of Thue & Siegel, many more. Much material not readily available elsewhere. 239 problems. Index. I figure. viii + 183pp. 5⅜ x 8.
60342-3 Paperbound $1.75

MATHEMATICAL TABLES AND FORMULAS, *compiled by Robert D. Carmichael and Edwin R. Smith*

Valuable collection for students, etc. Contains all tables necessary in college algebra and trigonometry, such as five-place common logarithms, logarithmic sines and tangents of small angles, logarithmic trigonometric functions, natural trigonometric functions, four-place antilogarithms, tables for changing from sexagesimal to circular and from circular to sexagesimal measure of angles, etc. Also many tables and formulas not ordinarily accessible, including powers, roots, and reciprocals, exponential and hyperbolic functions, ten-place logarithms of prime numbers, and formulas and theorems from analytical and elementary geometry and from calculus. Explanatory introduction. viii + 269pp. 5⅜ x 8½. 60111-0 Paperbound $1.50

CHANCE, LUCK AND STATISTICS: THE SCIENCE OF CHANCE,
Horace C. Levinson

Theory of probability and science of statistics in simple, non-technical language. Part I deals with theory of probability, covering odd superstitions in regard to "luck," the meaning of betting odds, the law of mathematical expectation, gambling, and applications in poker, roulette, lotteries, dice, bridge, and other games of chance. Part II discusses the misuse of statistics, the concept of statistical probabilities, normal and skew frequency distributions, and statistics applied to various fields—birth rates, stock speculation, insurance rates, advertising, etc. "Presented in an easy humorous style which I consider the best kind of expository writing," Prof. A. C. Cohen, Industry Quality Control. Enlarged revised edition. Formerly titled *The Science of Chance*. Preface and two new appendices by the author. xiv + 365pp. 5⅜ x 8. 21007-3 Paperbound $2.00

BASIC ELECTRONICS,
prepared by the U.S. Navy Training Publications Center

A thorough and comprehensive manual on the fundamentals of electronics. Written clearly, it is equally useful for self-study or course work for those with a knowledge of the principles of basic electricity. Partial contents: Operating Principles of the Electron Tube; Introduction to Transistors; Power Supplies for Electronic Equipment; Tuned Circuits; Electron-Tube Amplifiers; Audio Power Amplifiers; Oscillators; Transmitters; Transmission Lines; Antennas and Propagation; Introduction to Computers; and related topics. Appendix. Index. Hundreds of illustrations and diagrams. vi + 471pp. 6½ x 9¼.
61076-4 Paperbound $2.95

BASIC THEORY AND APPLICATION OF TRANSISTORS,
prepared by the U.S. Department of the Army

An introductory manual prepared for an army training program. One of the finest available surveys of theory and application of transistor design and operation. Minimal knowledge of physics and theory of electron tubes required. Suitable for textbook use, course supplement, or home study. Chapters: Introduction; fundamental theory of transistors; transistor amplifier fundamentals; parameters, equivalent circuits, and characteristic curves; bias stabilization; transistor analysis and comparison using characteristic curves and charts; audio amplifiers; tuned amplifiers; wide-band amplifiers; oscillators; pulse and switching circuits; modulation, mixing, and demodulation; and additional semiconductor devices. Unabridged, corrected edition. 240 schematic drawings, photographs, wiring diagrams, etc. 2 Appendices. Glossary. Index. 263pp. 6½ x 9¼.
60380-6 Paperbound $1.75

GUIDE TO THE LITERATURE OF MATHEMATICS AND PHYSICS,
N. G. Parke III

Over 5000 entries included under approximately 120 major subject headings of selected most important books, monographs, periodicals, articles in English, plus important works in German, French, Italian, Spanish, Russian (many recently available works). Covers every branch of physics, math, related engineering. Includes author, title, edition, publisher, place, date, number of volumes, number of pages. A 40-page introduction on the basic problems of research and study provides useful information on the organization and use of libraries, the psychology of learning, etc. This reference work will save you hours of time. 2nd revised edition. Indices of authors, subjects, 464pp. 5⅜ x 8.
60447-0 Paperbound $2.75

MATHEMATICAL PHYSICS, *D. H. Menzel*
Thorough one-volume treatment of the mathematical techniques vital for classical mechanics, electromagnetic theory, quantum theory, and relativity. Written by the Harvard Professor of Astrophysics for junior, senior, and graduate courses, it gives clear explanations of all those aspects of function theory, vectors, matrices, dyadics, tensors, partial differential equations, etc., necessary for the understanding of the various physical theories. Electron theory, relativity, and other topics seldom presented appear here in considerable detail. Scores of definition, conversion factors, dimensional constants, etc. "More detailed than normal for an advanced text . . . excellent set of sections on Dyadics, Matrices, and Tensors," *Journal of the Franklin Institute*. Index. 193 problems, with answers. x + 412pp. 5⅜ x 8. 60056-4 Paperbound $2.50

THE THEORY OF SOUND, *Lord Rayleigh*
Most vibrating systems likely to be encountered in practice can be tackled successfully by the methods set forth by the great Nobel laureate, Lord Rayleigh. Complete coverage of experimental, mathematical aspects of sound theory. Partial contents: Harmonic motions, vibrating systems in general, lateral vibrations of bars, curved plates or shells, applications of Laplace's functions to acoustical problems, fluid friction, plane vortex-sheet, vibrations of solid bodies, etc. This is the first inexpensive edition of this great reference and study work. Bibliography, Historical introduction by R. B. Lindsay. Total of 1040pp. 97 figures. 5⅜ x 8. 60292-3, 60293-1 Two volume set, paperbound $6.00

HYDRODYNAMICS, *Horace Lamb*
Internationally famous complete coverage of standard reference work on dynamics of liquids & gases. Fundamental theorems, equations, methods, solutions, background, for classical hydrodynamics. Chapters include Equations of Motion, Integration of Equations in Special Gases, Irrotational Motion, Motion of Liquid in 2 Dimensions, Motion of Solids through Liquid-Dynamical Theory, Vortex Motion, Tidal Waves, Surface Waves, Waves of Expansion, Viscosity, Rotating Masses of Liquids. Excellently planned, arranged; clear, lucid presentation. 6th enlarged, revised edition. Index. Over 900 footnotes, mostly bibliographical. 119 figures. xv + 738pp. 6⅛ x 9¼. 60256-7 Paperbound $4.00

DYNAMICAL THEORY OF GASES, *James Jeans*
Divided into mathematical and physical chapters for the convenience of those not expert in mathematics, this volume discusses the mathematical theory of gas in a steady state, thermodynamics, Boltzmann and Maxwell, kinetic theory, quantum theory, exponentials, etc. 4th enlarged edition, with new material on quantum theory, quantum dynamics, etc. Indexes. 28 figures. 444pp. 6⅛ x 9¼. 60136-6 Paperbound $2.75

THERMODYNAMICS, *Enrico Fermi*
Unabridged reproduction of 1937 edition. Elementary in treatment; remarkable for clarity, organization. Requires no knowledge of advanced math beyond calculus, only familiarity with fundamentals of thermometry, calorimetry. Partial Contents: Thermodynamic systems; First & Second laws of thermodynamics; Entropy; Thermodynamic potentials: phase rule, reversible electric cell; Gaseous reactions: van't Hoff reaction box, principle of LeChatelier; Thermodynamics of dilute solutions: osmotic & vapor pressures, boiling & freezing points; Entropy constant. Index. 25 problems. 24 illustrations. x + 160pp. 5⅜ x 8. 60361-X Paperbound $2.00

A SOURCE BOOK IN MATHEMATICS,
D. E. Smith

Great discoveries in math, from Renaissance to end of 19th century, in English translation. Read announcements by Dedekind, Gauss, Delamain, Pascal, Fermat, Newton, Abel, Lobachevsky, Bolyai, Riemann, De Moivre, Legendre, Laplace, others of discoveries about imaginary numbers, number congruence, slide rule, equations, symbolism, cubic algebraic equations, non-Euclidean forms of geometry, calculus, function theory, quaternions, etc. Succinct selections from 125 different treatises, articles, most unavailable elsewhere in English. Each article preceded by biographical introduction. Vol. I: Fields of Number, Algebra. Index. 32 illus. 338pp. 5⅜ x 8. Vol. II: Fields of Geometry, Probability, Calculus, Functions, Quaternions. 83 illus. 432pp. 5⅜ x 8.
60552-3, 60553-1 Two volume set, paperbound $5.00

FOUNDATIONS OF PHYSICS,
R. B. Lindsay & H. Margenau

Excellent bridge between semi-popular works & technical treatises. A discussion of methods of physical description, construction of theory; valuable for physicist with elementary calculus who is interested in ideas that give meaning to data, tools of modern physics. Contents include symbolism; mathematical equations; space & time foundations of mechanics; probability; physics & continua; electron theory; special & general relativity; quantum mechanics; causality. "Thorough and yet not overdetailed. Unreservedly recommended," *Nature* (London). Unabridged, corrected edition. List of recommended readings. 35 illustrations. xi + 537pp. 5⅜ x 8. 60377-6 Paperbound $3.50

FUNDAMENTAL FORMULAS OF PHYSICS,
ed. by D. H. Menzel

High useful, full, inexpensive reference and study text, ranging from simple to highly sophisticated operations. Mathematics integrated into text—each chapter stands as short textbook of field represented. Vol. 1: Statistics, Physical Constants, Special Theory of Relativity, Hydrodynamics, Aerodynamics, Boundary Value Problems in Math, Physics, Viscosity, Electromagnetic Theory, etc. Vol. 2: Sound, Acoustics, Geometrical Optics, Electron Optics, High-Energy Phenomena, Magnetism, Biophysics, much more. Index. Total of 800pp. 5⅜ x 8.
60595-7, 60596-5 Two volume set, paperbound $4.75

THEORETICAL PHYSICS,
A. S. Kompaneyets

One of the very few thorough studies of the subject in this price range. Provides advanced students with a comprehensive theoretical background. Especially strong on recent experimentation and developments in quantum theory. Contents: Mechanics (Generalized Coordinates, Lagrange's Equation, Collision of Particles, etc.), Electrodynamics (Vector Analysis, Maxwell's equations, Transmission of Signals, Theory of Relativity, etc.), Quantum Mechanics (the Inadequacy of Classical Mechanics, the Wave Equation, Motion in a Central Field, Quantum Theory of Radiation, Quantum Theories of Dispersion and Scattering, etc.), and Statistical Physics (Equilibrium Distribution of Molecules in an Ideal Gas, Boltzmann Statistics, Bose and Fermi Distribution. Thermodynamic Quantities, etc.). Revised to 1961. Translated by George Yankovsky, authorized by Kompaneyets. 137 exercises. 56 figures. 529pp. 5⅜ x 8½.
60972-3 Paperbound $3.50

CELESTIAL OBJECTS FOR COMMON TELESCOPES,
Rev. T. W. Webb
Classic handbook for the use and pleasure of the amateur astronomer. Of inestimable aid in locating and identifying thousands of celestial objects. Vol I, The Solar System: discussions of the principle and operation of the telescope, procedures of observations and telescope-photography, spectroscopy, etc., precise location information of sun, moon, planets, meteors. Vol. II, The Stars: alphabetical listing of constellations, information on double stars, clusters, stars with unusual spectra, variables, and nebulae, etc. Nearly 4,000 objects noted. Edited and extensively revised by Margaret W. Mayall, director of the American Assn. of Variable Star Observers. New Index by Mrs. Mayall giving the location of all objects mentioned in the text for Epoch 2000. New Precession Table added. New appendices on the planetary satellites, constellation names and abbreviations, and solar system data. Total of 46 illustrations. Total of xxxix + 606pp. 5⅜ x 8. 20917-2, 20918-0 Two volume set, paperbound $5.00

PLANETARY THEORY,
E. W. Brown and C. A. Shook
Provides a clear presentation of basic methods for calculating planetary orbits for today's astronomer. Begins with a careful exposition of specialized mathematical topics essential for handling perturbation theory and then goes on to indicate how most of the previous methods reduce ultimately to two general calculation methods: obtaining expressions either for the coordinates of planetary positions or for the elements which determine the perturbed paths. An example of each is given and worked in detail. Corrected edition. Preface. Appendix. Index. xii + 302pp. 5⅜ x 8½. 61133-7 Paperbound $2.25

STAR NAMES AND THEIR MEANINGS,
Richard Hinckley Allen
An unusual book documenting the various attributions of names to the individual stars over the centuries. Here is a treasure-house of information on a topic not normally delved into even by professional astronomers; provides a fascinating background to the stars in folk-lore, literary references, ancient writings, star catalogs and maps over the centuries. Constellation-by-constellation analysis covers hundreds of stars and other asterisms, including the Pleiades, Hyades, Andromedan Nebula, etc. Introduction. Indices. List of authors and authorities. xx + 563pp. 5⅜ x 8½. 21079-0 Paperbound $3.00

A SHORT HISTORY OF ASTRONOMY, *A. Berry*
Popular standard work for over 50 years, this thorough and accurate volume covers the science from primitive times to the end of the 19th century. After the Greeks and the Middle Ages, individual chapters analyze Copernicus, Brahe, Galileo, Kepler, and Newton, and the mixed reception of their discoveries. Post-Newtonian achievements are then discussed in unusual detail: Halley, Bradley, Lagrange, Laplace, Herschel, Bessel, etc. 2 Indexes. 104 illustrations, 9 portraits. xxxi + 440pp. 5⅜ x 8. 20210-0 Paperbound $2.75

SOME THEORY OF SAMPLING, *W. E. Deming*
The purpose of this book is to make sampling techniques understandable to and useable by social scientists, industrial managers, and natural scientists who are finding statistics increasingly part of their work. Over 200 exercises, plus dozens of actual applications. 61 tables. 90 figs. xix + 602pp. 5⅜ x 8½.
61755-6 Paperbound $3.50

PRINCIPLES OF STRATIGRAPHY,
A. W. Grabau
Classic of 20th century geology, unmatched in scope and comprehensiveness. Nearly 600 pages cover the structure and origins of every kind of sedimentary, hydrogenic, oceanic, pyroclastic, atmoclastic, hydroclastic, marine hydroclastic, and bioclastic rock; metamorphism; erosion; etc. Includes also the constitution of the atmosphere; morphology of oceans, rivers, glaciers; volcanic activities; faults and earthquakes; and fundamental principles of paleontology (nearly 200 pages). New introduction by Prof. M. Kay, Columbia U. 1277 bibliographical entries. 264 diagrams. Tables, maps, etc. Two volume set. Total of xxxii + 1185pp. 5⅜ x 8. 60686-4, 60687-2 Two volume set, paperbound $6.25

SNOW CRYSTALS, *W. A. Bentley and W. J. Humphreys*
Over 200 pages of Bentley's famous microphotographs of snow flakes—the product of painstaking, methodical work at his Jericho, Vermont studio. The pictures, which also include plates of frost, glaze and dew on vegetation, spider webs, windowpanes; sleet; graupel or soft hail, were chosen both for their scientific interest and their aesthetic qualities. The wonder of nature's diversity is exhibited in the intricate, beautiful patterns of the snow flakes. Introductory text by W. J. Humphreys. Selected bibliography. 2,453 illustrations. 224pp. 8 x 10¼. 20287-9 Paperbound $3.25

THE BIRTH AND DEVELOPMENT OF THE GEOLOGICAL SCIENCES,
F. D. Adams
Most thorough history of the earth sciences ever written. Geological thought from earliest times to the end of the 19th century, covering over 300 early thinkers & systems: fossils & their explanation, vulcanists vs. neptunists, figured stones & paleontology, generation of stones, dozens of similar topics. 91 illustrations, including medieval, renaissance woodcuts, etc. Index. 632 footnotes, mostly bibliographical. 511pp. 5⅜ x 8. 20005-1 Paperbound $2.75

ORGANIC CHEMISTRY, *F. C. Whitmore*
The entire subject of organic chemistry for the practicing chemist and the advanced student. Storehouse of facts, theories, processes found elsewhere only in specialized journals. Covers aliphatic compounds (500 pages on the properties and synthetic preparation of hydrocarbons, halides, proteins, ketones, etc.), alicyclic compounds, aromatic compounds, heterocyclic compounds, organophosphorus and organometallic compounds. Methods of synthetic preparation analyzed critically throughout. Includes much of biochemical interest. "The scope of this volume is astonishing," *Industrial and Engineering Chemistry*. 12,000-reference index. 2387-item bibliography. Total of x + 1005pp. 5⅜ x 8. 60700-3, 60701-1 Two volume set, paperbound $4.50

THE PHASE RULE AND ITS APPLICATION,
Alexander Findlay
Covering chemical phenomena of 1, 2, 3, 4, and multiple component systems, this "standard work on the subject" (*Nature*, London), has been completely revised and brought up to date by A. N. Campbell and N. O. Smith. Brand new material has been added on such matters as binary, tertiary liquid equilibria, solid solutions in ternary systems, quinary systems of salts and water. Completely revised to triangular coordinates in ternary systems, clarified graphic representation, solid models, etc. 9th revised edition. Author, subject indexes. 236 figures. 505 footnotes, mostly bibliographic. xii + 494pp. 5⅜ x 8. 60091-2 Paperbound $2.75